AF362055

Cartography and Geomedia

Cartography and Geomedia

Editors

Beata Medyńska-Gulij
David Forrest
Thomas P. Kersten

Basel • Beijing • Wuhan • Barcelona • Belgrade • Novi Sad • Cluj • Manchester

Editors

Beata Medyńska-Gulij
Department of Cartography
and Geomatics
Adam Mickiewicz University
Poznań
Poland

David Forrest
School of Geographical &
Earth Sciences
University of Glasgow
Glasgow
UK

Thomas P. Kersten
Photogrammetry & Laser
Scanning Laboratory
HafenCity University Hamburg
Hamburg
Germany

Editorial Office
MDPI
St. Alban-Anlage 66
4052 Basel, Switzerland

This is a reprint of articles from the Special Issue published online in the open access journal *ISPRS International Journal of Geo-Information* (ISSN 2220-9964) (available at: https://www.mdpi.com/journal/ijgi/special_issues/geomedia).

For citation purposes, cite each article independently as indicated on the article page online and as indicated below:

Lastname, A.A.; Lastname, B.B. Article Title. *Journal Name* **Year**, *Volume Number*, Page Range.

ISBN 978-3-0365-9006-6 (Hbk)
ISBN 978-3-0365-9007-3 (PDF)
doi.org/10.3390/books978-3-0365-9007-3

Cover image courtesy of Beata Medyńska-Gulij

Contents

International Journal of
Geo-Information

Editorial

Cartography and Geomedia in Pragmatic Dimensions

Beata Medyńska-Gulij [1,*], David Forrest [2] and Thomas P. Kersten [3]

[1]   Adam Mickiewicz University, Poznań, Poland, Department of Cartography and Geomatics, 61-712 Poznań, Poland
[2]   University of Glasgow, School of Geographical & Earth Sciences, Glasgow G12 8QQ, UK;
     david.forrest@glasgow.ac.uk
[3]   HafenCity University Hamburg, Photogrammetry & Laser Scanning Laboratory, 20457 Hamburg, Germany;
     thomas.kersten@hcu-hamburg.de
*   Correspondence: bmg@amu.edu.pl

Abstract: This article summarizes the Special Issue of *Cartography and Geomedia.* Here, *Cartography and Geomedia* presents a view of cartography as a combination of technology, science, and art, with a focus on the development of geomedia in a geomatic and design-based context. Individual considerations are presented according to the following topics: efficiency of mapping techniques; historical cartographic works in a geomedial context; cartographic pragmatics for cultural heritage, teaching, and tourism; and pragmatism in gaming cartography. The main conclusion is that the two approaches to learning, revealing, and understanding geographic phenomena—starting from a specific geographical phenomenon and starting from maps and geomedia to understand geographical space—have their pragmatic strengths.

Keywords: cartography; geomedia; pragmatics; cartography in virtual space; symbol efficiency of visualization; information potential of geomedia; map; cartographic symbol

1. Introduction

If geography describes geographic space, and cartography illustrates this space, one can identify the media that both disciplines use to transfer knowledge: text, photography, graphics, animation, video, sound, music, virtual environments, and computer games, among others [1]. Geomedia are therefore means of transmitting information about geographical space that the creator uses separately or jointly. *Cartography and Geomedia* holds the view of cartography as a combination of technology, science, and art, with a focus on the development of geomedia with regard to geomatic and design aspects. Modern cartography applies analogue, digital, and virtual maps, visualizations, and media to objects and landscapes, the evolution of map design, events in the historio-geographical space, topographical change representation socioeconomic transformations, geomedia for teaching, pragmatic geo-applications, social media, and immersion in geographic space, among others. In the context of *Cartography and Geomedia*, we can take two approaches to learn, reveal, and understand geographic phenomena. The first starts from a specific geographical phenomenon and searches for contemporary maps, historic maps, topographical and statistical visualizations, and other uses geomedia to describe and explain this phenomenon. The alternative starts with maps and geomedia to understand historical and contemporary geographical space. In this study, we focus on the use of individual media as information carriers in the context of the role of a map or cartographic symbols as the core of a pragmatic geomedia product. This Special Issue on *Cartography and Geomedia* takes a broad approach to the complementarity of cartography and geomedia, showing how many different problems may be addressed.

2. Efficiency of Mapping Techniques

Ban and Kim [2] propose a new application for the relative motion analysis and visualization of maritime traffic in space and time. They argue that an ideal maritime

Citation: Medyńska-Gulij, B.; Forrest, D.; Kersten, T.P. Cartography and Geomedia in Pragmatic Dimensions. *ISPRS Int. J. Geo-Inf.* **2023**, *12*, 326. https://doi.org/10.3390/ijgi12080326

Received: 30 July 2023
Accepted: 1 August 2023
Published: 4 August 2023

traffic control system should implement specific characteristics, including knowledge of the relative motion of vessels in the same areas, to support navigators in decision making. User evaluation shows it is crucial to apply the principles of cartographic symbolization for the effective visualization of the navigational information of many vessels to effectively support the maritime traffic control system.

The study by Wielebski and Medyńska-Gulij [3] yield interesting results based on research concerning six thematic maps made with various mapping techniques and related to various aspects of the activities of rescue helicopters. Users ranked these maps in terms of four subjective evaluation criteria, which were the graphical attractiveness of the maps, the readability of the maps, the usefulness and importance of the information, and the complexity of the information presented on the maps. In the case of the usefulness and importance of the information, the map's topic—important for saving health and life from the user's point of view—was of the greatest importance, while the amount of information in the legend significantly influenced the evaluation of information complexity.

Eye tracking tests are of great importance for checking the effectiveness of geomedia designs. Cybulski [4] apply eye-tracking in a controlled laboratory experiment with 15 animated choropleth maps illustrating various spatial and temporal changes. The results of this experiment show that effective pattern or trend recognitions can be influenced by user' visual behavior, but animated choropleth maps are more suitable for presenting temporal trends than spatial patterns.

An interesting example of the parametric modeling method for 3D symbols in geology is Li and Chen's proposal [5]. According to the complexity and diversity of the fold structures, they discuss the generation of 3D fold symbols which abstractly express the morphologies of fold structures developed in the experimental area and can express most fold structure subdivision types. The parametric modeling method for the 3D symbols of fold structures follows the traditional cartographic symbol creation rules, but the modeling method avoids excessive dependence on geological survey data, and fits well with the abstract character of the geological symbol.

Research by Lyu et al. [6] suggest that cartographic generalization of road networks cannot be considered in isolation as there is a strong geographical correlation between roads and settlements. They describe a method for generalizing a road network, with settlements as nodes, which removes redundant paths between settlements and verifies the visual continuity and topological connectivity of the network.

3. Historical Cartographic Works in Geomedial Dimension

In a new methodological approach, Medyńska-Gulij et al. [7] employ a complementary combination of the knowledge of historical geography, map design principles, general good practices of writing a short film script, and the use of media. In one shared video, they use as many different types of geomedia as possible to tell a story, e.g., historical printed maps, digital political maps, hypsometric maps, old maps, 3D maps, and cartographic animations. They offer potential designers recommendations for filming historical geography with analog maps and maps created specifically for films published on online video-sharing and social media platforms.

Justová and Cajthmal [8] develop a process for adapting originally printed historical maps for presentation online which goes beyond zoomable, scanned analogues or default GIS maps. They discuss how the additional functionality the new geomedia dimension of these historical maps on the web allows for the simplification of the originally complex map and to increase the information potential of the maps. They go onto set out a complex methodology for effective presentation of printed historical maps on the web.

Globes are a special cartographic product, but cannot simply be scanned or photographed like a flat map to create a digital visualization. Stefanova et al. [9] develop a model for creating a virtual copy. In this process, the key aspects in digitizing the globe were the appropriate photogrammetric parameters, new metadata from the bibliographic description, the specificity of the old engraving in the copperplate, and web presentation.

They also compare web browsing performance and offer good practice guidelines that can be applied more generally.

The selection of appropriate geomedia attributes for constructing natural and suggestive perspective visualizations can be a new option for 3D digitizing historical watercolor topographic maps. Medyńska-Gulij [10] selected several 18th century manuscript maps with specific painterly means of expression, which she subjected to rectification and vectorization of contour lines, and the transformation to a GRID model. After applying parameter variations such as elevation rise, azimuth and altitude, contrast of illumination, and the creation of the final bird's-eye-view visualization, she concludes that the proposed parameters for the three maps work well for creating the general static bird's-eye-view visualization, with the natural and suggestive perception of the landscapes' relief. However, it is crucial to maintain their complementarity.

4. Cartographic Pragmatics for Cultural Heritage, Teaching and Tourism

An example of using cartographic methods in cultural heritage research is identifying relatively stable boundaries of dialect and subdialect areas, allowing language surveys to focus on boundaries. Duan and Zhou [11] note that scholars often use the watershed boundary as the boundary of dialect areas. In the Tibetan case they discovered that road breakpoint lines and other features can be reliable and effective indicators in determining the boundaries of (sub)dialect areas, and that such data can highlight issues with language survey samples.

In order to promote the rich cultural and natural heritage of an area to tourists not aware of it, Luppichini et al. [12] propose the use of Web Mapping and Real–Virtual Itineraries based on geological, biological, and archaeological data. Based on a multidisciplinary approach, they create palaeoenvironment maps and real and virtual itineraries with the aid of a web application, more specifically web mapping, developed with free and open-source libraries.

Due to their text-based presentation, local chronicles recording the urban characteristic and culture of a location are limited in their accessibility and geographical context. He et al. [13] create visualizations of city-specific culture extracted from local chronicles is based on designing various types of content-framework construction with the most important being the cartographic approach. Evaluation of the four sample designs shows that despite some limitations resulting from the size and content of the chronicles, it is possible to integrate local chronicles with maps to create meaningful visualizations of spatial and temporal connections. The case study can be treated as a universal model for creating visualizations of cultural, didactic, and touristic significance.

A similar cultural and teaching dimension is found in the considerations of Janovský et al. [14] concerning the development of a visualization of a historic river valley before the construction of the so-called Vltava Cascade. Historical cadastral maps were the input data for creating a virtual reality (VR) application using Unreal Engine software. Digital transformations and programming result in 3D scenes in a form approximately corresponding to the state of the river valley in the 19th century, which is available for viewing via a web application, and a VR scene used for demonstration at exhibitions for a wider audience.

Creating a virtual tour of the site of an onshore wind farm can be counted as a specific use of stitching images with multi-row panorama and multimedia sources. Lai et al. [15] propose improving image stitching quality by calculating the root mean square error (RMSE) of tie point matching and adjustment. Once the images are stitched together effectively, they are combined with multimedia materials to provide information on wind turbine attributes to establish a virtual reality tour platform accessible on a smartphone.

Mercier et al. [16] describe a comparative test to assess system usability and understand the impact of different geolocation techniques on the usability of Augmented Reality (AR) applications. Their cartographic authoring tool for the visualization of geolocated media in augmented reality (AR) has been tested to extend education technology towards biodiversity education. The use of the UCD method (User-Centered Design) allowed for adaption to the needs and expectations of teachers and students. However, the variable positional accuracy of mobile devices is an issue. While an external RTK (Real-Time Kinematic Positioning) module solves the problem of geolocation data inaccuracy, without its usability, issues remain.

Halik and Wielebski [17] suggest a new method of generating AR with a new name plane-based augmented geovisualizations (PAGs). Based on the embedding of several models of mountain peaks, they tested them as geovisualizations on any plane detected with the AR device. Three age groups of users pointed out that the AR mode was preferable against all compared criteria; only with regard to the criterion of ease of use of the AR mode, was the result less definitive.

5. Pragmatics in Gaming Cartography

Gaming cartography is a developing trend in geomedia research. Therefore, in this Special Issue, there are studies which feature players' opinions about the characteristics of cartographic symbols and the interface and the use of maps in games.

Medyńska-Gulij et al. [18] investigate map design and usability issues when exploring topographical space in landscape versus portrait orientation in mobile phone games. The study incorporated an appropriate research methodology, including establishing conceptual assumptions, developing map applications with gaming elements, user testing, and visualizing results. The results reveal the main differences and similarities in participant performance when using a simplified topographic 2D Map. The different phone orientations solicited different visual strategies, which in turn influenced decisions regarding path selection.

Based on the analysis of 100 popular video games, Zagata and Medyńska-Gulij [19] identify features of mini-map design used as navigational support in the virtual geographical space. The analytical study revealed eight features of mini-map design and their popular parameters and attributes: projection—orthographic; centering—player-centered; base layers—artificial; shape—circular; orientation—camera view; position—bottom left; proportions—2–3%; additional navigational element—north arrow. They are able to confirm the feasibility of designing a mini-map according to traditional cartographic design principles when these attributes were considered separately, complementarily, and holistically.

The interpretation of app symbols in the context of gamers' age and experience is investigated by Horbiński and Zagata [20]. Using a popular survival game as their focus, they explored the effects that players' age and the time spent playing has on their interpretation of map symbols used in a game. Based on the outcomes of the study, it is concluded that there is correlation between the age of the players, the time spent playing, and the interpretations of symbols within it (for individual symbols but not the entire symbol set).

6. Conclusions

Critical reflections on deviance in representation—in the context of the relationship between aesthetics and cartography—is a key aspect of the discourse on the relationship between cartography and geomedia. Edler and Kühne [21] treat cartographic representations as being subject to sensory perception, and these representations rely on the translation of sensory perceptions into cartographic symbols. On the one hand, Edler and Kühne accept traditional cartography, on the other hand, they see great potential in augmented and virtual environments to generate aesthetically constructed cartographic representations.

The studies presented above show the increasingly strong links between cartography and geomedia—as it is broadly understood—in primarily pragmatic terms. It is worth noting that maps and cartographic symbols are the elements that bind the representation of geographical phenomena in new media. Additionally, the adaptation of cartographic forms to the virtual environment creates new opportunities for the visualization of space using photogrammetric and geomatic techniques [22] as well as the design–programming approach [18].

Considering the above mentioned studies, we can conclude that the two approaches to learning, revealing, and understanding geographic phenomena—starting from a specific geographical phenomenon and starting from maps and geomedia to understand geographical space—are advantageous. Starting from a specific geographical phenomenon and using maps and geomedia to understand geographical space focuses on the pragmatic dimension [23]. Cartographic pragmatism which emphasize the principles of map design, is highly advantageous, because it guarantees the effectiveness of the transmission of quantitative and quantitative geographic information [24].

Author Contributions: Conceptualization, Beata Medyńska-Gulij, David Forrest and Thomas P. Kersten; methodology, Beata Medyńska-Gulij, David Forrest and Thomas P. Kersten; writing—original draft preparation, Beata Medyńska-Gulij, David Forrest and Thomas P. Kersten. All authors have read and agreed to the published version of the manuscript.

Funding: This research received no external funding.

Conflicts of Interest: The authors declare no conflict of interest.

References

1. Kersten, T.P.; Trau, D.; Tschirschwitz, F. The Four-masted Barque Peking in Virtual Reality as a New Form of Knowledge Transfer. *ISPRS Ann. Photogramm. Remote Sens. Spatial Inf. Sci.* **2020**, *2020*, 155–162. [CrossRef]
2. Ban, H.; Kim, H.-j. Analysis and Visualization of Vessels' RElative MOtion (REMO). *ISPRS Int. J. Geo-Inf.* **2023**, *12*, 115. [CrossRef]
3. Wielebski, Ł.; Medyńska-Gulij, B. User Evaluation of Thematic Maps on Operational Areas of Rescue Helicopters. *ISPRS Int. J. Geo-Inf.* **2023**, *12*, 30. [CrossRef]
4. Cybulski, P. An Empirical Study on the Effects of Temporal Trends in Spatial Patterns on Animated Choropleth Maps. *ISPRS Int. J. Geo-Inf.* **2022**, *11*, 273. [CrossRef]
5. Li, A.-B.; Chen, H.; Du, X.-F.; Sun, G.-K.; Liu, X.-Y. Parametric Modeling Method for 3D Symbols of Fold Structures. *ISPRS Int. J. Geo-Inf.* **2022**, *11*, 618. [CrossRef]
6. Lyu, Z.; Sun, Q.; Ma, J.; Xu, Q.; Li, Y.; Zhang, F. Road Network Generalization Method Constrained by Residential Areas. *ISPRS Int. J. Geo-Inf.* **2022**, *11*, 159. [CrossRef]
7. Medyńska-Gulij, B.; Tegeler, T.; Bauer, H.; Zagata, K.; Wielebski, Ł. Filming the Historical Geography: Story from the Realm of Maps in Regensburg. *ISPRS Int. J. Geo-Inf.* **2021**, *10*, 764. [CrossRef]
8. Justová, P.; Cajthaml, J. Cartographic Design and Processing of Originally Printed Historical Maps for Their Presentation on the Web. *ISPRS Int. J. Geo-Inf.* **2023**, *12*, 230. [CrossRef]
9. Štefanová, E.; Novotná, E.; Čábelka, M. Digitization, Visualization and Accessibility of Globe Virtual Collection: Case Study Jüttner's Globe. *ISPRS Int. J. Geo-Inf.* **2023**, *12*, 122. [CrossRef]
10. Medyńska-Gulij, B. Geomedia Attributes for Perspective Visualization of Relief for Historical Non-Cartometric Water-Colored Topographic Maps. *ISPRS Int. J. Geo-Inf.* **2022**, *11*, 554. [CrossRef]
11. Duan, M.; Zhou, S. Geographic Approach: Identifying Relatively Stable Tibetan Dialect and Subdialect Area Boundaries. *ISPRS Int. J. Geo-Inf.* **2022**, *11*, 280. [CrossRef]
12. Luppichini, M.; Noti, V.; Pavone, D.; Bonato, M.; Ghizzani Marcìa, F.; Genovesi, S.; Lemmi, F.; Rosselli, L.; Chiarenza, N.; Colombo, M.; et al. Web Mapping and Real–Virtual Itineraries to Promote Feasible Archaeological and Environmental Tourism in Versilia (Italy). *ISPRS Int. J. Geo-Inf.* **2022**, *11*, 460. [CrossRef]
13. He, X.; Liu, C.; Wu, L.; Wang, Y.; Tian, Z. Thematic Content and Visualization Strategy for Map Design of City-Specific Culture Based on Local Chronicles: A Case Study of Dengfeng City, China. *ISPRS Int. J. Geo-Inf.* **2022**, *11*, 542. [CrossRef]
14. Janovský, M.; Tobiáš, P.; Cehák, V. 3D Visualisation of the Historic Pre-Dam Vltava River Valley—Procedural and CAD Modelling, Online Publishing and Virtual Reality. *ISPRS Int. J. Geo-Inf.* **2022**, *11*, 376. [CrossRef]
15. Lai, J.-S.; Tsai, Y.-H.; Chang, M.-J.; Huang, J.-Y.; Chi, C.-M. A Technical and Operational Perspective on Quality Analysis of Stitching Images with Multi-Row Panorama and Multimedia Sources for Visualizing the Tourism Site of Onshore Wind Farm. *ISPRS Int. J. Geo-Inf.* **2022**, *11*, 362. [CrossRef]
16. Mercier, J.; Chabloz, N.; Dozot, G.; Ertz, O.; Bocher, E.; Rappo, D. BiodivAR: A Cartographic Authoring Tool for the Visualization of Geolocated Media in Augmented Reality. *ISPRS Int. J. Geo-Inf.* **2023**, *12*, 61. [CrossRef]
17. Halik, Ł.; Wielebski, Ł. Usefulness of Plane-Based Augmented Geovisualization—Case of "The Crown of Polish Mountains 3D". *ISPRS Int. J. Geo-Inf.* **2023**, *12*, 38. [CrossRef]
18. Medyńska-Gulij, B.; Gulij, J.; Cybulski, P.; Zagata, K.; Zawadzki, J.; Horbiński, T. Map Design and Usability of a Simplified Topographic 2D Map on the Smartphone in Landscape and Portrait Orientations. *ISPRS Int. J. Geo-Inf.* **2022**, *11*, 577. [CrossRef]

19. Zagata, K.; Medyńska-Gulij, B. Mini-Map Design Features as a Navigation Aid in the Virtual Geographical Space Based on Video Games. *ISPRS Int. J. Geo-Inf.* **2023**, *12*, 58. [CrossRef]
20. Horbiński, T.; Zagata, K. Interpretation of Map Symbols in the Context of Gamers' Age and Experience. *ISPRS Int. J. Geo-Inf.* **2022**, *11*, 150. [CrossRef]
21. Edler, D.; Kühne, O. Aesthetics and Cartography: Post-Critical Reflections on Deviance in and of Representations. *ISPRS Int. J. Geo-Inf.* **2022**, *11*, 526. [CrossRef]
22. Kersten, T.P.; Edler, D. Methods and Applications of Virtual and Augmented Reality in Geo-Information Science. *PFG—J. Photogramm. Remote Sens. Geoinf. Sci.* **2020**, *88*, 119–120. [CrossRef]
23. Medyńska-Gulij, B. Map compiling, map reading and cartographic design in "Pragmatic pyramid of thematic mapping". *Quaest. Geogr.* **2010**, *29*, 57–63. [CrossRef]
24. Forrest, D. Thematic Maps in Geography. *Int. Encycl. Soc. Behav. Sci.* **2015**, *24*, 260–267.

International Journal of
Geo-Information

Article

Analysis and Visualization of Vessels' RElative MOtion (REMO)

Hyowon Ban [1] and Hye-jin Kim [2,*]

[1] Department of Geography, California State University Long Beach, Long Beach, CA 90840, USA
[2] Korea Research Institute of Ships & Ocean Engineering, Daejeon 34103, Republic of Korea
* Correspondence: hjk@kriso.re.kr; Tel.: +82-42-866-3114

Abstract: This research is a pilot study to develop a maritime traffic control system that supports the decision-making process of control officers, and to evaluate the usability of a prototype tool developed in this study. The study analyzed the movements of multiple vessels through automatic identification system (AIS) data using one of the existing methodologies in GIScience, the RElative MOtion (REMO) approach. The REMO approach in this study measured the relative speed, delta-speed, and the azimuth of each vessel per time unit. The study visualized the results on electronic navigational charts in the prototype tool developed, V-REMO. In addition, the study conducted a user evaluation to assess the user interface (UI) of V-REMO and to future enhance the usability. The general usability of V-REMO, the data visualization, and the readability of information in the UI were tested through in-depth interviews. The results of the user evaluation showed that the users needed changes in the size, position, colors, and transparency of the trajectory symbols in the digital chartmap view of V-REMO for better readability and easier manipulation. The users also indicated a need for multiple color schemes for the spatial data and more landmark information about the study area in the chartmap view.

Keywords: maritime transportation; expert evaluation; nautical charts; relative motion; situation awareness system; geographic information science; cartographic visualization

Citation: Ban, H.; Kim, H.-j. Analysis and Visualization of Vessels' RElative MOtion (REMO). *ISPRS Int. J. Geo-Inf.* **2023**, *12*, 115. https://doi.org/10.3390/ijgi12030115

Academic Editors: Beata Medynska-Gulij, David Forrest, Thomas P. Kersten and Wolfgang Kainz

Received: 1 January 2023
Revised: 28 February 2023
Accepted: 7 March 2023
Published: 8 March 2023

1. Introduction

Maritime transportation in port control areas can be complex when several vessels operate, anchor, or ship. Therefore, detecting unusual maritime traffic situations becomes easier for traffic controllers if they can observe the spatio-temporal characteristics of the transportation and operation of vessels—such as speed and azimuth—in terminal control areas.

It is necessary for vessel traffic service (VTS) operators to constantly be informed about the status of multiple vessels' operations and make decisions to prevent any possible maritime traffic accidents. However, it is challenging for traffic controllers to constantly observe large amounts of information that changes through space and time, so they may generate errors in their decision making about maritime traffic. Therefore, it is vital to develop a system that supports users—for example, traffic controllers—to recognize the characteristics of maritime traffic patterns and visualize marine traffic information from a large amount of data. The system can use information on the course over ground (COG) and speed over ground (SOG) of the vessels [1,2].

Recently, maritime transportation accidents have increased for various reasons. Some accidents have occurred because of overloading; technical problems; collisions with other ships; weather and water conditions; and social, economic, and political structures [3,4]. Examples include the Le Joola disaster in 2002, the Shariatpur-1 disaster in 2012, and the Sewol Ferry disaster in 2014 [4–7]. Especially in the case of South Korea, the number of maritime traffic accidents increased from 1093 cases (1306 vessels involved) to 2307 cases (2549 vessels involved) between 2013 and 2016 [8].

In the vicinity of ports, vessels are on course for entry and departure to and from the ports. Buoys only define the entry and departure courses, and vessels need to sail within the extent of these courses. The direction of currents and the motion of seawater, such as tidal movements in port areas, influence the sailing of the vessels. Operators of vessels understand such characteristics of seawater, and they sail by manipulating the speed and azimuth of the vessels based on this knowledge. Additionally, traffic controllers monitor the operation of vessels in port areas, often controlling them by providing restrictions or advice [9].

Outside of ports, port control centers do not control vessels. Instead, they operate by themselves based on consideration of the seawater, seafloor geomorphology, and other vessels nearby. Therefore, it is necessary to manage information on traffic situations every time a vessel moves. Automatic identification system (AIS) data are helpful in analyzing marine traffic because they provide information on high temporal resolutions collected per second. AIS data consist of a vessel's position, identity, speed, course with other nearby ships, satellites, and so on. Usually, the resolution of the AIS data may range from low (sampling rate in minute unit) to high (sampling rate in second unit) [10]. The current study uses AIS data to deal with maritime transportation.

Recently, there has been much research regarding the risk analysis of maritime transportation, some of which has dealt with the methodology and applications of the analysis of maritime transportation [11]. For example, AIS data have been used to locate vessels by exchanging data with each other, including vessels' position, identity, speed, course with other nearby ships, AIS base stations, and satellites. Further, AIS data have been widely studied in relation to the topics of spatio-temporal distributions, anomaly detection, the prediction of the routes of vessels, the spatial domain of ships, the statistical analysis of traffic patterns and the collision risk of vessels, the emission estimation of ships, the uncertainty of the AIS data, and so on [12–18].

There exist decision support systems for maritime traffic. Many of them display the current locations of vessels on digital charts and generate an alarm sound when distances among the vessels decrease or signals regarding the locations of the vessels show abnormal patterns. Additionally, the systems show the locations of the vessels on charts. However, few provide more detailed information, such as analyses of each vessel's operations or maritime traffic in areas. Due to this, it is challenging for traffic controllers to recognize and control complex maritime traffic situations in the relevant areas [1,2].

The current research is a pilot study aiming to develop a maritime traffic control system that supports control officers' decision-making process and to conduct a usability test of the system. We analyze multiple vessels' movements using one of the existing approaches in GIScience, the RElative MOtion (REMO) approach, and visualize the results on two-dimensional, electronic navigational charts. Furthermore, we assess the tool via user interviews, which is the novelty of this study. Ultimately, the study aims to answer the following questions: (i) What are the crucial characteristics of a maritime traffic control system to support control officers in making decisions? (ii) How can multiple vessels' navigation information at a port be effectively analyzed and visualized for control officers? To answer these questions, we analyze AIS data and visualize the results using the REMO approach and GIScience following Laube and Imfeld [19]. The AIS data used in this study were collected from the Korea Research Institute of Ships and Ocean Engineering (KRISO).

Background

AIS data have been widely used to analyze and report vessels' movements. As a result, maritime traffic management has improved information processing, integration, and presentation using AIS data [1]. However, a large number of vessels' existing control systems rely on experienced officers' mental capacity for information analysis and decision making [20]. Additionally, the number of skilled workforces is limited, so officers can be prone to errors in decision making that may lead to marine traffic accidents [21]. Therefore, to increase the safety of marine traffic, it is necessary to develop enhanced control systems that support the analysis of vessels' spatio-temporal distributions, and the prediction of routes and collision risk [12].

Several works have been conducted to aid the guidance and decision making of vessels. For instance, by using AIS data, the drifting of ships can largely influence marine traffic accidents [22]. Moreover, a reference model of port information systems was developed consisting of various types of data and analyses; however, the model does not deal with the analysis or visualization of AIS data [23]. Another method was developed to enhance responses to maritime emergencies by using the Electronic Chart Display and Information System. The system analyzes and displays optimized situation-dependent maneuvering plans for maritime traffic emergencies [24]. Further, hydrographic geospatial standards for marine data and information and their effective representation were suggested for communities, including marine science and maritime technology [25]. Additionally, a scenario for shipping industry was developed for stakeholders that employs analytics, stream processing, monitoring, alerting, and vessel route optimization over big data [26].

Some countries or cities have provided information for warning and guiding marine traffic by operating authorities. For example, according to the U.S. Coast Guard Navigation Center, the Coast Guard operates 12 vessel traffic centers (VTC) and 200 very-high-frequency (VHF) receiver sites located throughout the coastal areas of the United States [27]. In the case of Hong Kong, its Marine Department has 13 radars employed in the system to provide radar surveillance coverage of Hong Kong's navigable waters. The radar system is designed to automatically track a maximum of 10,000 targets at any one time [28]. The Swedish Maritime Administration has a network of land-based AIS base stations to receive AIS information from vessels and transmit safety-related information. The network uses AIS information to improve maritime safety information, search and rescue missions, and icebreaking operations [29]. Further, the Port of London Authority has a team of 44 VTS personnel that oversees its VTS area on a 24/7 basis, 365 days a year. A VTS supervisor with the delegated powers of the harbor master leads each VTS Center, and a team of VTS officers and shipping coordinators supports the VTS supervisor [30]. In South Korea, the Maritime Transportation Control Center of the Korea Cost Guard operates 20 VTS centers [31].

Many works have dealt with spatial and temporal approaches to analyzing AIS data on vessels. For instance, AIS data were utilized to measure the safety of a vessel's bridge and showed their potential impact on the safety of marine navigation. Furthermore, issues related to uncertainty that may be generated by different regulations to supervision for proper use, training, and management of AIS users were investigated [13]. AIS data were used for creating charts and analyzing individual vessel movements and general traffic patterns to quantify air pollutant emissions from the vessels [16]. Moreover, a prediction model of fuel oil consumption and a weather routing algorithm were used together to propose an optimal route for a vessel for fuel oil consumption efficiency using data collected from AIS, Sensors, Noon Reports, and Weather Service APIs [32]. An algorithm was developed for three-dimensional space–time density analysis and the visualization of AIS data collected from the Gulf of Finland [33]. An open-source toolbox was also developed and released that provides modules to support AIS data processing and visualizing traffic density analyzed on charts [34]. Additionally, a linear regression model was developed for AIS data to avoid collision by identifying the correlation of the closest point of approach, which is a crucial indicator for collision avoidance, to the vessel's size, speed, and course [14]. A methodology was developed to extract traffic routes, detect low-likelihood behaviors of vessels, and predict the routes of vessels from a large amount of AIS data. Moreover, the authors visualized the predicted destination of vessels and the probability of the observed tracks [15]. A comfortable navigational distance was estimated for vessels by analyzing the intensity plots generated from AIS data [12]. The risk of the collision of vessels was analyzed by using developed software and AIS data. The tool calculated the collision risk from traffic patterns in the data and visualized the results [18].

A user test on visualization tools is necessary to support user-oriented communication and decision making [35,36]. An agenda for empirical research on user interactive design for digital maps was developed [37]. Several works have conducted user studies on the use of maps and the cartographic visualization of spatial data. For instance, data from historical sites were visualized in a virtual reality application, and users tested the effectiveness and attractiveness of the application [38]. An online atlas contained usability research and adopted a user-centered design methodology for better decision making [39]. The usability of some mapping techniques for visualizing spatial accessibility was evaluated in terms of effectiveness, efficiency, graphical attractiveness, and user-perceived effectiveness [40].

However, most of these previous works have limitations regarding providing information to officers about multiple vessels moving concurrently, so that the officers can understand quickly because of their complex representations of information [21]. Therefore, in this research, we apply the REMO analysis method and visualization, the existing approaches in GIScience, to extensive marine traffic data, in order to improve the analysis and visualization of vessel movements and to help users better understand such information. The REMO analysis measures relative speed, change of speed, and the azimuth of moving objects to detect constancy, concurrence, dispersion, and so on. Studies have used REMO analysis to detect and measure constancy, concurrence, trendsetting, turn, opposition, and dispersion of moving objects such as herds, athletes, and dancers [19,41–44]. Further, several studies have extended the REMO approach. For example, the REMO approach has been used in studies to discover locational and temporal patterns of moving objects, such as caribou, soccer players and dancers [41,43–45]. Moreover, methods were developed to detect spatio-temporal patterns of flock, leadership, convergence, and encounter by following the REMO approach and using computational geometry. The researchers improved existing algorithms by decreasing running time bounds [42]. A framework was developed to detect the motion anomalies of vessels by conducting a statistical analysis of real-time AIS data [17]. In addition, a semantic recognition method was proposed to analyze ship motion patterns entering and leaving a port based on a probabilistic topic model. The analysis used a large amount of trajectory data in an unsupervised manner [46]. Moreover, a visual analytics tool was proposed that used spatial segmentation to identify vessels' anomalous trips and measure the degree of unusual vessel behavior [47]. Furthermore, the motion patterns of moving objects, such as dancers using REMO matrices and dynamic time warping, were visualized and analyzed. The motion attributes of speed, motion azimuth, vertical angle over time, and time-series analysis of the attributes were measured [45]. However, little research has investigated vessels' movements using the REMO approach. Therefore, using the REMO approach, we attempt to fill the gap in the literature by analyzing and visualizing the characteristics of multiple vessels' movements at a port based on the empirical AIS datasets of the vessels. Further, we develop a prototype software tool, "V-REMO", to interactively analyze the vessels' relative movements based on the AIS data and display the results on digital charts. For instance, the digital charts in V-REMO show changes in the azimuth, speed, and δ-speed—or change of speed—of each vessel per time unit and location based on the user's manipulation. Finally, we conduct user evaluations of V-REMO to enhance the tool in the future.

2. Materials and Methods

2.1. Study Areas and Data

The study areas included the Port of Yeosu and nearby areas located in South Jeolla Province, South Korea, and the southwestern areas of the Korean Peninsula (Figure 1). The Port operates passenger terminals and cargo docks [48].

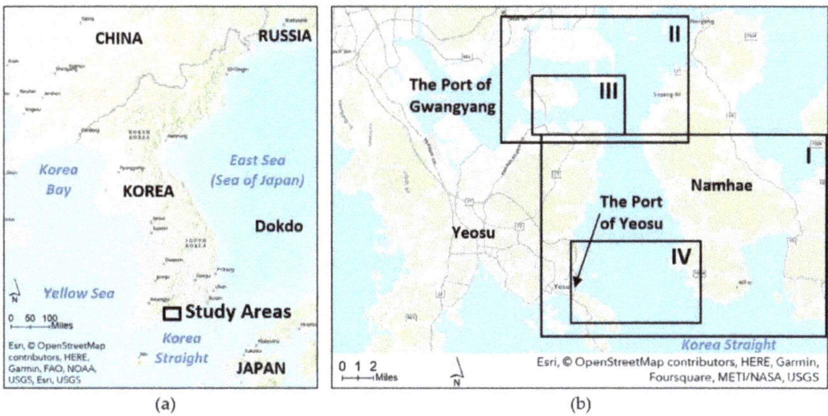

(a)   (b)

Figure 1. (**a**) Location of the study areas; (**b**) the study areas enlarged from (**a**): area I includes the Port of Yeosu and City of Namhae, area II includes Myo Island and Cities of Gwangyang, Namhae, and Yeosu, area III includes Myo Island and City of Yeosu, and area IV includes the Port of Yeosu and Cities of Yeosu and Namhae.

AIS prevents the collision of vessels by providing real-time navigation information such as the location, course, and velocity of each vessel. The International Maritime Organization (IMO) regulation requires AIS to be mounted on vessels of certain types [49] (p. 2). AIS can identify a vessel's existence, name, navigation direction, and velocity, even when the naked eye or marine radars cannot. Thus, maritime safety management activities, such as preventing vessel collisions, controlling broad areas, and searching and rescuing wrecked vessels, can be more efficient using AIS. In general, AIS data help to analyze vessels' navigation because they consist of multiple types of information on a vessel. Since AIS messages are encoded in binary format, they should be separated from other non-binary types of data in a database (Figure 2).

```
!AIVDM,1,1,,A,B6SWVSP002@p2o5FojNc<vkUoM06,0*4D,5,END
!AIVDM,1,1,,A,16SlPM0P00a2doBE;V8ev0E>2D3l,0*19,2,END
!AIVDM,1,1,,B,8>lEJr00Bj8C4kWNieId0000042P00000001HL0000000000000000h0000,2*0D,0,END
!AIVDM,1,1,,A,16SWbsPP00a8QuVAKcOjM?w>25AP,0*71,5,END
!AIVDM,1,1,,A,10aucQ0000a1vbPEY9a52HW@2@GV,0*6C,5,END
```

Figure 2. An example AIS message encoded in a binary format. Data from [50].

Generally, a VTS center operates multiple AIS-receiving stations that collect all AIS signals within the maritime territory, including the control areas. Therefore, a vessel's course characteristics per time can be studied by analyzing its AIS data accumulated over a certain period. AIS data primarily consist of the items in Table 1.

AIS receivers collect messages from vessels within signal reception areas per the order of the reception. Thus, AIS data are not distinguished per vessel. Additionally, AIS data are generated every few seconds or minutes according to the velocity of vessels. Therefore, the amount of AIS data received within a control area can be large when multiple vessels move concurrently. Table 2 shows the transmission frequency of AIS messages. AIS data have different time intervals since the data are not collected by a regular time unit. Therefore, the density of the data collection becomes higher when the velocity of a vessel increases.

Table 1. Items in AIS data.

Items	Descriptions
Static information	- Identifies information about a vessel. - Is transmitted every 6 min or per request. - Includes information such as Maritime Mobile Service Identity, International Maritime Organization number, name of vessel, maritime call sign, beam and length, and type of vessel.
Dynamic information	- Is related to a vessel's movement. - Transmits a vessel's frequency per velocity. - Includes the location of the vessel, course over ground, speed over ground, heading, and angular velocity of the vessel's head.
Voyage related data	- Include vessel draft, types of cargo, destination, and estimated time of arrival.

Table 2. Transmission frequency of AIS messages.

Status of Vessel	Transmission Frequency
Ship at anchor	3 min
Ship 0–14 knots	12 s
Ship 0–14 knots and changing course	4 s
Ship 14–23 knots	6 s
Ship 14–23 knots and changing course	2 s
Ship > 23 knots	3 s
Ship > 23 knots and changing course	2 s

KRISO provided AIS data on the study areas. The data included 31,708,973 records of 212 vessels collected during 11–20 January 2012, in the Port of Yeosu. The AIS data were decoded for use by the REMO software tool.

2.2. REMO Analysis

In the study of moving objects in space and time, it is important to reveal information on movement attributes, such as direction and speed [33]. We adopted the REMO approach [19] to analyze the patterns of the relative movements of vessels, including the azimuth (or direction), speed, and δ-speed of vessels. Figure 3 introduces example concepts of the REMO approach for vessels following Laube and Purves [43]. For example, in Figure 3a, vessel one moves towards the south with a constant motion azimuth of 135°. The movement of vessel one occurs during an interval from t_1 to t_3 (Figure 3b,c). The vessel's movement includes three discrete time steps of length δt and shows a constancy of motion azimuth (Figure 3d). The vessel's speed is measured based on the distance and duration between the two adjacent locations of the vessel's trajectory. The vessel's δ-speed is calculated based on the differences between the speed of the two adjacent locations of the vessel.

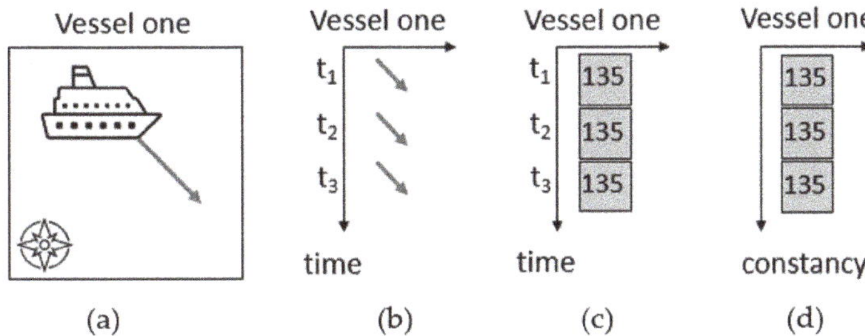

Figure 3. Example concepts of the REMO approach for vessels following [43]: (**a**) a vessel moves toward the south with a constant motion azimuth; (**b**) the constant motion azimuth occurs per time unit; (**c**) values of the motion azimuth, 135°, are coded per time unit; (**d**) the vessel's movement shows a constancy of motion azimuth.

The patterns of REMO can be graphically recognized in a "REMO matrix", a two-dimensional conceptual space consisting of a time axis and another axis of individual moving objects [19,44]. Based on the steps of the example in Figure 3, three REMO matrices of the vessels were generated from the database in this study. The REMO matrices consist of information about the motion azimuth, speed, and δ-speed—or change of speed—of the vessels, respectively, following Laube and Imfeld [19]. Then, the motion azimuth of the vessel was measured based on the x and y coordinates between two adjacent point locations of the vessel from the dataset. The motion azimuth of the vessel was classified as the following: N (either 0~22.5° or 337.5~360°), NE (22.5~67.5°), E (67.5~112.5°), SE (112.5~157.5°), S (157.5~202.5°), SW (202.5~247.5°), W (247.5~292.5°), and NW (292.5~337.5°). Further, "O" was used for no movement in location. However, the "sorites paradox" in classifying angles may exist because of the crisp classification used in this study. For instance, two values, 22.5° and 22.51°, were almost identical. Nonetheless, they were classified into two classes, N and NE, respectively (Section 3.1). Finally, charts that showed empirical locations of the vessel's REMO were created to provide spatial patterns of the vessel's REMO (Section 3.2). The REMO matrices and the charts of the REMO analysis were constructed using ArcGIS Pro (10.8) [51].

In the next section, we introduce V-REMO, a software tool that supports the REMO analysis on the AIS data.

2.3. Development of Software, V-REMO

This section describes the development process of V-REMO, a prototype software tool for marine traffic. V-REMO was developed by the project team at KRISO to analyze multiple vessels' real-time movements from the decoded AIS data. The tool provides information for safe navigation based on the Microsoft NET framework. The tool aims to help the decision-making process of vessel traffic control officers and reduce their workload. V-REMO consists of a menu pane and a chart view pane to visualize analysis results on digital charts. The digital chart is displayed as the background on the chart view pane because such charts are familiar to navigation officers and controllers, and they know how to use them. The digital chart provides data on topics such as the fairway, aid to navigation, and the water depth, so that information affecting a vessel's navigation can be grasped. The digital chart is displayed as raster data on the user interface (UI) of V-REMO to shorten the data loading time.

A list of the vessels' Maritime Mobile Service Identity (MMSIs) from the database is displayed on the left-hand side of the UI. When the user selects a vessel, the vessel's trajectory is displayed in a different color from other vessels' trajectories on the digital chart. The user can display changes in the selected vessel's trajectory per time by using a time slider at the bottom of the UI. The user can select SOG, COG, δ-SOC (or change of SOC), and δ-COG (or change of COG) in the UI to visualize the analysis results. SOG and δ-SOG refer to information about the vessel's velocity, and COG and δ-COG refer to information about the vessel's course. Usually, AIS data include information on the heading of a vessel and its rate of turning (ROT). However, AIS data were not used to analyze the vessels' trajectories in this study because the ROT values represent the intention of a vessel's navigation rather than the results of its navigation. Finally, the user can select one of the analysis results from V-REMO and adjust its graphic symbology, such as the outline and transparency, to enhance the display. Figure 4 shows the process of the REMO analysis of a vessel's movement (Figure 4a) and the visualization of the results (Figure 4b) in V-REMO.

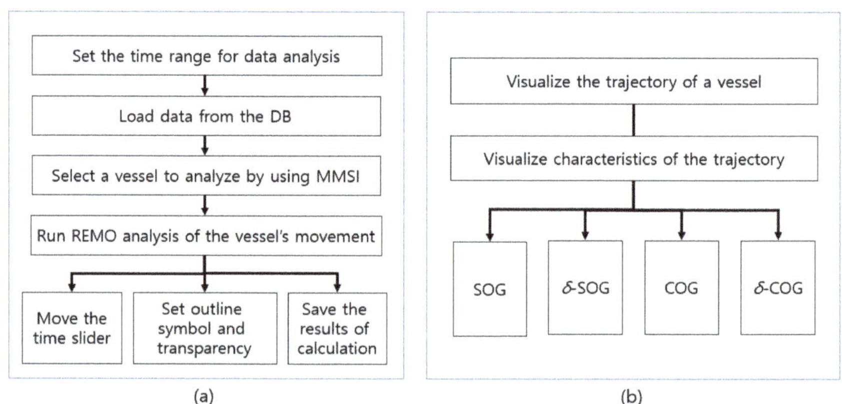

Figure 4. Functions of V-REMO ((**a**): flow of user's manipulation; (**b**): information view).

Figure 5 shows the flow of the development and assessment of V-REMO in this study. It consists of data preparation, the development of V-REMO, the analysis of results, and tool assessment. In the next section, we describe the assessment of V-REMO in detail.

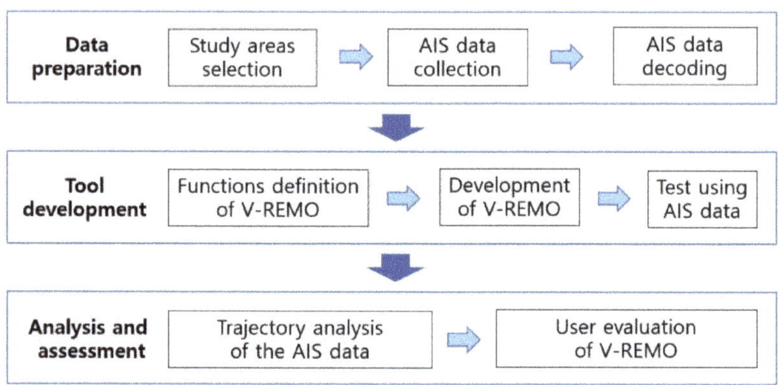

Figure 5. Structure of the study.

2.4. User Evaluation of V-REMO

The usability of an application for maritime transportation should be tested in close collaboration with the intended users, including the analysts at VTS stations and maritime authorities [33]. We conducted user evaluations of V-REMO to test its usability. We examined the prototype of V-REMO by conducting in-depth interviews with stakeholders in South Korea to determine whether V-REMO's UI supports data analysis and trajectory recognition activities, and to identify its existing limitations for its further development.

The participants in the interviews were six professionals in the maritime domain, including researchers and navigation officers. The participants were recruited because they had diverse specializations, including marine safety, marine transportation, maritime standards, product specification, e-navigation, and ergonomics. The participants' ages varied from those in their 20s to those in their 40s. Their work experience in the industry varied from 1 to 18 years. None of them was involved in any processes of the development of V-REMO. All subjects gave their informed consent for inclusion before participating in the study. The study was conducted in accordance with the Declaration of Helsinki, and the protocol was approved by the Institutional Review Board of California State University, Long Beach (Project number: 1146800-2, Reference number: 18-149).

The user evaluation consisted of practical demonstrations of V-REMO and in-depth interviews. First, each participant was informed about the purpose of the user evaluation and the concept of the REMO analysis of the vessels' trajectories. Then, examples of demonstrations using V-REMO were provided, and each participant was asked to answer questions about their own experience of using V-REMO. Finally, the in-depth interviews comprised 26 questionnaires regarding demographic information on the participants, the general usability of V-REMO, the data visualization, and the readability of information in the UI. For example, the participants were asked to evaluate the UI's graphic design and the usability of the menus and buttons, the chart design of the REMO analysis results on the digital chart, and the readability of the real-time traffic of vessels in the UI (figures are provided in Section 3). The questionnaires used in the user evaluation of this study are provided in Supplementary File S1.

3. Results

In this section, we provide results of the REMO analysis on AIS data. Then, we show the developed V-REMO and describe the results of the user evaluations.

3.1. REMO Analysis

First, we introduce the hypothetical results of the REMO analysis in Section 3.1. The AIS data used in this section were produced arbitrarily to provide an example of the REMO analysis. Then, we provide the results of the REMO analysis using the empirical AIS data collected in the study areas in Section 3.2.

Figure 6 presents the initial results of the REMO analysis regarding the azimuth (Figure 6a), speed (Figure 6b), and δ-speed (Figure 6c) of vessels 1, 2, 3, and 4 in REMO matrices. The four rows in each REMO matrix in Figure 6 visualize the relative movements of the corresponding vessels. Since the four vessels started their navigation at different times, the starting time of their navigation and the lengths of their data in the REMO matrices are not identical. Figure 6d–f show a part of the REMO matrices enlarged from Figure 6a–c, respectively. In general, the motion azimuth of the vessels shows a variety of their moving directions ("divergence" in [19]) during the navigation (Figure 6a). Some vessels changed their directions (the color hue of a row changes in Figure 6d ("turn" in [19])), while others did not (the same color hue continues to appear in the row in Figure 6d ("independence" in [19])). Blank cells represent no data, since some vessels often had no difference of azimuth. Sometimes, vessels' movements that are not clear in the azimuth matrix (Figure 6d) become more explicit in the speed matrix (Figure 6e) or the δ-speed (Figure 6f). For instance, area I in Figure 6d does not show any changes in the azimuth of the vessels. However, area II in Figure 6e and area III in Figure 6f show variations in the speed and δ-speed for the same data as Figure 6d, respectively.

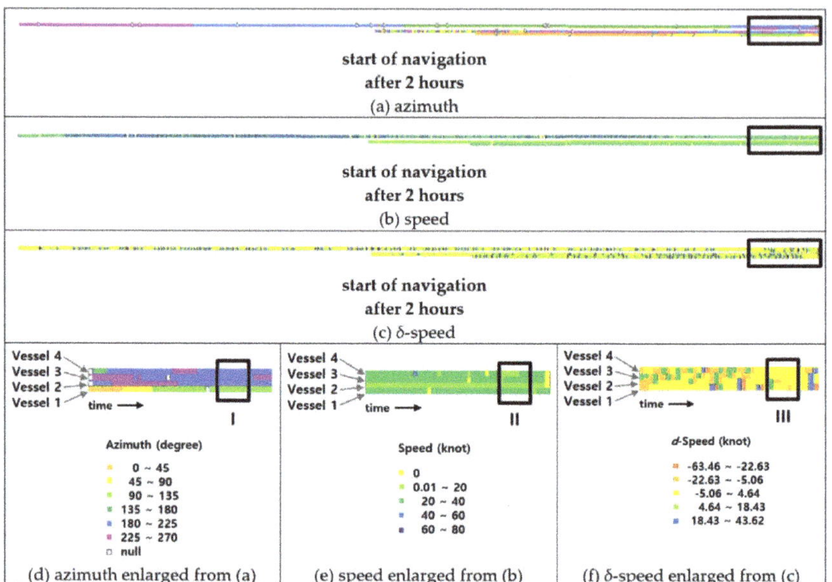

Figure 6. Example REMO matrices of the four vessels during two hours of navigation: (**a**) REMO matrix of azimuth, (**b**) REMO matrix of speed, (**c**) REMO matrix of δ-speed, (**d**) part of (**a**) enlarged, (**e**) part of (**b**) enlarged, (**f**) part of (**c**) enlarged.

Figure 7 spatially visualizes the REMO analysis results of the four vessels in terms of azimuth (a), speed (b), and δ-speed (c) during the whole navigation. Figure 7a shows the relative azimuth of each vessel at its location. In Figure 7a, yellow-green, green, or blue colors indicate a substantial change in azimuth, and yellow, orange, or purple colors indicate a small change in azimuth. The color hues in Figure 7a change where the azimuth of each vessel changes during their navigation. Figure 7b shows the relative speed of each vessel at its location. The relative speed of the data clearly shows that the speed of the vessel "MMSI: 636091308" changed, although its azimuth was maintained while navigating in area I in Figure 7b. Finally, Figure 7c shows the relative δ-speed of each vessel at its location. The relative δ-speed, or change of the speed, of the data visualizes an increase (in green or blue color hue) or a decrease (in red or orange color hue) in the speed of each vessel while navigating (Figure 7c).

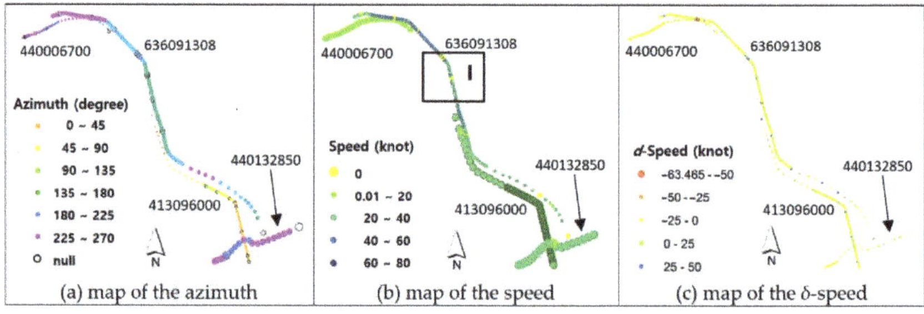

Figure 7. Charts showing REMO analysis of the four vessels during the navigation in terms of azimuth (**a**), speed (**b**), and δ-speed (**c**).

3.2. Application: Empirical REMO Analysis Using V-REMO Software

This section provides the results of the REMO analysis on empirical AIS data collected in the study areas as an application. The REMO analysis was conducted by using the V-REMO software developed in this study. The UI of V-REMO primarily consists of a data processing section and a chart visualization section. First, a user can load the trajectory database and specify the time range to query (A of Figure 8). Once the trajectory data are loaded, a list of vessels is displayed, including each vessel's MMSI number (B of Figure 8). Then, the user can select a vessel from the list and move the time slider to visualize the vessel as a symbol on the electronic navigational chart along the timeline (G of Figure 8). Thus, the user can see the trajectories of each vessel and select a vessel to analyze and visualize the vessel on the electronic navigational chart (C of Figure 8). The user can preprocess the trajectory data on the selected vessel and change the size and color of the vessel's symbol by using the "Calc" button. Additionally, the user can adjust the transparency of the electronic navigational chart to enhance the visibility of information. Figure 8 shows the UI of the V-REMO software.

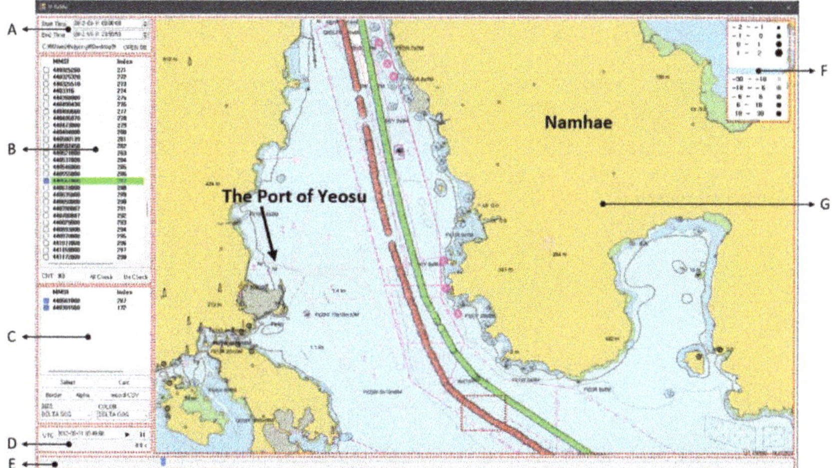

Figure 8. The UI of V-REMO showing area I in Figure 1b: (**A**) the vessels' empirical AIS data are loaded from a trajectory database, and the time range for a query is set; (**B**) vessels in the database are selected; (**C**) vessels to be analyzed are selected, and visualization attributes are set; (**D**) the playback of trajectory data is controlled; (**E**) the slider bar is controlled; (**F**) the symbols of the vessels are displayed in the legend; and (**G**) the electronic navigational chart is shown in the chart view.

One of the unique characteristics of the V-REMO software is it can efficiently visualize information for the REMO analysis of the vessel's trajectory data. AIS collects the trajectory data. The data consist of MMSI numbers, the location information (longitude and latitude) of the vessel, the SOG and COG of the vessel, time, and so on. Among these, the values of SOG and COG are closely related to the vessel's movement characteristics. The V-REMO software measures the vessel's movement over time and visualizes the results using symbols on the digital chart. A vessel's symbols are located on the vessel's trajectory and are displayed in different sizes and colors depending on SOG, COG, δ-SOG, and δ-COG values. Figure 9 provides an example of SOG changing along with the vessel's movement and the amount of change.

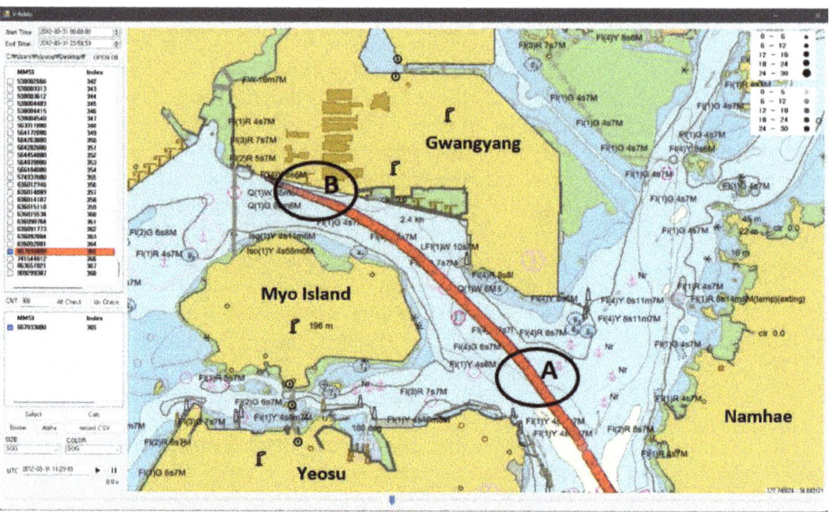

(A) The UI of V-REMO showing the area II in Figure 1(b)

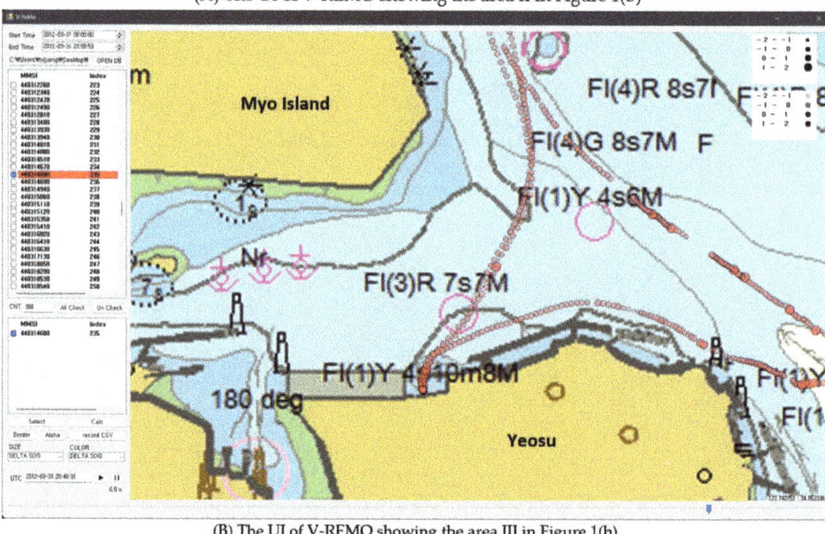

(B) The UI of V-REMO showing the area III in Figure 1(b)

Figure 9. Example of vessel trajectory visualization: (**A**) a vessel's SOG while approaching the pier at Gwangyang Bay; (**B**) a vessel's symbols showing the change of SOG during the movement.

Figure 9A visualizes information on the trajectory of a vessel during its pier approach: the size and color of the vessel's symbol change following its speed change. For example, the symbol becomes more prominent and has higher color saturation when the vessel's speed increases (Area A in Figure 9A). The user can see the vessel decreasing its speed as it approaches the pier (Area B in Figure 9A). Figure 9B displays the trajectory of a vessel moving to a pier within a port: the size and the color of the vessel's symbol change following its speed change in five classes. The user can easily recognize the location where the vessel's speed suddenly changed and the time when it changed.

Additionally, the user can replay the vessel's movements in the trajectory data through time using the V-REMO UI. The user can control the playback time of the trajectory data by moving the slider bar beneath the UI and by visualizing the trajectory data on the chart. By doing so, the user can analyze the vessel's trajectory per time and compare it with that of other vessels.

Figure 10A,B displays the trajectories of the two vessels MMSI: 440144600 (in green) and MMSI: 440132190 (in red) at t_1 and t_2, respectively. The two vessels may appear to meet in Figure 10B; however, their trajectories per time in Figure 10A indicate that they did not meet.

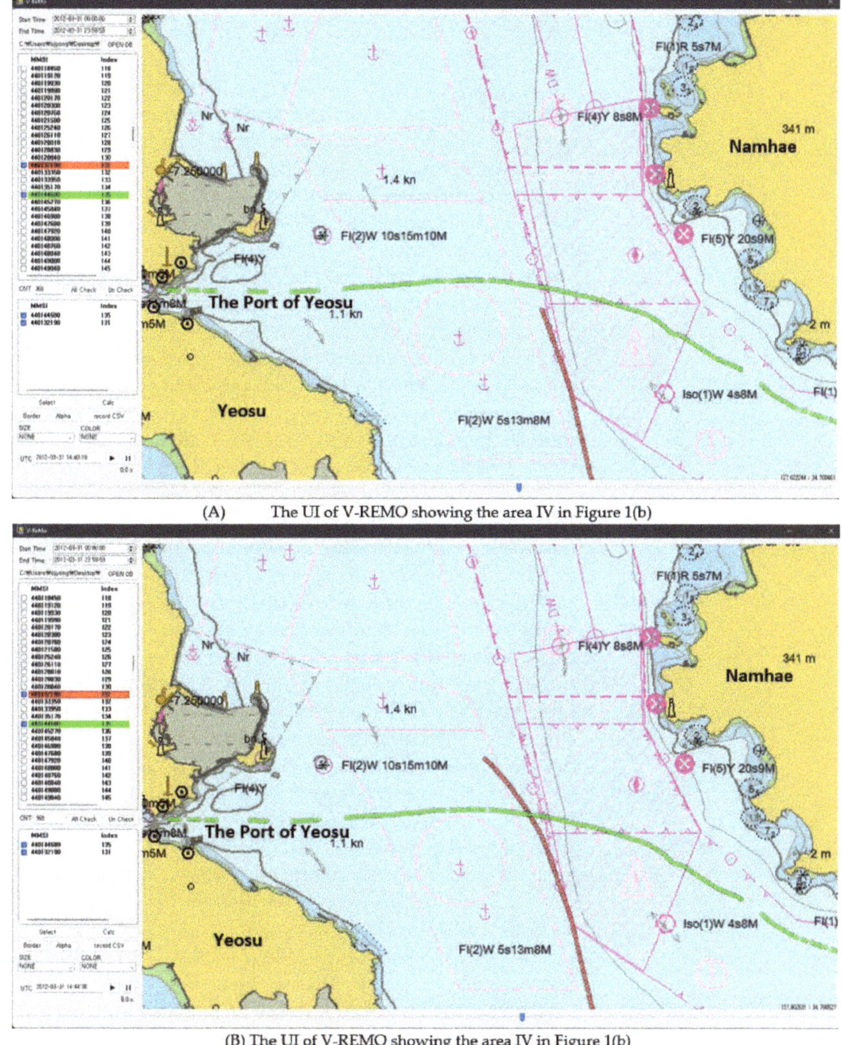

(A)   The UI of V-REMO showing the area IV in Figure 1(b)

(B) The UI of V-REMO showing the area IV in Figure 1(b)

Figure 10. Example visualization of the vessel's trajectories: (**A**) trajectories of two vessels, MMSI: 440144600 (in green) and MMSI: 440132190 (in red) at t_1 = 14:40:19; (**B**) the same trajectories at t_2 = 14:44:36.

3.3. User Evaluations of V-REMO

The usability of the V-REMO software was assessed in the user evaluations. The in-depth interviews consisted of a few questionnaires on the usability of V-REMO. First, regarding the general usability of the UI, the interviewees responded that the V-REMO software tool was easy to manipulate and the electronic navigational chart in the UI was user-friendly. However, the interviewees also pointed out that the querying functions were too limited to represent a vessel's trajectories. For instance, one user stated that querying vessels' trajectories was performed quickly using only time ranges in the current V-REMO. However, finding a single vessel to analyze its trajectories from the data was challenging. This was because the user misunderstood the characteristics of V-REMO. For example, the user was trying to query a specific vessel's trajectory throughout the timeline from the dataset. However, the UI was designed to compare multiple vessels' movements during a specific time range rather than a vessel's movements. This suggests that users of V-REMO may need to have good knowledge of REMO analysis and V-REMO before using the software tool. Further, the interviewees mentioned that the menus in the UI were not easy to find and too many menus were shown. The interviewees also pointed out that querying specific information for individual vessels was not possible.

Second, regarding the issues in visualizing analysis results in the V-REMO UI, the interviewees responded that the visualization of the tool is convenient for understanding the trajectories of a vessel. However, most interviewees could not recognize the analysis results easily. Namely, they found it difficult to differentiate the colors and sizes of the multiple vessels' symbols. They specified that the circular symbol representing the vessels looked identical to another type of symbol in the electronic navigational chart. Therefore, it is necessary to examine the standard of the electronic navigational chart for symbolization.

Finally, regarding the readability of information on the real-time traffic of the vessels in the UI, some interviewees responded that they could recognize moving vessels because their trajectories and attribute information were displayed together on charts. However, most of them pointed out that it was challenging to understand the vessel's traffic in the UI when multiple vessels appeared or when the vessels' trajectories were cluttered on the electronic navigational chart.

The interviewees suggested the following for improving V-REMO. First, they recommended providing more options for querying the trajectory of a vessel in the UI. For example, the user should be able to query the trajectory data on a vessel by using the name, MMSI, type, and size. Additionally, the user should be able to view specific information about the vessel, including specifications and nationality, by selecting its symbol on the digital chart. The interviewees also recommended adding a function to display only a vessel's trajectories on the digital chart. Often, a single vessel may have multiple trajectories to display for a particular time range that the user defines. Thus, the digital chart of V-REMO can quickly become cluttered when multiple vessels are selected and their trajectory histories are displayed for the same time range.

Finally, the interviewees provided multiple suggestions for visualizing results in V-REMO. For instance, symbols of vessels should be able to represent the speed and course of each vessel. Additionally, the size and color of the symbol showing the analysis results should be clearly distinguished from other symbols. In particular, users can confuse circular symbols with other symbols in the digital chart that are not appropriate to represent the speed and direction of navigation. More studies are necessary to symbolize vessels following IMO's Guidelines for the Presentation of Navigation-Related Symbols [49]. The interviewees also proposed showing the navigation data on all vessels using a regular time unit. However, this suggestion does not apply to the V-REMO UI because periodic intervals of the transmission of navigation data can be an essential factor for marine transportation analysis.

The user evaluation examined the UI design and the software functionality of V-REMO. In terms of the UI design, the users wanted to see changes in the size, position, colors, and transparency of the trajectory symbols in the digital chart for better readability and easier manipulation. The users also indicated a need for multiple color schemes for the spatial data and more landmark information about the study area in the chart view. The responses from the user evaluation participants of this study are provided in Supplementary File S2. Additionally, the user feedback contributed to discovering existing limitations of V-REMO that needed to be rectified, including supporting users in understanding how it works and designing some functions and visual displays in the UI.

4. Discussion

This study demonstrated how the REMO approach and visualization could analyze multiple vessels' movements from empirical AIS data. The significance of this research is that it expands the range of GIScience to the domains of ocean engineering and maritime transportation. Our research introduced a new application of the REMO approach to enhance the work environment of maritime control officers. In South Korea, maritime control officers at all six selected VTS centers out of eighteen centers have a high workload that may lead to human errors in recognizing information about vessels' status and making decisions in real-time [5]. Therefore, it is necessary to develop a system to decrease officers' workload so that they can analyze complex maritime information.

Despite the novel approach of this study, there exist a few limitations. We discuss some major issues in three categories as examples. First, issues of the data: This study analyzed and visualized the AIS data in only two dimensions. Analysis and visualization of the data in three or higher dimensions in V-REMO—i.e., by including time as an additional dimension—might provide more insight about the data than the current study. Moreover, in general, AIS can provide data in a high resolution (i.e., record per second). The high resolution often makes AIS data useful for maritime transportation analysis [10]. However, other types of data related to maritime transportation, such as climate (i.e., 30 m), topography (i.e., 30 m), and weather (i.e., 10 km) have lower resolution than AIS data [52–54]. For this reason, the current study did not address the data of the climate, topography, and weather of the study areas. In addition, the results of this study might not be useful in understanding the movement of small vessels if they are not equipped with AIS [55]. Therefore, additional data types—for instance, very-high-frequency (VHF) radio data—should be considered in the analysis for more inclusive research.

Second, issues of V-REMO: The tool developed in this study did not deal with real-time AIS data due to the limitation of the pilot version of the software. Data collection, analysis, visualization, and traffic control for vessels should be in real-time for the better prevention of maritime transportation accidents than analyzing non-real-time data. Therefore, the functionality of V-REMO needs to be enhanced to deal with real-time data through interactive analysis and visualization.

Third, applying the REMO approach: The REMO approach in this study could be challenging for stakeholders in the maritime domain to understand if they do not have knowledge of the approach. The user evaluations of this study revealed that users should first understand the REMO approach before they use V-REMO. Moreover, users' needs for the analysis of AIS data should be investigated before a tool is developed to address their needs in the UI of the tool.

5. Conclusions

In this article, we present a new application of the REMO analysis and visualization on moving objects in space and time. This is a new approach in the research of the maritime transportation domain to deal with the movement of vessels. Further, it should be considered how the approach of this study compares with a similar approach for the analysis of maritime traffic data. To demonstrate the utility of the new application, we tested the REMO approach on an empirical dataset of moving objects representing maritime traffic in the Port of Yeosu. The results attempted to answer the research questions of this study as follows. First, an ideal maritime traffic control system should implement specific characteristics, including knowledge of the relative motion of vessels in the same areas, to support control officers in making decisions. Second, it is crucial to effectively visualize multiple vessels' navigation information to support the maritime traffic control system effectively. Specifically, the user evaluation results in this study demonstrated the importance of human factors in software development for maritime traffic control. For instance, users' understanding of the REMO analysis of AIS data can be largely affected by the symbolization in the software UI. Finally, the study provided new insights into the collaboration between geography and ocean engineering disciplines dealing with maritime traffic safety.

Future extensions of this study will include the development of a decision-making system based on real-time maritime traffic data by embedding artificial intelligence (AI) and the V-REMO functionality to reduce the labor of maritime traffic controllers. Therefore, it is necessary to develop an ontology-based context-aware system for maritime traffic accidents to embed AI in the decision-making system. In addition, the UI design of the V-REMO will be enhanced by applying the user evaluation results of this study. Finally, the variables of the closest point of approach (CPA)/time to closest point of approach (TCPA) between two adjacent vessels will be included in the dataset to measure the probability of collision between them. The CPA/TCPA data in the V-REMO will be able to support maritime transportation safety.

Supplementary Materials: The following supporting information can be downloaded at: https://www.mdpi.com/article/10.3390/ijgi12030115/s1, Supplementary File S1: the questionnaires used in the user evaluation of this study translated in English; Supplementary File S2: the responses from the user evaluation participants of this study translated in English.

Author Contributions: Hyowon Ban and Hye-jin Kim contributed to the conceptualization; methodology; software; validation; formal analysis; investigation; resources; data curation; writing—original draft preparation, review and editing; visualization; supervision; and project administration of the study. Hye-jin Kim contributed to the funding acquisition. All authors have read and agreed to the published version of the manuscript.

Funding: This research was supported by a grant from the Korean National Research and Development Project, "Development of Technology of VTS Big Data and Safety Information System," funded by the Korea Coast Guard, Republic of Korea (1535000332-201904962).

Data Availability Statement: Restrictions apply to the availability of these data. Data were obtained from KRISO. Please contact http://www.kriso.re.kr (accessed on 7 March 2023) for inquiries.

Acknowledgments: The authors acknowledge the contribution made by the project team members at KRISO, who provided technical support for the development of V-REMO.

Conflicts of Interest: The authors declare no conflict of interest. The funders had no role in the design of the study; in the collection, analyses, or interpretation of data; in the writing of the manuscript; or in the decision to publish the results.

References

1. Chang, S.J. Development and Analysis of AIS Applications as an Efficient Tool for Vessel Traffic Service. In Proceedings of the Oceans '04 MTS/IEEE Techno-Ocean '04 (IEEE Cat. No.04CH37600), Kobe, Japan, 9–12 November 2004; Volume 4, pp. 2249–2253. [CrossRef]
2. Mazzarella, F.; Vespe, M.; Damalas, D.; Osio, G. Discovering Vessel Activities at Sea Using AIS Data: Mapping of Fishing Footprints. In Proceedings of the 17th International Conference on Information Fusion (FUSION), Salamanca, Spain, 7–10 July 2014; pp. 1–7.
3. Jin, J.; Song, G. Bureaucratic Accountability and Disaster Response: Why Did the Korea Coast Guard Fail in Its Rescue Mission During the Sewol Ferry Accident? *Risk Hazards Crisis Public Policy* **2017**, *8*, 220–243. [CrossRef] [PubMed]
4. Rothe, D.; Muzzatti, S.; Mullins, C.W. Crime on the High Seas: Crimes of Globalization and the Sinking of the Senegalese Ferry Le Joola. *Crit. Crim.* **2006**, *14*, 159–180. [CrossRef]
5. Kim, T.; Nazir, S.; Øvergård, K.I. A STAMP-Based Causal Analysis of the Korean Sewol Ferry Accident. *Saf. Sci.* **2016**, *83*, 93–101. [CrossRef]
6. Abdul Rahman, N.S.F.; Rosli, Z. An Innovation Approach for Improving Passenger Vessels Safety Level: Overload Problem. *Int. J. Bus. Tour. Appl. Sci.* **2014**, *2*, 61–74.
7. Won, D.; Yoo, S.; Yoo, H.; Lim, J. Complex Adaptive Systems Approach to Sewol Ferry Disaster in Korea. *J. Open Innov. Technol. Mark. Complex.* **2015**, *1*, 22. [CrossRef]
8. Korea Maritime Safety Tribunal. Available online: https://www.kmst.go.kr/eng/page.do?menuIdx=227 (accessed on 31 December 2022).
9. Gionet, P. *Public Vessel Operator's Study Guide*; New York State Office of Parks, Recreation and Historic Preservation, Marine Services Bureau: Albany, New York, USA, 2019.

10. Mestl, T.; Tallakstad, K.T.; Castberg, R. Identifying and Analyzing Safety Critical Maneuvers from High Resolution AIS Data. *TransNav Int. J. Mar. Navig. Saf. Sea Transp.* **2016**, *10*, 69–77. [CrossRef]

11. Goerlandt, F.; Montewka, J. Maritime Transportation Risk Analysis: Review and Analysis in Light of Some Foundational Issues. *Reliab. Eng. Syst. Saf.* **2015**, *138*, 115–134. [CrossRef]

12. Hansen, M.G.; Jensen, T.K.; Lehn-Schiøler, T.; Melchild, K.; Rasmussen, F.M.; Ennemark, F. Empirical Ship Domain Based on AIS Data. *J. Navig.* **2013**, *66*, 931–940. [CrossRef]

13. Harati-Mokhtari, A.; Wall, A.; Brooks, P.; Wang, J. Automatic Identification System (AIS): Data Reliability and Human Error Implications. *J. Navig.* **2007**, *60*, 373–389. [CrossRef]

14. Mou, J.M.; Tak, C.v.d.; Ligteringen, H. Study on Collision Avoidance in Busy Waterways by Using AIS Data. *Ocean. Eng.* **2010**, *37*, 483–490. [CrossRef]

15. Pallotta, G.; Vespe, M.; Bryan, K. Vessel Pattern Knowledge Discovery from AIS Data: A Framework for Anomaly Detection and Route Prediction. *Entropy* **2013**, *15*, 2218–2245. [CrossRef]

16. Perez, H.M.; Chang, R.; Billings, R.; Kosub, T.L. Automatic Identification Systems (AIS) Data Use in Marine Vessel Emission Estimation. In Proceedings of the 18th Annual International Emission Inventory Conference, Baltimore, MD, USA, 14–17 April 2009; Volume 14, p. e17.

17. Ristic, B.; La Scala, B.; Morelande, M.; Gordon, N. Statistical Analysis of Motion Patterns in AIS Data: Anomaly Detection and Motion Prediction. In Proceedings of the 2008 11th International Conference on Information Fusion, Cologne, Germany, 30 June–3 July 2008; pp. 1–7.

18. Silveira, P.A.M.; Teixeira, A.P.; Soares, C.G. Use of AIS Data to Characterise Marine Traffic Patterns and Ship Collision Risk off the Coast of Portugal. *J. Navig.* **2013**, *66*, 879–898. [CrossRef]

19. Laube, P.; Imfeld, S. Analyzing Relative Motion within Groups of Trackable Moving Point Objects. In *Geographic Information Science*; Egenhofer, M.J., Mark, D.M., Eds.; Lecture Notes in Computer Science; Springer: Berlin, Germany, 2002; pp. 132–144. [CrossRef]

20. Rydstedt, L.W.; Lundh, M. An Ocean of Stress? The Relationship between Psychosocial Workload and Mental Strain among Engine Officers in the Swedish Merchant Fleet. *Int. Marit. Health* **2010**, *62*, 168–175.

21. Chauvin, C.; Lardjane, S.; Morel, G.; Clostermann, J.-P.; Langard, B. Human and Organisational Factors in Maritime Accidents: Analysis of Collisions at Sea Using the HFACS. *Accid. Anal. Prev.* **2013**, *59*, 26–37. [CrossRef] [PubMed]

22. Gao, X.; Makino, H.; Furusho, M. Analysis of Ship Drifting in a Narrow Channel Using Automatic Identification System (AIS) Data. *WMU J. Marit. Affairs* **2017**, *16*, 351–363. [CrossRef]

23. Lambrou, M.A.; Rødseth, Ø.J.; Foster, H.; Fjørtoft, K. Service-Oriented Computing and Model-Driven Development as Enablers of Port Information Systems: An Integrated View. *WMU J. Marit. Affairs* **2013**, *12*, 41–61. [CrossRef]

24. Baldauf, M.; Benedict, K.; Fischer, S.; Gluch, M.; Kirchhoff, M.; Klaes, S.; Schröder-Hinrichs, J.-U.; Meißner, D.; Fielitz, U.; Wilske, E. E-Navigation and Situation-Dependent Manoeuvring Assistance to Enhance Maritime Emergency Response. *WMU J. Marit. Affairs.* **2011**, *10*, 209. [CrossRef]

25. Graff, J. E-Maritime: A Framework for Knowledge Exchange and Development of Innovative Marine Information Services. *WMU J. Marit. Affairs.* **2009**, *8*, 173–201. [CrossRef]

26. Varelas, T.; Plitsos, S. Real-Time Ship Management through the Lens of Big Data. In Proceedings of the 2020 IEEE 6th International Conference on Big Data Computing Service and Applications (BigDataService), Oxford, UK, 3–6 August 2020; pp. 142–147. [CrossRef]

27. United States Coast Guard, U.S. Department of Homeland Security. Navigation Center. Available online: https://www.navcen.uscg.gov/?pageName=vtsMain (accessed on 31 December 2022).

28. Marine Traffic Control—Marine Department. Available online: https://www.mardep.gov.hk/en/pub_services/ocean/vts.html (accessed on 31 December 2022).

29. Swedish Maritime Administration. Shipping of the Future. Available online: https://www.sjofartsverket.se/en/ (accessed on 31 December 2022).

30. Port of London Authority. About London VTS. POLA2012. Available online: http://www.pla.co.uk/Safety/Vessel-Traffic-Services-VTS-/About-London-VTS (accessed on 31 December 2022).

31. Korea Coast Guard. Available online: https://www.kcg.go.kr/kcg/vts/main.do (accessed on 31 December 2022).

32. Kaklis, D.; Eirinakis, P.; Giannakopoulos, G.; Spyropoulos, C.; Varelas, T.J.; Varlamis, I. A Big Data Approach for Fuel Oil Consumption Estimation in the Maritime Industry. In Proceedings of the 2022 IEEE Eighth International Conference on Big Data Computing Service and Applications (BigDataService), Newark, CA, USA, 15–18 August 2022; pp. 39–47. [CrossRef]

33. Demšar, U.; Virrantaus, K. Space–Time Density of Trajectories: Exploring Spatio-Temporal Patterns in Movement Data. *Int. J. Geogr. Inf. Sci.* **2010**, *24*, 1527–1542. [CrossRef]

34. Troupiotis-Kapeliaris, A.; Spiliopoulos, G.; Vodas, M.; Zissis, D. Navigating through Dense Waters: A Toolbox for Creating Maritime Density Maps. In Proceedings of the 12th Hellenic Conference on Artificial Intelligence, SETN '22, New York, NY, USA, 7–9 September 2022; pp. 1–4. [CrossRef]

35. Hurni, L. Multimedia Atlas Information Systems. In *Encyclopedia of GIS*; Shekhar, S., Xiong, H., Eds.; Springer: Boston, MA, USA, 2008; pp. 759–763. [CrossRef]

36. Elzakker, C.P.J.M.v.; Ooms, K. Understanding Map Uses and Users. In *The Routledge Handbook of Mapping and Cartography*; Routledge: Oxfordshire, UK, 2017.
37. Roth, R.E.; Çöltekin, A.; Delazari, L.; Filho, H.F.; Griffin, A.; Hall, A.; Korpi, J.; Lokka, I.; Mendonça, A.; Ooms, K.; et al. User Studies in Cartography: Opportunities for Empirical Research on Interactive Maps and Visualizations. *Int. J. Cartogr.* **2017**, *3* (Suppl. S1), 61–89. [CrossRef]
38. Medyńska-Gulij, B.; Zagata, K. Experts and Gamers on Immersion into Reconstructed Strongholds. *ISPRS Int. J. Geo-Inf.* **2020**, *9*, 655. [CrossRef]
39. Kramers, E.R. Interaction with Maps on the Internet—A User Centred Design Approach for The Atlas of Canada. *Cartogr. J.* **2008**, *45*, 98–107. [CrossRef]
40. Wielebski, Ł.; Medyńska-Gulij, B. Graphically Supported Evaluation of Mapping Techniques Used in Presenting Spatial Accessibility. *Cartogr. Geogr. Inf. Sci.* **2019**, *46*, 311–333. [CrossRef]
41. Ban, H.; Ahlqvist, O. Geographical Counterpoint to Choreographic Information Based on Approaches in GIScience and Visualization. *Int. J. Geospat. Environ. Res.* **2020**, *7*, 4.
42. Gudmundsson, J.; van Kreveld, M.; Speckmann, B. Efficient Detection of Patterns in 2D Trajectories of Moving Points. *Geoinformatica* **2007**, *11*, 195–215. [CrossRef]
43. Laube, P.; Purves, R.S. An Approach to Evaluating Motion Pattern Detection Techniques in Spatio-Temporal Data. *Comput. Environ. Urban Syst.* **2006**, *30*, 347–374. [CrossRef]
44. Laube, P.; Imfeld, S.; Weibel, R. Discovering Relative Motion Patterns in Groups of Moving Point Objects. *Int. J. Geogr. Inf. Sci.* **2005**, *19*, 639–668. [CrossRef]
45. Chavoshi, S.H.; De Baets, B.; Neutens, T.; Ban, H.; Ahlqvist, O.; De Tré, G.; Van de Weghe, N. Knowledge Discovery in Choreographic Data Using Relative Motion Matrices and Dynamic Time Warping. *Appl. Geogr.* **2014**, *47*, 111–124.
46. Li, G.; Liu, M.; Zhang, X.; Wang, C.; Lai, K.; Qian, W. Semantic Recognition of Ship Motion Patterns Entering and Leaving Port Based on Topic Model. *J. Mar. Sci. Eng.* **2022**, *10*, 2012. [CrossRef]
47. Abreu, F.H.O.; Soares, A.; Paulovich, F.V.; Matwin, S. A Trajectory Scoring Tool for Local Anomaly Detection in Maritime Traffic Using Visual Analytics. *ISPRS Int. J. Geo-Inf.* **2021**, *10*, 412. [CrossRef]
48. Yeosu Hang. Available online: http://encykorea.aks.ac.kr/Contents/Index?contents_id=E0036407 (accessed on 31 December 2022).
49. International Maritime Organization. Guidelines for the Installation of a Shipborne Automatic Identification System (AIS) 2003. Available online: https://wwwcdn.imo.org/localresources/en/OurWork/Safety/Documents/AIS/SN.1-Circ.227.pdf (accessed on 31 December 2022).
50. *Automatic Identification System Message*; Secure Data Repository; Dataset; Korea Research Institute of Ships & Ocean Engineering: Daejeon, Republic of Korea, 2020.
51. ESRI. *ArcGIS Pro: Release 2.8*; Environmental Systems Research Institute: Redlands, CA, USA, 2021.
52. Hettiarachchige, C.; Cavallar, S.v.; Lynar, T.; Hickson, R.I.; Gambhir, M. Risk Prediction System for Dengue Transmission Based on High Resolution Weather Data. *PLoS ONE* **2018**, *13*, e0208203. [CrossRef] [PubMed]
53. Davy, R.; Kusch, E. Reconciling High Resolution Climate Datasets Using KrigR. *Environ. Res. Lett.* **2021**, *16*, 124040. [CrossRef]
54. La Rosa, A.; Pagli, C.; Hurman, G.L.; Keir, D. Strain Accommodation by Intrusion and Faulting in a Rift Linkage Zone: Evidences From High-Resolution Topography Data of the Afrera Plain (Afar, East Africa). *Tectonics* **2022**, *41*, e2021TC007115. [CrossRef]
55. van Westrenen, F.; Baldauf, M. Improving Conflicts Detection in Maritime Traffic: Case Studies on the Effect of Traffic Complexity on Ship Collisions. *Proc. Inst. Mech. Eng. Part M: J. Eng. Marit. Environ.* **2020**, *234*, 209–222. [CrossRef]

International Journal of
Geo-Information

Article

User Evaluation of Thematic Maps on Operational Areas of Rescue Helicopters

Łukasz Wielebski * and Beata Medyńska-Gulij

Adam Mickiewicz University, Poznań, Poland; Department of Cartography and Geomatics,
61-712 Poznań, Poland
* Correspondence: lukwiel@amu.edu.pl; Tel.: +48-61-829-6310

Abstract: This article presents the results of research on users concerning six thematic maps made with various mapping techniques and related to various aspects of the activities of the Helicopter Emergency Medical Service. The aim of the survey was to determine how the respondents rank these maps in terms of the four subjective evaluation criteria, which were the graphical attractiveness of maps, the readability of maps, the usefulness and importance of information, and the complexity of information presented on the maps. The greatest discrepancies were noted for the dot map, while the flow map obtained the most consistent evaluations. To check what the respondents were guided by while building the ranking for each criterion, a catalog of factors was created, the importance of which was assessed using the Likert scale. In the case of graphical attractiveness, users attach particular importance to the arrangement of objects visible on the map. The speed of reading the information is particularly important for map readability. In the case of the usefulness and importance of the information, the map topic, important for saving health and life from the user's point of view, was of the greatest importance, while the amount of information in the legend significantly influenced the evaluation of information complexity.

Keywords: thematic maps; ranking; subjective evaluation; mapping techniques; graphical attractiveness of map; map readability; usefulness and importance of information; information complexity; user study; Helicopter Emergency Medical Service

Citation: Wielebski, Ł.; Medyńska-Gulij, B. User Evaluation of Thematic Maps on Operational Areas of Rescue Helicopters. *ISPRS Int. J. Geo-Inf.* **2023**, *12*, 30. https://doi.org/10.3390/ijgi12020030

Academic Editors: Florian Hruby and Wolfgang Kainz

Received: 29 October 2022
Revised: 12 January 2023
Accepted: 14 January 2023
Published: 18 January 2023

1. Introduction

Research on users is highly important when it comes to controlling the effectiveness of cartographic products and giving recommendations to create good maps [1]. Knowledge of map design is passed in the form of more or less precisely described rules [2–4]; it has been worked out over the years and has been formalized in classic cartography textbooks, and further research continues to expand it [4–9]. Some of those rules result from the cartographic tradition, and some are justified by empirical scientific studies, frequently carried out with the application of new and advanced research techniques [10]. Such studies may be based on professional or public users [11] and may be carried out on-site or online, mainly via online surveys whose advantage is that they may reach a wide audience.

Depending on the type of the map, rules and recommendations on map design are centered around different problems related to their design, and the testing of alternative solutions may take place with the application of different research techniques, whose choice is crucial to the results obtained, their analysis opportunities, and formulating conclusions later [12,13]. As far as thematic maps [14] are concerned, it is significant to select the mapping technique and visual variables adequate for the nature of the phenomenon demonstrated. Such studies often consist of comparing graphical solutions used for presenting the same spatial phenomena, but worked out by means of different mapping techniques [11,15,16] that consider different levels of complexity of the phenomenon [17], or in the context of specific tasks of different complexity that can be carried out based on those maps [18], and of the number of pieces of information being read [19], as well as, e.g., the role of numbers in quantitative information reading [20]. The research may include criteria of evaluation that constitute objective or subjective measurements [17].

Valuation may consist in evaluating each map separately and independently, such as in the case of rating. Rating means that different maps may receive the same notes, whereas ranking consists of arranging maps in the defined order. According to Oldendick [21], "Ranking is a question response format used when a researcher is interested in establishing some type of priority among a set of objects". In surveys, the ranking method is applied less frequently than the rating method [22,23]. Advantages and disadvantages of rating and ranking were analyzed in detail by Rokeach [24], Alwin and Krosnick [25], Ovadia [26] or Tarka [27], among others. In rating, there is a tendency of the respondents to evaluate all objects or traits relatively high or low, which may result in the data obtained from two different respondents not being comparable with each other [27]. Ranking, as opposed to rating, is not as susceptible to the impact of the respondent's lack of motivation, which, when it comes to rating, results in the lack of diversity in evaluation. A disadvantage of the ranking method pointed out by many authors (among others, Oldendick [21], Alwin and Krosnick [25], Hino and Imai [28], Chiusole and Stefanutti [29], Heyman and Sailors [30]) is a greater cognitive load and the effort that the respondent has to put into the evaluation, which is also associated with longer time needed to complete the task. Rating scales are easier to use, and more convenient, which determines their popularity [29,31], whereas ranking favors the diversification of results [30], while being more difficult at the same time.

Rankings can be created for a small or large number of stimuli (items) [32], which in the latter case often means that "ranking datasets often contain missing or tie rankings, which are impossible to analyze without computers"[32]. The increase in the number of stimuli increases the degree of difficulty, and sometimes it also requires reaching for solutions in which the respondent focuses only on, e.g., the three best items, in their opinion, out of many available [33,34]. In the case of rankings in which the possibility of taking an equivalent place is allowed (rankings built in a secondary manner on the basis of objective criteria, e.g., cases where the final ranking is created based on the number of points, or time, as in in sport competitions), the problem becomes which strategy to adopt to rank the equal measurements [35].

According to McCarty et al. [36], the method by which respondents rank the items before the rating provides more differentiation and lower final scores than a method with simple rating only. Similar recommendations are made by Chiusole and Stefanutti [29], whose research results suggest that "when a subject performs both tasks, ranking should come before rating", and this makes the reliability of the responses better.

The solutions used in the graphical user interface of the online survey are not without significance for the course of the ranking creation process. Various solutions are known that allow the creation of a ranking, such as drag and drop, numeric entry, radio button, text box or select box [37,38]. The ranking method can also be used in e-mail surveys [39], although this is associated with greater difficulties for respondents and researchers.

The strategy of comparing and evaluating cartographic visualizations may also take a different form. It may consist in comparing pairs of maps (Paired Comparison Technique), but the analysis of results collected in this way may be problematic; therefore, this method is rarely used in practice [40]. The visual analogue scale may be another example of a research method that allows one to compare and evaluate maps [41]. The visual analogue scale (VAS) takes the form of a horizontal scale with opposite definitions on the opposite ends. In many studies, VAS is frequently presented as a remedy for the flaws of the Likert scale and the ranking, as well as the Paired Comparison Technique [41]. The study by Cawthon and Moere [42], related to cartographic visualization, in which respondents used such a scale to evaluate the visualization in terms of esthetics, is an example of that solution.

This research focuses on the problem of evaluation by the method of subjectively ranking several thematic maps perceived by the user in a single thematic visualization. The study is concentrated on the ranking of several cartographic visualizations related to a geographical phenomenon that is important and comprehensible to each citizen, i.e., different aspects of the activity of the Helicopter Emergency Medical Service (HEMS). The issue of formulating the criteria of subjective evaluation of a cartographic visualization of such aspects by respondents still remains to be solved. Researching factors that influence the decisions of survey participants was crucial to the analysis of the ranking results. Suggested special measurements and graphical ways of presenting results of the research were necessary for the analysis and synthesis in terms of drawing general and detailed conclusions.

The objective of the research was to study several thematic maps by means of the subjective ranking method in terms of one geographical phenomenon (here, the collection of data on different aspects of the HEMS' activity) in the context of graphical attractiveness, map readability, usefulness and importance of information, and information complexity.

To meet the study objective, the following questions were raised:

- How often did respondents indicate particular rankings for specific thematic maps and criteria of subjective evaluation?
- What is the level of unambiguity in respondents' answers?
- What was the final ranking of mapping techniques in terms of particular criteria, considering all rankings and importance attributed to them?
- What factors do users take into consideration in creating the ranking according to specific criteria?
- What techniques of mapping spatial data on HEMS are preferred by users?

2. Materials and Methods

To meet the objective and answer the above questions, four main research stages were adopted:

- — to create thematic maps (Section 2.1);
- — to create the procedure and carry out surveys among map users (Section 2.2);
- — to conduct statistical and graphical analysis of the results: users' rankings and final ranking (Sections 3.1 and 3.2); respondents' motivations (Section 3.3).

2.1. Cartographical Materials

Thematic maps presenting different information about the work of the HEMS in Poland have been used as the research material. Each of six maps marked with the letters A–F has been made in a different mapping technique and presents a different set of data on the HEMS. Maps are visible in Figure 1 and in such form as they were presented to respondents during the survey.

Six "island" thematic maps were worked out at the same scale in square frames making tiles with the gray, neutral background [25]. Above them there was the main title of the entire map composition: Information on the activities of the Polish Medical Air Rescue (PMAR). The size of the figure, along with the set of maps, was 1185 × 811 px and was adjusted in such a way as to display the entire map in the survey window in the Internet browser on screens with the most popular definition 1920 × 1080 (Full HD). Thus, it was possible to compare six maps on the screen at the same time without the necessity of scrolling the page down.

Above each frame there was the title of the phenomenon presented, whereas in empty spaces around island maps there were linear scales and map legends. Apart from the common topic centered around HEMS, the same scale and common graphical elements that repeat themselves on all maps constitute the common denominator for all the maps, making it possible to compare them. Administrative borders of separate provinces [voivodships], drawn in dark graphite, are one of those graphical elements. Locations of HEMS bases, along with the names of those bases, constitute a common reference object.

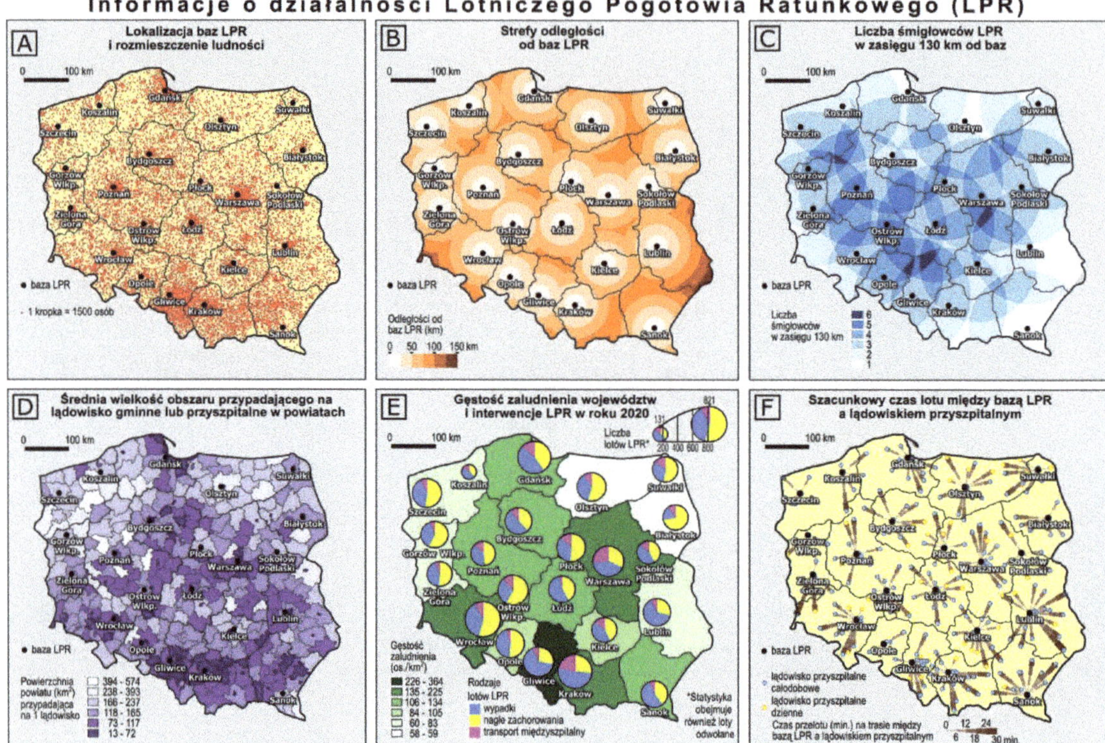

Figure 1. Six thematic maps connected with the topic of Helicopter Emergency Medical Service evaluated in the study (Polish original spelling used in the survey; for details see Figure 2).

During the process of cartographic symbolization, a different color variant was selected for each map to graphically code each piece of geographical information differently [10]. The type of symbols used on a particular map determines which graphical elements are highlighted and also has an impact on the decisions made by users based on those elements [43]. What is crucial to drawing conclusions is the choice of colors and other graphic means, from classic graphical variables [44,45] as well as new visual variables [46–48]. The rule of using the maximum of six classes was adopted in order to ensure good communication and, at the same time, demonstrate a highly diverse level of a given phenomenon. Six maps differed in how much the legend and the title were developed, depending on the topic, and the number of text labels placed in the cartographic content of each map was identical. The most significant traits of thematic maps were collected in Figure 2. Each of the thematic maps is briefly characterized in the following paragraphs.

Map A—a dot map—presents diverse distribution of population on the background of the administrative division of a country by provinces [voivodships] and HEMS bases' locations. A single dot was assigned the number of 1500 people. Intensely red dots contrast with the neutral, light yellow background of cartographic content. Map A has the fewest colors out of all the maps. It has the least number of map symbols and comments in the legend, which makes it rank 6th (last) in terms of complexity of the legend.

Thematic map	Map title in English	Mapping techniques	Color scheme (coloring)	Number of colors	Number of objects (excluding reference objects)	Map title length in the native survey language
A 	Location of HEMS bases and population distribution	dot map	dots: red; cartographic background: light yellow	5	points: 2477	6 words; 45 characters with spaces
B 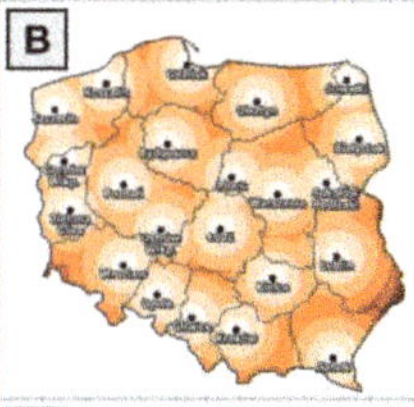	Distance zones from HEMS bases	isoline map - equidistances (isochromatic map)	sequential, single hue: oranges	9	polygons: 65 polygon classes: 6	5 words; 28 characters with spaces
C 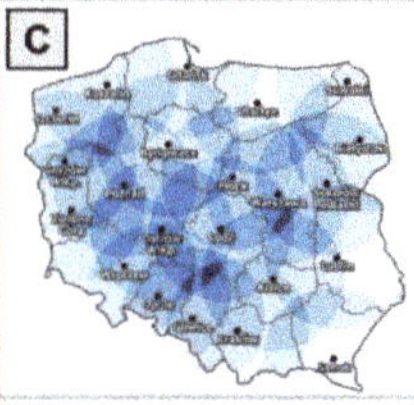	Number of HEMS helicopters within 130 km from the bases	spot ranges map	sequential, single hue: blues	9	polygons: 184 polygon classes: 6	9 words; 45 characters with spaces
D 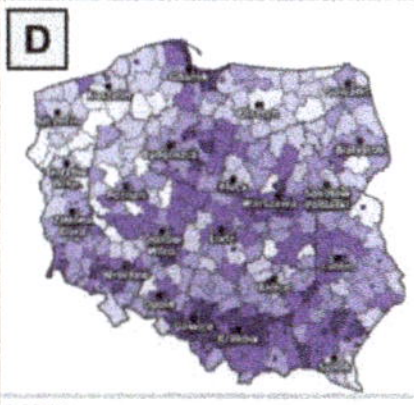	Average size of the area per commune or hospital landing site in counties	choropleth map	sequential, single hue: violetes	9	polygons: 380 polygon classes: 6	11 words; 89 characters with spaces
E 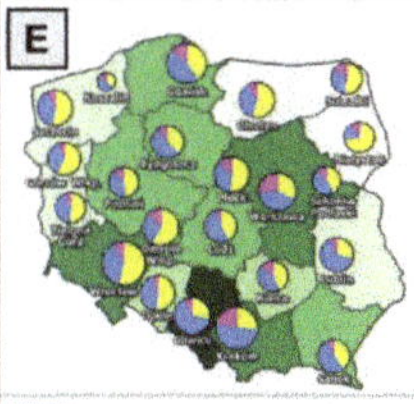	Population density in provinces [voivodships] and HEMS interventions in 2020	choropleth map and proportional point symbol map formatted as a pie chart	choropleth map: sequential, single hue: greens; pie chart: qualitative (cyan, yellow, magenta)	12	polygons: 16 polygon classes: 6 pie charts: 23 pie chart classes: 3	9 words; 60 characters with spaces
F 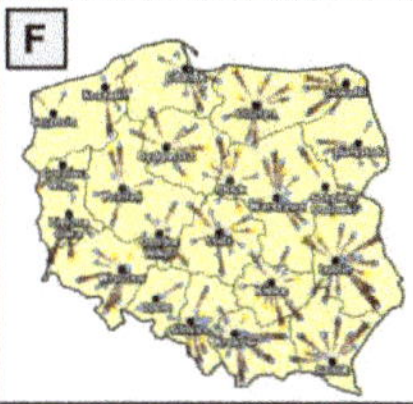	Estimated flight time between HEMS base and hospital landing site	flow map and point signature map	flow map: sequential, single hue: browns; cartographic background: light yellow	11 (+ 2 point outlines colors)	lines: 670 (colored segments) points: 272 line classes: 5 points: 272 point classes: 2	9 words; 65 characters with spaces

Figure 2. Comparison of the characteristics of the six thematic maps evaluated in the study (compare with Figure 1).

Map B—an isoline map—equidistances (isochromatic map)—demonstrates the distances to HEMS bases that define physical (measured by the distance) accessibility of the territory of Poland to helicopters that provide emergency medical service. Map B has the shortest title out of all analyzed titles. The phenomenon presented is graphically coded by means of 6 shades of orange. The map is ranked 5th in terms of the complexity of the legend.

Map C—a spot ranges map—shows the number of HEMS helicopters available within a radius of 130 km. Construction-wise, the map is based on partially overlapping ranges of HEMS bases. Apart from the optimal range of action, the number of overlapping areas that, at the same time, indicates the number of helicopters reaching a particular region is the main focus of the map. In this case, a blue color variant has been adopted.

Map D—a choropleth map—presents the average size of the area that is assigned to a single landing site for HEMS helicopters, divided according to respective administrative units. In this case, the purple color variant has been used. Due to untypical relation, the opposite order of colors has been adopted for those relative data. The data classification of choropleth map classes was established according to the Jenks (natural breaks) method [49]. It is the map with the longest title.

Map E—a choropleth map and proportional point symbol map including pie charts—shows the density of population in provinces (the choropleth map) and the number of the Polish Medical Air Rescue's interventions in 2020 (the proportional point symbol map), formatted as a pie chart showing three types of emergencies: accidents, sudden illnesses, and inter-hospital transfer of patients (pie chart slices distinguishing the types of categories). Proportional point symbols represent the locations of HEMS bases. The area of a given proportional symbol corresponds with the number of interventions shown by the nomogram in the legend. In this case, the units of reference on the choropleth map are maintained at the sequential single-tone green scale, whereas colors used for highlighting the categories of emergencies are sets consisting of three colors, selected from the complementary color wheel (magenta, yellow, cyan) [50]. The data classification into 6 classes of the choropleth map was made by the Jenks classification method. Map E depicts two phenomena and has the largest number of colors and the most complex legend.

Map F—a flow map and point signature map—demonstrates time distances (in 6 min intervals) from Polish Medical Air Rescue bases to the closest hospital landing fields prepared for HEMS helicopters. Lines of the flow map have a variable width that is increasing discontinuously along with the distance to the base, and a sequential color scheme in different shades of brown. The colors were achieved by means of an online app, a sequential color scheme generator [51], allowing one to create sequential color schemes with an easily definable intensity of difference between individual shades, based on the CIEDE2000 method [52]. Landing sites for HEMS helicopters are categorized according to whether they are open 24/7 (blue dots) or just during the day (yellow dots). The map has the same light yellow background as map A, which denotes the area of the country.

2.2. Procedure and Carrying out of Surveys among Map Users

The survey was carried out by the Limesurvey app and the survey form was posted on the website. The evaluation of the map within the experiment consisted of creating a map ranking by each respondent, starting with the map that would meet a given criterion best (ranking 1st) and ending with the one that meets the criterion least (ranking 6th). A classic version of ranking was adopted, in which the user has no opportunity to rank two maps ex aequo, which means that each map had to be separately included in the ranking.

Apart from learning the respondents' subjective evaluation of six specific island maps, factors that respondents took into consideration while evaluating maps were also studied. The information was obtained at the Likert scale with five response categories. Apart from the set of questions related to possible factors, respondents had an opportunity to add their own comments. The results were analyzed with the use of a Libre Calc spreadsheet. Maps were worked out in software QGIS and Inkscape, based on the data by the Polish Medical Air Rescue available online (www.lpr.com.pl/pl/; first access date: 20 April 2021) and the spatial analyses carried out.

The research was conducted on a sample of 104 respondents (45 female, 58 male, 1 person refused to answer), students majoring in different subjects related to land surveying, cartography, and geomatics. The average age of respondents was 22. The participation in the research was voluntary and respondents received no remuneration for it. To be able to present the way the study had been conducted in a more vivid way, a scheme of the survey was prepared [17] on which the way the survey had been organized, and the order of questions (Q) along with numbers (Figure 2) are demonstrated. The online survey included 14 questions (Q1–Q14). The time for completing the survey had no limits and respondents had an opportunity to return to previous questions. Firstly, respondents were asked to fill in the questionnaire, in which they were asked about sex (Q1) and age (Q2). On the page that preceded the proper survey a short film manual (a looped gif file) was placed. It was related to the way of creating rankings and the opportunities of managing particular ranks by means of the interface of the Limesurvey application, based on an intuitive drag and drop solution [37]. It was arranged in this manner as to make sure that all respondents would be able to create a ranking correctly and exhaust all opportunities that the interface of the survey offers. Next, questions (Q3–Q14) were organized in four questions blocks connected with analyzed criteria: graphical attractiveness (GA), readability (MR), usefulness and importance of information (UI), and information complexity (IC) (Figure 2). Each block of questions contained three questions (two mandatory and one optional). The first question in each block of questions was related to the creation of the ranking of six thematic maps prepared with different mapping techniques according to a given criterion (Q3, Q6, Q9, Q12). An example command for the criterion of graphic attractiveness was as follows: "Order the maps according to their graphical (visual) attractiveness. Assign the map with the highest graphical attractiveness to the first position in the ranking (1), and the least graphically attractive map to the last position (6)". The second question was linked to respondents defining to what extent the factors mentioned in the survey affected their choice (Q4, Q7, Q10, Q13). In the case of questions about the evaluation of the impact of different factors on the ranking arranged by each respondent, the variants of answers were constructed according to the Likert scale, with five response categories [definitely YES/rather YES/hard to say/rather NOT/definitely NOT (1)]. For each criterion, 8 sub-questions were asked. In these sub-questions, respondents were asked whether and to what extent they were guided by the certain factors when creating the ranking. The question connected with graphical attractiveness (Q4) was as follows: "Have you taken into account the following factors when assessing the graphical attractiveness of each map (visualization)?":

- GA-1: type of mapping technique used (method of visualization of spatial data);
- GA-2: knowledge of the mapping technique (method of visualization) used;
- GA-3: originality of the graphic solution (unusual way of presenting a given topic);
- GA-4: visualization colors (color variant—type of colors used, e.g., shades of green, blue, purple, etc.);
- GA-5: the method of arranging the objects visible on the map;
- GA-6: visualization complexity (amount of graphically presented information);
- GA-7: beauty and aesthetics of visualization (harmony, information ordering, symmetry, visually pleasing reception);
- GA-8: selected graphic means (visual variables that attract attention, e.g., shape, color, orientation, color brightness, size) and their differentiation (number and variety of colors, shapes, sizes of graphic elements, etc.).

For the map readability (MR) criterion, respondents were obligated to rate following these factors (Q7):

- MR-1: speed of reading information (no need to analyze the transmitted information for a long time);
- MR-2: correct reading of the information (no need to search for it and check it on the map again);
- MR-3: clarity of visualization (graphic distinguishability of details);

- MR-4: the simplicity of the graphic solutions used (e.g., simple figures, instead of complex shapes, 2D instead of 3D);
- MR-5: getting used to a specific mapping technique (common, popular method, commonly used, experience in its use);
- MR-6: legend readability;
- MR-7: visibility of details on the screen with a given scale (100%); no need to rescale the page view;
- MR-8: overlapping of graphic elements (objects creating cartographic content).

On the other hand, for the criterion of usefulness and importance of information (UI), respondents were obligated to rate following factors (Q10):

- UI-1: interesting and absorbing topic of cartographic visualization;
- UI-2: a topic important for saving life and health from one's own point of view (as a potential victim);
- UI-3: a topic important for saving life and health from the point of view of emergency services (as an aid in the organization of rescue);
- UI-4: location of the visualization in the figure (specified order of maps);
- UI-5: selected graphic means accentuating specific information (visual variables that attract attention);
- UI-6: type and amount of information presented on the visualization;
- UI-7: suggestiveness of the information presented (affecting the imagination, the strength of the influence of the information presented on the visualization or the way of perceiving the phenomenon);
- UI-8: new, previously unknown knowledge about the phenomenon.

In the case of the information complexity (IC) criterion, the suggested factors were as follows (Q13):

- IC-1: number of graphic elements in the cartographic content (on the map);
- IC-2: amount of information in the legend;
- IC-3: overlapping graphic elements (objects);
- IC-4: the shape and size of the elements constituting the reference for the data (e.g., borders and areas of voivodeships or counties, the size of the point objects that represent HEMS bases);
- IC-5: number and variety of visual variables used;
- IC-6: length of the title that identifies the topic of the map;
- IC-7: the number of mapping methods used within one visualization;
- IC-8: deviation of the shapes of objects from the reference elements (i.e., the location of HEMS bases and the boundaries of administrative units).

The third question allowed respondents to add their own information about other possible factors that they were led by (Q5, Q8, Q11, Q14). Questions received codes, were adjusted to the graphical representation of results, and received references to the attached scheme of the survey, presented in Figure 3.

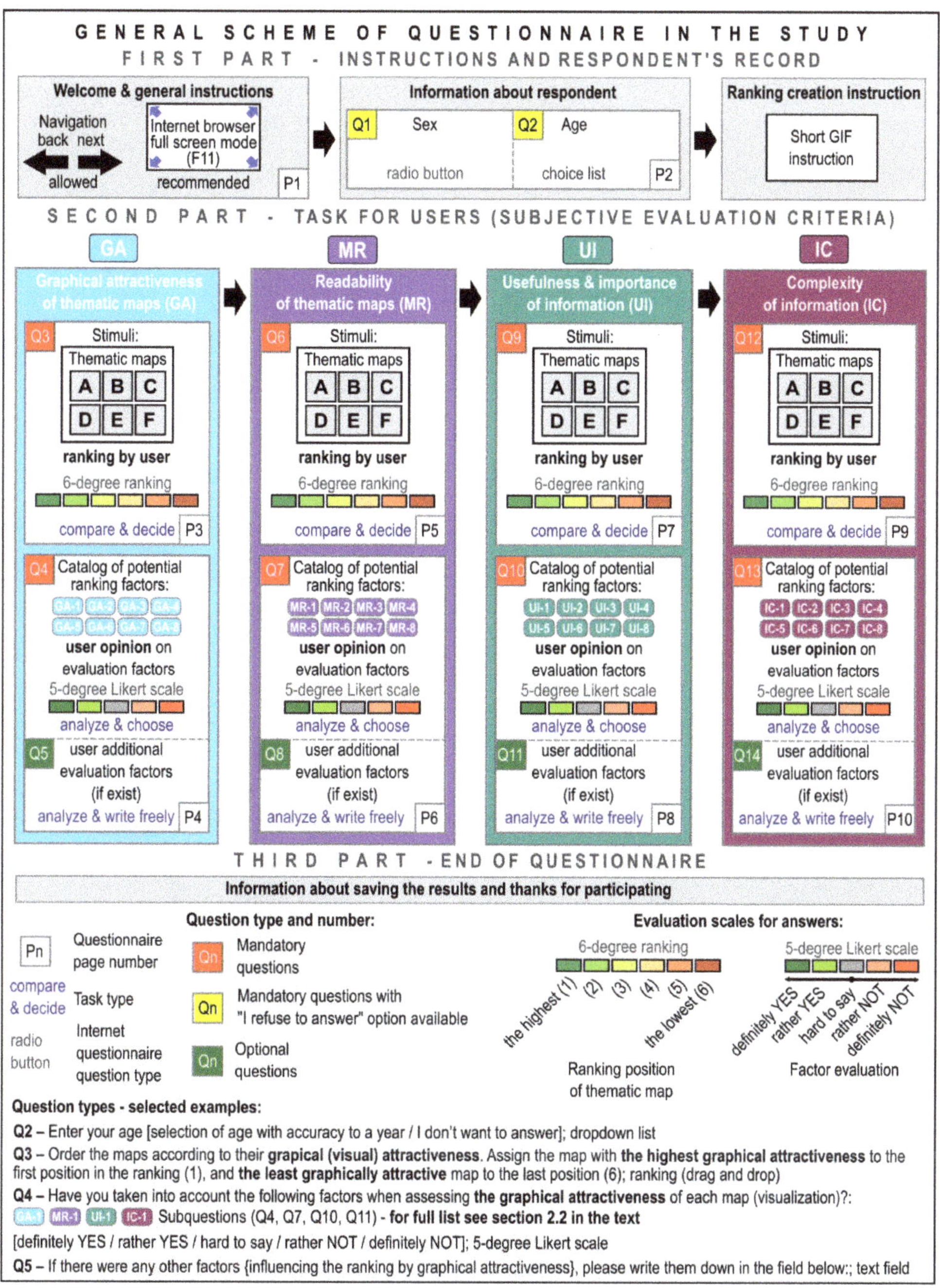

Figure 3. Scheme of the survey—the construction of the online questionnaire used in the study.

3. Results

Figure 4 shows the results of the evaluation of six thematic maps in the graphical form. On four structural bar diagrams, the percentage of consecutive ranks assigned to particular thematic maps as a part of the analyzed criteria of subjective evaluation was demonstrated.

The percentages are arranged according to the ranking. At the top of each bar diagram the highest rank (rank 1—dark green) was placed, and at the bottom the lowest rank (rank 6—red) was placed.

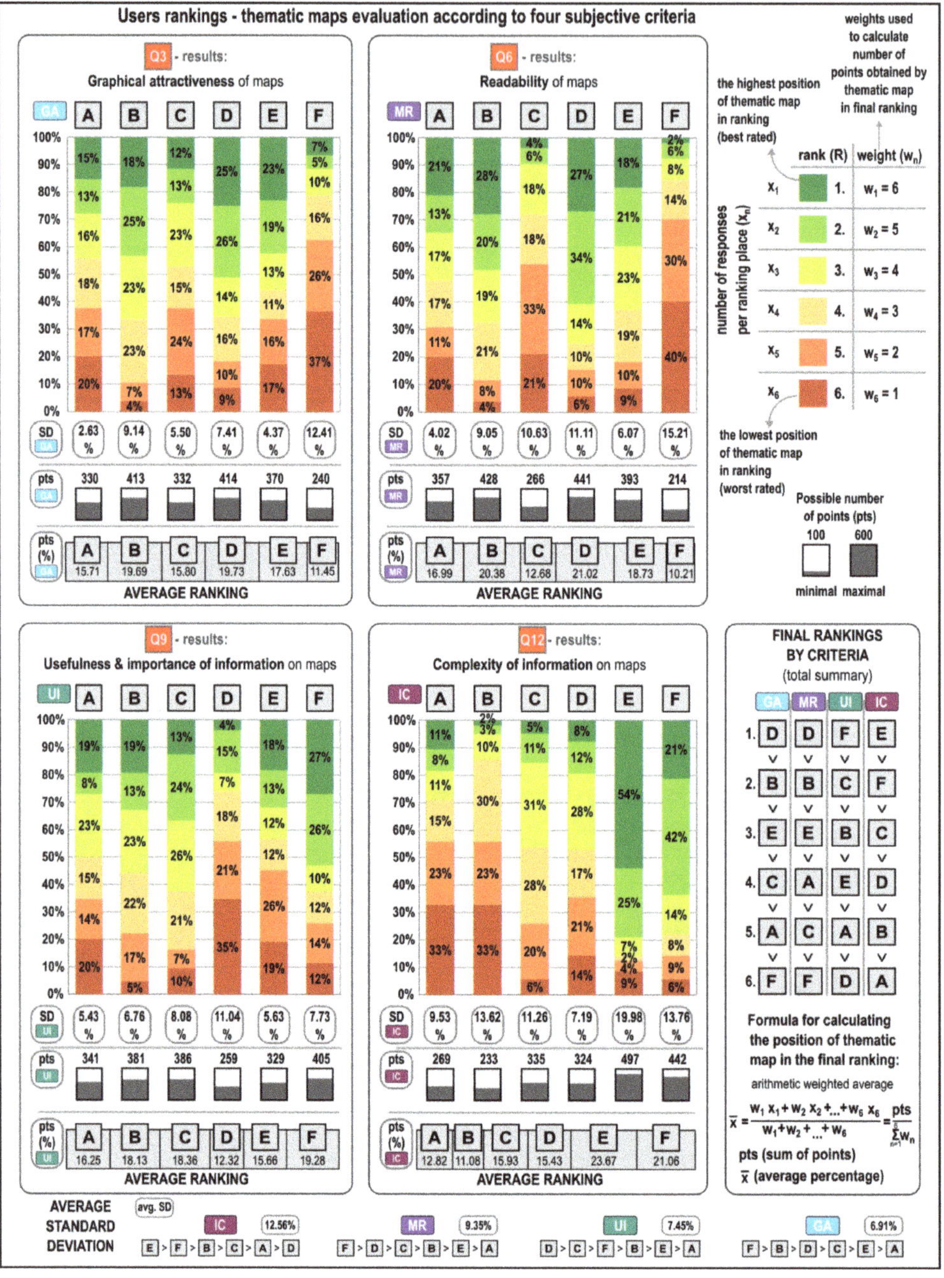

Figure 4. Results for the rankings of thematic maps created by individual participants of the study according to the subjective evaluation criteria and the final rankings created, taking into account all positions in particular rankings and assigned weights according to the formula provided.

Under bar diagrams, the value of standard deviation (SD) that tells by what proportional value the results obtained for particular ranks deviate, on average, was placed. Smaller, gray diagrams present the final number of points for each thematic map (pts). The importance is assigned in the order that is opposite to the ranks in the ranking, i.e., the best rank (with the value of 1) is assigned the importance of 6, and the worst rank (with the value of 6) is assigned the importance of 1. The way importance was assigned was visible at the top of the figure on the right side. In such an approach, each out of the six maps may obtain between the minimum of 100 points (assuming that all respondents would rank a given map the lowest) to the maximum of 600 points (assuming that all respondents would rank a given map the highest). The final ranking of the six thematic maps (A–F) according to four evaluation criteria GA, MR, UI, IC is demonstrated in the graphics next to the diagrams.

3.1. User Rankings

In terms of the first criterion, graphical attractiveness (GA), map D was ranked highest (the choropleth map) because as much as 25% of respondents decided the map should rank first (Figure 4). Slightly fewer respondents (23%) considered map E to be the most graphically attractive. Map F ranked the lowest in that category (ranked the lowest by 37% of all answers). Considering the distribution of answers, the greatest differences between percentages obtained for particular ranks occurred for map F (SD = 12.73%), which proves that respondents were unanimous about the evaluation of that map. A smaller disproportion in the distribution of evaluations was observed for map B (SD = 9.14%), then map D (7.41%), C (5.50%) and E (4.37%). Map A had the most similar distribution of evaluations in terms of rank and the lowest value of standard deviation (2.63%), which proved that the evaluation of a dot map was not unambiguous.

When it comes to the map readability (MR) criterion, map B was ranked first (28%), and map D ranked second (27%). Visibly, the lowest result was achieved by map F, ranked last by 40% of respondents and last but one by 30% of respondents. Furthermore, map C was considered not very transparent either (ranked last according to 21% of responses and last but one according to 20% of responses). In terms of map readability, the highest unanimity occurred for map F (SD = 15.21, and the slightly lower value of standard deviation was achieved by map D (11.11%), then C (10.63%), B (9.05%), and E (6.07%). Answers were distributed most evenly for map A (standard deviation for the results achieved for particular ranks was 4.02%). That method was considered the most readable by 21% of respondents and the least readable by 20% of respondents. Opinions about the evaluation of that method were, thus, the most divided.

As far as usefulness and importance of information (UI) are concerned, map D was ranked last or last but one in the ranking (34.62% and 21.15%, respectively) whereas map F had a high percentage of responses, indicating the first and the second rank. Map D had the most diverse evaluations among all respondents (SD = 11.04%), which also means the most unambiguous evaluation. Map A received the most evenly distributed evaluations (SD = 5.43%). Other maps received the SD value of 8.08% for map C, 7.73% for map F, 6.76% for map B, and 5.63% for map E.

In terms of information complexity (IC), map E was considered to be the most complex (ranked first according to 54% of respondents and second according to 25%, which gives 78.85% in total). Map F also ranked highly (ranked first according to 21% of respondents and second according to 42% of respondents). Maps B and A were considered the most simple in terms of information complexity, ranking first by 33% of respondents and second by 23% of them. The standard deviation of map E is 19.98%, and that is the highest result obtained in the entire study, proving that respondents were strongly convinced that the level of information complexity in that map was high. The smallest differences were noted for map D in that category (SD = 7.19%).

On the graphics on the right side in Figure 3 it is visible that map D ranked first in the final ranking of graphical attractiveness (GA) (414 pts), map B received only one point less (413 pts), and map F, with a visibly smaller number of points, ranked last (240 pts). Map D turned out to be the most readable (MR) (441 pts), and map F the least readable (214 pts). In the final ranking of maps in terms of usefulness and importance of information (UI), map F ranked the highest, presenting time distances from HEMS bases to the closest hospital landing sites prepared for HEMS helicopters (405 pts). The information presented on map D (the choropleth map), related to the size of the area assigned to one local or hospital landing site, was considered the least useful and relevant (259 pts). Users decided that, as far as the criterion of information complexity (IC) was concerned, map E was the most frequently indicated thematic map (442 pts). At the same time, it was also a map in which two mapping techniques had been used and the range of the information included in the map was the widest (compare Figure 2). Map F, demonstrating time distances, in which two types of mapping techniques were also included, and the range of information presented also included qualitative data (the type of hospital landing sites), ranked second. Map C ranked third in terms of information complexity (335 pts). Map B ranked the lowest in the entire ranking (233 pts).

3.2. Multicriteria Final Ranking—Four Subjective Evaluation Criteria in Total

By combining the points obtained for all four subjective criteria, map E (1589 points) would rank first, map B would rank second (1455 pts), then map D (1438 pts), C (1318 pts), F (1302 pts), and map A would rank last (1297 pts). Figure 5 demonstrates this ranking. A comparison of the number of points obtained in each category is presented in the column diagram, whereas the total number of points obtained in four categories, along with the information about the percentage contribution of each of them in the final result, is visible on summary-structural diagrams. The maximum possible number of points to be obtained is 2100, and the minimum number 600.

The best result achieved by map E was influenced by the evaluation of information complexity (31.28% of points), then by map readability (24.74%), and also by graphical attractiveness (23.29%) and information usefulness and importance (20.69%). In the case of map B, which ranked second, the overall result is as follows: 29.41% of it is readability of the map, 28.42% graphical attractiveness, 26.17% information usefulness and importance, and 15.99% information complexity. Map D, which ranked third, had the following results of particular categories that determined the overall result: 30.68%—map readability, 28.81%—graphical attractiveness, 22.53%—information complexity and 17.98%—information usefulness and importance. As far as map C was concerned, it ranked fourth, with 29.25% being the information usefulness and importance, 25.38% being information complexity, 25.16% graphical attractiveness, and 20.20% map readability. Map F, ranking fifth, achieved the following results: 33.97%—information complexity, 31.09%—information usefulness and importance, 18.46%—graphical attractiveness, and 16.47%—map readability. Map A, ranking last, obtained: 27.5% in terms of map readability, 26.32% for information usefulness and importance, 25.43% for graphical attractiveness, and 20.76% for information complexity.

3.3. Respondents' Motivations

Figure 6 demonstrates the results related to answers to the question of what respondents were guided by when evaluating thematic maps and to what extent they considered the factors mentioned that were distinguished by means of codes (e.g., GA-1). Apart from detailed results presenting the percentages of answers for particular levels of the Likert scale, the average results were also used. Particular answers from the Likert scale were assigned importance, allowing the quantification of qualitative variables [53] (1 for the answer "Definitely not", 2—"Rather not", 3—"Hard to say", 4—"Rather yes", 5—"Definitely yes"). Calculating the average importance allowed for making it possible to draw general conclusions about the importance of the factor and create the final ranking of those factors (1 to 1.8—Definitely not, 1.9 to 2.6—"Rather not", 2.7 to 3.4—"Hard to say", 3.5 to 4.2—"Rather yes", 4.3 to 5—"Definitely yes"). The factors were specified under diagrams and arranged in order from the most important to the least important, based on the average results. In the case of additional factors indicated by respondents, the following answers appeared for graphical attractiveness (GA): (1) "interesting content, theme of the map"; (2) "Maps on which there is a lot going on discourage from focusing on them. Firstly, I analyzed maps with fewer details"; (3) "easily readable"; (4) "none" and for readability of map (MR): (1) "none". There were no extra factors for other criteria (IC and UI).

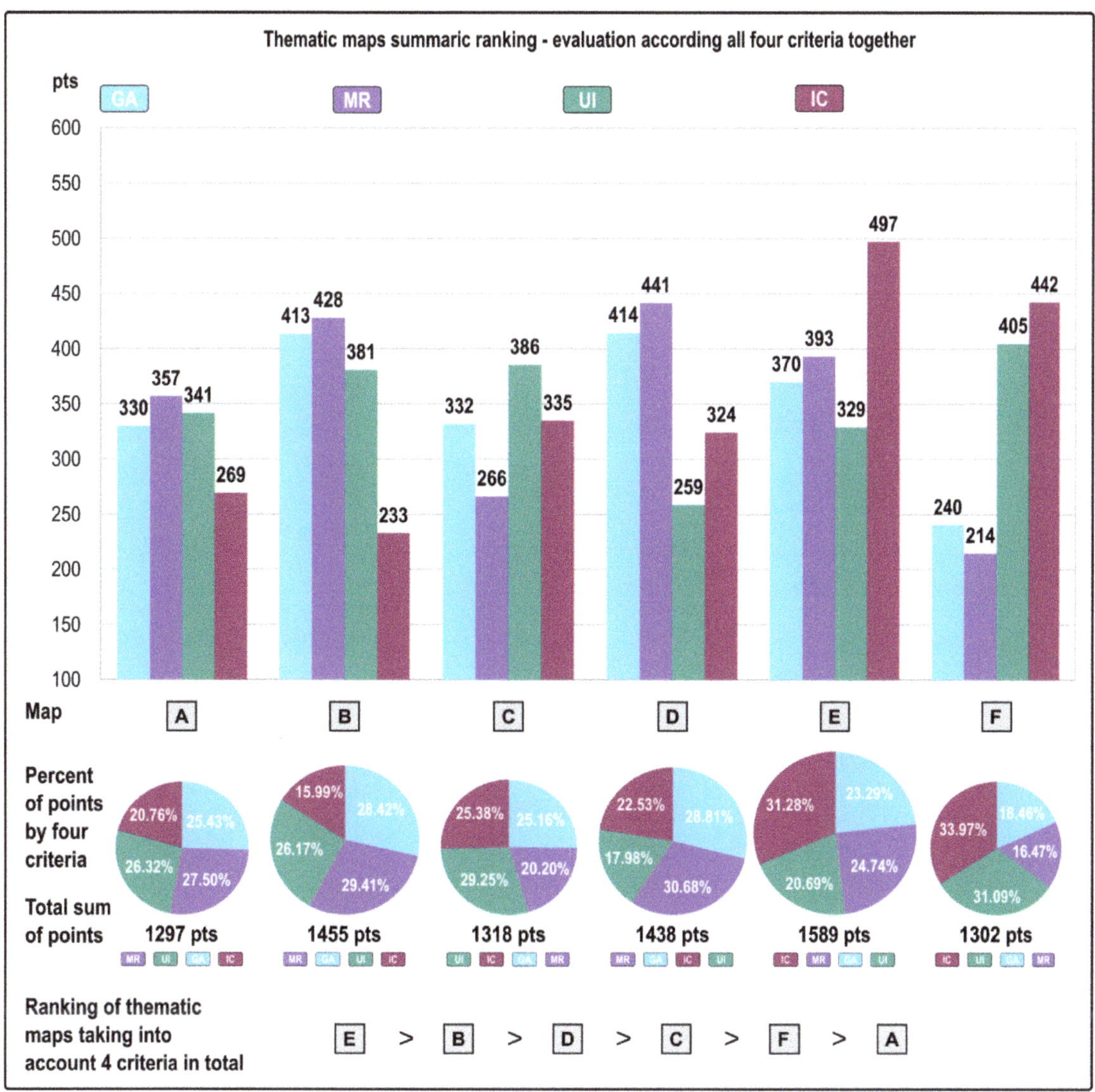

Figure 5. Thematic maps ranking according to four subjective evaluation criteria in comparative and summary terms—total number of points and the percentage share of individual criteria in the overall result.

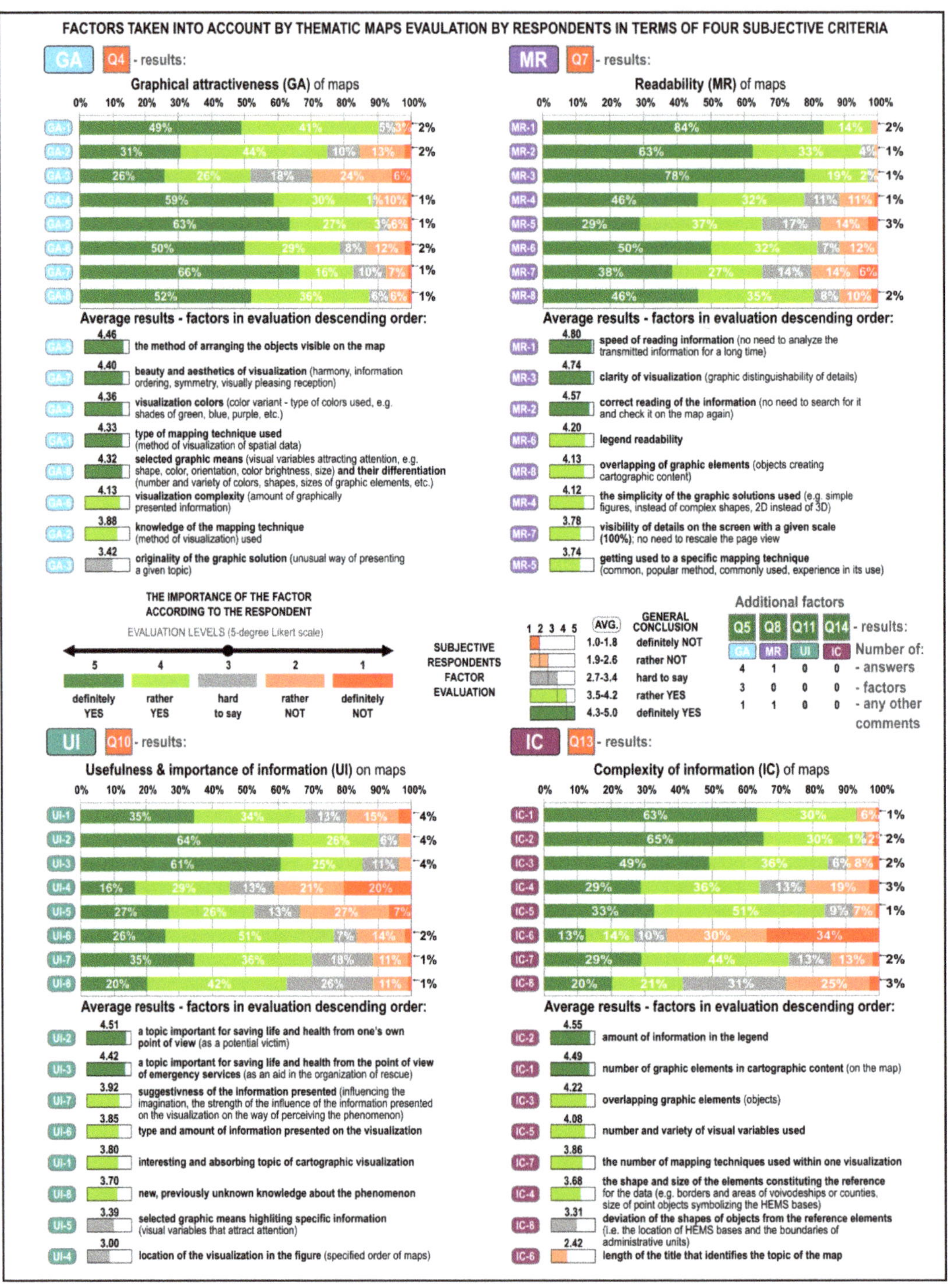

Figure 6. Detailed and generalized (averaged) results of users' evaluation of factors that may affect the way of assessing thematic maps in terms of graphical attractiveness (GA), map readability (MR), information usefulness and importance (UI), and information complexity (IC).

As far as graphical attractiveness (GA) for respondents was concerned, the method of arranging objects visible on the map (GA-5), with the average of 4.46, turned out to be the most relevant to respondents. The type of mapping technique used (GA-1) ranked fourth (4.33). At the same time, it was the factor that obtained the largest total number of positive votes, near 90%. Selected graphic means (GA-8) ranked fifth, with the slightly lower average (4.32). Considering the distribution of answers, it may be concluded that the impact of those five factors was relevant, in respondents' opinion. In each case, nearly 50% of respondents selected the answer "Definitely yes". In the case of visualization complexity (GA-6), the average result was 4.13 due to the relatively larger number of negative votes. The knowledge of the mapping technique used (GA-2) with the average of 3.88 ranked last but one. Generalizing the two last results, it is possible to conclude that they do have an impact on the manner of arranging maps according to the criterion of graphical attractiveness. The originality of the graphical solution ranked last (the average of 3.42; GA-3), which is equivalent to the result "Hard to say".

The largest number of "Definitely yes" answers was given for the factor of map readability (MR), specifically for how fast the information was read (84%), and along with "Rather yes" answers the result was 98% (the average of 4.8). Clarity of visualization (MR-3), understood as the opportunity to distinguish between details, was a factor of map readability that ranked second, with the average of 4.74. Correct reading of the information (MR-2) with the average of 4.57 ranked third. The result for those factors may be generally defined as "Definitely yes". Legend readability (MR-6) ranked fourth (4.20), followed by overlapping graphic elements (4.13; MR-8), simplicity of the graphic solutions used (4.12; MR-4), visibility of details on the screen with a given scale of 100% (3.78; MR-7), and getting used to a specific mapping technique (3.74; MR-5). Generally, it is reasonable to say that those factors are rather relevant in the creation of the ranking of maps according to their readability.

When it comes to usefulness and importance of information (UI), users would indicate a topic important for saving life and health from one's own point of view (UI-2) the most often, and secondly a topic important for saving life and health from the point of view of emergency services (UI-3). In terms of selected graphic means highlighting specific information (UI-5), the average was 3.39, whereas the location of the visualization in the figure (UI-4) was 3.0, which is equivalent to "Hard to say".

As far as information complexity (IC) was concerned, respondents would pick the "Definitely yes" answer in the case of the amount of information in the legend (the average of 4.55; IC-2), the number of graphic elements in the cartographic content (on the map), with the average of 4.49, ranked second (IC-1). Those two factors could be generalized as the "Definitely yes" answer. Overlapping graphic elements (objects) (the average of 4.22) ranked third (IC-3), then the number and variety of visual variables used (4.08; IC-5), the number of mapping techniques used within one visualization (3.86; IC-7), the shape and size of the elements constituting the reference (3.68; IC-4), the deviation of the shape of objects from reference elements (3.31; IC-8—that question may have been misunderstood), and the length of the title that identifies the theme of the map (2.42; IC-6). The last factor generally was "Rather not", and the highest of all results "Definitely not" (34%).

4. Discussion

The study was carried out in order to examine how public users perceive such traits of cartographic visualizations as graphical attractiveness, map readability, usefulness and importance of information, and complexity of the information presented on the map in the context of specific thematic maps. Researchers can choose from different research strategies that can be adopted for comparing maps. To evaluate maps the ranking method was employed, whose potential in cartographic studies had not been used enough so far, in our opinion. The method of analyzing data becomes a significant problem in interpreting data obtained by means of this method. The result obtained by ranking is a complex result, which makes the analysis much more difficult than in the case of rating. The approach, in which instead of a single result a few partial results were achieved, may hinder their interpretation, particularly a comparison with other criteria of evaluation tested the same way. In order to draw the right conclusions from the ranking of mapping techniques, it is not enough just to look at the first and last rank but one needs to consider all the ranks. Assigning importance to particular ranks and calculating the weighted arithmetic mean results in achieving individual numeric results that allow the creation of the final ranking.

The ranking method benefits a stronger differentiation of maps according to the criteria established, and even the smallest details may determine the order adopted. It increases the accuracy of the research, as even the smallest differences need to be taken into account [25], which is actually highly desirable in terms of cartographic research. It provides a lot more than just the first impression, because the respondent is forced to make a decision. Thus, the quality of the subjective evaluation rises, at the same time increasing the cognitive load of the respondent and the effort that needs to be put in to filling in the survey [21,25]. However, the method used also has its flaws, one of them being the lack of knowledge of how big those differences are, because they can be either very small or very big between consecutive ranks. This results from the orderly character of that measurement scale [54]. The lack of opportunity to rank the compared thematic maps the same may also be a disadvantage. Subjectively, it seems that the respondent should have such an opportunity. However, if one were to consider measurable criteria, as well as the plurality and diversity of factors that may determine the evaluation, the probability of maps achieving exactly the same result seems to be very low, particularly considering the fact that already in the assumption of the research the decision has been made to embrace a diversity of mapping techniques, colors and shapes. It is also confirmed by quantitative attempts to describe the diversity of particular maps analyzed in this research. VAS, already employed in studies of cartographic visualizations, e.g., by Cawthon and Moere [42], constitutes the alternative solution.

Examining which factors affect respondents' choices is as important as obtaining information about users' preferences. The ranking method is conducive to gaining insight into criteria according to which it is possible to make a subjective evaluation. Respondents are forced to make an in-depth analysis, i.e., they have to dive deeper into the factors that differentiate evaluated maps in terms of the set criteria. The method makes users examine maps in more detail and analyze criteria they consider in making a decision. All the presented techniques of analyzing rankings complement one another in a way and bring factual, complex results. The employment of labels and colors, and maintaining accuracy and consistency, was truly significant to facilitate the understanding of graphical presentation of the results. The ranking method worked well for the realization of the adopted goal. It was possible to learn about priorities that respondents are guided by because the ranking was also created to analyze factors that respondents select while building the ranking.

5. Conclusions

The research also allowed the formulation of the following conclusions regarding particular mapping techniques for representing phenomena that are relevant to citizens (here, the Helicopter Emergency Medical Service). The dot map turned out to be the most problematic and, relatively, the most difficult one to evaluate by users. Opinions of respondents were the most divided on that map, and the consensus was the lowest in assigning ranks. The dot map has also received one of the lowest evaluations in terms of graphical attractiveness, usefulness and importance of information and complexity of information. Map A was, on the one hand, distinguished by great simplicity and a small number of colors used. On the other hand, it was characterized by certain discomfort of touching and overlapping dots that is a well-known flaw of this method. Such dichotomy supposedly may be the source of very diverse evaluations by respondents.

The isoline map (Isochromatic map) was the one that was judged the lowest in terms of complexity of information. At the same time, it was evaluated highly in terms of graphic attractiveness, ranking second. Map B turned out to be graphically attractive, readable and the least complex in terms of complexity of information, with usefulness evaluated high. In the ranking considering all criteria in total, map B ranked second.

The spot ranges map received highly diverse ranks in the final ranking in terms of our four criteria: it ranked second in terms of usefulness, third for complexity of information, fourth in terms of graphical attractiveness and fifth in terms of readability. However, considering all criteria, the map ranked fourth.

The choropleth map has turned out to be the most graphically attractive and readable; however, it ranked last in terms of usefulness of information. That mapping technique was, according to respondents, the most readable and attractive despite the large number of enumeration units, the unusual assignment of values, and a cool violet color. It may result from the common usage of that technique for representing statistical data. In the ranking that considers all four criteria, map D ranked third.

The choropleth map, with the proportional point symbol map including pie charts distinguishing the types of emergencies, is characterized by the greatest complexity of information and the highest resoluteness of users, ranking third in graphical attractiveness and readability and fourth in information usefulness. Map E ranked the highest in the total ranking that considered all four criteria.

The map with a flow map and point signature map has turned out to be the least graphically attractive and also the least readable and the most usable. Map F was also the most unambiguously evaluated by users, i.e., respondents would more often indicate a selected rank than other ranks. For that map, high usefulness could be related to the most sensitive piece of information to the health and life of the citizen, i.e., approximate time of the flight between the helicopter base and the hospital landing site. Considering all maps, it is possible to draw the conclusion that complexity of information was evaluated by respondents in the most unambiguous way, map readability ranked second, usefulness and importance of information ranked third, and consensus between respondents was the lowest in terms of the evaluation of graphical attractiveness.

Out of 32 factors highlighted that may have affected the process of map ranking creation (eight per each criterion), 12 of them can be considered definitely significant, the majority of them (five) being related to graphical attractiveness. The speed of reading information was the highest score achieved (criterion: map readability). When it comes to factors that matter in terms of building the ranking, 15 were identified, with the majority (five) being for the map readability criterion. For that criterion, all the predefined factors were treated as the average (answer "Definitely yes" or "Rather yes"). A factor defined as the one that did not matter that much was related to the length of the title that identified the topic of the map and it was related to the complexity of information criterion. None of the factors obtained such a low average to make a statement that it definitely did not matter in establishing the order of the maps evaluated. The authors realize that they have not exhausted the entire issue of the subjective ranking of several mapping techniques that the user can see in parallel in a single view on the computer screen. The research has shown the necessity of undertaking in-depth studies of the properties of particular thematic maps for a separate or complementary analysis of phenomena in geographic space that are relevant to the health and life of society.

Author Contributions: Conceptualization, Beata Medyńska-Gulij and Łukasz Wielebski; methodology, Łukasz Wielebski; software, Łukasz Wielebski; validation, Łukasz Wielebski and Beata Medyńska-Gulij; formal analysis, Łukasz Wielebski and Beata Medyńska-Gulij; investigation, Łukasz Wielebski; resources, Łukasz Wielebski; data curation, Łukasz Wielebski; writing—original draft preparation, Łukasz Wielebski; writing—review and editing, Beata Medyńska-Gulij; visualization, Łukasz Wielebski; supervision, Beata Medyńska-Gulij; project administration, Łukasz Wielebski. All authors have read and agreed to the published version of the manuscript.

Funding: This research received no external funding.

Institutional Review Board Statement: Not applicable.

Informed Consent Statement: Not applicable.

Data Availability Statement: Not applicable.

Conflicts of Interest: The authors declare no conflict of interest.

References

1. Robinson, A.H. *Elements of Cartography*, 6th ed.; Wiley: New York, NY, USA; Chichester, UK, 1995; ISBN 9780471555797.
2. Dent, B.D. *Cartography: Thematic Map Design*, 5th ed.; WCB/McGraw-Hill: Boston, MA, USA, 2007; ISBN 9780072319026.
3. Medynska-Gulij, B. How the Black Line, Dash and Dot Created the Rules of Cartographic Design 400 Years Ago. *Cartogr. J.* **2013**, *50*, 356–368. [CrossRef]
4. Medyńska-Gulij, B. Map compiling, map reading, and cartographic design in "Pragmatic pyramid of thematic mapping". *Quaest. Geogr.* **2010**, *29*, 57–63. [CrossRef]
5. Kraak, M.-J.; Ormeling, F. *Cartography: Visualization of Geospatial Data*, 4th ed.; CRC Press, Taylor & Francis Group: Boca Raton, FL, USA, 2021; ISBN 1138613959.
6. Hake, G.; Grünreich, D.; Meng, L. *Kartographie: Visualisierung Raum-Zeitlicher Informationen*, 8th ed.; Walter de Gruyter: Berlin, Germany, 2002; ISBN 978-3-11-016404-6.
7. Slocum, T.A.; McMaster, R.B.; Kessler, F.C.; Howard, H.H. *Thematic Cartography and Geovisualization*, 3rd ed.; Pearson Prentice Hall: Hoboken, NJ, USA, 2010; ISBN 0138010064.
8. Tyner, J.A. *Principles of Map Design*; Guilford Press: New York, NY, USA, 2010; ISBN 978-1-60623-544-7.
9. Ratajski, L. *Metodyka Kartografii Społeczno-Gospodarczej*, 2nd ed.; Państwowe Przedsiębiorstwo Wydawnictw Kartograficznych im. Eugeniusza Romera: Warszawa, Poland; Wrocław, Poland, 1989; ISBN 9788370000554.
10. Krassanakis, V.; Cybulski, P. Eye Tracking Research in Cartography: Looking into the Future. *IJGI* **2021**, *10*, 411. [CrossRef]
11. Medyńska-Gulij, B.; Wielebski, Ł.; Halik, Ł.; Smaczyński, M. Complexity Level of People Gathering Presentation on an Animated Map—Objective Effectiveness Versus Expert Opinion. *IJGI* **2020**, *9*, 117. [CrossRef]
12. Cybulski, P.; Wielebski, Ł. Effectiveness of Dynamic Point Symbols in Quantitative Mapping. *Cartogr. J.* **2019**, *56*, 146–160. [CrossRef]
13. Horbiński, T.; Cybulski, P.; Medyńska-Gulij, B. Graphic Design and Button Placement for Mobile Map Applications. *Cartogr. J.* **2020**, *57*, 196–208. [CrossRef]
14. Forrest, D. Thematic Maps in Geography. *Int. Encycl. Soc. Behav. Sci.* **2015**, *24*, 260–267. [CrossRef]
15. Forrest, D.; Medyńska-Gulij, B. Which mapping technique for population density is effective, attractive, and suggestive? *Abstr. Int. Cartogr. Assoc.* **2021**, *3*, 1. [CrossRef]
16. Besançon, L.; Cooper, M.; Ynnerman, A.; Vernier, F. An Evaluation of Visualization Methods for Population Statistics Based on Choropleth Maps. *arXiv* **2020**, arXiv:Abs/2005.00324. [CrossRef]
17. Wielebski, Ł.; Medyńska-Gulij, B. Graphically supported evaluation of mapping techniques used in presenting spatial accessibility. *Cartogr. Geogr. Inf. Sci.* **2019**, *46*, 311–333. [CrossRef]
18. Korycka-Skorupa, J.; Gołębiowska, I. Multivariate mapping for experienced users: Comparing extrinsic and intrinsic maps with univariate maps. *Misc. Geogr.* **2021**, *25*, 259–271. [CrossRef]
19. Šašinka, Č.; Stachoň, Z.; Čeněk, J.; Šašinková, A.; Popelka, S.; Ugwitz, P.; Lacko, D. A comparison of the performance on extrinsic and intrinsic cartographic visualizations through correctness, response time and cognitive processing. *PLoS ONE* **2021**, *16*, e0250164. [CrossRef] [PubMed]
20. Korycka-Skorupa, J.; Gołębiowska, I. Numbers on Thematic Maps: Helpful Simplicity or too Raw to Be Useful for Map Reading? *IJGI* **2020**, *9*, 415. [CrossRef]
21. Oldendick, R.W. *Ranking. Encyclopedia of Survey Research Methods*; Lavrakas, P., Ed.; Sage Publications, Inc: Thousand Oaks California, CA, USA, 2008; ISBN 9781412918084.
22. Moors, G.; Vriens, I.; Gelissen, J.P.T.M.; Vermunt, J.K. Two of a Kind. Similarities between Ranking and Rating Data in Measuring Values. *Surv. Res. Methods* **2016**, *10*, 15–33. [CrossRef]
23. Langville, A.N.; Meyer, C.D. *Who's #1?* Princeton University Press: Princeton, NJ, USA, 2012; ISBN 9781400841677.
24. Rokeach, M. *The Nature of Human Values*; The Free Press: New York, NY, USA, 1973; ISBN 9780029267509.
25. Alwin, D.F.; Krosnick, J.A. The Measurement of Values in Surveys: A Comparison of Ratings and Rankings. *Public Opin. Q.* **1985**, *49*, 535–552. [CrossRef]
26. Ovadia, S. Ratings and rankings: Reconsidering the structure of values and their measurement. *Int. J. Soc. Res. Methodol.* **2004**, *7*, 403–414. [CrossRef]
27. Tarka, P. Statistical choice between rating or ranking method of scaling consumers' values. *Stat. Transit.* **2010**, *11*, 177–186.
28. Hino, A.; Imai, R. Ranking and Rating: Neglected Biases in Factor Analysis of Postmaterialist Values. *Int. J. Public Opin. Res.* **2019**, *31*, 368–381. [CrossRef]
29. Chiusole, D.; Stefanutti, L. Rating, ranking, or both? A joint application of two probabilistic models for the measurement of values. *Test. Psychom. Methodol. Appl. Psychol.* **2011**, *18*, 49–60.
30. Heyman, J.; Sailors, J. A Respondent-friendly Method of Ranking Long Lists. *Int. J. Mark. Res.* **2016**, *58*, 693–710. [CrossRef]

31. van Herk, H.; van de Velden, M. Insight into the relative merits of rating and ranking in a cross-national context using three-way correspondence analysis. *Food Qual. Prefer.* **2007**, *18*, 1096–1105. [CrossRef]
32. Lee, P.H.; Yu, P.L.H. An R package for analyzing and modeling ranking data. *BMC Med. Res. Methodol.* **2013**, *13*, 269. [CrossRef] [PubMed]
33. Harzing, A.-W.; Baldueza, J.; Barner-Rasmussen, W.; Barzantny, C.; Canabal, A.; Davila, A.; Espejo, A.; Ferreira, R.; Giroud, A.; Koester, K.; et al. Rating versus ranking: What is the best way to reduce response and language bias in cross-national research? *Int. Bus. Rev.* **2009**, *18*, 417–432. [CrossRef]
34. Finch, H. An introduction to the analysis of ranked response data. *Pract. Assess. Res. Eval.* **2022**, *27*, 1–20. [CrossRef]
35. Barwa, T.M. How Rankings Superiorly Differ Than Ratings? *Int. J. Bus. Manag. Invent.* **2016**, *5*, 9–10.
36. McCarty, J.; Shrum, L.J. Measuring the Importance of Positive Constructs: A Test of Alternative Rating Procedures. *Mark. Lett.* **1997**, *8*, 239–250. [CrossRef]
37. Genter, S.; Trejo, Y.G.; Nichols, E. Drag-and-Drop versus Numeric Entry Options: A Comparison of Survey Ranking Questions in Qualtrics. *J. Usability Stud.* **2022**, *17*, 117–130.
38. Blasius, J. Comparing Ranking Techniques in Web Surveys. *Field Methods* **2012**, *24*, 382–398. [CrossRef]
39. Smyth, J.D.; Olson, K.; Burke, A. Comparing survey ranking question formats in mail surveys. *Int. J. Mark. Res.* **2018**, *60*, 502–516. [CrossRef]
40. Lavrakas, P. Paired Comparison Technique. In *Encyclopedia of Survey Research Methods*; Lavrakas, P., Ed.; Sage Publications, Inc.: Thousand Oaks California, CA, USA, 2008; ISBN 9781412918084.
41. Hoag, J.R.; Kuo, C.-L. Ranking question designs and analysis methods. *J. Med. Stat. Inform.* **2016**, *4*, 6. [CrossRef]
42. Cawthon, N.; Moere, A.V. The Effect of Aesthetic on the Usability of Data Visualization. In Proceedings of the 2007 11th International Conference Information Visualization (IV '07), Zurich, Switzerland, 4–6 July 2007; pp. 637–648.
43. Medynska-Gulij, B. The Effect of Cartographic Content on Tourist Map Users. *Cartography* **2003**, *32*, 49–54. [CrossRef]
44. Medyńska-Gulij, B.; Cybulski, P. Spatio-temporal dependencies between hospital beds, physicians and health expenditure using visual variables and data classification in statistical table. *Geod. Cartogr.* **2016**, *65*, 67–80. [CrossRef]
45. Bertin, J. *Semiologie Graphique: Les Diagrammes, Les Réseaux, Les Cartes*; The Royal Geographical Society: London, UK, 1967.
46. Kraak, M.-J.; Roth, R.; Ricker, B.; Kagawa, A.; Sourd, G. *Mapping for a Sustainable World*; Mouton: New York, NY, USA, 2020; ISBN 9789216040468.
47. Halik, Ł. The analysis of visual variables for use in the cartographic design of point symbols for mobile Augmented Reality applications. *Geod. Cartogr.* **2012**, *61*, 19–30. [CrossRef]
48. Medyńska-Gulij, B. *Kartografia i Geomedia*, 1st ed.; Wydawnictwo Naukowe PWN: Warszawa, Poland, 2021; ISBN 8301215542.
49. Całka, B. Comparing continuity and compactness of choropleth map classes. *Geodesy and Cartography* **2018**, *67*, 21–34. [CrossRef]
50. *Lexikon der Kartographie und Geomatik*; Spektrum, Akad. Verl.: Heidelberg/Berlin, Germany, 2002; ISBN 3827410568.
51. Brychtova, A.; Doležalová, J. Designing Usable Sequential Color Schemes for Geovisualizations. In Proceedings of the 1st ICA European Symposium on Cartography, Vienna, Austria, 10–12 November 2015.
52. Sharma, G.; Wu, W.; Dalal, E. The CIEDE2000 color-difference formula: Implementation notes, supplementary test data, and mathematical observations. *Color Res. Appl.* **2005**, *30*, 21–30. [CrossRef]
53. Kukuła, K. Propozycja budowy rankingu obiektów z wykorzystaniem cech ilościowych oraz jakościowych. *Metod. Ilościowe W Bad. Ekon.* **2012**, *13*, 5–16.
54. Borsboom-van Beurden, J.A.M.; de Jong, K.; De Niet, R.; de Nijs, A.C.M.; Hagen, A.; CGM, K.G.; Verburg, P. *The MAP COMPARISON KIT: Methods, Software and Applications*; Rijksinstituut voor Volksgezondheid en Milieu RIVM: De Bilt, Netherlands, 2004.

International Journal of
Geo-Information

Article

An Empirical Study on the Effects of Temporal Trends in Spatial Patterns on Animated Choropleth Maps

Paweł Cybulski

Department of Cartography and Geomatics, Faculty of Geographical and Geological Sciences, Adam Mickiewicz University Poznan, 61-712 Poznań, Poland; p.cybulski@amu.edu.pl

Abstract: Animated cartographic visualization incorporates the concept of geomedia presented in this Special Issue. The presented study aims to examine the effectiveness of spatial pattern and temporal trend recognition on animated choropleth maps. In a controlled laboratory experiment with participants and eye tracking, fifteen animated maps were used to show a different spatial patterns and temporal trends. The participants' task was to correctly detect the patterns and trends on a choropleth map. The study results show that effective spatial pattern and temporal trend recognition on a choropleth map is related to participants' visual behavior. Visual attention clustered in the central part of the choropleth map supports effective spatio-temporal relationship recognition. The larger the area covered by the fixation cluster, the higher the probability of correct temporal trend and spatial pattern recognition. However, animated choropleth maps are more suitable for presenting temporal trends than spatial patterns. Understanding the difficulty in the correct recognition of spatio-temporal relationships might be a reason for implementing techniques that support effective visual searches such as highlighting, cartographic redundancy, or interactive tools. For end-users, the presented study reveals the necessity of the application of a specific visual strategy. Focusing on the central part of the map is the most effective strategy for the recognition of spatio-temporal relationships.

Keywords: geomedia; cartographic animation; choropleth map; temporal trend; spatial pattern; pattern recognition; eye tracking

Citation: Cybulski, P. An Empirical Study on the Effects of Temporal Trends in Spatial Patterns on Animated Choropleth Maps. *ISPRS Int. J. Geo-Inf.* **2022**, *11*, 273. https://doi.org/10.3390/ijgi11050273

Academic Editor: Wolfgang Kainz

Received: 1 March 2022
Accepted: 15 April 2022
Published: 20 April 2022

Publisher's Note: MDPI stays neutral with regard to jurisdictional claims in published maps and institutional affiliations.

1. Introduction

Dynamic maps, so-called cartographic animations, are part of multimedia cartography [1–3]. The advancement of digital technology has brought significant developments. Existing animated maps have basic time navigation features such as a time sliders, using audio narration instead of a legend, or taking advantage of modern APIs [4,5] and JavaScript libraries [6,7]. Cartographic animation is consistent with the concept of geomedia, which means communicating about geographic space through multimedia related to spatio-temporal dimensions [8]. Owing to this, cartographic animation finds application not only in geovisualization [9], but also in augmented reality systems [10], movies [11], and computer games [12].

However, as Harrower [13,14] correctly noted, the problem with animated maps is their perception. Changes in space on maps require the user's memorization of spatial patterns, and not only comparison, observation, and interpretation. The problem of dynamic maps' perception relates to a multitude of changes such as disappearance, value changes, or movement [15]. Nevertheless, dynamic maps play an important role in visual analysis [16]. Hegarty et al. [17] and Lee et al. [18] gave evidence that people process animated maps through a series of static snapshots rather than a dynamic representation. Juergens [19] suggests that design issues such as map projections could be also a source of misleading interpretation of temporal geographic data.

Much scientific research has dealt with the animated map perception issue. Some research is concerned with the perception of smooth and abrupt changes [20–22]. Other studies are concerned with comparing effectiveness against static maps [23,24]. Part of the research is related to searching for the most efficient solution in animated map design [25,26]. These studies focus on perception issues, e.g., change blindness [27–30].

Eye tracking technology allows one to record eye movements in real time. The main two components of this movement are saccades and fixations [31]. Saccades are rapid gaze movements from one location to another. The second component is fixation, which relates to a relatively stationary gaze on a single location. There is also a specific eye movement, i.e., smooth pursuit. This is slow eye tracking of a moving objects [32]. On dynamic maps, it concerns, among other things, the movement of point objects along the static route [33]. However, the dominant change in the choropleth map is the change in enumeration units [34]. Therefore, the user's visual strategy could rely on the specific sequence of fixations and saccades (e.g., fixation duration and saccadic amplitude). When users search for trends and patterns on a map, the visual strategy could be used to determine the effective performance. The connection between eye tracking measures and the cognitive processing of a map lies in the idea that fixating objects correlate with the cognitive processing of the fixated objects. This assumption is well established in cognitive studies such as Just and Carpenter [35] or Rayner [36] and also in recent cartographic research [37]. Thus, the following research questions arise: What is an effective visual strategy in pattern searching on a dynamic map? Can different spatial patterns be easier to detect? Are components of spatio-temporal relationships recognized with the same level of effectiveness?

Both spatial patterns and temporal trends are the subject of research in the context of the methodology adopted by Andrienko et al. [38]. In this regard, the main aim of this research was to assess the effectiveness of pattern and trend recognition on animated choropleth maps. Eye tracking could be used to help understand the differences between effective and ineffective recognition among study participants. Knowing that a specific spatial pattern or temporal trend is difficult to detect, end-users should be aware of a specific visual strategy that supports correct recognition. Additionally, we state the hypothesis that effective spatial pattern and temporal trend recognition on a choropleth map are related to participants' visual behavior. To verify this research assumption, a map-based empirical study with participants and eye tracking technology was conducted. Therefore, in Section 3, the adopted methodology is presented, in particular: participants, equipment, materials, procedure, and analysis.

2. Related Research

Spatio-temporal data include data classifications according to changes occurring over time [39,40]. One can distinguish existential changes (disappearance/appearance), spatial property changes (visual variables), or thematic property changes (increase/decrease). Earlier, Peuquet [41] distinguished the space (where), time (when), and object (what) components of spatio-temporal data. Andrienko et al. [38] extended this issue. Identification and comparison tasks could serve as elementary searches or general searches since there is the possibility of distinguishing spatio-temporal map-based tasks. Such cartographic tasks do not contradict the phenomena taking place in geographical space. Moreover, many phenomena that change in space and time show regularities [42]. The correct identification of regularities is possible, among others, owing to an appropriate cartographic presentation. To estimate the correctness of the users' perception of these regularities, one needs to design appropriate map-based tasks [43,44].

Regularities in geographical space can be viewed as spatial patterns or temporal trends [45]. The spatial pattern is related to the distribution of a phenomenon in physical or administrative space [46]. In geographic research, one can distinguish several patterns: clustered, peripheral, grid, linear, radial, or dispersed patterns [47–50]. In animated maps, these patterns change over time. However, despite the changes, several trends occur. The most basic temporal trends are an increase in phenomena, a decrease in a phenomenon, or the absence of a trend [51,52]. Of course, a more detailed breakdown of temporal trends considers increases and decreases, e.g., linear or logarithmic. In addition, it can also be cyclical.

Blok et al. [53] presented research on the visualization of the relationship between spatial patterns in time. They focused on synchronization, which is the main dynamic variable in cartographic animation. In their research, they designed an alternative visual exploration of spatio-temporal relationships. The results showed that one synchronous animation is sufficient to display juxtaposed data. However, it was found that, in some cases, the relationship between spatial patterns requires two alternative visualizations. Therefore, they concluded that empirical research is lacking in the determination of the effectiveness of the proposed tools.

Edsall et al. [54] provided tools for the visualization of spatial and temporal periodicity in geographic data. The proposed tools used both idealized and real-world environmental space–time data. Decomposing geographic time series involved using a three-dimensional Fourier transformation into spatio-temporal periodicities. Owing to these tools patterns and trends were easily observed. However, this study did not involve users' visual cognition to understand the difficulties of pattern and trend recognition.

Griffin et al. [55] compared the visual identification of space–time clusters on animated maps and static small-multiple maps. Their research proved that animated maps contributed to the correct identification of patterns, contrary to small-multiple maps. The animation pace supported different types of clusters. Griffin et al. (2006) found a threshold in the animation pace. They identified that if the animation pace is too slow, correct spatial pattern identification decreases among users. However, in their study, clustered patterns moved in time and space, and they did not test the identification of other types of spatial patterns.

Schiewe [56] conducted empirical research on the visual perception of spatial patterns in choropleth maps. He examined the intuitive ranking of color lightness, the neglect of small areas, and the attention to data classification when interpreting spatial patterns. The study participants detected extreme values, central tendencies, and homogeneities. The results confirmed that a legend improves the effective identification of extreme values. Additionally, the detection rate of extreme values in small areas was increased. However, most participants relied on the color lightness distribution and not on the data classification during pattern interpretation.

Karsznia et al. [57] studied effective spatial pattern recognition in choropleth maps with an optimal hexagonal enumeration unit size. They examined nine sizes of hexagonal units and showed that the enumeration unit size impacts users' spatial pattern recognition abilities. In conclusion, they argued that the largest and the smallest units' were not suitable for indicating spatial patterns. Their research delivered the optimal range of EUS (enumeration size units) without indicating sharp boundaries. However, their experimental procedure did not include changes in the spatial pattern over time.

Changes in the detection of global trends and local outliers on an animated choropleth map were studied by Traun et al. [58]. In a user testing study, they found that outlier-preserving value generalization in space supports local outlier identification. However, generalization did not support the participants' ability to correctly detect the spatial pattern movement. They conclude that spatial pattern recognition is related more to personal features (e.g., age) rather than specific map design.

Summarizing this section, it should be mentioned that spatial pattern cognition was well studied based on cartographic and psychological basics. The research gaps exist in the context of temporal trends in spatial patterns. There has been some indication of the effectiveness of patterns in trend recognition. However, the presented study focused more on trends in patterns, especially their effectiveness in terms of correct recognition and, additionally, users' visual behavior during a map-based task.

3. Methodology

3.1. Participants

Eighteen students of the Faculty of Geographic and Geological Sciences at Adam Mickiewicz University participated in the experiment. The participants described their characteristics: age, gender, and education. A total of 10 of them indicated male gender and 8 indicated female gender. The participants' age range was 21–22 years, and they were in the third year of their studies, taking geodetic and cartographic courses. They did not receive payment and participated voluntarily. They could resign from participation at any time in the study.

3.2. Equipment

To display map stimuli, an 27″ ASUS LCD with a 1980 × 1080 px screen resolution was used. A Gazepoint GP3 HD eye tracker with a sampling frequency of 150 Hz recorded the participants' eye movements. The visual angle accuracy was 0.5°, allowing up to 35 cm (horizontal) × 22 cm (vertical) movement. The average distance between the participant and the monitor was approximately 62 cm. The calibration procedure used a 9-point method using Gazepoint Control software (www.gazept.com, accessed 12 December 2021).

3.3. Materials

Fifteen animated choropleth maps of the unemployment rate constituted the main cartographic material. Each choropleth had 63 enumeration units classified into 5 classes of a sequential single-hue greyscale according to ColorBrewer 2.0 (5-class Greys). Each of the animated maps lasted 30 seconds and showed changes over twelve months. The study participants viewed each map once.

In the first month of the animation (January), the maps legibly showed (using one color hue) one of the following patterns: radial, clustered, linear, grid, peripheral, or dispersed (Figure 1). In the following months, the enumeration units changed their hue randomly except for the pattern units. The pattern units did change their hue randomly, but changed in a specific direction. There were two possible directions of this temporal trend: an increase (from light to dark) or a decrease (from dark to light). This indicated the phenomenon intensity. The pattern units did not change their hue altogether, but consequently changed in one of the directions of the trend.

The only exception was the dispersed pattern. In this case, all of the enumeration units changed their hue randomly, even the pattern units. As a consequence, the pattern visible at the beginning of the animation was lost and did not follow any trend. Finally, in the last month of the animation (December), each pattern was again clearly visible with one color hue. Figure 2 present thes frames of the animated maps of linear patterns with an increasing trend and dispersed patterns with no trend.

3.4. Procedure

The procedure consisted of presenting all fifteen maps in a random order to each participant. Before the presentation, each study participant was given an introduction to the purpose of the study. The experiment began with a 9-point calibration reaching a gaze sample score of 90% on average. Secondly, participants performed a set of familiarization tasks. Next, they watched fifteen animated maps. After each map, there were three questions regarding spatial pattern and temporal trend (Figure 3 presents an interactive response protocol that appeared after the animation ended):

- Was there a spatial pattern in (random month except for January and December)? (Possible answers: Yes or No).
- Which spatial pattern does it represent? (One of three maps to choose from; only one was correct).
- Was there a temporal trend in this pattern? (Possible answers: Yes or No; if Yes, then the possible answers were Increase or Decrease)

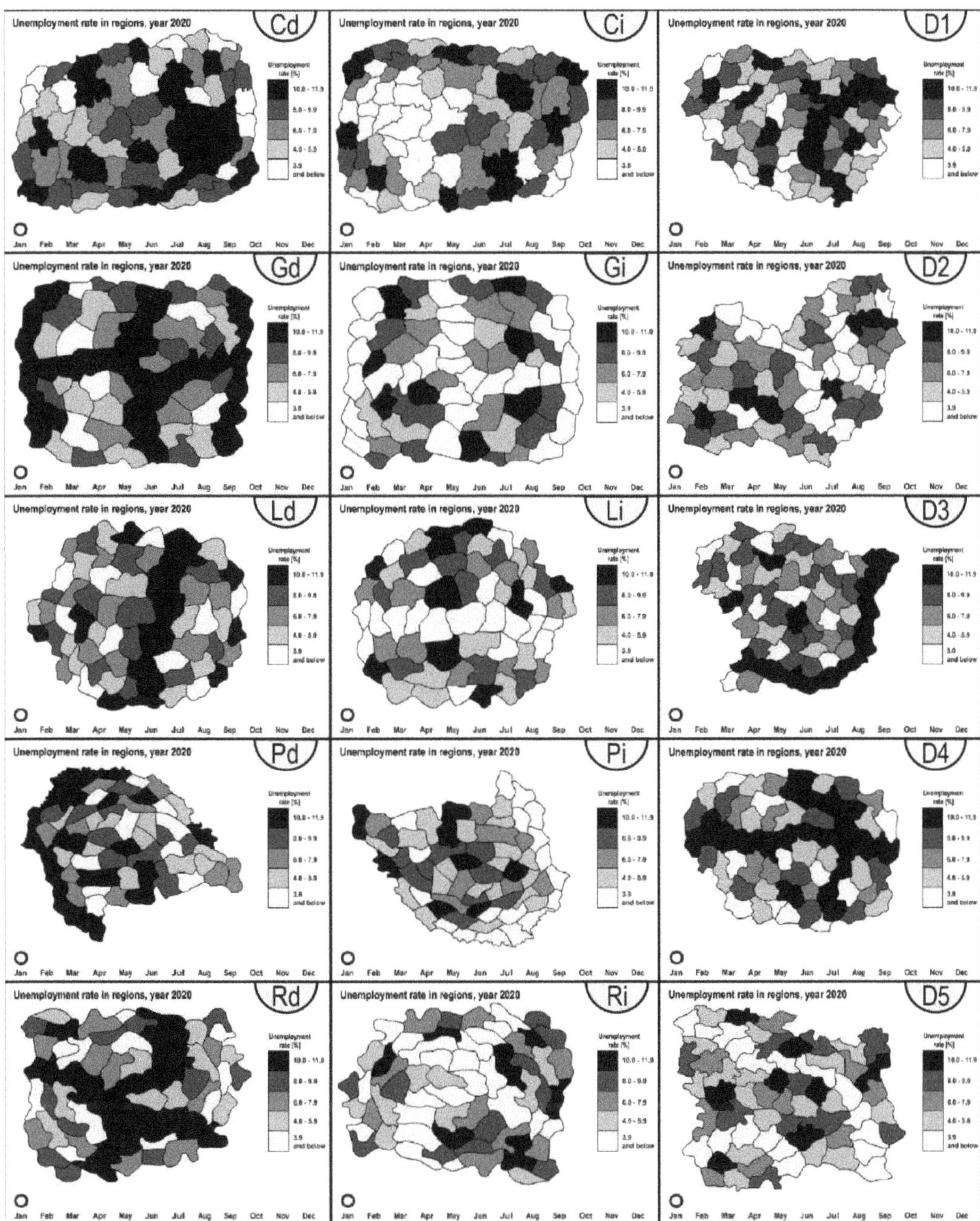

Figure 1. Map stimuli used in the research. Spatial patterns: C: clustered, G: grid, L: linear, P: peripheral, R: radial, D: dispersed. Temporal trends: d: decrease, i: increase.

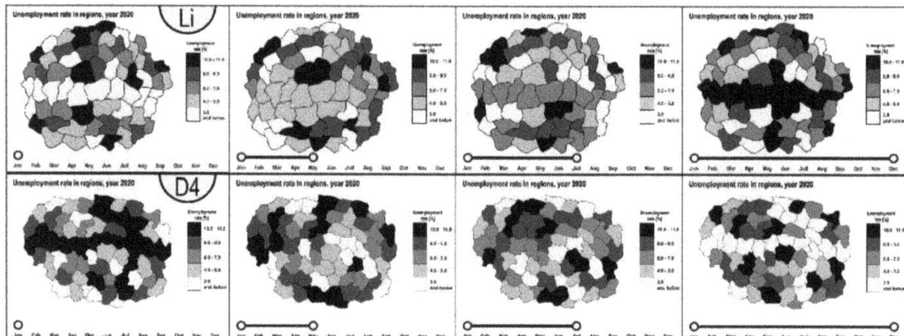

Figure 2. Selected frames of a linear and dispersed pattern. An increasing trend is in a linear pattern. The dispersed pattern has no trend. In D4, the spatial pattern is only visible at both ends of animation and did not appear in other months. Li—linear increasing, D4—dispersed pattern map number 4.

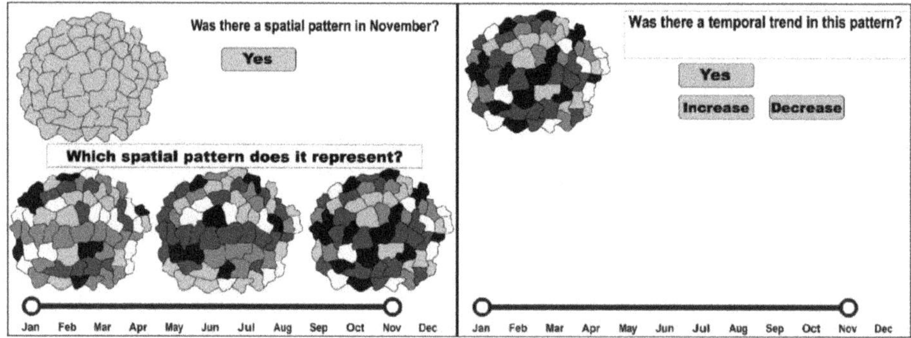

Figure 3. An interactive protocol for the participants' responses was presented after each animated map ended.

3.5. Analysis

The analysis of participants' visual behavior included eye tracking metrics such as fixation location, fixation count, fixation duration, and saccadic magnitude. Fixation count describes how many times the participants gazed at a specific location. This might indicate the ease in understanding the spatio-temporal data. The more fixations on the timeline, the more focused attention it required. This is related to the fixation duration. Longer durations might indicate a longer duration of attention. Therefore, these issues are related to perception. Saccadic magnitude describes the length of rapid eye movement from one fixation to another. This metric might suggest the difficulty in map interpretation. Participants' shorter saccades correlated with a high number of long-duration fixations, which might result in ineffective spatio-temporal dependencies. The statistical analysis involved 6*2 Friedman's test since the data had a non-parametrical distribution. The within-subject factors were the spatial pattern type (six conditions) and the correct recognition of patterns and temporal trends (two conditions). Correct pattern recognition effectiveness was based on the number of correct spatial pattern indications after the animation was viewed. It was a binary response, where 0 was assumed when the participants incorrectly identified a spatial pattern in a given month. Additionally, analyzing the spatial distribution of fixations was based on fixation coordinates recorded with eye tracking and spatial statistics such as Average Nearest Neighbor and Kernel Density.

4. Results

4.1. Spatial Pattern and Temporal Trend Recognition Effectiveness

Figure 4 presents the results of spatial pattern and temporal trend recognition effectiveness. The participants recognized the lack of a pattern the best (dispersed pattern); 73% of their answers were correct. There was a similar result in the case of recognizing the absence of a trend; 84% of the participants responded correctly. Correct identification at a random point (month) in the animation time was recorded for most of the studied patterns. The pattern recognition effectiveness ranged between 64% and 55%. The only exception was the radial pattern. Its effectiveness achieved 28%. The effectiveness of identifying temporal trends in spatial patterns ranged between 53% and 75%.

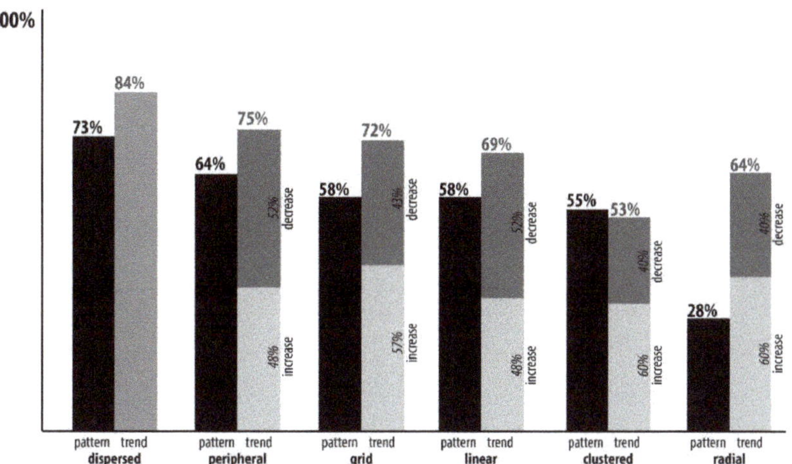

Figure 4. Results of spatial pattern and temporal trend recognition effectiveness calculated as a percentage of participants' correct responses.

However, in some cases, the participants recognized the increasing trend more effectively than the decreasing trend. This was the case for the clustered pattern (recognition of 60% of the increasing trends vs. 40% of the decreasing trends), radial pattern (recognition of 60% of the increasing trends vs. 40% of the decreasing trends), and grid pattern (recognition of 58% of the increasing trends vs. 43% of the decreasing trends). In the case of a linear pattern and peripheral pattern, the trends' recognition rates were similar (recognition of 48% of the increasing trends vs. 52% of the decreasing trends).

4.2. Participants' Visual Behavior

Figure 5 presents Freidman's test results in the context of three eye tracking metrics describing the participants' visual behavior. Friedman's test detected statistically significant differences between all three eye tracking metrics. Fixation count varied substantially among the selected factors. The highest differences were among clustered patterns juxtaposed with correct pattern and trend recognition. The second large significant difference between correct and incorrect answers was in the peripheral and linear patterns. In both patterns, participants who answered correctly had more fixation on the map in contrast to participants who recognized them incorrectly. Additionally, the radial pattern had significant differences in the fixation count among participants who responded correctly or incorrectly. However, in the case of grid patterns, the participants who correctly recognized spatio-temporal dependencies had fewer fixations than those who responded incorrectly ($\chi^2 = 82.4$; $p < 0.000001$). This shows that, in most cases, if participants had to fixate more on the animated map, there was a higher probability of incorrect recognition of spatial patterns and temporal trends.

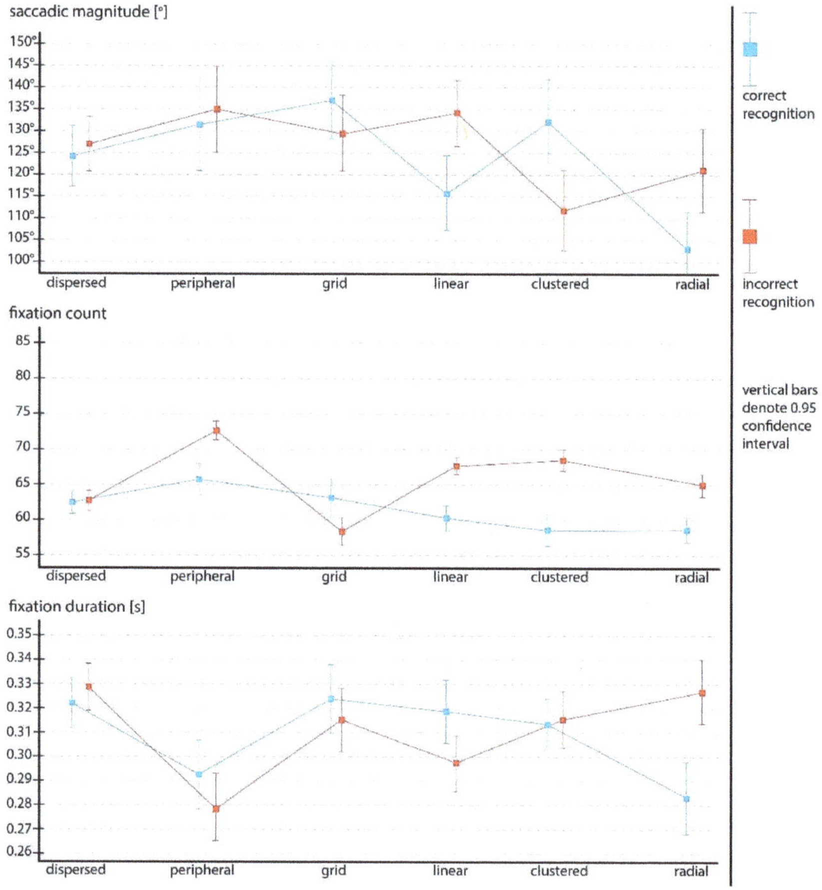

Figure 5. 3 Spatial analysis of participants' visual behavior.

Fixation duration was differed significantly among participants' pattern and trend recognition in radial pattern conditions ($\chi^2 = 9.6$; $p < 0.00193$). Interestingly, the radial pattern was the least effective recognized pattern by participants. Participants who incorrectly recognized spatio-temporal dependencies had longer fixations, while participants who responded correctly had a statistically significantly shorter fixation duration.

Saccadic magnitude expressed in visual angle degrees was statistically significant among the selected within-subject factors ($\chi^2 = 3.28$; $p < 0.04$). The saccadic magnitude was similar in the case of dispersed and peripheral patterns. The most crucial variation between participants' correct and incorrect recognition of spatio-temporal relations was visible in the case of the clustered, radial, and linear patterns. It is significant that these patterns had the lowest effectiveness in temporal trend detection. In the linear and radial pattern conditions, participants who incorrectly responded had longer saccades, while in the clustered pattern it was the opposite.

Inconsistency between the spatial patterns regarding correct and incorrect responses among the selected eye tracking metrics is visible. It should be noted that different spatial patterns require various visual behaviors for effective performance. The main factor crucial for determining effective visual behavior is the fixation count. Statistically significant differences between the saccadic magnitude in the linear, clustered, and radial patterns might indicate that these are more complex patterns for visual analysis. Since the dispersed, peripheral, and grid patterns had similar saccadic magnitude differences, shorter saccades could be crucial for effective recognition of these patterns. This might be confirmed by the finding that these patterns had similar visual strategies according to the fixation count.

Based on Euclidean distance Average Nearest Neighbor was calculated for correct and incorrect recognition participants. The results are presented in Table 1, which contains all related values such as Observed Mean Distance (OMD), Expected Mean Distance (EMD), Nearest Neighbor Index (NNI), z-score (z), and p-value (p). Most fixations, both correct and incorrect recognition, showed statistically significant clustering visual behavior. The only exception was a dispersed pattern with incorrect recognition participants (random distribution of fixations). The calculated distance included the local coordinate system of the recorded screen. Therefore, the X and Y axes of the coordinate system were a percentage of the monitor height and width.

Table 1. Average Nearest Neighbor Summary of the fixation distribution of participants' correct and incorrect recognition.

	Average Nearest Neighbor Summary									
	Incorrect Recognition					Correct Recognition				
	OMD	EMD	NNI	z	p	OMD	EMD	NNI	z	p
D1	0.0235	0.0299	0.785	−4.460	<0.0000	0,0099	0.0167	0,593	−16.922	<0.0000
D2	0.0148	0.0223	0.663	−7.327	<0.0000	0.0103	0.0159	0.652	−13.643	<0.0000
D3	0.0267	0.0285	0.936	−0.794	0.4272	0.0112	0.0168	0.667	−13.192	<0.0000
D4	0.0254	0.0292	0.871	−2.722	0.0065	0.0086	0.0136	0.637	−17.654	<0.0000
D5	0.0153	0.0237	0.646	−8.800	<0.0000	0.0157	0.0219	0.716	−8.255	<0.0000
Pd	0.0195	0.0309	0.630	−7.148	<0.0000	0.0121	0.0173	0.702	−11.623	<0.0000
Pi	0.0158	0.0202	0.782	−5.799	<0.0000	0.0112	0.0140	0.796	−8.267	<0.0000
Gd	0.0114	0.0176	0.647	−9.854	<0.0000	0.0126	0.0173	0.727	−9.382	<0.0000
Gi	0.0167	0.0251	0.664	−8.596	<0.0000	0.0137	0.0186	0.737	−8.942	<0.0000
Ld	0.0145	0.0196	0.741	−6.947	<0.0000	0.0136	0.0189	0.720	−7.926	<0.0000
Li	0.0114	0.0168	0.678	−11.092	<0.0000	0.0140	0.0230	0.609	−12.306	<0.0000
Cd	0.0149	0.0214	0.696	−8.868	<0.0000	0.0166	0.0232	0.713	−7.757	<0.0000
Ci	0.0196	0.0287	0.682	−7.423	<0.0000	0.0099	0.0154	0.640	−16.880	<0.0000
Rd	0.0104	0.0158	0.662	−13.700	<0.0000	0.0173	0.0265	0.655	−7.640	<0.0000
Ri	0.0113	0.0159	0.710	−11.433	<0.0000	0.0142	0.0211	0.675	−8.584	<0.0000

This clustering of participant fixations was confirmed and located on a map stimulus with Kernel Density. In most cases, participants who correctly recognized the spatial patterns and temporal trends had different fixation concentration areas from those of participants with incorrect responses. As can be seen in Figure 6, correct recognition corresponds to a specific visual behavior. This consisted of the concentration of fixations near the spatial pattern. However, in particular, what distinguished correct recognition from incorrect recognition was that, if the recognitions were correct, the fixation clusters covered a larger area of the map. It is significant that the correct recognition fixation cluster covers a larger part of the spatial pattern than in the case of incorrect recognition. The only exception was the clustered pattern. For the decreasing temporal trends, the participants' correct recognition fixation cluster covers a larger area. However, it is slightly shifted from the pattern. Therefore, the cluster area of fixations is outside the pattern on the map. In the case of increasing temporal trends, the incorrect recognition fixation cluster covers a larger area than that of the correct recognition cluster. Therefore, there is a difference between the participants' visual behavior.

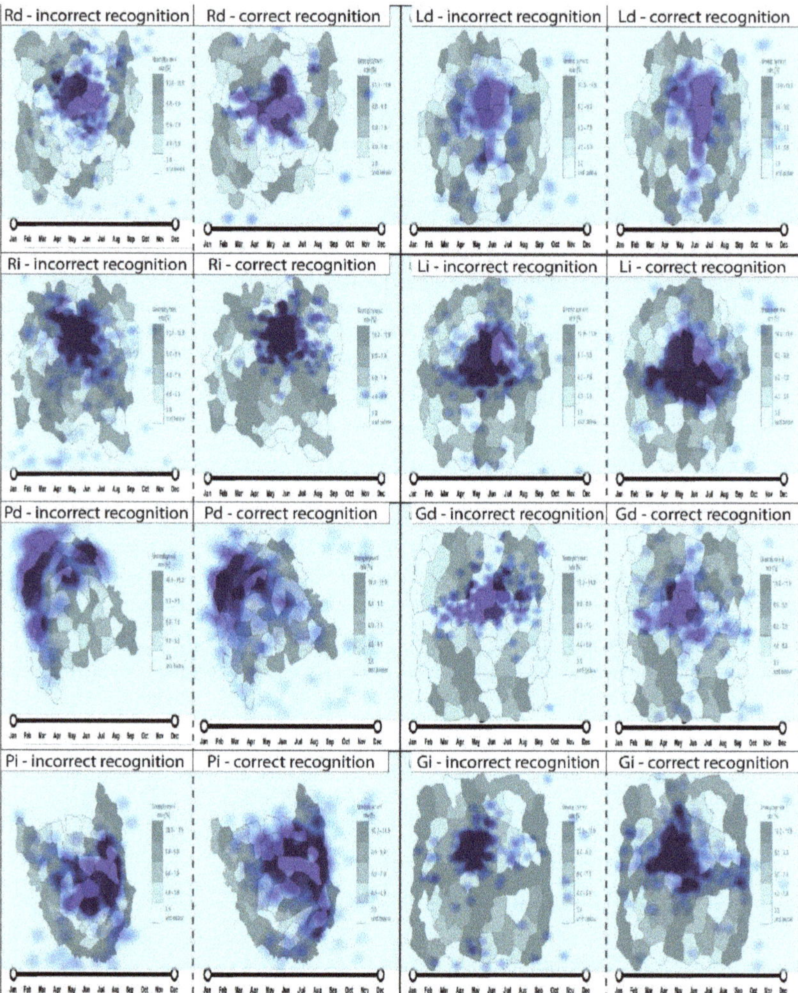

Figure 6. Fixation density clusters for correct and incorrect recognition of spatial patterns and temporal trends.

When analyzing the lack of patterns (dispersed pattern), the results are not as clear as when there was a presence of a spatial pattern. Participants' fixation clusters occur in both incorrect and correct recognition. In the case of the D1, D2, and D4 maps, the cluster distribution is similar to the case of the occurrence of the spatial patterns and temporal trends. Correct recognition fixation clusters cover a larger area. The situation is different when analyzing the D3 and D5 maps. In both cases, the incorrect recognition fixation clusters cover a larger area than the correct recognition fixation clusters. Figure 7 presents the result of Kernel Density based on the participants' fixation and recognition.

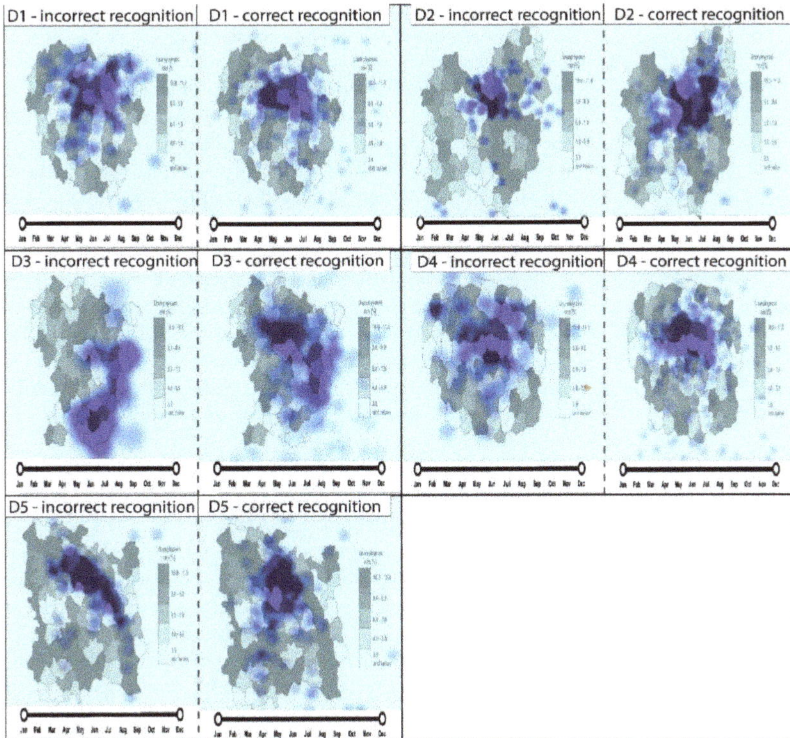

Figure 7. Fixation density clusters for correct and incorrect recognition of spatial patterns and temporal trends when analyzing dispersed patterns.

5. Discussion

The effectiveness, measured as the correct recognition of temporal trends and spatial patterns, differed among the analyzed stimuli. It appears that the study participants were better at recognizing temporal trends for most of the maps. This is especially beneficial for animated maps related to injury prevention. Cinnamon et al. [59] showed that participants could detect temporal trends relatively quickly, in comparison to static and interactive maps. Therefore, this study consists of empirical evidence supporting the previous hypothesis regarding the usefulness of animated products for the visualization of temporal trends [14,60]. Choropleth maps are more effective when the data have a peripheral spatial pattern or the lack of one (effectiveness above 64%). For the visualization of data with a radial pattern, one should search for other mapping methods due to the low choropleth map effectiveness (below 29%). The presented research had only one animation speed. However, according to Griffin et al. [55], different animation speeds may support the recognition of some patterns. Additionally, following Roth's [4] suggestions of including interactive tools could increase animated map effectiveness. Considering hexagonal enumeration units might also help in the recognition of less effective patterns (e.g., radial) [57]. Even some techniques of highlighting the most crucial information might be worthy of consideration [25].

The recognition of spatial patterns was found to be more difficult than the recognition of temporal trends. Based on the results, the most effective was the dispersed pattern. This proves previous assumptions that animated conditions support both the lack of pattern recognition and strong pattern recognition [55]. This is also true for the recognition of temporal trends, which did not appear while displaying dispersed pattern maps. It was found that most of the patterns analyzed in this study had similar effectiveness (from 55%

to 64%). The only exception was the radial pattern, which had the lowest effectiveness (28%). This type of spatial pattern is related to road and railroad mapping [61].

The study results report statistically significant differences between correct (effective) and incorrect (ineffective) recognition of temporal trends and spatial patterns. These differences are related to participants' visual behavior. As Shiewe [56] showed, different choropleth mapping techniques could support the successful interpretation of cartographic data. However, this study used only one mapping method. Therefore, the spatial pattern classification is related to the color intensities, and a legend was rarely fixated upon. Differences in the participants' correct and incorrect recognition were visible in the saccadic magnitude in the case of the clustered, radial, and linear patterns. Two of these patterns were the only ones where the participants recognized more increasing trends than decreasing ones. In Cybulski's [33] study, different saccade lengths are related to variability in spatial abilities or other personal features. Additionally, Dong et al. [62] presented the role of previous experiences in differences between the spatial behavior of novices and experts. Finally, Trauns et al.'s [58] study confirmed the reported results regarding the role of map users' personal features, which may determine map effectiveness to a greater extent than the specific map design itself.

The spatial analysis of the visual behavior showed that the study participants fixated mostly on specific areas. Fixation clusters appeared among correct and incorrect recognitions. The fixation clusters were located in the central part of the map. The only exceptions were maps with a peripheral pattern and some dispersed patterns (D3), which generated clusters that were shifted to near the pattern. Previous studies by Dong et al. [63] and Cybulski and Krassanakis [64] have explained this type of visual behavior. To observe the whole map during animation, the participants mainly observed its central part. The participants' sporadically observed peripheral parts of the map. A larger area covered by longer duration fixations characterized the participants' correct recognition fixation clusters. This might be related to change blindness when a participant fails to notice a change on a dynamic map [63,65].

6. Conclusions

The main aim of this research was to investigate the effectiveness of spatial pattern and temporal trend recognition assessment on a choropleth map. The presented study achieved this goal. The results show that spatial pattern recognition on a choropleth map varies, and that this depends on the pattern itself. The radial pattern was the least effective, and the peripheral pattern was the most effective. Despite this, the results were compared to those for maps with no pattern (dispersed pattern) and showed that recognizing the lack of a pattern is even more effective. Therefore, animated choropleth maps could support the visualization of specific spatial patterns. However, they are more suitable for presenting temporal trends. Temporal trends in spatial data had a recognition effectiveness of 64%.

The empirical results confirm the research hypothesis that effective spatial pattern and temporal trend recognition on a choropleth map is related to participants' visual behavior. Visual attention clustered in the central part of the choropleth map supports effective spatio-temporal relationship recognition. The larger the area covered by the fixation cluster, the higher the probability of correct temporal trend and spatial pattern recognition. This has specific implications for map makers and map users. For the visualization of temporal trends in spatial patterns using a choropleth map, one might consider a simple design since the interpretation of this type of data is more demanding than a static map. Understanding the difficulty in the correct recognition of spatio-temporal relationships might be a reason for implementing techniques that support effective visual searches such as highlighting, cartographic redundancy, or interactive tools. For end-users, the presented study reveals the necessity of the application of a specific visual strategy. Focusing on the central part of the map was found to be the most effective strategy for the recognition of spatio-temporal relationships.

The study limitations relate to the number of participants. Due to the COVID-19 pandemic, we experienced difficulties with user testing. This was related to legal restrictions and the necessity to maintain social distance and disinfect equipment and rooms. Therefore, it would be beneficial to increase the number of research participants in future studies.

In future studies, it would be interesting to compare the presented results with different spatial pattern and temporal trend mapping methods. This would help find the best cartographic solutions for spatio-temporal relationships, which are ineffective using a choropleth map. The presented study does not exhaust the topic of recognizing patterns and trends on maps. Users might experience more complex spatial patterns or more than one pattern on a map. Additionally, a temporal trend might initially be an increasing trend, and afterward become a decreasing trend. These more complicated spatio-temporal relationships require interdisciplinary studies, including cartography and psychology. Future cartographic visualization designers might benefit from these empirical results by taking into account the awareness that map users do not perceive space and time with the same level of effectiveness. The author believes that future temporal cartographic products need to be reconsidered in terms of temporal legend design to maintain effective communication with the user.

Funding: This research received no external funding.

Data Availability Statement: The data that support the findings of this study are openly available in the "Harvard Dataverse" at https://doi.org/10.7910/DVN/NFJW6B.

Acknowledgments: The author would like to thank all anonymous reviewers for their fruitful comments and recommendations. The paper is the result of research on visualization methods carried out within statutory research in the Department of Cartography and Geomatics, Faculty of Geographical and Geological Sciences, Adam Mickiewicz University Poznan, Poland.

Conflicts of Interest: The author declares no conflict of interest.

References

1. Gartner, G.; Bennet, D.A.; Morita, T. Towards ubiquitous Cartography. *Cartogr. Geogr. Inf. Sc.* **2007**, *34*, 247–257. [CrossRef]
2. Edler, D.; Dickmann, F. The Impact of 1980s and 1990s Video Games on Multimedia Cartography. *Cartographica* **2017**, *52*, 168–177. [CrossRef]
3. Medyńska-Gulij, B.; Forrest, D.; Cybulski, P. Modern Cartographic Forms of Expression: The Renaissance of Multimedia Cartography. *ISPRS Int. J. Geoinf.* **2021**, *10*, 484. [CrossRef]
4. Roth, R.E. Interactive maps: What we know and what we need to know. *J. Spat. Inf. Sci.* **2013**, *6*, 59–115. [CrossRef]
5. Roth, R.E.; Donohue, R.G.; Sack, C.M.; Wallace, T.R.; Buckingham, T.M.A. A Process for Keeping Pace with Evolving Web Mapping Technologies. *Cartogr. Perspect.* **2014**, *78*, 25–52. [CrossRef]
6. Sack, C.M.; Donohue, R.G.; Roth, R.E. Interactive and Multivariate Choropleth Maps with D3. *Cartogr. Perspect.* **2014**, *78*, 57–78. [CrossRef]
7. Horbiński, T.; Cybulski, P.; Medyńska-Gulij, B. Web Map Effectiveness in the Responsive Context of the Graphical User Interface. *ISPRS Int. J. Geoinf.* **2021**, *10*, 134. [CrossRef]
8. Medyńska-Gulij, B. *Cartography and Geomedia*; Wydawnictwo Naukowe PWN: Warsaw, Poland, 2021; p. 12.
9. Medyńska-Gulij, B.; Halik, Ł.; Wielebski, Ł.; Dickmann, F. Multiperspective visualization of spatial spare time activities—A smartphone-based approach to mapping tourist routes. *KN-J. Cartogr. Geogr.* **2015**, *65*, 323–329.
10. Halik, Ł.; Kent, A.J. Measuring user preferences and behaviour in a topographic immersive virtual environment (TopoIVE) of 2D and 3D urban topographic data. *Int. J. Digit. Earth* **2021**, *14*, 1835–1867. [CrossRef]
11. Medyńska-Gulij, B.; Tegeler, T.; Bauer, H.; Zagata, K.; Wielebski, Ł. Filming the Historical Geography: Story from the Realm of Maps in Regensburg. *ISPRS Int. J. Geoinf.* **2021**, *10*, 764. [CrossRef]
12. Horbiński, T.; Zagata, K. Interpretation of Map Symbols in the Context of Gamers' Age and Experience. *ISPRS Int. J. Geoinf.* **2022**, *12*, 150. [CrossRef]
13. Harrower, M. A look at the history and future of animated maps. *Cartographica* **2004**, *39*, 33–43. [CrossRef]
14. Harrower, M. The Cognitive Limits of Animated Maps. *Cartographica* **2007**, *42*, 349–357. [CrossRef]
15. Fish, C. Cartographic Challenges in Animated Mapping. In Proceedings of the ICA Workshop on Envisioning the Future of Cartographic Research, Curitiba, Brazil, 21 August 2015.
16. Andrienko, G.; Andrienko, N.; Demsar, U.; Dransch, D.; Dykes, J.; Fabrikant, S.I.; Jern, M.; Kraak, M.-J.; Schumann, H.; Tominski, C. Space, time and visual analytics. *Int. J. Geogr. Inf. Sci.* **2010**, *24*, 1577–1600. [CrossRef]

17. Hegarty, M.; Kriz, S.; Cate, C. The roles of mental animations and external animations in understanding mechanical systems. *Cogn. Instr.* **2003**, *21*, 325–360. [CrossRef]
18. Lee, P.U.; Klippel, A.; Tappe, H. The effect of motion in graphical user interfaces. In *Smart Graphics: Lecture Notes in Computer Science*; Butz, A., Krüger, A., Olivier, P., Eds.; Springer: Berlin/Heidelberg, Germany, 2003; Volume 2733, pp. 12–21.
19. Juergens, C. Trustworthy COVID-19 Mapping: Geo-Spatial Data Literacy Aspects of Choropleth Maps. *KN-J. Cartogr. Geogr.* **2020**, *70*, 155–161. [CrossRef] [PubMed]
20. von Wyss, M. The production of smooth scale changes in an animated map project. *Cartogr. Perspect.* **1996**, *23*, 12–20. [CrossRef]
21. Battersby, S.E.; Goldsberry, K.P. Considerations in Design of Transition Behaviors for Dynamic Thematic Maps. *Cartogr. Perspect.* **2010**, *65*, 16–32. [CrossRef]
22. Roth, R.E. Cartographic Design as Visual Storytelling: Synthesis and Review of Map-Based Narratives, Genres, and Tropes. *Cartogr. J.* **2021**, *58*, 83–114. [CrossRef]
23. Harrower, M. Tips for Designing Effective Animated Maps. *Cartogr. Perspect.* **2003**, *44*, 63–65. [CrossRef]
24. Lloyd, R.E. Attention on Maps. *Cartogr. Perspect.* **2005**, *52*, 28–57. [CrossRef]
25. Robinson, A.C. Highlighting in Geovisualization. *Cartogr. Geogr. Inf. Sci.* **2013**, *38*, 373–383. [CrossRef]
26. Maggi, S.; Fabrikant, S.I.; Imbert, J.-P.; Hurter, C. How Do Display Design and User Characteristics Matters in Animation? An Empirical Study with Air Traffic Control Displays. *Cartographica* **2016**, *51*, 25–37. [CrossRef]
27. Goldsberry, K.P.; Battersby, S.E. Issues of Change Detection in Animated Choropleth Maps. *Cartographica* **2009**, *44*, 201–215. [CrossRef]
28. Fish, C.; Goldsberry, K.P.; Battersby, S.E. Change Blindness in Animated Choropleth Maps: An Empirical Study. *Cartogr. Geogr. Inf. Sci.* **2011**, *38*, 250–362. [CrossRef]
29. Moon, S.; Kim, E.-K.; Hwang, C.-S. Effects of Spatial Distribution on Change Detection in Animated Choropleth Maps. *J. Korean Soc. Surv. Geod. Photogramm. Cartogr.* **2014**, *32*, 571–580. [CrossRef]
30. Cybulski, P.; Medyńska-Gulij, B. Cartographic Redundancy in Reducing Change Blindness in Detecting Extreme Values in Spatio-Temporal Maps. *ISPRS Int. J. Geoinf.* **2018**, *7*, 8. [CrossRef]
31. Fisher, B.; Ramsperger, E. Human express saccades: Extremely short reaction times of goal directed eye movements. *Exp. Brain Res.* **1984**, *57*, 191–195. [CrossRef] [PubMed]
32. Lisberger, S.G. Visual guidance of smooth pursuit eye movements: Sensation, action, and what happens in between. *Neuron* **2010**, *66*, 477–491. [CrossRef]
33. Cybulski, P. Effectiveness of Memorizing an Animated Route—Comparing Satellite and Road Map Differences in the Eye-Tracking Study. *ISPRS Int. J. Geoinf.* **2021**, *10*, 159. [CrossRef]
34. Harrower, M. Unclassed Animated Choropleth Maps. *Cartogr. J.* **2007**, *44*, 313–320. [CrossRef]
35. Just, M.A.; Carpenter, P.A. Eye Fixations and Cognitive Processes. *Cogn. Psychol.* **1967**, *8*, 441–480. [CrossRef]
36. Rayner, K. Eye Movements and Attention in Reading, Scene Perception, and Visual Search. *Q. J. Exp. Psychol.* **2009**, *62*, 1457–1506. [CrossRef] [PubMed]
37. Edler, D.; Keil, J.; Tuller, M.-C.; Bestgen, A.-K.; Dickmann, F. Searching for the 'Right' Legend: The Impact of Legend Position on Legend Decoding in a Cartographic Memory Task. *Cartogr. J.* **2020**, *57*, 6–17. [CrossRef]
38. Andrienko, N.; Andrienko, G.; Gatalsky, P. Exploratory spatio-temporal visualization: An analytical review. *J. Vis. Lang. Comput.* **2003**, *14*, 503–541. [CrossRef]
39. DiBiase, D.; MacEachren, A.M.; Krygier, J.B.; Reeves, C. Animation and the role of map design in scientific visualization. *Cartogr. Geogr. Inf. Sys.* **1992**, *19*, 201–214. [CrossRef]
40. Blok, C. Monitoring change: Characteristics of dynamic geo-spatial phenomena for visual exploration. In *Spatial Cognition II, Lecture Notes on Artificial Intelligence*; Freska, C., Habel, C., Brauer, W., Wender, K.F., Eds.; Springer: Berlin/Heidelberg, Germany, 2000; Volume 1849, pp. 16–30.
41. Peuquet, D.J. It's about time: A conceptual framework for the representation of temporal dynamics in geographic information systems. *Ann. Am. Assoc. Geogr.* **1994**, *84*, 441–461. [CrossRef]
42. Li, D.; Samet, H.; Varshney, A. Visualizing accessibility with choropleth maps. In Proceedings of the 5th ACM SIGSPA-TIAL International Workshop on Location-based Recommendations, Geosocial Networks and Geoadvertising, Bejijng, China, 2 November 2021.
43. Bertin, J. *Semiology of Graphics Diagrams Networks Maps*; The University of Wisconsin Press: Madison, WI, USA, 1986.
44. Medyńska-Gulij, B. The effect of cartographic content on tourist map users. *Cartography* **2003**, *32*, 49–54. [CrossRef]
45. Schroeder, J.P. Trends in patterns versus patterns in trends: A key distinction for visualizing geographic time-series data. In Proceedings of the 23rd International Cartographic Conference, Moskov, Russia, 4–10 August 2007.
46. Borregaard, M.K.; Hendrichsen, D.; Nachman, G. Spatial distribution patterns. In *Encyclopedia of Ecology*, 2nd ed.; Jørgensen, S.E., Fath, B.D., Eds.; Elsevier: Oxford, UK, 2009; pp. 3303–3310.
47. Ripley, B.D. Modelling Spatial Patterns. *J. R. Stat. Soc. Ser. B Stat. Methodol.* **1977**, *39*, 172–192. [CrossRef]
48. Anders, A.M.; Roe, G.H.; Hallet, B.; Montgomery, D.R.; Finnegan, N.J.; Putkonen, J. Spatial patterns of precipitation and topography in the Himalaya. In *Tectonics, Climate, and Landscape Evolution*; Willett, S.D., Hovius, N., Brandon, M.T., Fisher, D.M., Eds.; The Geological Society of America: Colorado, CO, USA, 2006; pp. 39–53.
49. Sapiro, P. Explaining geographic patterns of small group internal migration. *Popul. Space Place* **2017**, *23*, e0278. [CrossRef]

50. Sinnott-Armstrong, M.; Downie, A.E.; Federman, S.; Valido, A.; Jordano, P.; Donoghue, M.I. Global geographic patterns in the colours and sizes of animal-dispersed fruits. *Glob. Ecol. Biogeogr.* **2018**, *27*, 1339–1351. [CrossRef]
51. Kraak, M.-J.; MacEachren, A.M. Visualization of spatial data's temporal component. In Proceedings of the Spatial Data Handling, Advances in GIS Research, Edinburgh, Scotland, UK, 5–9 September 1994.
52. Guo, D.; Chen, J.; MacEachren, A.M.; Liao, K. A Visualization System for Space-Time and Multivariate Patterns. *IEEE Trans. Vis. Comput. Graph.* **2006**, *12*, 1461–1474. [PubMed]
53. Blok, C.; Köbben, B.; Cheng, T.; Kuterema, A.A. Visualization of Relationship Between Spatial Patterns in Time by Cartographic Animation. *Cartogr. Geogr. Inf. Sci.* **1999**, *26*, 139–151. [CrossRef]
54. Edsall, R.M.; Harrower, M.; Mennis, J.L. Tools for visualizing properties of spatial and temporal periodicity in geographic data. *Comput. Geosci.* **2000**, *26*, 109–118. [CrossRef]
55. Griffin, A.L.; MacEachren, A.M.; Hardisty, F.; Steiner, E.; Li, B. A Comparison of Animated Maps with Static Small-Multiple Maps for Visually Identifying Space-Time Clusters. *Ann. Am. Assoc. Geogr.* **2006**, *96*, 740–753. [CrossRef]
56. Schiewe, J. Empirical Studies on the Visual Perception of Spatial Patterns in Choropleth Maps. *KN-J. Cartogr. Geogr.* **2019**, *69*, 217–228. [CrossRef]
57. Karsznia, I.; Gołębiowska, I.M.; Korycka-Skorupa, J.; Nowacki, T. Searching for an Optimal Hexagonal Shaped Enumeration Unit Size for Effective Spatial Pattern Recognition in Choropleth Maps. *ISPRS Int. J. Geoinf.* **2021**, *10*, 576. [CrossRef]
58. Traun, C.; Schreyer, M.L.; Wallentin, G. Empirical Insight from a Study on Outlier Preserving Value Generalization in Animated Choropleth Maps. *ISPRS Int. J. Geoinf.* **2021**, *10*, 208. [CrossRef]
59. Cinnamon, J.; Rinner, C.; Cusimano, M.D.; Marshall, S.; Bekele, T.; Hernandez, T.; Glazier, R.H.; Chipman, M.L. Evaluating web-based static, animated and interactive maps for injury prevention. *Geospat. Health* **2009**, *4*, 3–16. [CrossRef]
60. Koussoulakou, A.; Kraak, M.-J. Spatio-temporal maps and cartographic communication. *Cartogr. J.* **1992**, *29*, 101–108. [CrossRef]
61. Monmonier, M. *Mapping It Out: Expository Cartography for the Humanities and Social Sciences*; University of Chicago Press: Chicago, IL, USA, 2015; p. 12.
62. Dong, W.; Zheng, L.; Liu, B.; Meng, L. Using Eye Tracking to Explore Differences in Map-Based Spatial Ability between Geographers and Non-Geographers. *ISPRS Int. J. Geoinf.* **2018**, *7*, 337. [CrossRef]
63. Dong, W.; Liao, H.; Xu, F.; Liu, Z.; Zhang, S. Using Eye Tracking to Evaluate the Usability of Animated Maps. *Sci. China Earth Sci.* **2014**, *57*, 512–522. [CrossRef]
64. Cybulski, P.; Krassanakis, V. The Role of Magnitude of Change in Detecting Fixed Enumeration Units on Dynamic Choropleth Maps. *Cartogr. J.* **2021**, 1–17. [CrossRef]
65. Rensink, R.A. Change Detection. *Annu. Rev. Psychol.* **2002**, *53*, 245–277. [CrossRef] [PubMed]

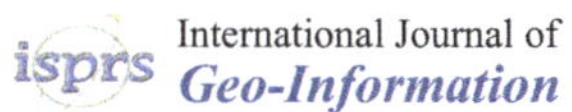

International Journal of
Geo-Information

Article

Parametric Modeling Method for 3D Symbols of Fold Structures

An-Bo Li [1,2,3,*], Hao Chen [1], Xiao-Feng Du [4], Guo-Kai Sun [4] and Xian-Yu Liu [1]

[1] Key Laboratory of Virtual Geographic Environment (Nanjing Normal University), Ministry of Education, Nanjing 210023, China
[2] Jiangsu Center for Collaborative Innovation in Geographical Information Resource Development and Application, Nanjing 210023, China
[3] State Key Laboratory Cultivation Base of Geographical Environment Evolution (Jiangsu Province), Nanjing 210023, China
[4] Shandong Rail Transit Engineering Consulting Co., Ltd., Jinan 250101, China
*   Correspondence: mrlab@njnu.edu.cn; Tel.: +86-136-4516-1205

Citation: Li, A.-B.; Chen, H.; Du, X.-F.; Sun, G.-K.; Liu, X.-Y. Parametric Modeling Method for 3D Symbols of Fold Structures. *ISPRS Int. J. Geo-Inf.* **2022**, *11*, 618. https://doi.org/10.3390/ijgi11120618

Academic Editors: Wolfgang Kainz, Beata Medynska-Gulij, David Forrest and Thomas P. Kersten

Received: 3 November 2022
Accepted: 12 December 2022
Published: 13 December 2022

Abstract: Most fabrication methods for three-dimensional (3D) geological symbols are limited to two types: directly increasing the dimensionality of a 2D geological symbol or performing appropriate modeling for an actual 3D geological situation. The former can express limited vertical information and only applies to the three-dimensional symbol-making of point mineral symbols, while the latter weakens the difference between 3D symbols and 3D geological models and has several disadvantages, such as high dependence on measured data, redundant 3D symbol information, and low efficiency when displayed in a 3D scene. Generating a 3D geological symbol is represented by the process of constructing a 3D geological model. This study proposes a parametric modeling method for 3D fold symbols according to the complexity and diversity of the fold structures. The method involves: (1) obtaining the location of each cross-section in the symbol model, based on the location parameters; (2) constructing the middle cross-section, based on morphological parameters and the Bezier curve; (3) performing affine transformation according to the morphology of the hinge zone and the middle section to generate the sections at both ends of the fold; (4) generating transition sections of the 3D symbol model, based on morphing interpolation; and (5) connecting the point sets of each transition section and stitching them to obtain a 3D fold-symbol model. Case studies for different typical fold structures show that this method can eliminate excessive dependence on geological survey data in the modeling process and realize efficient, intuitive, and abstract 3D symbol modeling of fold structures based on only a few parameters. This method also applies to the 3D geological symbol modeling of faults, joints, intrusions, and other geological structures and 3D geological modeling of typical geological structures with a relatively simple spatial morphology.

Keywords: fold structure; 3D geological symbols; parametric modeling; 3D geological modeling; 3D GIS; visual suggestiveness; symbols efficiency

1. Introduction

In recent years, traditional geographic information system (GIS) technology and its applications have undergone revolutionary changes [1,2], and it has transformed from 2D GIS to 3D GIS, with a primary focus on 3D modeling, 3D scene display, and 3D spatial analysis [3]. In particular, 3D GIS has become one of the iconic contents of GIS technology today and for the future [4,5]. At present, Digital Earth systems, including Google Earth, NASA WorldWind and Cesium, have been widely embraced by geoscientists in various disciplines. These 3D GIS-related technologies are convenient tools for promoting the scientific research of different application scenarios and contribute to integrating geospatial information, improving visualization capabilities, exploring spatio-temporal changes, and communicating scientific results [6–8]. In this context, traditional 2D map symbols cannot easily meet the demand for the intuitive expression of complex geological objects in 3D

GIS. The research and application of 3D symbols have gradually become a hot topic in the current GIS field [9,10].

Similar to traditional 2D symbols, 3D symbols are also the language of map visualization, which aims to express the 3D spatial distribution and attribute information of geographical entities [11,12]. The essence of a 3D symbol is that it is a 3D model that directly reflects the spatial morphology of the corresponding geographical object. Therefore, the production of 3D symbols is the process of generating 3D models. Given the growing demand for 3D symbols in 3D GIS applications, such as virtual reality (VR)\augmented reality (AR)\mixed reality (MR), scholars have proposed various methods for creating 3D symbols. These methods can be roughly divided into dimension-raising methods for planar map symbols [13], 3D model construction methods [14–17], and parametric modeling methods [18,19]. In the following, a summary of the modeling methods for 3D geological symbols, from these three aspects, is provided.

Few studies have been conducted on 3D symbol-making methods based on the dimension-raising of planar map symbols. Zhu et al. [13] proposed an automatic dimension-raising scheme from 2D vector mineral point symbols to 3D symbols, based on the expression characteristics of the 2D vector point symbols. This method has a high degree of automation and is mainly applied to the 3D symbol fabrication of mineral point symbols. However, because of the failure to entirely use the essential geometric parameter information of related geological objects, this method is unsuitable for generating symbols of 3D geological structures.

Fabricating 3D symbols based on a 3D model construction method is the primary method of rapidly generating 3D symbols. Modeling methods can be further subdivided into manual 3D symbol construction methods, based on general 3D modeling software, and automatic 3D symbol construction methods, based on particular modeling software. Early research has primarily focused on manual modeling methods based on general 3D modeling software. Xu et al. [20] used computer aided design (CAD) and MDL models to build 3D solid model symbols for 3D urban scenes, and Gu et al. [21] used Sketchup to build a 3D solid symbol library containing point, line, face, and body symbols. With the continued maturation and development of 3D modeling technology, an automatic construction method based on special modeling software has become the primary method of performing 3D modeling in different fields. According to the source of modeling data, the methods of 3D geological modeling mainly include borehole-based modeling [14,22], section-based modeling [23–25], planar geological map-based modeling [26], and multi-source data modeling [22,27–29].

Furthermore, based on the application of knowledge in the modeling process, 3D geological modeling methods can be divided into explicit modeling methods [16,30–32] and implicit modeling methods [17,33–35]. All of the 3D geological modeling methods described above can be used to rapidly construct 3D geological symbols. However, these methods require sufficient modeling data, and the model results are too elaborate. Directly applying the relevant 3D geological model to a 3D scene as a 3D symbol leads to low display efficiency, which contradicts the abstract principle of the symbol. Implicit modeling methods, which have developed rapidly in recent years, have dramatically reduced the dependence on modeling data by implicitly defining the geological interface as the isosurface of one or more scalar fields of 3D space [32,34,35]. However, owing to its refined modeling requirements, this method is unsuitable for creating many 3D geological symbols.

In the parametric modeling method, the characteristic parameters of natural objects are used to automatically control and construct 3D models based on parametric technology [18]. Parametric modeling was first invented by Rhino, which is a 3D draughting software that evolved from AutoCAD. The key advantage of parametric modeling is that, when setting up a 3D geometric model, the shape of the model's geometry can be changed as soon as the parameters, such as the dimensions or curvatures, are modified [19,36,37]. Parametric modeling has long been the primary method of machinery modeling in the CAD field. This method is suitable for structured entity modeling, owing to its low data dependency and high modeling efficiency. In recent years, parametric modeling methods have been promoted and applied to 3D tree modeling [38,39], 3D human body modeling [40], 3D building (structure) modeling [36], and many other fields. However, owing to the heterogeneity and non-parametric characteristics of geological bodies, parametric modeling methods have not played a role in 3D geological modeling. Using a geological map sketch and symbol information, Amorim et al. [34] realized the rapid construction of 3D models of folds, faults, and other geological structures by quantitatively calculating the topological structure and stratigraphic contact relationship of geological maps. Through parametric analysis and 2D geological symbol application, this method shows a specific idea of parametric modeling and effectively reduces the dependence on modeling data. However, this method is not designed for 3D geological symbols; therefore, its relatively refined modeling requirements do not entirely eliminate the dependence on geological maps and fail to fully reflect the abstract and 3D characteristics of 3D geological symbols.

Map symbols have prominent parametric characteristics because of their high abstraction as either a 2D or multidimensional symbol [41]. The parametric modeling method should become the primary method of 3D symbol modeling in the future because it requires only a small amount of feature parameter data to support the construction of 3D models with abstract expression characteristics. Similarly, according to the characteristics and requirements of 3D geological symbols based on parametric modeling methods, studying targeted 3D geological symbol fabrication methods should be the primary trend of future 3D geological symbol modeling research and application. In addition, although geological bodies have various types and complex structures, they still have apparent spatial distribution characteristics, and their geometric shapes can be mathematically simulated [41]. Implicit modeling methods that have been developed in recent years, particularly geological structure 3D modeling methods based on geological map sketches and geological symbols proposed by Amorim et al. [34] have effectively verified the feasibility of this idea.

As one of the most common geological structures in the crust, folds are of great significance for studying oil and gas traps and other mineral resources [42]. Furthermore, their geometric shapes are complex, and their types are diverse. Therefore, this study intends to take fold structure as the study object and use the parametric modeling method to discuss the rapid construction method of three-dimensional geological structure symbols. The remainder of this paper is organized as follows: Section 2 introduces the methodology; Section 3 presents the experimental results; Section 4 presents a discussion; and Section 5 presents the conclusions and future work.

2. Methodology

2.1. Modeling Parameters of the Fold Structures

In general, fold structures not affected by strong denudation are wavy structures in shape [34] whose spatial morphology can be described by fold elements, such as limbs, hinges, and axial surfaces [42–44]. The limb refers to the rock strata on both sides of the core, the hinge is the connecting line of the maximum bending point on each cross-section of the fold, and the axial surface refers to the surfaces formed by connecting the hinge lines on each adjacent fold stratum (Figure 1).

The formation mechanism of the folds is closely related to their stress mode and deformation environment and the deformation behavior of the rock stratum [44]. The diversity of fold formation mechanisms determines the complexity of the fold structures and the diversity of the fold types. Different types of folds are mainly classified according to their position, axial occurrence, and section morphology (Table 1).

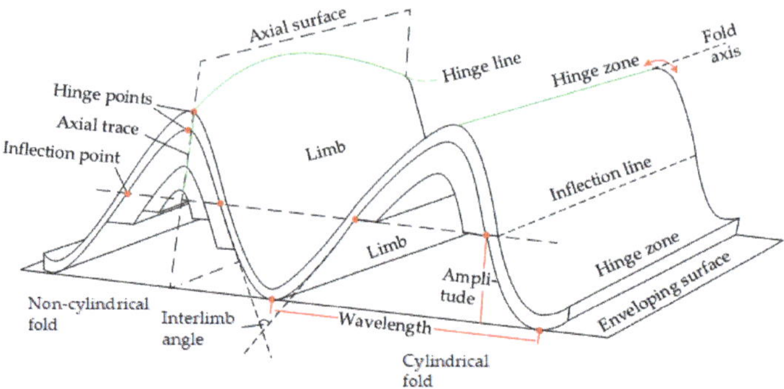

Figure 1. Diagram of fold structure characteristics.

Table 1. Main fold types and their characteristic parameters [41].

Classification Basis	Fold Type	Morphological Characteristics	Parameters
Vertical section morphology	Upright fold	Axial surface is nearly vertical (80~90°); the inclination of the two limbs are opposite, and their dip angles are nearly equal.	Strike of axial surface; Strike of limbs; Interlimb angle
	Oblique fold	Axial surface inclines; the inclination of two limbs tends to be opposite, and their dip angles are unequal.	
	Overturned fold	Axial surface inclines; the inclination of two limbs tends to be the same, and their dip angles are unequal.	
	Recumbent fold	Axial surface is nearly horizontal (1~10°), and one limb of the rock stratum overturns.	
	Circular fold	Folded surface is curved in a circular arc.	
	Chevron fold	Two limbs intersect straightly; the hinge zone is sharp; the interlimb angle is generally less than 30°.	
	Box fold	Two limbs are steep, the hinge zone is straight, and a pair of conjugate axial surfaces occurs.	
	Fan fold	Folded surface is fan-shaped, and both limbs are overturned.	
Cross-section morphology	Gentle fold	Interlimb angle: 120~180°	Interlimb angle
	Broad fold	Interlimb angle: 70~120°	
	Closed fold	Interlimb angle: 30~70°	
	Tight fold	Interlimb angle < 30°	
	Isoclinal fold	Interlimb angle ≈ 0°, the two limbs are parallel, and their occurrences are the same.	
Longitudinal section morphology	Horizontal fold	Hinge is a horizontal line.	The occurrence of hinge; The occurrence of the axial surface.
	Hinge fold	Hinge plunges.	
	Vertical fold	Both the hinge and the axial surface are nearly vertical (inclination angle: 80~90°).	
Stratum thickness and morphological changes	Isopach fold	Stratum thickness is equal.	Curvature center, curvature radius, and thickness of each stratum.
	Similar fold	One curvature radius, different curvature centers.	
	Harmonic fold	Bending morphology of each stratum is roughly the same.	
	Others	Including disharmonious folds, intestinal folds, and other folds whose strata morphology changes greatly.	

Due to the lack of borehole data in the fold structure development area, generating a series of cross-cutting sections based on geological maps, and using the contour comparison algorithm [45] to build a 3D model of the fold structure, is the most feasible method. The two critical points of this method are to reasonably plan the location of the section lines based on the location parameters and to efficiently generate a series of cross-sections based on the section morphology parameters. First, the location parameter of the cross-section (Table 2) was mainly used to determine the location and strike of the cross-sections. The relevant parameters can be directly obtained from geological maps or geological structure maps. Second, the morphological parameters of the cross-section (Table 2) represent the primary information that controls the cross-sectional morphology. The relevant parameters mainly include the axial dip angle, interlimb angle, limb occurrence, and stratigraphic information. These parameters control the local details of each stratum in the cross-sections of the fold structure. The geometric characteristics of the different fold types can be constrained and controlled by the different value ranges of the fold morphology parameters.

Table 2. Modeling parameters of the fold structure.

Parameter Type	Parameter Subdivision	Abbreviation	Unit	Parameter Description
Location parameters of cross-section	Center point of middle section	D	m	Determine the approximate position of the whole fold with the lower vertex
	Fold width	d	m	Width range of the middle fold cross- section
	Distance between sections	dist	m	Control the overall length of folds
	Dip angle of hinge line	δ	°	Angle between the hinge line and the horizontal line in the axial surface
	Strike of hinge line	μ	°	Strikes of the hinge lines between the middle section and two end sections
	Dip angle of axial surface	γ	°	Dip angles of the two end sections
Morphological parameters of cross-section	Dip angle of axial surface	γ	°	Dip angle of middle section
	Dip angles of inflection points	α, β	°	Dip angles of inflection points on both limbs of the reference stratum boundary in the middle section
	Interlimb angle	θ	°	Interlimb angle of the middle section
	Morphological parameter	cur	-	Control the morphology of the hinge zone of each section
	Collection of strata thickness	Dep	m	Thickness of each fold stratum

Based on the modeling parameters outlined above, this study presents a parametric 3D modeling method for fold symbols (Figure 2). The specific modeling steps primarily include: (1) the parametric generation of the cross-section lines based on the section location parameters; (2) the parametric generation of the middle cross-section based on the section morphology parameters; (3) the generation of the cross-cutting section at both ends based on the affine transformation; (4) the generation of the transition sections based on the morphing interpolation; and (5) the stitching and attribute assignment of the fold model.

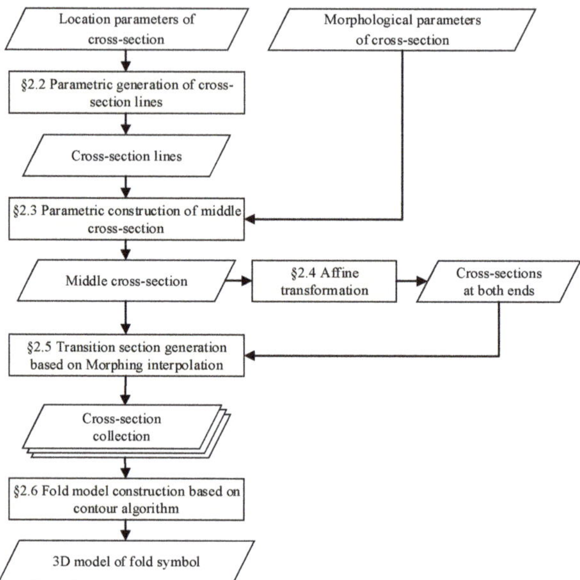

Figure 2. Parametric modeling process of 3D fold symbols.

2.2. Generation of Cross-Section Lines

The premise of generating a 3D fold-symbol model based on a contour comparison algorithm is to generate cross-sections by locating each cross-section of the fold. The stratigraphic information in the cross-section of the fold structure should be comprehensive and reflect the typical characteristics of the fold structure. Therefore, the position of the cross-sections should be determined according to the distribution of various strata of the fold structure and the direction of the hinge line (Figure 3a). This process makes the symbol model closer to the actual situation of the structure, which implies a higher degree of model restoration.

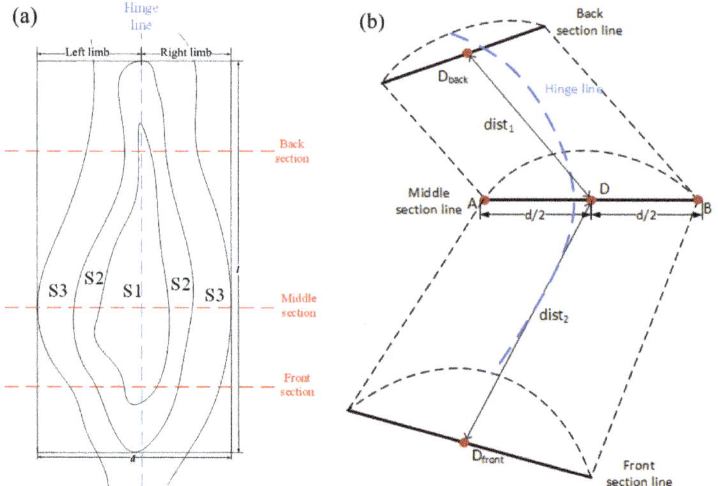

Figure 3. Determination of cross-section locations: (**a**) extraction of location parameters based on a geological map and (**b**) location of sections based on location parameters. S1–S3 are different stratum codes.

Furthermore, to break through the excessive dependence on geological survey data, this study applies the premise of extracting section location parameters based on geological maps (Figure 3a) and mainly determines the location of cross-sections according to the modeling parameters input by users (Figure 3b). The specific steps are as follows: (1) take the center point D of the middle section, according to the position parameters as the midpoint, and extend, perpendicularly, to the average hinge line strike (hinge line in Figure 3a) to obtain the positions of inflection points A and B of the middle section at d/2; and (2) calculate the affine transformation coefficient and generate the positions of the section lines at both ends, according to the distances ($dist_1$ and $dist_2$) from the front and back sections to the middle section, the occurrence of the hinge line (including front and back strikes of hinge lines μ_1 and μ_2 and front and back dip angles of hinge lines δ_1 and δ_2) and dip angle of the axial surface of the front and back sections γ_1 and γ_2.

2.3. Construction of Middle Cross-Section

The construction of the middle cross-section is essentially the process of generating the boundary of each stratum based on the morphological parameters of the cross-section. According to the morphological consistency of different stratum boundaries in the cross-section, and based on the premise of generating one stratum boundary based on the parameters, this boundary can be used as the reference stratum boundary to deduce and generate other stratum boundaries. Therefore, the stratum boundary of the middle section can be divided into two types: the reference stratum boundary and the general stratum boundary (Figure 4). The boundary of the outermost stratum was generally selected as the reference stratum boundary. The generation of the stratum boundary mainly includes three parts: the generation of the reference stratum boundary; the deduction of the general stratum boundary; and the trimming of the stratum boundary.

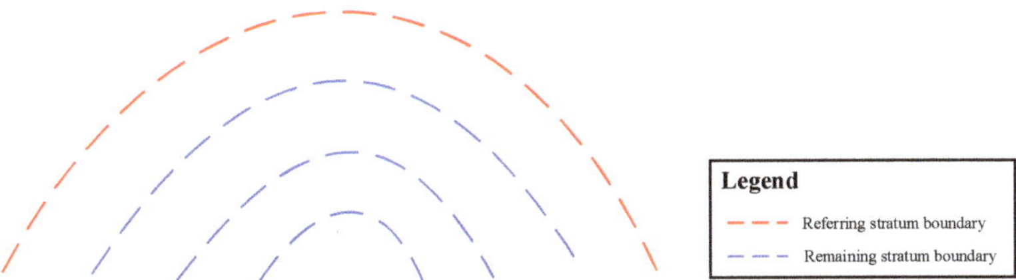

Figure 4. Diagrammatic sketch of stratum boundary.

2.3.1. Generation of Reference Stratum Boundary

The stratum morphology of the fold structure is generally smooth and regular and shows specific function laws, and it can be described in terms of amplitude and wavelength [42]. Therefore, it is feasible to use a quadratic Bessel curve to simulate the stratum boundary of the fold structure. In addition, folds are usually formed by the hinge zone and its two connected limbs, and the stratum morphologies of the two limbs tend to be different. Therefore, dividing the boundary line of the fold stratum into two segments according to the hinge zone is appropriate, and the Bezier curve is used to express this division [46,47]. In addition, the fitting of the Bezier curve was realized through the endpoints and control points. Suppose that only one segment of the Bezier curve is used for the fitting; in this case, obtaining the Bezier control points according to the shape of the interlimb angle is difficult, which causes the morphology of the hinge zone to not be precisely controlled. Therefore, the generation of the reference stratum boundary can be transformed into the selection of the Bezier endpoints and control points of the two curves.

The generation of the reference stratum boundary mainly includes the following steps: (1) calculate the intersection point C of the tangent lines at the inflection points, according to the inflection points (A and B) of the middle section, and the dip angle of the two limbs (α and β) mentioned in the section morphological parameters; (2) find the intersection point D between the angle bisector of angles ACB and AB, and CD is the axial surface line of the middle section; (3) cut point E proportionally on CD as the hinge point, based on the morphological parameter curve; (4) extend the vertical line of CD through point E and intersect AC and BC to points F and G, respectively; and (5) take points A and E and points E and B as the endpoints, and F and G as the Bezier control points of the left and right curve segments, respectively, and then conduct segmented Bezier interpolation to obtain the reference stratum boundary (Figure 5).

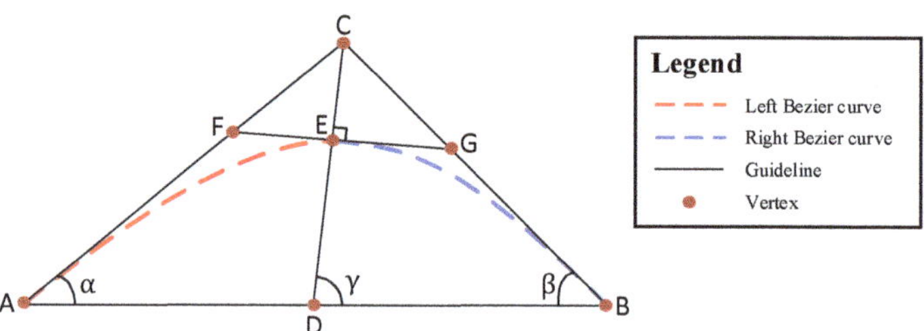

Figure 5. Diagrammatic sketch of reference stratum boundary generation.

2.3.2. Deduction of General Stratum Boundary

According to the principle of stratigraphic superposition [48], the bottom surface of each stratum in the fold structure is generally morphologically consistent. Therefore, if the bottom deduction rules of the reference stratum are determined, all the general stratum boundaries can be obtained. However, the deduction rules for the general stratum boundaries differ for different types of fold structures. Taking the isopach fold and similar folds as examples, the deductive method of the general stratigraphic boundary is briefly described below.

The isopach fold has typical characteristics and the "stratum thickness is equal everywhere" [42]. Deducing the general stratum boundaries in such a structure can be realized by moving a certain distance along the normal direction, according to the stratum thickness based on the points in the reference stratum boundary (Figure 6a).

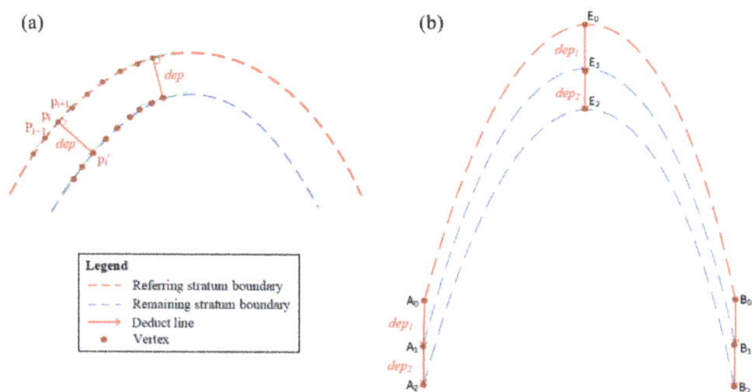

Figure 6. Deduction of general stratum boundaries: (**a**) isopach fold and (**b**) similar fold.

The characteristic of a similar fold is that the curvature radius of each stratum boundary is similar, although a common curvature center is not observed [42]. In other words, the morphology of the stratum boundaries is the same, but their locations are different. Therefore, the general stratum boundaries can be obtained by translating the reference stratum boundary in the vertical direction, and the translation distance can be determined by the stratum thickness at the hinge zone (Figure 6b).

2.3.3. Trimming of Stratum Boundary

As with 2D geological symbols, 3D fold symbols should also be designed to convey geological information as simply as possible and by focusing on the aesthetic expression of symbols. During stratum deduction, the strata of the isopach fold are inferred according to the normal direction of the reference stratum boundary, resulting in uneven stratum boundaries. Therefore, for anticlinal folds with low openness, the bottom of the entire section must be trimmed based on the inflection points of the two limbs in the outermost stratum (Figure 7a,b). For the syncline, the topographic line at the section location was used as a trimming line to trim the top of each stratum to truly express the morphology of the stratum surface (Figure 7c,d). In addition, the core stratum should be calculated according to the principle of stratum superposition.

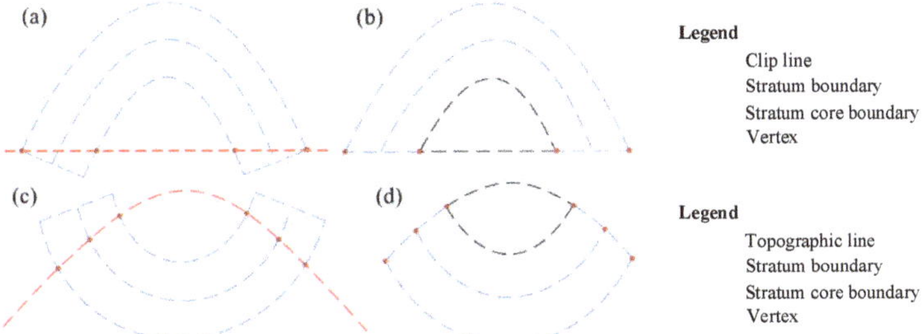

Figure 7. Trimming and optimization of the middle section, (**a**) bottom trimming line of the anticline, (**b**) supplement the core stratum of the anticline after trimming, (**c**) topographic line at the top of the syncline, and (**d**) supplement the core stratum of the syncline after trimming.

The specific steps of this method are as follows: (1) calculate the trimming line equation according to the inflection points of the middle section; (2) eliminate the points outside the trimming line in the stratum boundary of each stratum; (3) supplement the points on the trimming line in the stratum boundary of each stratum by linear interpolation; and (4) generate the core stratum boundary according to the area enclosed by the trimming line and the bottom stratum (Figure 7c,d).

2.3.4. Section Treatment for Special Fold Types

The parametric modeling method of the cross-section described above mainly applies to types of folds where no inversion occurs. Thus, the distorted and deformed morphology of overturned or recumbent folds must be optimized on this basis.

(1) Overturned fold

The two limbs of the overturned fold inclined in the same direction, and one of the limbs was overturned. The stratum boundary generation method in Section 2.3.1 can be used to construct the stratum boundaries by treating the dip angle at the overturned limb as an obtuse angle (Figure 8a). However, owing to the compression and deformation between strata, most of the strata of the overturned folds will be distorted to varying degrees [49]. Therefore, when constructing the middle cross-section, the rotation angle of the section point set around vertex D can be calculated according to Formula (1). Distortion and deformation processing are then carried out to realize an approximate simulation of the overturned fold morphology.

$$\theta = amax \times (R - r)/R \tag{1}$$

where amax is the maximum rotation angle, R is the maximum rotation radius, r is the distance from the current point to the rotation center D, and θ is the angle of rotation required for the current point.

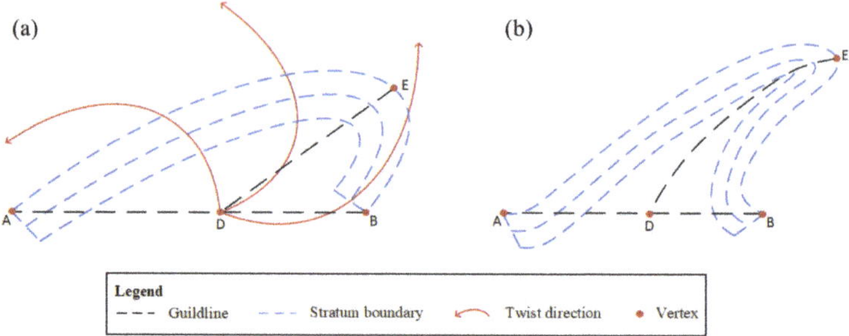

Figure 8. Distortion treatment for overturned folds: (**a**) before distortion and (**b**) after distortion.

(2)   Continuous fold belt

In general, typical geological structures with prominent characteristics are often accompanied by large-scale folding activities. One single symbol model of a fold cannot easily reflect the structural characteristics of a continuous fold belt. To enable fold symbols to effectively realize the abstract expression of geological characteristics under different geological scales, this study uses the periodic cycle method to repeat the middle section construction steps presented in Section 2.3 to generate the cross-section of continuous fold belts (Figure 9).

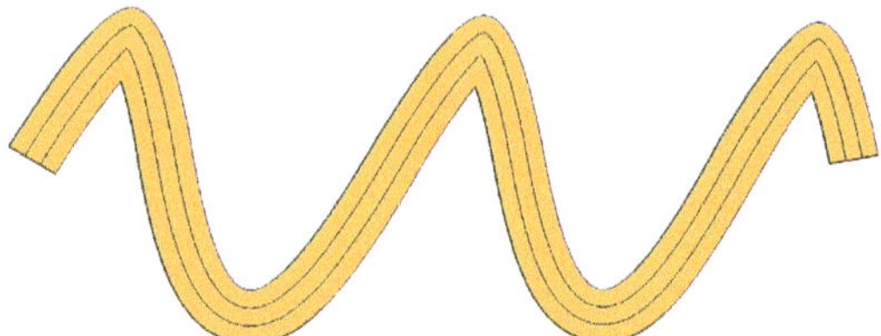

Figure 9. Periodic treatment for continuous fold belts.

2.4. Generation of Cross-Sections at Both Ends

Based on the morphological consistency among cross-sections of the fold structure, the cross-sections at both ends can be obtained by affine transformation from the middle cross-section. The affine transformation in this section includes translation, magnification, and rotation. The strike of the hinge affects the offset distance of the sections at both ends along the X-axis, the dip angle of the hinge affects the magnification ratio of the sections at both ends, and the dip angle of the axial surfaces of the sections at both ends affects the rotation angle. The specific affine transformation parameters and corresponding calculation methods are listed in Table 3.

Table 3. Affine transformation factor.

Factor Name	Computing Formula	Variable Description
Magnification scale	$scale = (h + \tan\delta \times dist)/h$	δ is the front or back dip angle of the hinge in the location parameter, dist is the distance from the current section to the middle section, and h is the height from the actual upper vertex E to the lower vertex D in the middle section (Figure 5).
Offset distance along X axis	$distX = \tan\mu \times dist$	μ is the front or back strike of the hinge in the location parameters, and dist is the distance from the current section to the middle section.
Rotation angle	$\Delta\gamma = \gamma' - \gamma$	γ' is the front or back section dip angle of the axial surface in location parameters, and γ is the dip angle of the axial surface in the middle section.

Note: Please refer to Figure 3b for the relevant parameters.

2.5. Generation of Transition Sections

The cross-sections, including the front, middle, and back, were obtained. Theoretically, these cross-sections can be stitched directly to complete the rough construction of the entire fold model. However, the spatial morphology of the hinge line is difficult to express using this method, which affects the accuracy and aesthetics of the model results to a large extent. Therefore, many transition sections need to be generated through morphing interpolation according to these cross-sections to realize the fine modeling of the fold symbol.

Morphing is an interpolation technique that smoothly transitions an initial object into a target object. This technology has been widely used in animation production, virtual reality, image compression, and image reconstruction [50–52]. Three types of boundaries are necessary for morphing interpolation (Figure 10): (1) the starting boundary with direction (SRC); (2) the target boundary (DEST) with a direction that must be consistent with that of SRC; and (3) the first and last constraint boundaries with directions (FC and LC) [53].

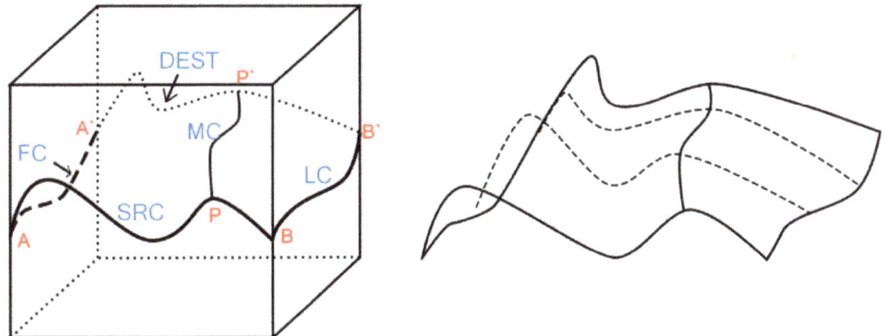

Figure 10. Principle of morphing interpolation.

2.5.1. Selection of Constraint Boundaries

Considering that the monotony of a single fold element in a folded section will experience one change at the hinge zone [54], this study considers the hinge line as the middle constraint and divides the strata of the fold sections into left and right parts. Morphing interpolation with four boundaries intersected by the hinge line was performed for the two limbs. This method allows the hinge information to participate effectively in the morphing interpolation process, thus ensuring the accuracy of the overall interpolation result.

Among the four constraint boundaries of the morphing interpolation method, the starting and ending stratum boundaries, SRC and DEST, can be obtained from the stratum boundary on the corresponding limb [46]. To make the section transition smoother, the constrained stratum boundaries, FC and LC, must be obtained by interpolation according to the corresponding characteristic points of each section. By connecting the points generated by the Lagrange interpolation results of A, E, and B (i.e., left inflection point, hinge point, and right inflection point, respectively), three smooth curves ($A_{front}A_{mid}A_{back}$, $E_{front}E_{mid}E_{back}$, and $B_{front}B_{mid}B_{back}$) of each stratum surface can be obtained (Figure 11a).

2.5.2. Transition Section Generation Based on Morphing Interpolation

Calculate any point $M(u, v)$ inserted in the transition section according to the morphing formula (2), and then obtain the point set of the transition section.

$$\begin{cases} M(u, v) = H(u, v) + L(u, v) \\ H(u, v) = (1 - v)F(u) + vG(u) \\ L(u, v) = (1 - u)a + ub \\ a = P(v) - H(0, v) \\ b = Q(v) - H(1, v) \end{cases} \tag{2}$$

where $M(u, v)$ represents the point to be inserted, and it is jointly controlled by the transverse path interpolation function $H(u, v)$ and longitudinal constraint boundary function $L(u, v)$. $0 \le u \le 1, 0 \le v \le 1$. Moreover, as shown in Figure 10, u represents the close degree between the point to be inserted and the FC or LC constraint lines and v represents the close degree between the point to be inserted and SRC or DEST lines. In addition, $F(u)$ is the curve function of the SRC, $G(u)$ is the curve function of the DEST, $P(v)$ is a function of FC and $Q(v)$ is a function of LC, a is the adjustment factor corresponding to the FC, and b is the adjustment factor corresponding to the LC.

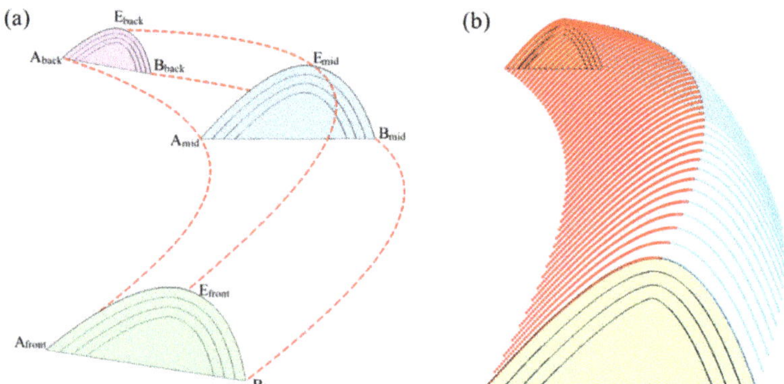

Figure 11. Interpolation of transition sections based on morphing, (**a**) constraint boundaries for morphing interpolation, and (**b**) point set of transition sections. The point sets of two colors are the results of morphing interpolation, respectively.

2.6. Fold Model Construction and Attribute Assignment

After the construction of all of the transition sections is completed, the boundary representation (Brep) surface model of the fold symbol can be built based on the contour comparison algorithm [55]. The difficulty of the contour algorithm lies in the treatment of branching, correspondence, meshing, and smoothing problems [56]. For the method used in this study, because the point sets between adjacent sections are all in one-to-one mode, the branching problem and corresponding problem are not observed. The smoothing problem can also be solved by generating transition sections.

Considering the facet complexity and algorithm efficiency of the modeling result, the shortest diagonal method is selected [57] (Figure 12b). This is a locally optimal objective function method. The fold symbol model was finally obtained based on the contour comparison algorithm, as shown in Figure 12c.

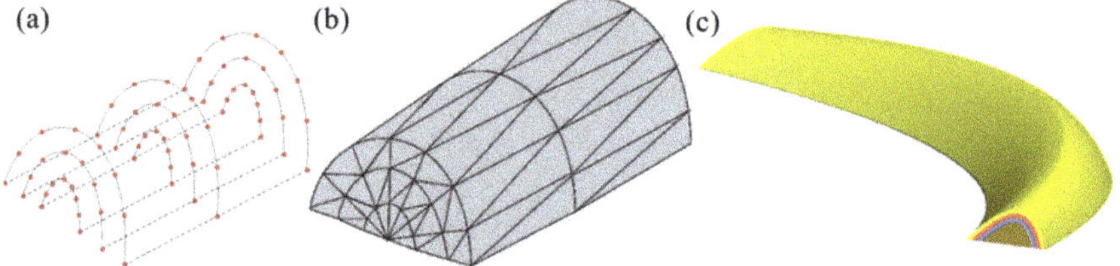

Figure 12. Fold surface model constructed based on the contour algorithm, (**a**) adjacent sections after morphing, (**b**) surface model constructed using the contour algorithm, and (c) fold symbol model result. Entities with different colors represent different strata.

After the model stitching is completed, attribute information, such as age, the lithological description of each stratum, and contact relationship between strata, can be stored in the corresponding JSON format model attribute file. The association query between the 3D model and the attribute information, 3D information annotation, and other functions can be effectively performed by binding the ID information of each object.

3. Case Studies

This study selects four folds with different types for testing the performance of the 3D modeling of fold-structure symbols. The experiments were carried out using a computer with the configuration of a 3.20 GHz Intel(R) Core (TM) i7-8700 CPU, 16 GB RAM. The operation system is Windows 10 Professional. All algorithms are implemented using GDAL3.2 and Dotspatial 1.8 and compiled using Microsoft visual C# 2017 compiler. The number of control and transition sections in all cases is three and eight, respectively. Information for each fold is presented in Table 4. The parameters of the different cases are summarized in Table 5.

Table 4. Details of each experimental area.

ID	Name	Fold Type	Source
1	Dayue Mountain in Mount Lu	Vertical horizontal anticline	[58]
2	Danaobo Mountain in Mount Lu	Oblique horizontal anticline	[59]
3	Fangdou Mountain in East Sichuan	Chevron fold	[60]
4	Scottish Highlands	Recumbent fold	[61]

Table 5. Modeling parameters of each experimental area.

Type	Name	Unit	Dayue Mountain	Danaobo Mountain	Fangdou MOUNTAIN	Scottish Highlands
Location parameters	Center point of middle section	coordinate	(1000, 0, 0)	(1000, 0, 0)	(1000, 0, 0)	(1000, 0, 0)
	Distance from front and back sections to middle section	m	2000, 2000	2000, 2000	6000, 6000	2000, 2000
	Fold width	m	2000	2000	2000	1000
	Front and back dip angles of hinge line	°	−10, −10	−5, −5	0, 0	0, 0
	Front and back strikes of hinge line	°	0, 0	−10, −5	0, 0	0, 0
	Axial dip angles of front and back sections	°	80, 80	70, 70	95, 95	20, 20
Morphological parameters	Dip angles of axial surface	°	80	70	95	20
	Dip angles of inflection points	°	55, 35	40	120	150, 10
	Morphological parameter		0.8	0.4	0.95	0.6
	Collection of strata thickness	M	50, 50, 50	50, 50, 50	30, 30, 30	40, 40, 40
	Maximum rotation angle of distortion	°	0	0	0	30

3.1. Case 1: Anticline Modeling

Dayue Mountain is located southeast of Mount Lu in Jiangxi Province and represents the second highest peak in the Mount Lu region. The measured data from geological records of Mount Lu [58] indicate that the 3D symbol model of Dayue Mountain (Figure 13b), generated based on the above parameters, reflects the structural characteristics of Dayue Mountain, such as equal stratum thickness, vertical anticline, and rounded hinge zone. The number of vertices and triangles is 16,755 and 28,418, respectively.

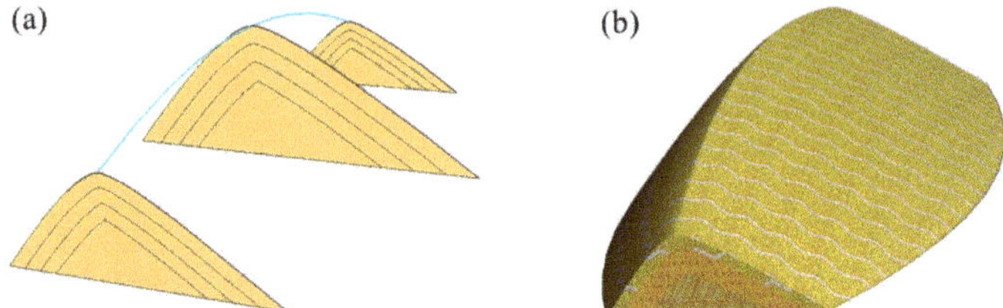

Figure 13. 3D symbol modeling process of Dayue Mountain: (**a**) sections and hinge line of Dayue Mountain and (**b**) 3D symbol model of Dayue Mountain upright anticline.

3.2. Case 2: Modeling of Syncline

Danaobo Mountain, as an oblique horizontal fold, is located in the Xifeng Cave and Danaobo mountain area at the northeast corner of the survey area (Mahui Ridge sheet (H-50-88-D)). The axial trace is in the northeast direction, and the axis surface inclines to the southeast at a dip angle of approximately 70°. The hinge was nearly horizontal, the fold length was approximately 9 km, and the width was approximately 2.5 km. Among the strata involved, the core is the second and fourth member of the Hanyangfeng Formation and the two limbs are the first and second members of the Hanyangfeng Formation. The generated 3D symbol model of Danaobo Mountain is shown in Figure 14. The number of vertices and triangles is 14,539 and 29,062, respectively.

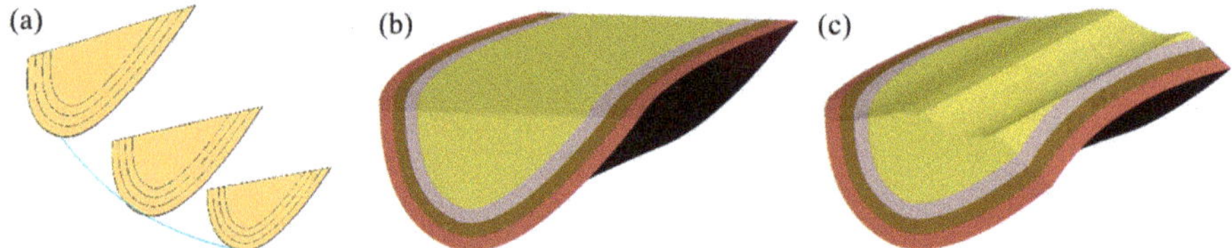

Figure 14. 3D symbol modeling process of the Danaobo mountain: (**a**) sections and hinge line of Danaobo mountain, (**b**) 3D symbol model of the Danaobo mountain upright syncline before trimming, and (**c**) 3D symbol model of the Danaobo mountain upright syncline after trimming.

3.3. Case 3: Modeling of Chevron Anticline

The Fangdou Mountain chevron anticline is located in the East Sichuan Fold Belt. The mountain is a narrow, high, and steep anticline in the Jura-type folds of eastern Sichuan. The Fangdou Mountain structure is a composite anticline composed of fault-bend folds and longitudinal-bend folds controlled by detachment, and its surface section is an asymmetric fold with a sharp hinge zone, which gradually widens and slows to form an asymmetric box fold [62]. According to the 95-23.5 survey line section data of Fangdou Mountain [60], the limb dip angles of the anticline are approximately 20°–25° and symmetrical and the axial surface is nearly vertical. Based on these parameters (Table 5), a 3D symbol model of Fangdou Mountain (Figure 15) is generated, which reflects the structural characteristics of its anticline, sharp edge, upright, and symmetrical characteristics. The number of vertices and triangles is 12,843 and 24,748, respectively.

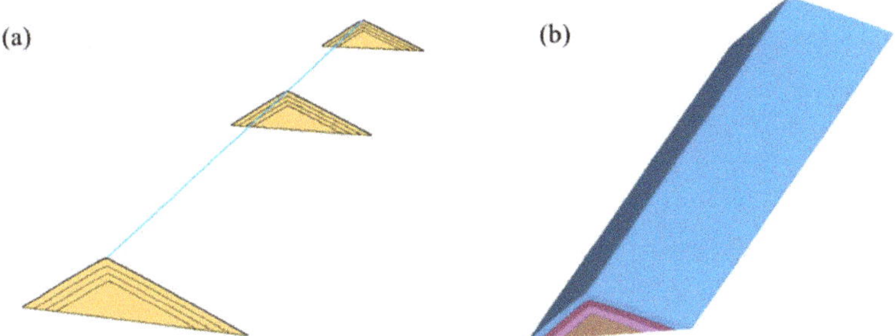

Figure 15. 3D symbol modeling process of Fangdou Mountain: (**a**) sections and hinge line of Fangdou Mountain, and (**b**) 3D symbol model of the Fangdou Mountain chevron fold.

3.4. Case 4: Modeling of Recumbent Folds

A large number of overturned folds are observed in the Scottish Highlands [61]. This study selects a recumbent fold to demonstrate symbol modeling in the central south zone of the area. The approximate fold information can be obtained from the geological section map (Figure 16a): the dip angle of the left limb was approximately 150° and the right is at approximately 10°. The generated 3D symbol model is shown in Figure 16b. The number of vertices and triangles is 7020 and 14,024, respectively.

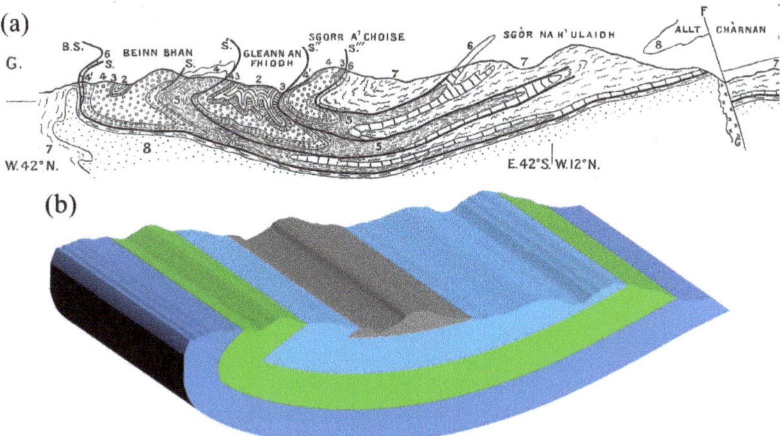

Figure 16. Section map and modeling result of recumbent fold in the Scottish Highlands: (**a**) geological section map of the recumbent fold and (**b**) symbol model of the recumbent fold. Entities with different colors represent different strata.

3.5. Case 5: Other Types of Folds

In addition to the fold structures in the specific areas mentioned above, this method can also be applied to a more general fold symbol modeling based on the inclusion of the appropriate modeling parameters. When the morphological parameter cur is set to 0.8, 0.5, and 0.2, the symbol models of the chevron fold, circular fold, and box fold can be generated. When the interlimb angle θ is set to 30°, 50°, 100°, and 160°, the symbol models of a tight fold, closed fold, broad fold, and gentle fold can be generated, respectively (Figure 17).

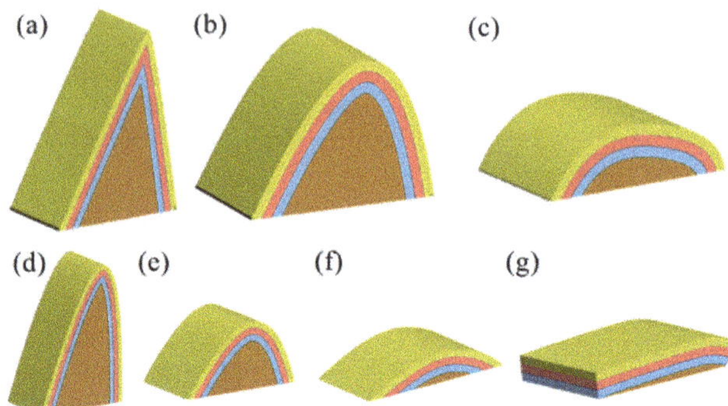

Figure 17. Other types of fold symbols obtained by setting the cur and θ parameters to different values, (**a**) cur = 0.8, chevron fold, (**b**) cur = 0.5, circular fold, (**c**) cur = 0.2, box fold, (**d**) θ = 30°, tight fold, (**e**) θ = 50°, closed fold, (**f**) θ = 100°, broad fold, and (**g**) θ = 160°, gentle fold. For a description of parameter cur and θ, please refer to Table 2. Entities with different colors represent different strata.

4. Discussion

4.1. Analysis of the Influencing Factors

The use of a greater number of control sections increases the accuracy of the generated model and ensures that the model is able to better restore the morphological characteristics of the fold structure. For fold structures with small-scale and prominent characteristics, fewer control sections are required to reflect the hinge morphology. In contrast, for fold types with large-scale and relatively complex hinge morphologies, more control sections are required. The typical fold structure selected in this experiment has a small scale; therefore, only three sections in the middle and at both ends were generated.

The use of a greater number of transition sections leads to the generation of a smoother model and requires a greater amount of model data. However, the running efficiency tends to be lower when a model with a large amount of data displays a large scene. For the fold type with a relatively smooth hinge zone, fewer transition sections are required to effectively reflect the 3D morphology of the fold structure. Otherwise, more transition sections would be required to effectively reflect the 3D morphology of the fold structure.

The different selection methods of the location parameters of the cross-sections significantly influence the model morphology. The location parameters of the cross-sections can be measured based on the outer rectangle of the fold distribution range or the outer rectangle of the core formation. These methods can be applied as long as they can effectively reflect the morphological characteristics of the fold structures.

The quality of the middle cross-section is a critical factor for determining the modeling quality of the fold symbol. The parametric modeling method is effective for the middle cross-section with a relatively simple stratum boundary morphology. However, the effect of parametric modeling is poor for a more complex middle cross-section. In this case, a better modeling quality can be obtained by replacing the cross-sections generated by the parameters with the measured cross-section.

4.2. Factors That Influence the Symbol Display Effect

The spatial information of the constructed models is stored in an OBJ format file, while the attribute information is in JSON. The constructed 3D fold symbols can be placed on a 3D Earth display platform (such as Cesium or Google Earth). When importing a model into Cesium, the process mainly includes: (1) converting the OBJ format model file to FBX format file, which contains information such as material and lighting; (2) converting the FBX format model files to GLB format file, which is the 3D model data format supported by the Cesium platform; (3) directly loading the glb file into the Cesium scene.

Consistent with the expression requirements of 2D symbols [63], when the constructed 3D folding symbols are placed on the 3D earth display platform, it is also necessary to set the appropriate symbol size to optimize the display effect. When the scale of the fold structure is small and the internal characteristics of the fold elements are apparent, the effect is better when the ratio of the symbol to the actual fold is 1:1. When the scale of the fold structure is large, the internal characteristics of the fold elements will not be sufficiently prominent, and the effect of the proportional magnification will be better. In addition, in 2D electronic maps, the size of related symbols usually needs to be adaptively adjusted at different scales. This rule also applies to symbol-size settings in 3D scenes. For example, when displayed on the Cesium platform, the size of the corresponding model can be changed for adaptive scaling and adjustment to ensure that the relative size of the symbol model remains unchanged at different scales (Figure 18).

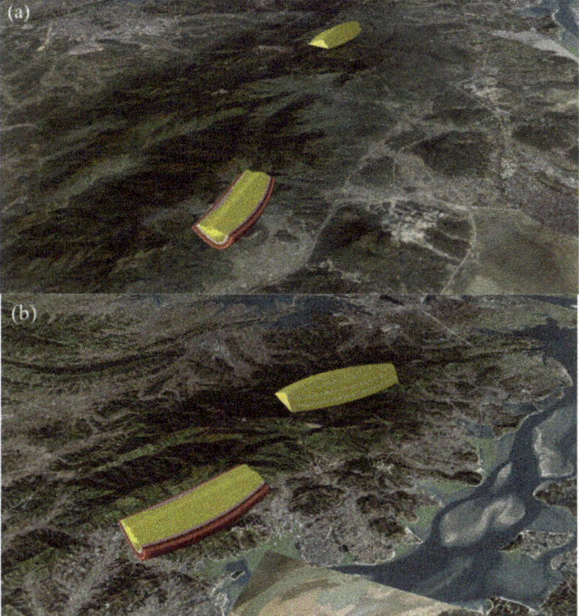

Figure 18. Display effect of fold symbols with different scaling ratios: (**a**) scaling down and (**b**) equal ratio.

When the symbol is smaller than the actual geological structure, it should be placed directly above the center point of the actual geological structure. When the scale is 1:1, the symbol should be placed in the actual position of the structure and the height should be consistent with the actual structure. In addition, when symbols are placed, the strike of the symbols must be consistent with that of the actual structure.

Different proportions may occur between the actual length along the hinge direction and the width of the fold structure. According to the ratio of longitudinal length to transverse width, folds can be divided into linear, long-axis, short-axis, or equiaxed folds (dome structure and tectonic basin) [64]. However, if the symbol model is constructed according to the actual proportion of the geological structures, then the structural information conveyed by the fold elements is weakened and the information about the hinge zone is highlighted. Moreover, the typical geological characteristics of some ample linear folds are likely to be reflected in the spatial information of the hinge; thus, comprehensive restoration of the entire fold is also required. Therefore, intercepting a part of the fold in the hinge direction or reflecting the 3D morphology of the entire fold depends on the characteristics of the fold and preferences of the user (Figure 19).

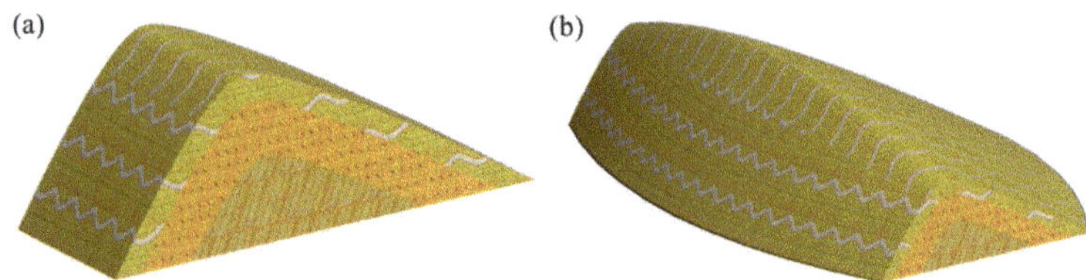

Figure 19. Examples of fold symbol models with different hinge lengths: (**a**) local morphology and (**b**) whole morphology.

Stratigraphic legends usually use color to indicate age and texture to indicate lithology. Therefore, by using an appropriate legend to decorate the corresponding strata, the fold symbols can better express and transmit the age and lithology information.

4.3. Analysis of Applicability

The premise of parametric modeling is to obtain relevant modeling parameters effectively. In this method, the modeling parameters of the fold structure can be directly obtained from the regional geological survey report or measured from corresponding maps, such as geological maps, structural maps, or section maps. Under exceptional circumstances, the parameters required for fold modeling can also be empirically assigned according to the fold type and structural knowledge.

This study's 3D parametric modeling method of fold symbol construction can realize symbol model construction when the geological data are insufficient. Although the symbol has an abstract requirement for the model structure, less dependence on geological data also means that the model results will be simplified in terms of morphology and large discrepancies with the actual structure may occur. In the case of complex sections, if the measured cross-sections are available, they can be used to replace the cross-sections generated by the parameters, so that more accurate modeling of the fold symbol can be carried out [23–25]. For example, by selecting the cross-section (Figure 20a) of the Danaobo mountain syncline as the middle section, a more detailed symbol model of the Danaobo mountain syncline can be built (Figure 20b). This experiment shows that the symbol model of the fold structure based on the measured cross-section can restore the actual morphology of the current fold structure. This method also has some shortcomings, such as relying on geological section data and being unable to directly reflect the original structural morphology before being subjected to external forces. In specific applications, the appropriate method for the cross-section can be selected according to the geological data and requirements to achieve the best data utilization and model expression effect.

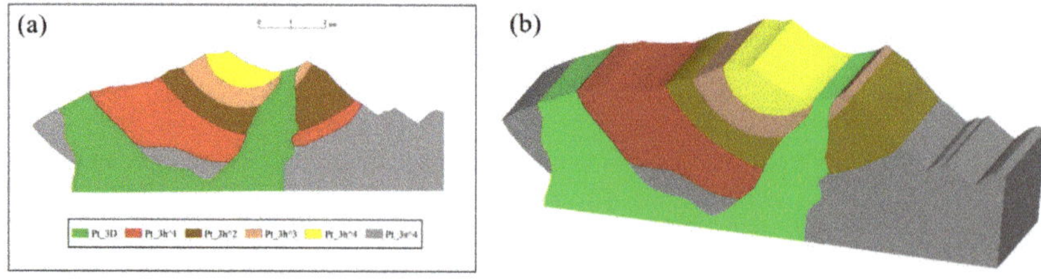

Figure 20. Fold structure modeling based on measured cross-section: (**a**) section of the Danaobo mountain syncline at Hanyang Peak of Mount Lu and (**b**) symbol model of the Danaobo mountain syncline.

A box fold refers to a fold in which the strata are all overturned, and the hinge zone is flat and wide, similar to a box [64]. The box fold is a component of the world-famous Jura-type folds [65], which were developed in East Sichuan in China as well as in the

Alps in Europe. In contrast to general folds, box folds typically have two hinges and a pair of conjugate axial surfaces. The parametric modeling method of fold symbols proposed in this paper cannot easily and directly realize the symbol modeling of box folds; therefore, a particular modeling method must be designed for box folds. According to the morphological characteristics of "two hinges and a pair of conjugate axial surfaces" of box folds, the middle cross-section of a box fold can be divided into left and right parts for modeling, and the topological transition at the middle junction should be guaranteed (Figure 21). The construction method of the stratum boundaries in the middle section of the box fold is as follows: (1) obtain the positions of inflection points A and B and the corresponding extension lines along the dip angles, according to the position of the inflection points and the occurrence of both limbs input by the user; (2) calculate the distance h from A to the extension line of B, take the midpoint K of AB, and make CK⊥AB so that the length of segment CK is equal to h; and (3) take point E on CK, according to the morphological parameter cur entered by the user, make FG//AB so that the dip extension lines of the two inflection points intersect with points F and G. After that, Bezier interpolation is carried out according to points A, F, E, E, G, and B to obtain the stratum boundary of the box fold; namely, the arc AEB.

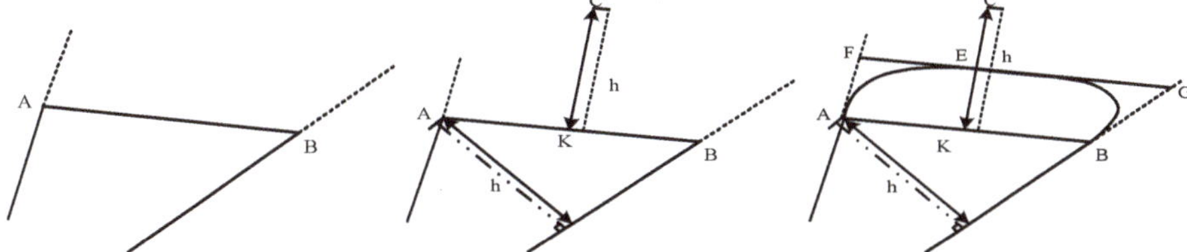

Figure 21. Construction process for the stratum boundaries of box folds.

A lack of apparent regularity is observed among stratum boundaries for more complex disharmonious folds. Rather than being deduced from a specific stratum boundary, the boundary of each stratum must be generated using the method described in Section 2.3.1. Therefore, the morphological parameters of each stratum of the disharmonious fold (such as the occurrence of inflection points, dip angles of the axial surfaces, and interlimb angles of each stratum) must be given in detail.

5. Conclusions

This study aims to develop a novel parametric modeling method for the 3D symbols of fold structures using cross-sections generated by fold parameters and a contour comparison algorithm. The cross-section generation method based on parameters and the Bezier curve effectively supports the parametric modeling of cross-sections of various fold structures. Transition section generation based on morphing satisfies the smooth requirement of the symbol model. The symbol-modeling method can control the model accuracy and data size by setting an appropriate number and quality of cross-sections, which may be measured or generated by fold parameters.

In this study, different types of fold structures were selected as experimental objects for modeling experiments. The experimental results show that the generated 3D fold symbol models can abstractly express the morphologies of fold structures developed in the experimental area and can express most fold structure subdivision types. The parametric modeling method for the 3D symbols of fold structures follows the traditional symbol creation rules [66] and cartographic rules [67]. The modeling method in this study is simple and efficient, avoids excessive dependence on geological survey data, and fits well with the abstract character of the geological symbol. Furthermore, the controllable characteristics of model accuracy and data size better meet the needs for efficient display in a wide range of 3D scenes.

In view of the inherent non-parametric geometric features of geological structures [41], this method breaks through the application limitation of parametric modeling methods in the field of 3D geological modeling. The parametric modeling method used for 3D fold-symbol modeling in this study can be applied for 3D symbol modeling of other geological structure types, such as faults, joints, and intrusions. It also applies for the 3D modeling of typical geological structures with relatively simple spatial morphology. The 3D geological symbols generated by this method have excellent research significance and application value for the 3D symbol expression of geological structures in digital Earth, digital cities, AR\VR\MR, geography teaching, science popularization, and other applications and research areas. In addition, the method in this paper has laid a sound data foundation for future research on the construction method of 3D spatio-temporal geological symbols (4D geological symbols).

Author Contributions: An-Bo Li conceived the original idea and is the primary author of the manuscript. Hao Chen developed the main modules of the prototype system and is the key contributor to the methodology. Xian-Yu Liu and Xiao-Feng Du developed partial modules of the prototype system. Guo-Kai Sun processed the relative data. All authors have read and agreed to the published version of the manuscript.

Funding: This study was supported by the National Key R&D Program of China (No. 2022YFB3904101, 2022YFB3904104) and by the National Natural Science Foundation of China (Project No. 41971068, 41771431).

Institutional Review Board Statement: Not applicable.

Informed Consent Statement: Not applicable.

Data Availability Statement: Not applicable.

Conflicts of Interest: The authors declare no conflict of interest.

References

1. Declan, B. Virtual globes: The web-wide world. *Nature* **2006**, *439*, 776–779.
2. Matrone, F.; Colucci, E.; De Ruvo, V.; Lingua, A.; Spanò, A. HBIM in a semantic 3D GIS database. *Int. Arch. Photogramm. Remote Sens. Spat. Inf. Sci.* **2019**, *42*, W11. [CrossRef]
3. Landeschi, G.; Dell'Unto, N.; Lundqvist, K.; Ferdani, D.; Campanaro, D.M.; Touati, A.-M.L. 3D-GIS as a platform for visual analysis: Investigating a Pompeian house. *J. Archaeol. Sci.* **2016**, *65*, 103–113. [CrossRef]
4. Petrov, V.A.; Veselovskii, A.V.; Kuz'mina, D.A.; Plate, A.N.; Gal'berg, T.V. Spatial-temporal three-dimensional GIS modeling. *Autom. Doc. Math. Linguist.* **2015**, *49*, 21–26. [CrossRef]
5. Zhu, Q. 3D GIS and its application in smart cities. *J. Earth Inf. Sci.* **2014**, *16*, 151–157.
6. Zhu, L.; Hou, W.; Du, X. Digital Earth—From surface to deep: Introduction to the Special issue. *Front. Earth Sci.* **2022**, *15*, 491–494. [CrossRef]
7. Goodchild, M.F.; Guo, H.; Annoni, A.; Bian, L.; de Bie, K.; Campbell, F.; Craglia, M.; Ehlers, M.; van Genderen, J.; Jackson, D.; et al. Next-generation Digital Earth. *Proc. Natl. Acad. Sci. USA* **2012**, *109*, 11088–11094. [CrossRef]
8. Gede, M.; Jeney, J. Thematische Kartierung mit Verwendung von Cesium Thematic Mapping with Cesium. *KN J. Cartogr. Geogr. Inf. Kartogr. Nachr.* **2017**, *67*, 210–213. [CrossRef]
9. Gual, J.; Puyuelo, M.; Lloveras, J. Three-dimensional tactile symbols produced by 3D Printing: Improving the process of memorizing a tactile map key. *Br. J. Vis. Impair.* **2014**, *32*, 263–278. [CrossRef]
10. Holloway, L.; Marriott, K.; Butler, M. Accessible maps for the blind: Comparing 3D printed models with tactile graphics. In Proceedings of the 2018 Chi Conference on Human Factors in Computing Systems, Montreal, QC, Canada, 21 April 2018; pp. 1–13.
11. Bandrova, T. Designing of Symbol System for 3d City Maps. In Proceedings of the 20th International Cartographic Conference, Beijing, China, 2016.
12. Patton, J.C. SOME Truth with Maps: A Primer on Symbolization and Design. *Cartogr. Perspect.* **1995**, *20*, 46–47. [CrossRef]
13. Zhu, S.R.; Liu, B.B.; Chen, F.X. Automatic 3D modeling of 2D geological vector point symbols. *Surv. Mapp. Sci.* **2017**, *42*, 132–138+146.
14. Lemon, A.M.; Jones, N.L. Building solid models from boreholes and user-defined cross-sections. *Comput. Geosci.* **2003**, *29*, 547–555. [CrossRef]
15. Wei, Y.J. Research on the Organization, Management and Symbolization of 3D GIS Data. Ph.D. Thesis, PLA Information Engineering University, Zhengzhou, China, 2006.

16. Fernández, O. Reconstruction of Geological Structures in 3D: An Example from the Southern Pyrenees. Ph.D. Thesis, Universitat de Barcelona, Barcelona, Spain, 2004.
17. Laurent, G.; Ailleres, L.; Grose, L.; Caumon, G.; Jessell, M.; Armit, R. Implicit modeling of folds and overprinting deformation. *Earth Planet. Sci. Lett.* **2016**, *456*, 26–38. [CrossRef]
18. Rife, D. Acoustic analysis and visualization using real time three-dimensional parametric modeling. *J. Acoust. Soc. Am.* **2010**, *128*, 2411. [CrossRef]
19. Badwi, I.M.; Ellaithy, H.M.; Youssef, H.E. 3D-GIS Parametric Modelling for Virtual Urban Simulation Using CityEngine. *Ann. GIS* **2022**, *28*, 325–341. [CrossRef]
20. Xu, M.; Liu, N.; Cong, F.B. Research on 3D Symbol Composition and Modeling Method. *Oceanogr. Surv. Mapp.* **2006**, *02*, 45–48.
21. Gu, G.W. Symbolization of Three-Dimensional Geospatial Models. Master Thesis, Xi'an University of Science and Technology, Xi'an, China, 2014.
22. Wu, Q.; Xu, H.; Zou, X.K. An effective method for 3D geological modeling with multi-source data integration. *Comput. Geosci.* **2005**, *31*, 35–43. [CrossRef]
23. Tipper, J.C. Computerized modeling in reconstruction of objects from serial sections. *AAPG Bul.* **1976**, *60*, 728.
24. Herbert, M.H.; Jones, C.B.; Tudhope, D.S. 3D reconstruction of geoscientific objects from serial sections. *Vis. Comput.* **1995**, *11*, 343–359. [CrossRef]
25. Thornton, J.M.; Mariethoz, G.; Brunner, P. A 3D geological model of a structurally complex Alpine region as a basis for interdisciplinary research. *Sci. Data* **2018**, *5*, 180238. [CrossRef]
26. Liu, X.-Y.; Li, A.-B.; Chen, H.; Men, Y.-Q.; Huang, Y.-L. 3D Modeling Method for Dome Structure Using Digital Geological Map and DEM. *ISPRS Int. J. Geo-Information* **2022**, *11*, 339. [CrossRef]
27. Mao, P.; Zhaoliang, L.I.; Zhongbo, G.; Yang, Y.; Gengyu, W. 3D Geological Modeling–Concept, Methods and Key Techniques. *Acta Geol. Sin.* **2012**, *86*, 1031–1036. [CrossRef]
28. Mallet, J.L. *Geomodeling*; Oxford University Press: New York, NY, USA, 2002; p. 599.
29. Hao, M.; Li, M.; Zhang, J.; Liu, Y.; Huang, C.; Zhou, F. Research on 3D geological modeling method based on multiple constraints. *Earth Sci. Informatics* **2020**, *14*, 291–297. [CrossRef]
30. Fernandez, O.; Muñoz, J.A.; Arbués, P.; Falivene, O.; Marzo, M. Three-dimensional reconstruction of geological surfaces: An example of growth strata and turbidite systems from the Ainsa basin (Pyrenees, Spain). *AAPG Bull.* **2004**, *88*, 1049–1068. [CrossRef]
31. Perrin, M.; Zhu, B.; Rainaud, J.-F.; Schneider, S. Knowledge-driven applications for geological modeling. *J. Pet. Sci. Eng.* **2005**, *47*, 89–104. [CrossRef]
32. Vidal-Royo, O.; Hardy, S.; Koyi, H.; Cardozo, N. Structural evolution of Pico del Águila anticline (External Sierras, Southern Pyrenees) derived from sandbox, numerical and 3D structural modeling techniques. *Geol. Acta* **2013**, *11*, 1–25.
33. De Kemp, E.A.; Sprague, K.B. Interpretive Tools for 3-D Structural Geological Modeling Part I: Bézier -Based Curves, Ribbons and Grip Frames. *GeoInformatica* **2003**, *7*, 55–71. [CrossRef]
34. Amorim, R.; Brazil, V.E.; Samavati, F.; Sousa, M.C. 3D geological modeling using sketches and annotations from geologic maps. In Proceedings of the 4th Joint Symposium on Computational Aesthetics, Non-Photorealistic Animation and Rendering, and Sketch-Based Interfaces and Modeling, Vancouver, BC, Canada, 8–10 August 2014; pp. 17–25.
35. Guo, J.; Wu, L.; Zhou, W.; Jiang, J.; Li, C. Towards Automatic and Topologically Consistent 3D Regional Geological Modeling from Boundaries and Attitudes. *Int. J. -Geo-Inf.* **2016**, *5*, 17. [CrossRef]
36. Kadi, H.; Anouche, K. Knowledge-based parametric modeling for heritage interpretation and 3D reconstruction. *Digit. Appl. Archaeol. Cult. Heritage* **2020**, *19*, e00160. [CrossRef]
37. Zhang, H.; Zhu, J.; Xu, Z.; Hu, Y.; Wang, J.; Yin, L.; Liu, M.; Gong, J. A rule-based parametric modeling method of generating virtual environments for coupled systems in high-speed trains. *Comput. Environ. Urban Syst.* **2016**, *56*, 1–13. [CrossRef]
38. Lintermann, B.; Deussen, O. Interactive modeling of plants. *IEEE Comput. Graph. Appl.* **1999**, *19*, 56–65. [CrossRef]
39. Ijiri, T.; Owada, S.; Okabe, M. Floral diagrams and inflorescences: Interactive flower modeling using botanical structural constraints. *Acm Trans. Graph.* **2005**, *24*, 720–726. [CrossRef]
40. Bastioni, M.; Re, S.; Misra, S. Ideas and methods for modeling 3D human figures: The principal algorithms used by Make Human and their implementation in a new approach to parametric modeling. In Proceedings of the 1st Bangalore Annual Compute Conference, Bangalore, India, 18–20 January 2008; pp. 1–6.
41. Li, A.B.; Zhou, L.C.; Lü, G.N. *Geological Information Systems*; Science Press: Beijing, China, 2013.
42. Fossen, H. *Structural Geology*; Cambridge University Press: New York, NY, USA, 2010.
43. Li, D.L.; Wang, E.L. *Structural Geology*; Jilin University Press: Changchun, China, 2001.
44. Billings, M. *Structural Geology*, 3rd ed.; Pearson College Div: New York, NY, USA, 1972.
45. Zyda, M.J.; Jones, A.R.; Hogan, P.G. Surface construction from planar contours. *Comput. Graph.* **1987**, *11*, 393–408. [CrossRef]
46. Wu, B.C. Research on 3D Geological Modeling and Visualization System Based on Triangular Prism. Master Thesis, Liaoning University of Engineering and Technology, Fuxin, China, 2007.
47. Yao, M.M. Research on Parametric 3D Modeling Method of Basic Geological Structure Types. Master Thesis, Nanjing Normal University, Nanjing, China, 2017.
48. Renfrew, C.; Bahn, P. *Archaeology: The Key Concepts*, 1st ed.; Routledge: London, UK, 2004.

49. Ray, S.K. Inverted fold hinge: An end member of hinge rotation by superposed buckle folding in the Precambrian terrain of western India. *J. Struct. Geol.* **2018**, *116*, 260–265. [CrossRef]
50. Wolberg, G. Image morphing: A survey. *Vis. Comput.* **1998**, *14*, 360–372. [CrossRef]
51. Gotsman, C.; Surazhsky, V. Guaranteed intersection-free polygon morphing. *Comput. Graph.* **2001**, *25*, 67–75. [CrossRef]
52. Deng, M.; Peng, D.L.; Xu, X.; Liu, H.M. A morphing method of linear features based on bending structure. *J. Cent. South Univ. (Nat. Sci. Ed.)* **2012**, *43*, 2674–2682.
53. Ming, J.; Yan, M. 3D Geological Surface Creation Based on Morphing. *Geogr. Geo-Inf. Sci.* **2014**, *30*, 37–40.
54. Carter, A.; Roques, D.; Bristow, C.; Kinny, P. Understanding Mesozoic accretion in Southeast Asia: Significance of Triassic thermotectonism (Indosinian orogeny) in Vietnam. *Geology* **2001**, *29*, 211–214. [CrossRef]
55. Stroud, I. *Boundary Representation Modeling Techniques*; Springer Science & Business Media: Berlin, Germany, 2006.
56. Meyers, D.; Skinner, S.; Sloan, K. Surfaces from contours. *ACM Trans. Graph.* **1992**, *11*, 228–258. [CrossRef]
57. Ekoule, A.B.; Peyrin, F.; Odet, C.L. A triangulation algorithm from arbitrary shaped multiple planar contours. *ACM Trans. Graph.* **1991**, *10*, 182–199. [CrossRef]
58. Xie, G.G. Specification of Lushan Sheet H-50-88-B 1/50000 Geological Map. Investigation and Research Team of Jiangxi Provincial Bureau of Geological and Mineral Exploration and Development. 1993.
59. Li, J.H. Specification of Mahuiling Sheet G-50-88-D 1/50000 Geological Map. Investigation and Research Team of Jiangxi Provincial Bureau of Geological and Mineral Exploration and Development. 1995.
60. Ding, D.D.; Guo, T.L.; Zhai, C.B.; Lv, J.X. Knee structure in West Hubei East Chongqing area. *Pet. Exp. Geol.* **2005**, 205–210.
61. Bailey, E.B. Recumbent Folds in the Schists of the Scottish Highlands. *Q. J. Geol. Soc.* **1910**, *66*, 586–620. [CrossRef]
62. Yan, D.P.; Wang, X.W.; Liu, Y.Y. Analysis of fold structural style and its genetic mechanism in Sichuan Hubei Hunan border area. *Mod. Geol.* **2000**, *01*, 37–43.
63. Halik, Ł.; Medyńska-Gulij, B. The differentiation of point symbols using selected visual variables in the mobile augmented reality system. *Cartogr. J.* **2017**, *54*, 147–156. [CrossRef]
64. Winterer, E.L. *Earth's Dynamic Systems*; Wiley Online Library: Hoboken, NJ, USA, 1998.
65. Jamison, W.R. Geometric analysis of fold development in overthrust terranes. *J. Struct. Geol.* **1987**, *9*, 207–219. [CrossRef]
66. Medyńska-Gulij, B. Geomedia Attributes for Perspective Visualization of Relief for Historical Non-Cartometric Water-Colored Topographic Maps. *ISPRS Int. J. Geo-Inf.* **2022**, *11*, 554. [CrossRef]
67. Medyńska-Gulij, B. Point Symbols: Investigating Principles and Originality in Cartographic Design. *Cartogr. J.* **2008**, *45*, 62–67. [CrossRef]

 International Journal of *Geo-Information*

Article

Road Network Generalization Method Constrained by Residential Areas

Zheng Lyu [1,2], Qun Sun [1], Jingzhen Ma [1,3,*], Qing Xu [1], Yuanfu Li [1] and Fubing Zhang [1]

[1] Institute of Geospatial Information, Information Engineering University, Zhengzhou 450000, China; lvzheng_xd@163.com (Z.L.); 13503712102@163.com (Q.S.); xq1982_no.1@163.com (Q.X.); alyf2011@sina.com (Y.L.); zhangfbing@163.com (F.Z.)
[2] Collaborative Innovation Center of Geo-Information Technology for Smart Central Plains, Zhengzhou 450000, China
[3] Key Laboratory of Spatiotemporal Perception and Intelligent Processing, Ministry of Natural Resources, Zhengzhou 450000, China
* Correspondence: zb50mjz@163.com; Tel.: +86-187-4945-3676

Abstract: Residential areas and road networks have a strong geographical correlation. The development of a single geographical feature could destroy the geographical correlation. It is necessary to establish collaborative generalization models suitable for multiple features. However, existing road network generalization methods for mapping purposes do not fully consider residential areas. Compared with road networks, residential areas have a higher priority in cartographic generalization. In this regard, this study proposes a road network generalization method constrained by residential areas. First, the roads and settlements obtained from clustering residential areas were classified. Next, the importance of the settlements was evaluated and certain settlements were selected as the control features. Subsequently, a geographical network with the settlements as the nodes was built, and the traffic paths between adjacent settlements were searched. Finally, redundant paths between the settlements were simplified, and the visual continuity and topological connectivity were checked. The data of a 1:50,000 road network and residential areas were used as the experimental data. The experimental results demonstrated that the proposed method preserves the overall structure and relative density characteristics of the road network, as well as the geographical correlation between the road network and residential areas.

Keywords: road network generalization; collaborative generalization; settlement; geographic correlation; divide-and-conquer

Citation: Lyu, Z.; Sun, Q.; Ma, J.; Xu, Q.; Li, Y.; Zhang, F. Road Network Generalization Method Constrained by Residential Areas. *ISPRS Int. J. Geo-Inf.* **2022**, *11*, 159. https://doi.org/10.3390/ijgi11030159

Academic Editors: Beata Medynska-Gulij, David Forrest, Thomas P. Kersten and Wolfgang Kainz

Received: 29 December 2021
Accepted: 18 February 2022
Published: 22 February 2022

Publisher's Note: MDPI stays neutral with regard to jurisdictional claims in published maps and institutional affiliations.

1. Introduction

Cartographic generalization is a technology and science that produces abstract and generalized spatial data when transferring large-scale map information to small-scale maps [1]. The concept of cartographic generalization was first proposed by Eckert in 1921. After years of development, it has experienced a transformation from "geometric perspective → computational perspective → functional perspective", and knowledge and semantic information have attracted increasing attention [2]. The map is a complex system formed by a variety of interactive geographical features. The spatial distribution characteristics of different geographical features, as well as the geographical correlation between features, need to be preserved in cartographic generalization. This is because the generalization of a single geographical feature is likely to destroy the geographical correlation between geographical features, resulting in spatial and semantic conflicts between the results [3]. Long experience has proven that the generalization method for a single type of geographical feature cannot adapt to the development of the scientific paradigm of cartographic generalization, so it is necessary to establish a collaborative generalization model directly facing multiple features.

Roads and residential areas are important socioeconomic features on maps. Residential areas are centers of human activity, and accordingly, the material and personnel between residential areas mainly flow through roads. Residential areas and the road network together constitute the main body of the human geographical activity network. In ancient times, people gathered to form settlements, and production activities naturally caused simple paths to appear around settlements. With the development of productive forces driven by political, economic, and military factors, the number and density of residential areas are gradually increasing, and the coverage and accessibility of road networks are also gradually increasing [4]. Although the essence of the relationship between the evolution of roads and residential areas has been controversial, the positive effect of residential area expansion on the development of the road network is recognized [5]. Residential area has a higher priority in cartographic generalization, which is dominant compared with roads, but the spatial information and attribute information of the road network need to be taken into account in residential area generalization [6]. The road network connects residential areas distributed in different spatial locations, which are usually selected after completing the generalization of residential areas. For closely related geographical features such as roads and residential areas, and rivers and contours, collaborative generalization should be carried out to maintain this geographical correlation [7]. The road network is an important part of the map skeleton and is the research object of this study. As such, we focus on the road network generalization constrained by residential areas.

In view of the geographical correlation between the road network and residential areas and the dominant position of residential areas in cartographic generalization, a road network generalization method constrained by residential areas is proposed. This method adopts the idea of divide-and-conquer, decomposes the road network with the settlements as the control features, and simplifies the complex road selection problem into a redundant path simplification problem between settlements. First, the roads and settlements obtained from residential clustering were classified. Next, the importance of the settlements was evaluated and part of the settlements are selected as the control features. Subsequently, a geographic network with settlements as nodes was built and the connecting paths between adjacent settlements were searched. Finally, the paths between settlements were compared and evaluated to eliminate redundant paths with high traffic costs.

The remainder of this paper is organized as follows. Section 2 describes related studies on road network generalization. Section 3 presents the proposed method. Section 4 validates the method with a real-world road network and residential area data. Section 5 offers conclusions and provides suggestions for future studies.

2. Related Studies

The goal of road network generalization is to preserve important roads in the scale reduction process, abandon redundant roads that do not fit the target scale, and preserve the overall form and connectivity of the road network [8]. Early road network generalizations tended to primarily depend on semantic information and ignore the topological and geometric information on the road network, thereby failing to establish an effective model. To improve the effectiveness of automatic selection, Mackaness and Beard introduced graph theory into road network generalization [9]. The general road network topology constructed with intersections as nodes and sections as edges can effectively maintain the topological characteristics of a road network. Recently, complex network theory has been gradually applied to transportation and geographic information systems, which can aid in analyzing the structure and functional characteristics of road networks [10]. Jiang and Harrie used a dual graph to express the topology of road networks, modeling road sections and intersections as nodes and edges, respectively [11]. They analyzed the importance of roads using complex network theory. Currently, the dual graph model is widely used in road network generalization research. For example, Ma et al. used the classic PageRank algorithm to calculate the importance of roads [12]. However, because graph-based road network generalization methods focus on the analysis of topological information, they ignore semantic and geometric information. Hence, Thomson and Richardson introduced the stroke model based on the principle of Good Continuity, which originated from the Gestalt theory [13]. A stroke is a road chain formed by connecting a group of topological and directional sections according to specific criteria. Stroke-based methods use a stroke as a unit, which can preserve the continuity characteristics of a road network. Subsequently, studies have continued to improve the importance evaluation indicators of the stroke and the selection process. For example, Liu et al. used the length, connectivity, and average local density of the sections that compose the stroke to comprehensively evaluate it [14]. Cao et al. introduced degree of centrality, clustering coefficient and geometric length to evaluate the stroke [15]. In addition, to more optimally preserve the density contrast of the road network, Chen et al. modelled the road network as a set of meshes and selected roads by iteratively merging high-density meshes [16]. To combine the advantages of stroke and mesh, Li and Zhou proposed an integrated method based on network hybrid hierarchies consisting of a linear hierarchy for strokes and an aerial hierarchy for meshes [17]. In addition to the continuous mining of the mode and character of the road network, other spatial data, such as trajectory points, points of interest, and important facilities, have also been introduced in road network generalization to improve importance evaluation. For example, Yu et al. proposed a road network generalization method that considers the geometric characteristics of the road network and traffic flow pattern based on taxi trajectory data [18]. Yamamoto et al. implemented an on-demand road generalization method that adapts to both the facility search results and stroke length [19]. Moreover, cartographic generalization is developing towards intelligence; accordingly, the data-driven road network generalization methods have been studied. Guo et al. proposed a road network selection method that uses ontology to organize cases and conduct knowledge reasoning [20]. In terms of deep learning, Zheng et al. proved that graph convolutional networks are appropriate tools for road network generalization [21]. Road network generalization is becoming increasingly mature and intelligent, but the above methods are insufficient to study the geographical correlation between geographical features.

Collaborative generalization is an important method considering geographical correlation. In terms of the collaborative generalization of road networks and residential areas, Wei proposed a collaborative displacement method, combining aggregation, elimination, and constrained reshape to solve all the possible spatial conflicts during urban building map generalization [22]. Xu summarized the available knowledge on collaborative selection of roads and buildings and used a multi-agent to control the process, which effectively avoided spatial conflict [23]. Touya divided roads into urban and rural area roads, using the shortest path between facilities to determine the importance of roads in rural areas [24]. The above method takes into account the geographical correlation between roads and residential areas, and mainly carries out process control through rules to avoid spatial conflict, but the correlation mode of the two geographical elements is not fully utilized. Therefore, we analyzed the geographical correlation mode of residential areas and the road network, and constructed a geographical network with residential areas and the road network as nodes and edges, respectively. On this basis, following the idea of divide-and-conquer, we adopt the mode of "road network-assisted residential area selection → residential area controlled road network generalization" to carry out road network generalization.

3. Materials and Methods

3.1. Road Network and Residential Area Maps

As presented in Figure 1, the 1:50,000 road network and residential area data around Ningbo City, Zhejiang Province, China were used as the experimental data, while an artificially produced 1:100,000 road network was used as the benchmark. The experimental area covered 1813 km². There were 10,770 entities in the residential area data. The settlements were dominated by small-scale villages, with a few medium-scale towns and discrete buildings. There were 2750 roads in the road network, with a combined length of 2876.94 km. There were 1959 roads in the benchmark, with a combined length of 2181.62 km.

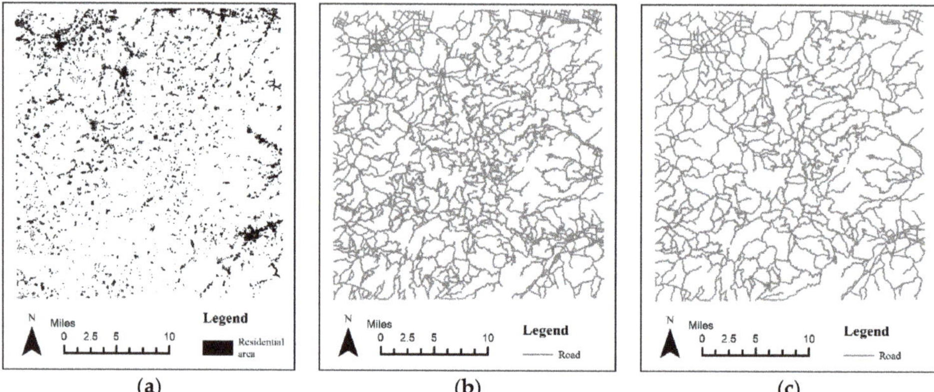

(**a**)   (**b**)   (**c**)

Figure 1. Experimental and benchmark data: (**a**) 1:50,000 residential areas, (**b**) 1:50,000 road network, and (**c**) 1:100,000 road network.

3.2. Framework

Our method aimed to constrain road network generalization by residential area to preserve the spatial distribution consistency and geographic correlation. This included the form of the road network, relative density characteristics, connectivity between settlements, and traffic efficiency. The settlements were used as the control features to decompose the complex road network generalization into relatively simple redundant paths between settlements. Figure 2 shows this method, which consisted of six basic steps:

- Cluster residential areas into settlements.
- Classify roads and settlements according to the topological relationship.
- Evaluate the importance of settlements and select certain settlements as control features.
- Build a geographic network with residential area and road network as nodes and edges, respectively, and search for the paths between settlements.
- Delete redundant paths between settlements.
- Check visual continuity and topological connectivity.

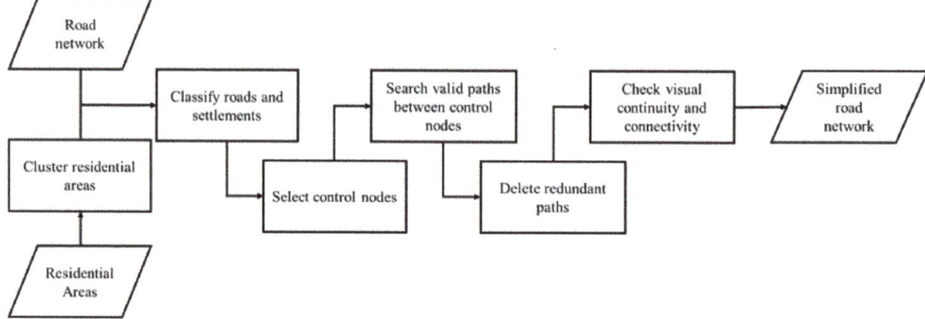

Figure 2. Flow of the road network generalization method constrained by residential areas.

3.3. Classifying Roads and Settlements

First, it was necessary to cluster the residential areas. For residential areas, density-based clustering methods are commonly used. For this study, the classic density-based spatial clustering of applications with noise (DBSCAN) algorithm [25] was used to cluster the residential areas. The rolling circle method [26] was used to extract the outline of the settlements.

In order to facilitate the subsequent construction of the geographical network and path search, roads and settlements needed to be classified. As presented in Table 1, according to the topological relationship between roads and settlements, roads can be divided into extending roads (E-Roads), passing roads (P-Roads), inner roads (I-Roads), and outer roads (O-Roads).

Table 1. Explanation of road types (examples are shown as red lines).

	Topological Characteristics	Example	Description
E-Road	One end is located inside the settlement and the other end is located outside the settlement.		E-Roads directly connect the interior and exterior of the settlement, which represent the starting and ending sections of the path between settlements.
P-Road	Both ends are outside the settlement with a section passing through the settlement.		P-Roads can also connect the interiors and exteriors of settlements. They are common in small settlements distributed along roads and occasionally in larger settlements.
I-Road	Both ends are contained within the settlement.		The function of an I-Road is to connect E-Roads and different areas within the settlement. The larger the settlement, the greater the number of I-Roads and the more complex the structure.
O-Road	Completely outside the settlement.		O-Roads are separated from the target settlement and do not participate in the construction of settlements.

Settlements can be divided into two categories, according to the road type. Type I settlements mainly include E-Roads and I-Roads, or only E-Roads. Figure 3a shows an ideal and typical settlement. We followed the principle of "must have access to the core of the settlement." Alternatively stated, the target road can only directly serve the settlement when it intersects with other roads within the settlement, or its range of influence covers the core of the settlement. The P-Road associated with the type I settlement shown in Figure 3b is an O-Road because it cannot directly serve the settlement. Type II settlements contain only P-Roads. As presented in Figure 3c, P-Roads, upon which settlements are completely dependent, penetrate most type II settlements. As shown in Figure 3d, some type II settlements contain multiple P-Roads. When evaluating which P-Roads are effective, we also followed the principle of "must have access to the core of the settlement."

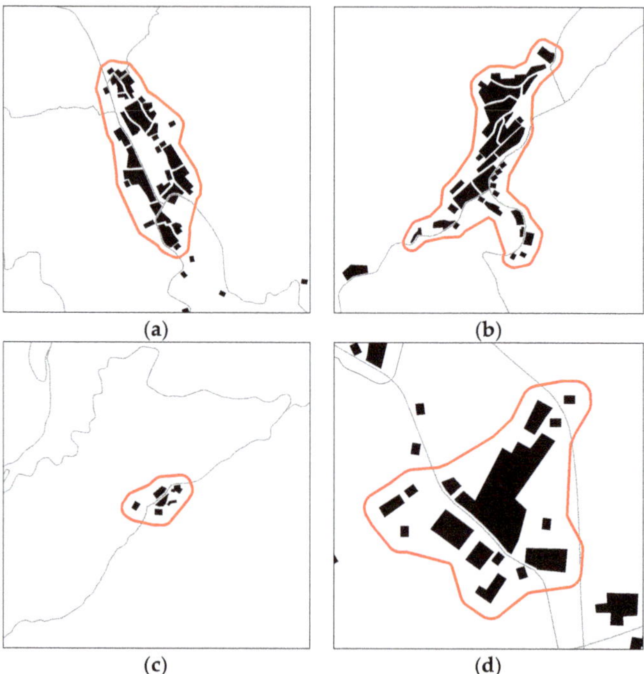

Figure 3. Classification of settlements: (**a**) typical type I settlement; (**b**) type I settlement with P-Road; (**c**) typical type II settlement; and (**d**) type II settlement with multiple P-Roads.

3.4. Evaluating Settlement Importance

As the service objects of road networks, residential areas have an important influence on the shape and structure of road networks. Compared with roads, residential areas have a higher priority, such that residential area generalization must be implemented first. Residential area generalization is also an important research topic in cartographic generalization, including selection, merging, and simplification. However, our focus was on road network generalization under the constraint of residential areas. Therefore, only some settlements were selected as the constraint conditions; no more steps were performed in residential areas. The importance of a settlement is usually evaluated in terms of semantic attributes, geometric characteristics, and location characteristics. Location characteristics refer to the positional relationships between settlements and roads. Hence, our method used the area (S), residential density (RD), Voronoi area (VS), and road index (RI) to evaluate the importance of settlements. A settlement set, i.e., $V = \{v_1, v_2, \ldots, v_n\}$, has the following information:

- S is the most intuitive embodiment of the importance of settlements. Generally, settlements with large areas are more important than those with small areas.
- RD reflects the density of residential areas within the settlement. When considering settlements with equal areas, settlements with more dense residential areas are more important. The RD was calculated as follows:

$$RD(v_i) = \frac{\sum_{j=0}^{n} S(p_j)}{S(v_i)},$$

(1)

where n is the number of residential areas in v_i and p_j denotes the residential area in v_i.

- *VS* reflects the spatial influence of the settlement. The larger the Voronoi area, the more representative the settlement and the greater the possibility of its being selected.
- *RI* reflects the connectivity of the settlement. Settlement importance is affected by the number and grade of roads. The more roads associated with a settlement, and the higher the road grade, the higher the importance of the settlement. Settlements primarily rely on E- and P-Roads to connect their interiors to distant locations. For this reason, the *RI* is calculated based on the number and grade of the E- and P-Roads. The *RI* was calculated as follows:

$$RI(v_i) = \sum_{j=0}^{n} Rank(r_j), \tag{2}$$

where n is the total number of effectively connected roads in v_i, r_j is the effectively connected road of v_i, and $Rank(r_j)$ denotes the numerical grade of r_j.

These measured values were normalized such that all of the values ranged from 0–1. Hence, we used an integrated indicator (*SP*) to evaluate the relative importance of settlements as follows:

$$SP(v_i) = \mu \times [w_1 \times S(v_i) + w_2 \times RD(v_i) + w_3 \times VS(v_i)] + w_4 \times RI(v_i), \tag{3}$$

where μ is an adjustment factor. When S, RD, and VS are constant, settlements located at road intersections have more geographical significance than settlements distributed along roads. Therefore, when the evaluation object is a type I settlement, $\mu = 1.0$. When the evaluation object is a type II settlement, $\mu < 1.0$. The variables w_1, w_2, w_3, and w_4 are the weights of S, RD, VS, and RI, respectively, and $w_1 + w_2 + w_3 + w_4 = 1.0$. These four weights and μ can be adjusted according to the requirements of the applications.

After the settlement importance evaluation was completed, some unqualified settlements were eliminated, according to the given rate threshold σ. All subsequent operations were based on settlements composed of the preserved settlements.

3.5. Searching Valid Paths

In a previous study, settlements were constructed and selected individually, rather than as parts of a neural network. The key to geographical networks is the proximity and connection between settlements. Cartographic generalization research usually uses the spatial distance, Delaunay triangulation, and Voronoi diagrams to express the spatial proximity relationships between geographical entities [27]. However, the spatial proximity relationship cannot effectively express the connected relationships of settlements in the road network. The effective path between settlements includes both direct and indirect connection paths. The entire direct connection path process involves only the start and end settlements, whereas the indirect connection path passes through other settlements in addition to the start and end settlements. If there is a direct connection path between two settlements, the two settlements are adjacent to each other along the road network.

The effective paths between settlements have two characteristics. (1) They connect from the E-Road of the beginning settlement to the end of the E-Road within the terminal settlement. Other E- and I-Roads in the beginning and ending settlements cannot be part of the path. The P-Road in the type II settlement, which effectively connects the interiors of two settlements, is an E-Road in the settlement. (2) The effective path is the least-cost path between the two E-Roads. This is because searching all of the paths between two nodes in a large network is time-consuming; people usually choose the shortest and fastest path when traveling. When planning a route, people pay more attention to time than distance, and tend to choose faster, high-grade roads. Therefore, travel time was used to express the cost. The travel time was primarily determined by the road length, road grade, and road tortuosity, as follows:

$$T = \frac{L \times ST}{V}, \tag{4}$$

where L is the length, V is the road design speed, and ST is the tortuosity.

The connection between settlements refers to all of the effective paths between settlements. A settlement may have multiple E-Roads beginning at varying locations within the settlement. Therefore, there may be multiple paths between a pair of settlements. When searching for an effective path between a pair of adjacent settlements, the E-Road of the start settlement and the E-Road of the end settlement are arranged and combined. The Dijkstra algorithm [28] was used to search for the least-cost path between all the E-Road pairs and record the most effective path based upon that requirement.

3.6. Simplifying Redundant Paths

The redundant paths simplification of a geographical network can be divided into the following two stages: path simplification of adjacent settlement pairs and path simplification of adjacent settlement groups. Generally, a path with a lower cost is more preservation-worthy. However, preservation also depends on whether the services it provides can be replaced.

In the first stage, direct connection paths between adjacent settlements were evaluated. When simplifying the path between a pair of adjacent settlements, we chose any path to compare with the optimal path. If the cost of the path was greater than the sum of the optimal path and the internal traffic costs, the path was substituted. This rule was defined as follows:

$$\alpha \times \left(\kappa_f + \kappa_s + \kappa_e \right) < \kappa_t, \tag{5}$$

where κ_f is the cost of the optimal path, κ_s is the internal travel cost of the optimal path and the current path in the start settlement, κ_e is the internal travel cost of the optimal path and the current path in the end settlement, κ_e is the cost of the current path, and α is the tolerable cost ratio, i.e., the service capacity of the optimal path in the target area is weaker than the current path (α ranges from 0–1).

As presented in Figure 4, there are three paths between settlements A and B. Path I begins at point $p1$ and extends along ea to point $p2$. Path II begins at point $p1$ and extends along ec and eb to point $p3$. Path III begins at point $p1$ and extends along ec and ed to point $p4$. Path I is the optimal path. The start and end points of path II are $p1$ and $p3$, respectively, and the travel cost is 4.1. However, the total cost of traveling from $p1$ along path I to $p2$ and then along internal road ib to $p3$ is 2.8. This is less than the cost of path II, such that path II can be replaced by path I. Decreasing the value of α can increase the intensity of simplification. For example, the cost of path III is 4.7. However, the total cost of traveling from $p1$ along path I to $p2$ and then along internal roads ib and ia to $p4$ is 5.0. When $\alpha = 1.0$, Equation (5) does not work and path I cannot replace path III. However, when $\alpha < 0.94$, Equation (5) is workable and path I can replace path III.

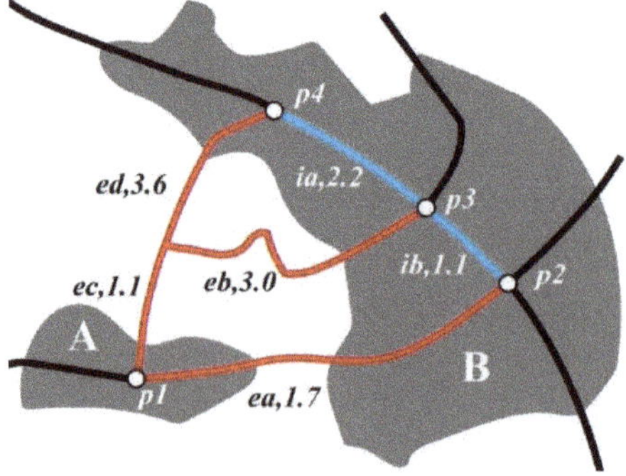

Figure 4. Schematic diagram of the path simplification between settlements (red lines are paths between settlements A and B; blue lines are I-Roads).

In the second stage, the proximity relationships between adjacent settlements are examined. All of the direct connection paths between adjacent settlements were treated as a whole, whose flow is shown in Figure 5. First, we constructed a graph with settlements and proximity relations as the nodes and edges, respectively, using the minimum cost of the direct connection paths between adjacent settlements as the edge weights. Next, we searched for the triangular structures in the graph, which were subgraphs composed of three adjacent settlements. Finally, we determined whether the edge with the largest weight in the triangular structure could be replaced by a combination of the other two edges. If so, all of the direct connection paths between the settlements connected by that edge were removed. This rule was defined as follows:

$$\alpha \times (\kappa_{e1} + \kappa_{e2}) < \kappa_{em}, \tag{6}$$

where α is the tolerable cost ratio, κ_{em} is the maximum weight value in the subgraph, and κ_{e1} and κ_{e2} are the weight values of the two other edges.

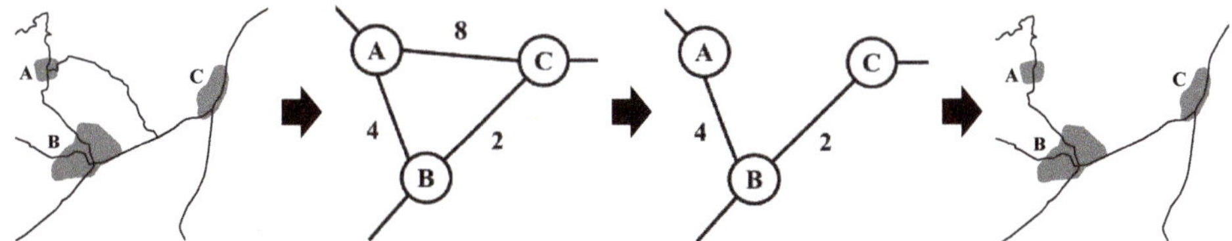

Figure 5. Settlement group path simplification process.

After the path simplification was completed, some additional roads were selected to supplement the results and check the connectivity. The roads to be supplemented were I-Roads and visually continuous roads. The method of selection for the I-Roads was different from the selection of roads between settlements. The larger the settlement, the more complicated the process. The focus of our method was the selection of roads between settlements, such that only the lowest-cost path that could satisfy the connectivity of the preserved E-Roads within the settlements was selected. Other roads were supplemented based upon visual continuity. Some roads lacked connectable settlements at one end, but were supplemented because they had good visual continuity with preserved roads. Another criterion for selecting a supplemented road was that the preserved road had no other visually continuous roads at the intersection.

4. Results and Discussion

4.1. Determination of Thresholds

The quantity and quality of the results were affected by many factors, including the clustering quality, settlement selection ratio, angle threshold for evaluating visual continuity, and tolerable cost ratio. The most strongly influential factors were the settlement selection ratio (σ) and the tolerable cost ratio (α). The number of roads in the road network generated by the method was recorded under different parameter values (Table 2). The data were visualized and the influence of σ and α on the number of roads was analyzed (Figure 6). The analysis showed that σ played a leading role in the selection and α played an auxiliary adjustment role, which corresponds to the concept that settlements constrain the road network generalization.

Table 2. Number of selected roads under varying σ and α values.

	$\sigma = 0.1$	$\sigma = 0.2$	$\sigma = 0.3$	$\sigma = 0.4$	$\sigma = 0.5$	$\sigma = 0.6$	$\sigma = 0.7$	$\sigma = 0.8$	$\sigma = 0.9$
$\alpha = 0.0$	1288	1523	1663	1738	1811	1883	1929	1957	1995
$\alpha = 0.1$	1288	1523	1663	1738	1811	1883	1929	1957	1995
$\alpha = 0.2$	1288	1523	1663	1738	1811	1883	1929	1957	1995
$\alpha = 0.3$	1288	1523	1664	1739	1812	1884	1930	1958	1996
$\alpha = 0.4$	1288	1523	1664	1739	1812	1884	1930	1958	1996
$\alpha = 0.5$	1288	1523	1664	1739	1817	1889	1935	1963	2001
$\alpha = 0.6$	1296	1543	1688	1765	1833	1902	1948	1975	2009
$\alpha = 0.7$	1354	1578	1717	1811	1873	1937	1984	2015	2039
$\alpha = 0.8$	1391	1639	1755	1863	1925	1976	2024	2058	2081
$\alpha = 0.9$	1403	1667	1815	1924	1972	2022	2067	2100	2120
$\alpha = 0.10$	1504	1768	1925	2021	2091	2139	2169	2195	2217
Ideal value	275	550	825	1100	1375	1650	1925	2200	2475

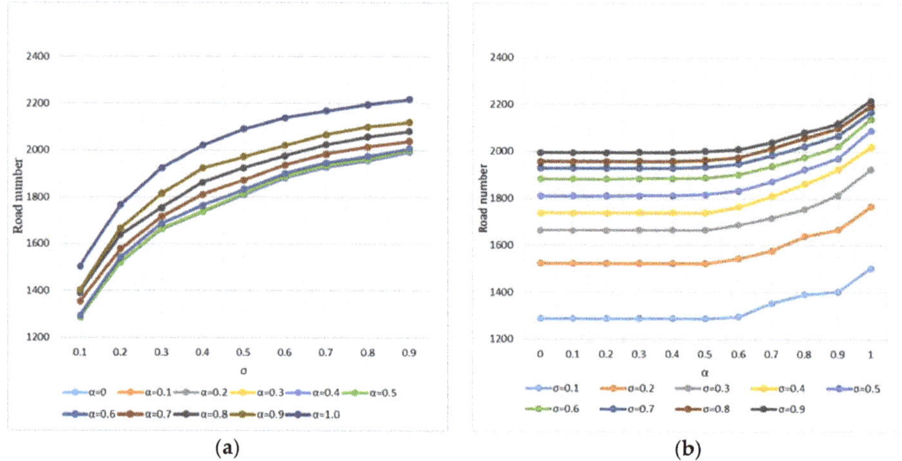

Figure 6. Number of selected roads under varying parameter values: (**a**) road number under varying σ values and (**b**) road number under varying α values.

The tolerable cost ratio (α) has limited adjustments. With respect to the experimental data, when the value of σ is fixed, the difference between the maximum and minimum values of the number of roads did not exceed 20%. The smaller the σ, the stronger the influence of α and the greater the difference between the extreme values of the number of roads. As the number of settlements decreased, there were more direct connection paths between settlements and more paths became redundant. In addition, when α was reduced to approximately 0.5, the maximum generalization intensity was reached. There was only one direct connection path between a pair of settlements and all the triangular structures on the graph were processed. Smaller values no longer affected the results. When $\alpha \geq 0.6$, different values could effectively adjust the number of roads, which could be flexibly determined according to the value of σ and the expected number.

The selection ratio of the roads and settlements should be the same for cartographic generalization. The analysis shows that the number of roads was relatively close to the ideal value when $\sigma \geq 0.6$. The smaller the value of σ, the greater the difference between the number of roads and the ideal value, indicating that our method is not suitable for road network generalization on a large scale because it is strongly influenced by residential areas. Road network and residential area generalization must be alternately implemented.

Residential area generalization includes selection, merging, and simplification. Selecting settlements only on a large scale while skipping the other steps is unreasonable. With the continuous generalization of roads and residential areas, the role of a particular settlement or road changes dynamically. Two E-Roads may merge into a P-Road, type I settlements may become type II settlements, and multiple adjacent settlements may merge into one settlement.

In addition, supplementing visually continuous roads also affects the number of roads, where the degree of impact varies depending on the dataset. To achieve the expected generalization, when applying the SSP method to other datasets, multiple experiments must be implemented on different parameter combinations, according to the target scale. For our data set and target scale, we set σ and α as 0.7 and 0.6, respectively.

4.2. Road Network Structure Analysis

The target scale was 1:100,000 and 70% of the roads and settlements were selected, according to the law of the open square root model. With 594 settlements as constraints, 1948 roads were selected, of which 1613 roads were preserved in the path simplification stage. Next, 306 visually continuous roads were supplemented, along with 29 roads supplemented during the connectivity check. Through visual inspection, we determined that the overall structure and relative density characteristics of the road network were well maintained and in good agreement with the spatial distribution of the settlements. As presented in Figure 7, the red lines were the result of path simplification, which was highly consistent with the spatial distribution of the residential areas and could meet the connectivity needs of the settlements. The green lines were visually continuous roads supplemented by the principle of good continuity. The added roads were mainly located at abrupt road turning points and at the ends of roads in the road network of the previous stage.

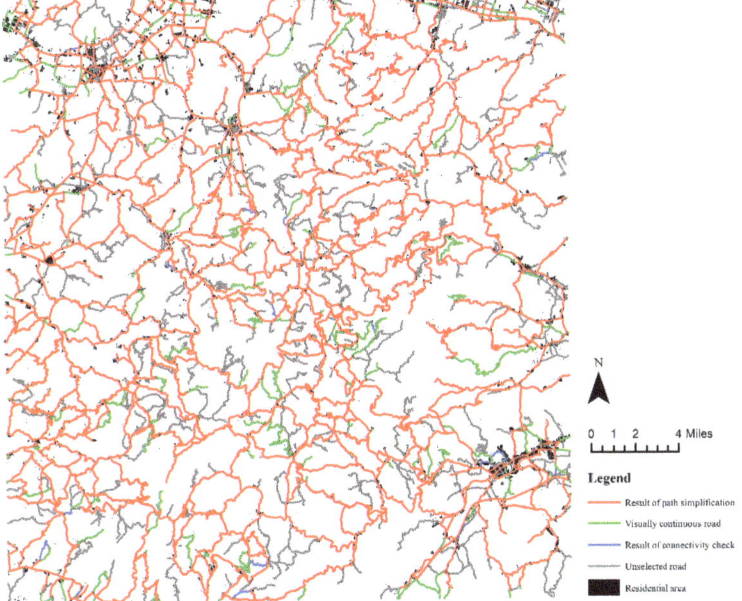

Figure 7. Results of generalization on a 1:100,000 scale (red lines are the results of path simplification, green lines are visually continuous roads, blue lines are the selected roads in the connectivity check, and gray lines are the unselected roads).

We performed a line density analysis on the road network before and after generalization. As presented in Figure 8, each density was divided into 10 sections and the relative density gradually increased (darkened) from low to high. After generalization, the high-density area of the road network was reduced, but the spatial distribution characteristics of the distinct density areas were comparable to those prior to the generalization. Areas with notable changes in the relative density were primarily concentrated in urban areas and sparse residential areas. The selection of I-Roads in towns was limited to those that represented the smallest sections that met the connectivity requirements. Sparse residential areas resulted in some roads being evaluated as having no connectivity values.

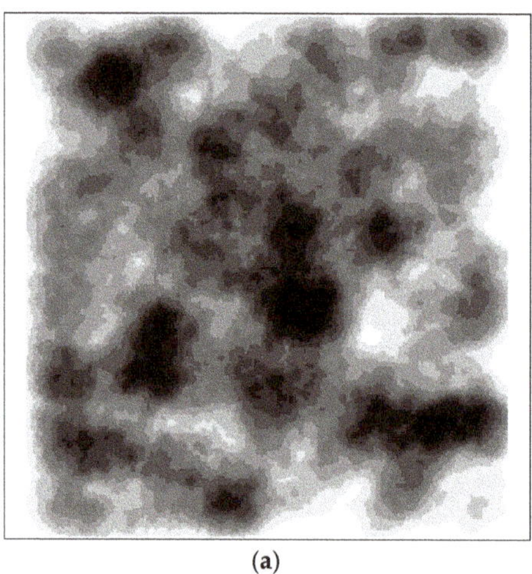
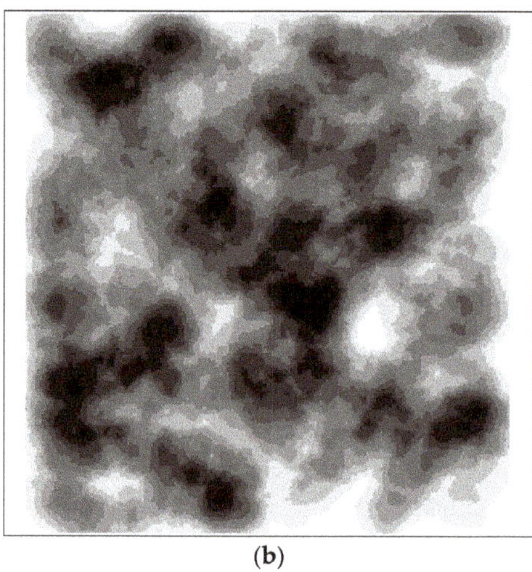

(a) (b)

Figure 8. Comparison of the relative road density before and after generalization: (**a**) road density before generalization and (**b**) road density after generalization.

4.3. Geographical Correlation Analysis

Our method focuses most on the functionality and geographical correlation of the generalized road network. As presented in Figure 9, the generalized road network is in good agreement with the residential areas as the control features on the 1:100,000 map. The two features are highly consistent in spatial distribution. Most settlements are connected by roads. At the same time, redundant paths and a large number of suspended roads were deleted. A small number of settlements are not directly connected by roads because they are not within the specified road service range on the original scale.

While maintaining geographical correlation, the redundant paths between settlements were also effectively simplified. In Figure 10a, there are three direct connection paths between the eastern and western settlements, including the main road and two partially overlapping trails. The two trails were removed because they could be replaced by the main road. In Figure 10b, there are three direct connection paths between the southern and northern settlements. The cost of the two gray roads in the middle was low, but they could only serve the start and end settlements. Although the path in the west was expensive, it was preserved because it was necessary for connecting the two western settlements with the other settlements. Figure 10c shows four suspended roads. The road connected with the settlement at the end was preserved, while the others were deleted. Therefore, the road network was generalized without damaging the connectivity of the settlements. However, our method performed poorly in sparse residential areas. In Figure 10d, gray roads are low in grade and weak in functionality, but they have an important impact on visual perception. Our method was not suitable for handling this type of road and requires further improvement.

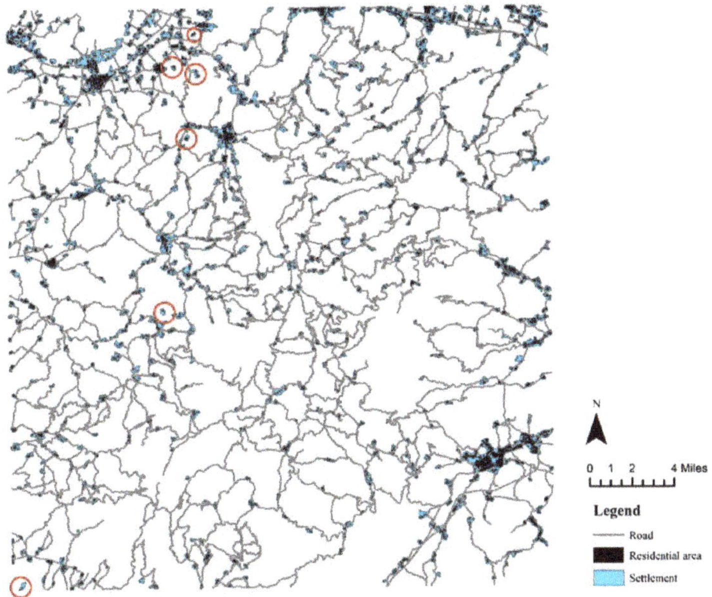

Figure 9. Generalized road network and residential areas in 1:100,000.

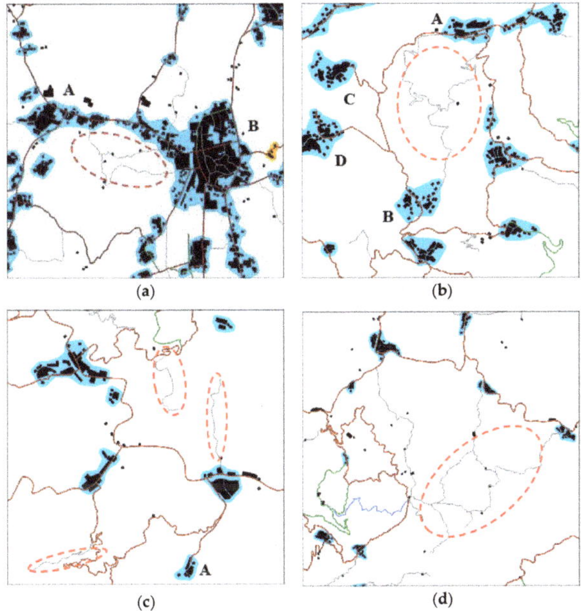

Figure 10. Partial details of the results: (**a–c**) show the advantages of SSP and (**d**) shows the shortcomings of SSP.

4.4. Comparison with Other Methods

In order to further verify the effectiveness of this method, taking the artificially produced 1:100,000 road network as the benchmark, this method was compared with the graph-based method, stroke-based method, and hybrid method combining graph and stroke. In Figure 11a, the graph-based method uses the PageRank algorithm to evaluate the importance of road sections, without considering the topological information and geometrical information of the road network, resulting in poor visual continuity in the selection results. In Figure 11b, stroke-based uses length and road grade to evaluate the importance of stroke. The overall visual continuity is good, but there is a large number of redundant roads. In Figure 11c, the hybrid method takes into account the advantages of graph and stroke, and comprehensively evaluates the importance of stroke by using parameters such as length, grade, degree, closeness, and betweenness. The selection effect is significantly improved, but some settlements cannot be effectively connected. In Figure 11d, the results of our method are significantly better than those of the other methods. The missed components were mainly long-suspended roads that did not end at settlements and low-grade roads in sparsely populated areas. These had an impact on visual perception, but did not affect the functionality of the road network. The extra components mainly consisted of short suspension roads connecting the settlements and some visually continuous roads. Short suspension roads were selected because the algorithm could not fully simulate human spatial perception when evaluating the service between roads and settlements. When selecting visually continuous roads, the self-best-fit stroke generation method was used. This method determined whether two roads were visually continuous based on the angle at the intersection. This method cannot consider the overall performance of roads. Therefore, although it can improve the visual effect, it reintroduces some redundant roads with poor global performance. In general, the results of this method were closer to the artificially produced results; the method has significant advantages for maintaining the structural characteristics and geographical correlation of road networks.

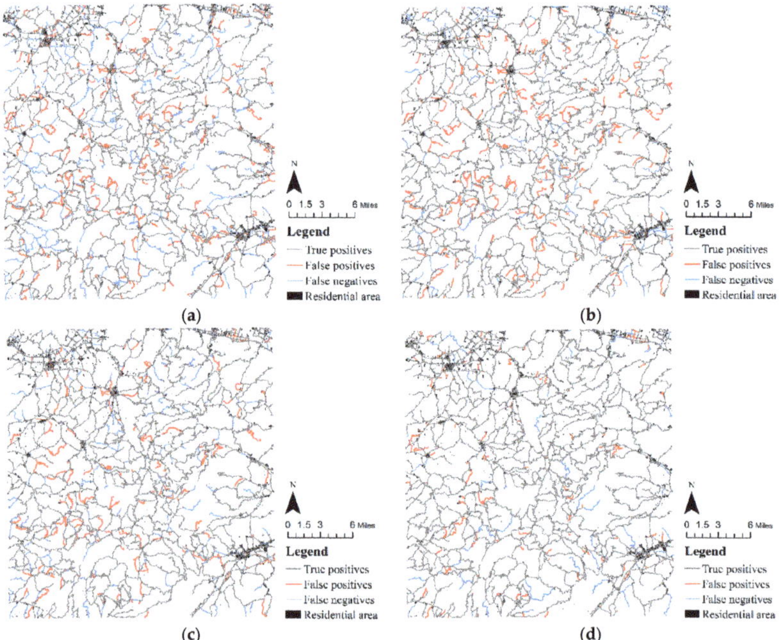

Figure 11. Comparisons between the benchmark and four methods: (**a**) Comparison between the results of the graph-based method and benchmark. (**b**) Comparison between the results of the stroke-based method and benchmark. (**c**) Comparison between the results of the hybrid method and benchmark. (**d**) Comparison between the results of our method and benchmark.

Further, the results were evaluated quantitatively. First, the consistency between the selection results and the benchmark was evaluated, and accuracy, precision, recall, and F1-Score were selected as the evaluation parameters. As presented in Figure 12, our method has the highest consistency with the benchmark (with accuracy, precision, recall, and F1-Score of 0.864, 0.907, 0.902, and 0.905, respectively), the closest relation to the artificially produced results, and the lowest difficulty and workload of subsequent processing.

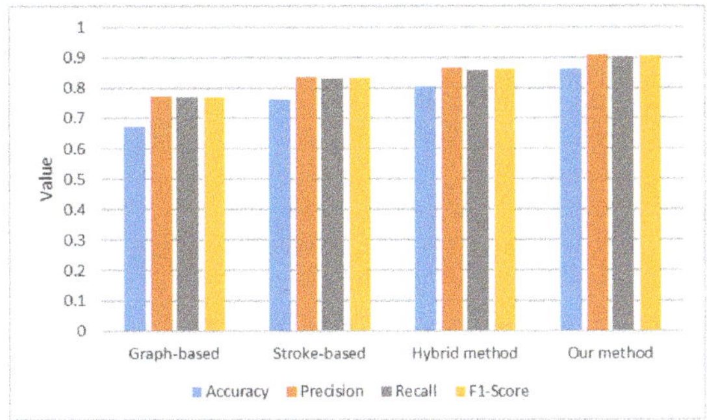

Figure 12. The consistency between results and benchmark.

In terms of geographical correlation, the global efficiency [29] and settlement coverage were used as evaluation parameters. The calculation object of global efficiency is the geographical network with settlements as nodes. Global efficiency reflects the traffic efficiency between the settlements in a network. It is defined as follows.

$$L = \frac{1}{N(N-1)} \sum_{i,j \in R, i \neq j} \frac{1}{c_{ij}},$$

(7)

where N is the number of selected settlements and c_{ij} is the path cost from settlement i to settlement j.

As presented in Figure 13, our method has the best traffic efficiency, the largest number of connected settlements, and the best geographical correlation between road networks and residential areas. Other methods do not take into account residential areas in the generalization, so their performance of geographical correlation is random. For example, although the hybrid method has high consistency with the benchmark, it is not as good as the stroke-based method in terms of geographical correlation.

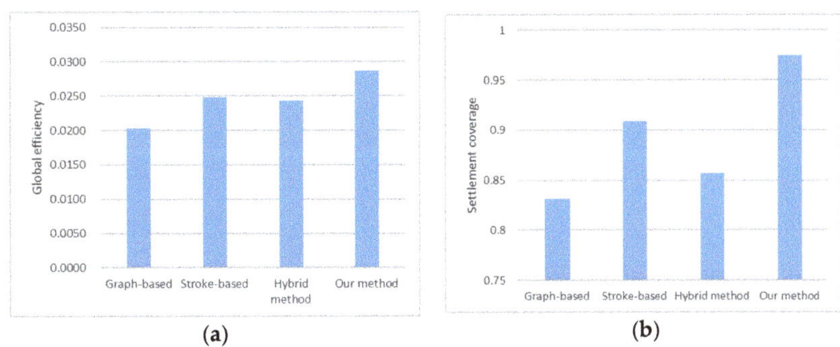

Figure 13. Performance of four methods in geographical correlation: (**a**) the global efficiency of the results; (**b**) the settlement coverage of the results.

5. Conclusions

Residential area and road network are a pair of map features with strong geographical correlation. When mapping, we should maintain their unique spatial distribution characteristics and geographical correlation between features. In this study, in order to ensure that the road network is in good agreement with residential areas, we propose a road network generalization method that is constrained by residential areas, which adopts the idea of divide-and-conquer, decomposes the road network with the settlements, and simplifies the complex road selection problem into a redundant path simplification problem. First, the roads and settlements obtained from residential clustering were classified. Subsequently, the importance of the settlements was evaluated and certain settlements were selected as the control features. Next, a geographical network with settlements as nodes was built. Finally, the redundant paths between settlements were deleted, and the visual continuity and topological connectivity were checked. This method makes full use of the geographical correlation of the two features. Using settlements as the constraint of road network generalization, the results were highly consistent with the spatial distribution of residential areas, which can ensure the connectivity between settlements and the functionality of road network.

In addition to the theoretical discussion, we also conducted empirical research. The experimental area was near Ningbo City, Zhejiang Province, China. The 1:50,000 road network and residential area data were used as the experimental data, while an artificially produced 1:100,000 road network was used as the benchmark. The experimental results show that this method performs well in terms of maintaining the spatial structure, density, and geographical correlation of road networks. Compared with the graph-based, stroke-based, and hybrid methods, the accuracy, precision, recall, and F1-Score of our method were as high as 0.864, 0.907, 0.902, and 0.905, respectively. The global efficiency and settlement coverage of our method were also better than other methods. The results showed that our method has significant advantages in the consistency of its benchmarking and geographical correlation with residential areas.

However, this study has some limitations. First, the proposed method represents only half of the full collaborative generalization method of the road network and residential area. Because it is incomplete, the proposed method is only suitable for road network generalization on a small scale. The complete method includes the residential area generalization method, considering the road network. Second, the proposed method highly depends on the spatial distribution of residential areas, and the effect is not good in areas with low residential density. To address this in the future, it will be necessary to identify and select additional roads in this type of area. Finally, the proposed method is not sufficiently automatic. To address this, it will be necessary to combine the data for multiple attempts to determine the parameter value. Road networks and geographical networks with settlements as nodes are typical complex-network structures, which are in good agreement with emerging graph neural networks. In the future, the collaborative generalization of roads and residential areas based on graph neural networks will be studied to improve the automation level of the algorithm.

Author Contributions: Conceptualization, Zheng Lyu; methodology, Zheng Lyu, Qun Sun, and Jingzhen Ma; software, Fubing Zhang; validation, Yuanfu Li; formal analysis, Zheng Lyu; investigation, Fubing Zhang; resources, Qun Sun; data curation, Yuanfu Li; writing—original draft preparation, Zheng Lyu.; writing—review and editing, Zheng Lyu, Jingzhen Ma, and Qing Xu; visualization, Zheng Lyu; supervision, Qun Sun; project administration, Qing Xu; funding acquisition, Qun Sun. All authors have read and agreed to the published version of the manuscript.

Funding: This study was supported by the National Natural Science Foundation of China (41571399, 42101458, 42101454, 42101455) and the Fund Project of Zhongyuan Scholar of Henan Province (202101510001).

Institutional Review Board Statement: Not applicable.

Informed Consent Statement: Not applicable.

Data Availability Statement: The data and codes that support the findings of this study are available at figshare.com with the following links: https://doi.org/10.6084/m9.figshare.17163533 (data, accessed on 12 December 2021) and https://doi.org/10.6084/m9.figshare.17163539 (code, accessed on 12 December 2021).

Acknowledgments: The authors are grateful to the editors and the anonymous referees for their valuable comments and suggestions.

Conflicts of Interest: The authors declare no conflict of interest.

References

1. Wang, J.; Wu, F. *Advances in Cartography and Geographic Information Engineering*, 1st ed.; Springer: Singapore, 2021; pp. 151–211. ISBN 978-981-16-0613-7.
2. Wu, F.; Gong, X.; Du, J. Overview of the Research Progress in Automated Map Generalization. *Acta Geod. Cartogr. Sin.* **2017**, *46*, 1645–1664. [CrossRef]
3. Steiniger, S.; Weibel, R. Relations among Map Objects in Cartographic Generalization. *Cartogr. Geogr. Inf. Sci.* **2007**, *34*, 175–197. [CrossRef]
4. Wang, C.; Wang, W.; Zhang, M.; Cheng, J. Evolution, Accessibility of Road Networks in China and Dynamics: From a Long Perspective. *Acta Geogr. Sin.* **2014**, *69*, 1496–1509. [CrossRef]
5. Zhou, Y.; Huang, H.; Liu, Y. The Spatial Distribution Characteristics and Influencing Factors of Chinese Villages. *Acta Geogr. Sin.* **2020**, *75*, 2206–2223. [CrossRef]
6. Gong, X. Research on Settlement Generalization Methods Considering Spatial Pattern and Road Networks. Ph.D. Thesis, PLA Strategic Support Force Information Engineering University, Zhengzhou, China, 2017.
7. Li, G.; Xu, W.; Long, Y.; Zhou, T.; Gao, C. A Multi-constraints Displacement Method for Solving Spatial Conflict between Contours and Rivers. *Acta Geod. Cartogr. Sin.* **2014**, *43*, 1204–1210. [CrossRef]
8. Šuba, R.; Meijers, M.; Oosterom, P.V. Continuous Road Network Generalization Throughout All Scales. *ISPRS Int. J. Geo-Inf.* **2016**, *5*, 145. [CrossRef]
9. Mackaness, W.A.; Beard, K.M. Use of Graph Theory to Support Map Generalization. *Cartogr. Geogr. Inf. Syst.* **1993**, *20*, 210–221. [CrossRef]
10. Liu, G.; Li, Y.; Yang, J.; Zhang, X. Auto-selection Method of Road Networks based on Evaluation of Node Importance for Complex Traffic Network. *Acta Geod. Cartogr. Sin.* **2014**, *43*, 97–104. [CrossRef]
11. Jiang, B.; Harrie, L. Selection of Streets from a Network Using Self-organizing Maps. *Trans. GIS* **2004**, *8*, 335–350. [CrossRef]
12. Ma, C.; Sun, Q.; Chen, H.; Xu, Q.; Wen, B. Application of Weighted PageRank Algorithm in Road Network Auto-selection. *Geomat. Inf. Sci. Wuhan Univ.* **2018**, *43*, 1159–1165. [CrossRef]
13. Thomson, R.C.; Richardson, D.E. The Good Continuation Principle of Perceptual Organization Applied to the Generalization of road networks. In Proceedings of the 19th International Cartographic Conference, Ottawa, ON, Canada, 14–21 August 1999; pp. 1215–1223.
14. Liu, X.; Zhan, F.; Ai, T. Road Selection Based on Voronoi Diagrams and "Strokes" in Map Generalization. *Int. J. Appl. Earth Obs. Geoinf.* **2010**, *12*, 194–202. [CrossRef]
15. Cao, W.; Zhang, H.; He, J.; Lan, T. Road Selection Considering Structural and Geometric Properties. *Geomat. Inf. Sci. Wuhan Univ.* **2017**, *42*, 520–524. [CrossRef]
16. Chen, J.; Hu, Y.; Li, Z.; Zhao, R.; Meng, L. Selective Omission of Road Features Based on Mesh Density for Automatic Map Generalization. *Int. J. Geogr. Inf. Sci.* **2009**, *23*, 1013–1032. [CrossRef]
17. Li, Z.; Zhou, Q. Integration of Linear and Areal Hierarchies for Continuous Multi-Scale Representation of Road Networks. *Int. J. Geogr. Inf. Sci.* **2012**, *26*, 855–880. [CrossRef]
18. Yu, W.; Zhang, Y.; Ai, T.; Guan, Q.; Chen, Z.; Li, H. Road Network Generalization Considering Traffic Flow Patterns. *Int. J. Geogr. Inf. Sci.* **2020**, *34*, 119–149. [CrossRef]
19. Yamamoto, D.; Murase, M.; Takahashi, N. On-demand Generalization of Road Networks Based on Facility Search Results. *IEICE Trans. Inf. Syst.* **2019**, *102*, 93–103. [CrossRef]
20. Guo, X.; Qian, H.; Wang, X.; Liu, J.; Zhong, J. A Method of Road Network Selection Based on Case and Ontology Reasoning. *Acta Geod. Cartogr. Sin.* **2021**, *50*, 1717–1727. [CrossRef]
21. Zheng, J.; Gao, Z.; Ma, J.; Shen, J.; Zhang, K. Deep Graph Convolutional Networks for Accurate Automatic Road Network Selection. *ISPRS Int. J. Geo-Inf.* **2021**, *10*, 768. [CrossRef]
22. Wei, Z.; He, J.; Wang, L. A Collaborative Displacement Approach for Spatial Conflicts in Urban Building Map Generalization. *IEEE Access* **2018**, *6*, 26918–26929. [CrossRef]
23. Xu, W. Collaborative Map Generalization Method of Roads and Buildings Based on Multi-Agent. Master's Thesis, Nanjing Normal University, Nanjing, China, 2014.
24. Touya, G.A. Road Network Selection Process Based on Data Enrichment and Structure Detection. *Trans. GIS* **2010**, *14*, 595–614. [CrossRef]

25. Ester, M.; Kriegel, H.P.; Sander, J.; Xu, X. A Density-Based Algorithm for Discovering Clusters in Large Spatial Databases with Noise. In Proceedings of the 2nd International Conference on Knowledge Discovery and Data Mining, Portland, OR, USA, 2–4 August 1996; pp. 226–231.
26. Zhi, L.; Yu, X.; Li, G. Spatial Point Clustering Analysis Based on the Rolling Circle. *Geomat. Inf. Sci. Wuhan Univ.* **2018**, *43*, 1193–1198. [CrossRef]
27. Brennan, J.; Martin, E. Spatial Proximity Is More than Just a Distance Measure. *Int. J. Hum. Comput. Stud.* **2012**, *70*, 88–106. [CrossRef]
28. Makariye, N. Towards Shortest Path Computation Using Dijkstra Algorithm. In Proceedings of the 2017 International Conference on IoT and Application (ICIOT), Nagapattinam, India, 19–20 May 2017; pp. 1–3.
29. Latora, V.; Marchiori, M. Efficient Behavior of Small-World Networks. *Phys. Rev. Lett* **2001**, *87*, 1–18. [CrossRef] [PubMed]

International Journal of
Geo-Information

Article

Filming the Historical Geography: *Story from the Realm of Maps in Regensburg*

Beata Medyńska-Gulij [1,*], **Tillmann Tegeler** [2], **Hans Bauer** [2], **Krzysztof Zagata** [1] and **Łukasz Wielebski** [1]

[1] Department of Cartography and Geomatics, Adam Mickiewicz University, 61-680 Poznan, Poland; krzysztof.zagata@amu.edu.pl (K.Z.); lukasz.wielebski@amu.edu.pl (Ł.W.)
[2] Leibniz Institute for East and Southeast European Studies, 93047 Regensburg, Germany; tegeler@ios-regensburg.de (T.T.); hbauer@ios-regensburg.de (H.B.)
*   Correspondence: bmg@amu.edu.pl

Abstract: Research on a specific topic requires the individualized cartographic methods of work that may be defined as the *Realm of Maps*. The double dimensionality in the *Realm of Maps* is understood here as a physical place—a studio workroom—and as a research method. In this study, we focused on the way of presenting a research method designed to study the topic of historico-geographical space in the form of a short film story. The second purpose is to indicate the legitimacy of combining two dimensions of working with maps, the real one and the virtual one, to be able to collect cartographic and descriptive sources in one scientific center. Our research on the *Story from the Realm of Maps in Regensburg: 'People Movement in Southeast Europe'* included a concept adopted by cartographical, historical, and geographical sources; the construction of a studio workroom; a script draft; individual sequences of the story in different types of media; editing the video, along with publishing it on an online video-sharing platform. We used as many different types of geomedia as possible, which, on the one hand, boosts the attractiveness of the film and, on the other, may hamper the proper perception of the main film plot. Finally, we recommend principles of map design for the film, with analog maps and maps created specifically for the short film, published using online video-sharing platforms.

Keywords: *Realm of Maps*; geomedia; map workroom; short film; cartographical story; medium means; map design; historico-geographical space; Southeast Europe; video-sharing platform

Citation: Medyńska-Gulij, B.; Tegeler, T.; Bauer, H.; Zagata, K.; Wielebski, Ł. Filming the Historical Geography: *Story from the Realm of Maps in Regensburg. ISPRS Int. J. Geo-Inf.* **2021**, *10*, 764. https://doi.org/10.3390/ijgi10110764

Academic Editor: Wolfgang Kainz

Received: 16 September 2021
Accepted: 8 November 2021
Published: 11 November 2021

Publisher's Note: MDPI stays neutral with regard to jurisdictional claims in published maps and institutional affiliations.

1. Introduction

The open access to multiple map collections changed their informational nature because users explore digital copies of maps in an interactive way, on computer screens [1]. More and more frequently, research methods are based on the analysis of map fragments on the computer screen without studying the entire map in paper form or without seeing a comparative analysis of multiple cartographic images at the same time. Old and modern maps play crucial roles in the study of changes that occur in geographical space in terms of their chronology and the topic analyzed [2].

Scientific institutes build map portals, the so-called geoportals, for their collections, along with rendering paper maps available in libraries [3]. Geoportals facilitate the interactive usage of maps through several functions such as enlarging the map contents, georeference, transparency, covering with transparency, etc. [4]. However, the resignation from looking at cartographic images in the original size may reduce the perception of spatial knowledge that a cartographer included at the first general level of perception and other levels as well. Since professional cartography began to develop in the 16th century, the principles of map design, related to publication technology, graphic pragmatism, and gestaltism, started to form [5]. Particularly, in the research on the artistry of manuscript maps, direct contact with cartography is indispensable [6].

A researcher, during a query in a map room, combines reading original maps with viewing digital copies and interactive access to map resources on their laptop screen in a complementary way [7]. Research on a specific topic requires individualized cartographic methods of work that may be defined as the *Realm of Maps*. In this study, double dimensionality in the *Realm of Maps* is understood as both a physical place, i.e., a studio/atelier, and a research method. Individualized stay in the *Realm of Maps* created can be compared with the presence of the gamer in a video game when he is immersed in the reality of the *Realm of Games* [8–10].

Methods of distributing historico-geographical research activities extend to more advanced audio–visual forms, apart from classic academic publications in journals or monographs [11,12]. Interactive multimedia websites, as well as online video sharing and social media platforms, are gaining popularity as media that transfer knowledge. Publishing the results of academic projects related to historico-geographical space in a traditional way is now complemented by multimedia websites [13,14]. Visual storytelling in cartography and map modification by using appropriate structural features of textual media more suitable for storytelling are considered interesting forms of an interdisciplinary approach [15–17]. This issue is also part of ongoing debates in the field of multimedia cartography, particularly in terms of established forms of expression [18–20]. Multiple studies touch upon the cognitive limits of animated maps [21] and the increasing role of story maps and interactive information graphics in spatial information [22].

We assume that map design should be based on fundamental general principles embedded in cartographic knowledge. In modern research of map design, it is a frequent practice to refer to classic principles of good map design formulated in the 1960s, 1970s, and 1980s. The fundamental principles of map design, adapted to modern ways of map publishing, include Bertin's visual variables [23], Hake's specific gestalt principles [24], Freitag's pragmatics of signs [25], Dent's focus of attention [26], Imhof's order of cartographic symbols [27], Robinson's figure–ground relation [28], Ratajski's primary symbol [29], and Keates' symbol generalization [30].

Short videos posted on online video-sharing and social media platforms constitute open forms of rendering a story widely available [31]. The most significant tips regarding the creation of one's own story from the world of maps are related to the length of the video and the construction of a coherent plot in a three-act script. The concept of the video includes one crucial idea related to space, e.g., in the form of a question, to which the viewer receives the answer in the last scene [32,33]. In terms of knowledge of geographical space based on maps, such a three-act script includes (1) the beginning, which introduces a narration built around the idea of a specific geographical phenomenon, sign, or cartographical shape; (2) the middle part, which employs the compilation of various geomedia; (3) the ending with conclusions and encouragement to undertake further studies [34]. The draft of the script is based on sequences written according to consecutive presentations of cartographic elements. A logical construction of the plot of the map story determines a good script: a coherent sequence of cause and effect events in space and time, focused on cartographic signs and symbols [7,26].

In general, in this study, we focused on the presentation of a topic from the historico-geographical field in the form of a short film that included maps. The creation of a map story according to a complementary combination of historical expertise in a specific political and physical region with traditional principles of map design, good practices of writing a short film script, and the use of media means was a methodological challenge. The work carried out in this study consisted of building the *Realm of Maps* in order to study historical events in geographical space, creating the draft of the script and individual sequences of the story in different types of media, and finally, formulating map design principles for cartographic visualizations in the short film.

2. Aims and Questions

Highlighting the essence and media quality of the presentation in the form of a short film story, we focused on the way of presenting research methods constructed to study the topic in historico-geographical space. In the process of creating a film story, we preferred using techniques and instruments for registering the way of moving around the *Realm of Maps*, both physically and digitally.

The first goal of the research was to suggest a method of storytelling in the form of a short video based on research methods, the so-called *Realm of Maps*, constructed particularly for the purpose dedicated to a specific research topic in historico-geographical space. The second purpose was to indicate the legitimacy of combining two dimensions of working with maps, the real one and the virtual one, to be able to collect cartographic and descriptive sources in one scientific center. Achieving both goals is linked to answering the following questions:

- What are the conditions of building a map workroom to study historical events in a geographical space?
- What elements should a script to a short film for stories from realms of maps include?
- How many and what medium means can be used for representing cartographic forms in specific sequences and generally in the entire film?
- What map design principles can be adapted for cartographic visualizations in the short film?

3. Materials and Methods

To achieve our objective and answer these questions, we adopted five main stages of research as follows:

- Pinpointing concepts adopted by the following types of approach: cartographical, historical, organization of library sources, and geographical, as well as adaptation of classic map design principles (Section 3.1.);
- Constructing the physical part of the *Story from the Realm of Maps in Regensburg*—a studio workroom according to the layout based on knowledge obtained from maps and publications in the Leibniz Institute for East and Southeast European Studies— IOS-Regensburg (Section 3.2, Figures 1 and 2);
- Creating the draft of the script for the *Story from the Realm of Maps in Regensburg: 'People Movement in Southeast Europe'* (Section 3.3, Figure 3);
- Creating individual sequences of the story in different types of media: animation, 3D models, animation of photos 180, 360, animation–immersion (Section 3.4, Figure 4);
- Editing the video and publishing it on an online video-sharing platform (Section 4).

3.1. Concept

The team working on the project consisted of scientists and experts in multimedia cartography, history of cartography, creation of old map databases and sharing cartographic resources, general physical and socio-economic geography, and the history of Central and Eastern Europe. Having discussed these interdisciplinary approaches, we adopted a common concept, including the following initial assumptions:

- The final media form of the story—a short film entitled the *Story from the Realm of Maps in Regensburg* for online video sharing;
- The viewer—a public user with general geographical and historical knowledge, an expert, or a researcher;
- The story structure—three acts consisting of a beginning (title, topic, and questions); a middle (the immersion in the *Realm of Maps*, individual sequences presenting historico-geographical space and specific events); an ending (answers and inspiration for future considerations; the closing of the *Realm of Maps*);
- The plot of the story—a single plot related to the most important historical events in one geographical space constituting a referential cartographic core with a brief textual narration, at a pace following the rhythm of a musical piece and with a wide range of geomedia;
- Two dimensions of the *Realm of Maps*—a physical dimension, i.e., a studio workroom (printed maps, atlases, wall maps, sketches in books, graphics, catalogs, etc.) and a dig-

ital dimension (maps, schemes, 3D models, cartographic visualizations, photographs, geoportals, map scans, etc.);

- A cartographic core of the story—a recognizable cartographic shape for a quick, intuitive spatial reference on all cartographic visualizations in all sequences;
- Medium means—the range was assumed as wide as possible, including textual (descriptions, dates, geographical names, dialogues); graphics (schemes, drawings, raster copies of old maps, photographs (traditional 180, 360); video (traditional, 360); animation; music, e.g., a piece of instrumental classical music representative of the specific historico-geographical space;
- Cartographic forms—cartographic pictograms, political maps, hypsometric maps, old maps, 3D maps, and cartographic animations [35];
- Principles of map design—e.g., high contrast [28], focus of attention [26], minimization of map elements [25,36], pictorial symbols [27,29,30], filling in the entire ratio of 16:9 with contents without a frame [11,26], and visual variables [23].

3.2. The Realm of Maps in Regensburg—Studio Workroom

The construction of the *Realm of Maps* as a workroom for creative activities, adjusted to the topic 'The People's Movement in Southeast Europe', occurred after the layout was prepared. Figure 1 shows the layout with a sketch map of Europe including the sea coastline, the line of the Danube, mountain chains, and circles of cities, along with their names. Around the contour of Europe, some historical events were placed, and they were matched to corresponding cartographic sources. The idea behind the workroom was to select the material from the Leibniz Institute for East and Southeast European Studies and place the selected sources that will be significant to our story in the workroom [37]. In the center of the workroom, there was a large table with maps, atlases, books, photographs, a notebook, and wall maps hung around, which 'enclosed' the physical dimension of the *Realm of Maps* (Figure 2) [38]. Access to digital map collection with metadata and georeference through the GeoPortOst portal was provided by laptops.

Figure 1. Layout with a sketch map of Europe including the sea coastline, the line of the Danube, mountain chains, and circles of cities, along with their names.

Figure 2. Physical dimension of the *Realm of Maps*—the workroom with a large table with maps, atlases, books, photographs, a notebook, and wall maps.

3.3. Drafting a Movie Script—The Story Plot: 'The People's Movement in Southeast Europe'

The layout (Figure 1) and detailed analysis of historico-cartographic materials in the workroom (Figure 2) allowed researchers to form the draft of the video script. When writing the draft of the script, we focused on the preservation of storytelling rights that are subject to the appropriate relationship between the narrator and viewer, as well as on a smooth plot in three acts [39]. The short film script includes ordered sequences in three acts with appropriate time proportions and matched time of duration for each sequence, with an accuracy of 5 s (Figure 3).

Following the specificity of creating the *Story from the Realm of Maps*, the draft of the script included the description of consecutive sequences according to the following points: text narration, different types of media, and cartographic form. Our audio effect was a single instrumental piece so that the viewer could focus solely on the visual aspects. In that stage of the research, we juxtaposed different types of media with corresponding cartographic forms to maintain a coherent plot of the story and show the full value of the *Realm of Maps* in IOS Regensburg.

3.4. Creating Story Sequences

A total of 13 sequences of the film, whose frames are graphically depicted in Figure 4, were created according to the concept described in Section 3.1 and the draft of the script in Figure 3. In the first sequence, viewers can see the white background, with the words of the title 'Story from the Realm of Maps in' appearing in turn. The last word of the title, 'Regensburg', appears at a distance near the circle symbolizing the location of the city. The blue line of the Danube, which becomes a cartographic core of the entire story from now on, 'flows' through the circle thanks to animation. In the next sequence, the Danube is the main element of the digitally created grayscale relief map, with the text of the narration being the topic and briefly asked questions in the form of interrogative pronouns. Additionally, in the third sequence, a special animation depicting spherical views of the workroom was employed to stimulate the opening of the workroom and cause the viewer to be immersed in the *Realm of Maps*.

Act 1: Beginning 25" 0:00 – 0:25

Sequence 1: Title   5"   0:00
Text narration: Story from Realm of Maps / Regensburg / Danube
Medium mean: text, animation, graphic, music
Cartographic form: line shape of Danube, circle symbol of Regensburg

Sequence 2: Topic and questions   20"   0:05
Text narration: PEOPLE MOVEMENT IN SOUTHEAST EUROPE / WHO? HOW? WHY? WHEN? WHERE? // ... to answer these questions YOU are welcome to our REALM OF MAPS
Medium mean: text, animation, graphic-map, music
Cartographic form: line shape of Danube, grayscale relief map with coast line

Act 2: Middle 3'15" 0:25 – 3:40

Sequence 3: Immersion in Realm of Maps   10"   0:25
Text narration: Immerse yourself in our Realm of Maps
Medium mean: text, animation, photos of workroom from spherical view, music
Cartographic form: folio maps and wall maps in the workroom

Sequence 4: Researchers in map workroom   60"   0:35
Text narration: [*Dialogue balloons*] The Homann Map of the Balkan Peninsula / ...it was published in Nuremberg around 1718 / Who moved the borders of the Empire south? / The genius strategist Prine Eugene of Savoy won the Battles of Petrovaradin 1716 and Belgrade 1717/ Here on the Danube / Eugene of Savoy took for Habsburgs the Banat
Medium mean: text in dialogue balloons, video 360, music
Cartographic form: folio maps and wall maps in the workroom

Sequence 5: Geographical space   10"   1:35
Text narration: Geographical space
Medium mean: text, animation, screenvideo: flyover with 3D hypsometric map, music
Cartographic form: grayscale relief map, hypsometric map 2D and 3D

Sequence 6: Historio-geographical space   15"   2:25
Text narration: Movement of state borders / 2021 / 1910 / 1800 / 1718 / 1647
Medium mean: text, animation, political map, music
Cartographic form: political maps

Sequence 7: Habsburg-Ottoman battles 1716–1717   10"   2:40
Text narration: Victory battles of Prince Eugene of Savoy
Medium mean: text, symbol of battle, political map, music
Cartographic form: political maps, old coloured coper map

Sequence 8: Settlement in Banat 1720–1820   10"   2:50
Text narration: Settlement organization by Habsburg in Banat 1720-1820
Medium mean: text, animation of boat symbol, political map, music
Cartographic form: political map, pictorial sign

Sequence 9: Settlement in Banat 18th-19th Centuries   20"   3:00
Text narration: Settlement organization by Habsburg in Banat 18.-19. Century
Medium mean: text, animation of cart symbol, political map, printscreen, music
Cartographic form: political map, pictorial sign, print screen of geoportal

Sequence 10: Economic migration to USA   20"   3:20
Text narration: Settlement Economic migration to U.S.A. / Prager Station
Medium mean: text, animation, train and port symbols, political map, video, music
Cartographic form: political map, pictorial signs

Act 3: End 50" 3:40 – 4:30

Sequence 11: Answers   20"   3:40
Text narration: Settlement organization by Habsburg in Banat 18.-19. Century
Medium mean: text, animation of cart symbol, political map, print screen, music
Cartographic form: political map, pictorial sign, print screen of geoportal

Sequence 12: Next ask, closing of Realm of Maps   20"   4:00
Text narration: Youth migration in the Danube Region? / to answer this question... visit our Realm of Maps again...
Medium mean: text, animation of cart symbol, political map, print screen, music
Cartographic form: political map, folio maps and wall maps in the workroom

Sequence 13: End credits   10"   4:20

Figure 3. The short film script with ordered sequences in three acts with text narration, different medium means, and cartographic form, and with appropriate time proportions and matched time of duration for each sequence, with an accuracy of 5 s.

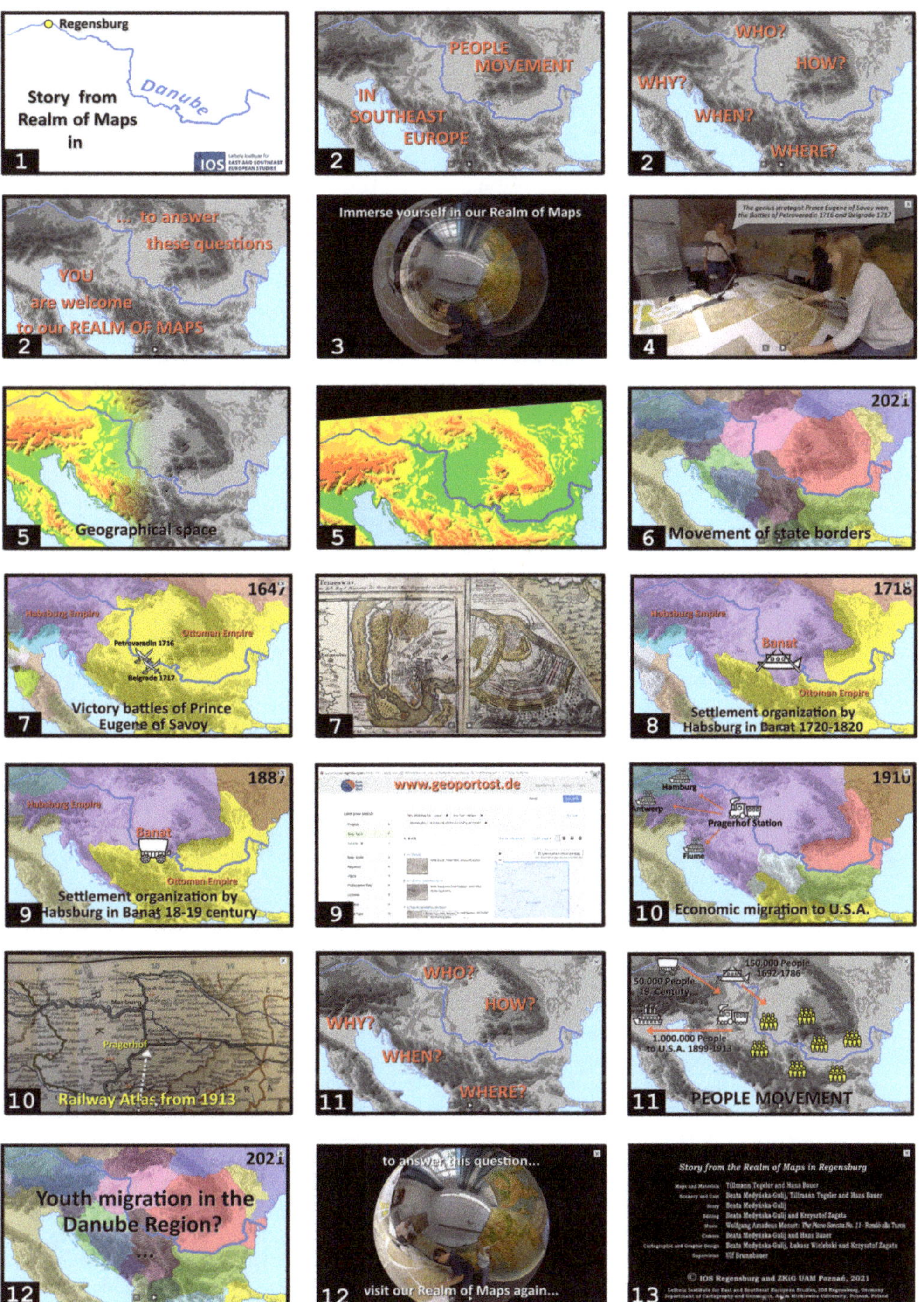

Figure 4. A total of 13 sequences of the film, whose frames are graphically depicted.

In the fourth section, a 360-degree camera shows three researchers walking in the workroom enclosed by wall maps, standing over maps on the table, and discussing historical events. The dynamics of the discussion are highlighted by speech balloons, stylized to resemble cartoon speech balloons, from which we can read crucial facts, dates, names, and geographical names.

In the fifth sequence, we can see the geographical space of Southeast Europe with a high generalization of the cartographic contents, leaving only the Danube, mountain chains, lowlands, and coastline on the map. The grayscale relief map from the first sequences transforms into a suggestive hypsometric map that becomes a 3D map before our eyes. We begin our flight over the Danube surrounded by mountain chains, which makes us perfectly aware of the geographical importance of the relief of this part of Europe. The presentation of the historico-geographical space occurs in the sixth sequence, in which colorful stains of individual states in the form of a political map with the animation of the changes in state borders from 2021 back to 1647 apply onto a relief map.

The next sequences of the film include four chronologically ordered events that were located on a political map by means of pictorial signs designed: a fight, boat, cart, train, and a port. The following additional media presentations of other cartographic materials from the *Realm of Maps* were added to this visualization:

- A photograph of a part of the copperplate map 'The War Theatrum' designed by Homann around 1718;
- A print screen of the internet service of map search from geoportost.de and a fragment of a map of Banat of 1858;
- Zooming the location of a junction station and a railway from the atlas of 1913 with a traditional camera.

To indicate the direction in which the population migrated, researchers used the animation of 'walking' pictorial symbols of boats and carts, as well as arrows that showed directions to seaports.

In the 11th sequence, the viewer can see the questions asked at the beginning of the film again and receives answers to them appearing on the map, with the symbols of residents and those representing the ways of migration, the number of people, and moving arrows indicating the directions of the people's movement. The 12th sequence is a political map with a question about modern migration around the Danube and the animation presenting spherical views of the workroom that imitates closing the gate to the *Realm of Maps*. The 13th sequence includes end credits following the convention of feature films.

4. Results—Editing the Film and Uploading It to the Online Video-Sharing Platform

To create suitable cartographic forms, the following types of media were employed: graphic software (Inkscape, Adobe Photoshop); geoinformation software (ArcGIS Pro); a screen recorder and video editor (Insta360Studio, Camtasia); a camera (Panasonic); a smartphone (Samsung GalaxyS10); a 360-degree camera (Insta360EVO); an MSI-VR laptop. The editing was performed in the Camtasia video editor and included the following parameters: an MP4 video file, 16:9 aspect ratio, encoded at the 1080p: 1920 × 1080 (Full HD) resolution. The film was uploaded to the YouTube channel 24 January 2021: https://www.youtube.com/watch?v=E-mChqmYn2A.

The length of the film with 13 sequences is 4 min and 30 s, with the first act lasting 25 s, the second one lasting 3 min and 13 s, and the last one, 50 s. Overall, 11 sequences are between 10 and 20 s long. The fourth sequence is an exception—it lasts 1 min and includes the dialogue of three researchers. According to the assumptions, in the film, there is only one piece of classical music with a fast tempo, directly linked to the history of Southeast Europe: The Piano Sonata No. 11 'Alla Turca' by Wolfgang Amadeus Mozart, from 1783 [7].

Simple, easily identifiable cartographic forms and brief subtitles are crucial to help the viewer follow the plot [17]. Neutral, black titles of individual sections, such as 'Movement of state borders', are centered and placed at the bottom of the film frame. The years in which the borders of the states changed were adjusted to the upper right corner of the frame to avoid the discomfort of watching 'jumping' years. Only in one case did we use speech balloons, created to resemble a dialogue from a cartoon, to make the exchange between three researchers preserve the dynamics of them moving around maps along with spontaneous dialogue. The ease of reading the subtitles results from the employment of a single-element, sans serif font [28].

To make our story coherent, we used two repetitions: two shots from the beginning of the film were used at the end. The first repetition is related to crucial questions: the 12th sequence uses the same questions and the same shot with interrogative pronouns appearing one by one along the Danube on the map. The other repetition is related to the opening of the *Realm of Maps* that imitates the immersion of the viewer in the world of maps in the second sequence. In the 12th sequence, the animation of the photograph 360 is reversed, which suggests the closing of the *Realm of Maps*.

To make moving from one cartographic visualization to another smoother, the study used a fade-out effect with a black frame that lasted for 1 s, and fade-through from the left to the right side for the scale of colors, and in the sixth sequence, 'Movement of state borders', for seven political maps. The line of the Danube is a stable graphical element of each shot because it is located in the same spot of the map, and it is always at the same scale and has the same cartographic symbol [25,26]. The cartographic shape of the Danube was preserved by designing a very thick blue line for the extending width from the river head in Swabia to the mouth of the river in the sea [24].

5. Discussion and Conclusions

To sum up our theoretical discussions, practical design, and the final version of the film, we examined the value of a suggested way of storytelling in the form of a film that applies certain specific research methods, i.e., the *Realm of Maps*, and is dedicated to a specific research topic related to historico-geographical space. Currently, in the time of digitalization and mapping websites, personal queries of cartographic sources in libraries and scientific institutes are frequently replaced with online studies of digital copies and the use of geoportals. Such a trend intensified during the pandemic; thus, we believe that the employment of geomedia is an excellent way to show the legitimacy of combining two dimensions of working with maps. It makes it possible to develop real and virtual methods of exploring the collections of cartographic and descriptive sources in one specific scientific center or in the library map department [40].

Returning to the question about the conditions of building a map workroom for the purpose of studying historical events in geographical space, we would recommend ordering old maps on a large table with possibly easy access to each map, and symbolically enclosing the working space with wall maps. A physical wall symbolically enclosed with historical maps allows one to grasp different perspectives on perceiving and combining facts. However, one needs to also consider individual preferences in terms of research methods based on maps [41].

The script that orders the plot of the story according to acts and sequences, including also the text of the narration, the medium, and the cartographic form, becomes crucial, especially when we assume that such film is the demonstration of the researcher's knowledge, but also a method of presenting the results or passing the knowledge to viewers. Such a combination of cartographic signs, symbols, and mapping techniques with media can be defined as a geomedia activity [42]. The legitimacy of creating content in the film for two types of viewers—typical users of online video sharing and social media platforms, with general knowledge of history and geography, and scientists, experts in the field—still remains open to debate [43].

The film is fast paced and includes very rhythmical classical piano music. The length of the film is also problematic. A 4.5 min video may seem too long for public users, while it may be considered too short for an expert and not discussing the topic profoundly enough [36].

For our film, we assumed two levels of knowledge perception. At the first, general level for public users, the user can see visually attractive sequences but may have a problem grasping the relation between pieces of information in individual sequences (e. g. due to the lack of geographical names). However, experts will easily associate historical facts with the location and, through such a comprehensive geomedia visualization, they may actually spot new links or be inspired to search for them [44]. At this point, we could ask a question about the potential of such a film for a better understanding of the presented phenomena. Although each map reader reads and interprets map signs individually, map designers may have an impact on reading and interpreting static and dynamic cartographic images in the right way thanks to the employment of classic principles [45,46]. One should also bear in mind that the appeal of the way of cartographic presentation does not need to go hand in hand with its effectiveness [20,36,47].

Hence, there is no single answer to the question about the right number of media and their types that should be used in the film [33]. In this case, we used as many different types as possible, which, on the one hand, boosts the appeal of the film and, on the other, may hamper the proper perception of the main film plot [20].

In our methodological approach, we decided to employ a complementary combination of the knowledge of historical geography, map design principles, general good practices of writing a short film script, and the use of media. A film script with sequences ordered in three acts with text narration, medium mean, and cartographic form, was vital to the creation of the *Story from the Realm of Maps in Regensburg*. However, building a workroom and drafting a layout together by historians and cartographers, as well as adapting classic map design principles in order to create new cartographic visualizations, proved to be crucial for visual storytelling in this case.

On the basis of the research, we can offer the following recommendations for filming historical geography with analog maps and maps created specifically for films published on online video-sharing and social media platforms:

- Adjust the map format to the ratio 16:9 in the horizontal orientation;
- Include an easily recognizable and repetitive referential element on maps (here, the Danube and the coastline);
- Fill in the entire film frame with the content;
- Do not use frames for cartographic content;
- Do not place a legend or a linear scale on the map (without a numerical scale!);
- Design pictorial map symbols of high contrast to the background (e. g., symbols filled in white with a black contour on colorful areas);
- Use line symbols with high width (such as on wall maps);
- Prioritize the animation of point symbols over the animation of area symbols on the map;
- Use the minimal number of subtitles and place them in a single spot on the film frame to prevent them from 'jumping' in consecutive shots;
- Present old folio and wall maps in the film along with the researcher to show their size and way of use.

We believe that availability on public platforms and an appealing form of sharing knowledge of maps in order to discover new interdisciplinary relations constitute the two greatest assets of short films about maps [48]. The informative value and appeal of such films are to be determined in further research. It may have an impact on the formulation of further recommendations for filming historical geography. Indeed, those alternative ways can serve better communication in the field of publishing research stages and results, as well as their individual scientific and professional methods [49]. Additionally, a question arises about the kind of impact such a film has on the perception of historico-geographical phenomena for different user groups, e.g., schoolchildren, the general public, experts, and researchers, and how it differs between them.

To summarize, the recommendations that we suggested could come to fruition thanks to the cooperation of historians, geographers, and cartographers who created this particular workroom (the *Realm of Maps*) while being also based on classic principles of map design. In fact, this is a good point for collecting some feedback from real users, i.e., the public, experts, and analysts. We look forward to opening a broader discussion on the limits of the use of the proposed framework and to making the differences in limits for designers and users part of the discourse, with the emphasis on various ways of using the proposed framework.

Author Contributions: Conceptualization, Beata Medyńska-Gulij; methodology, Beata Medyńska-Gulij, Tillmann Tegeler, and Hans Bauer; software, Beata Medyńska-Gulij, Krzysztof Zagata, and Łukasz Wielebski; investigation, Beata Medyńska-Gulij; resources, Tillmann Tegeler and Hans Bauer; data curation, Hans Bauer; writing—original draft preparation, Beata Medyńska-Gulij; writing—review and editing, Tillmann Tegeler and Hans Bauer; visualization, Beata Medyńska-Gulij, Krzysztof Zagata, and Łukasz Wielebski All authors have read and agreed to the published version of the manuscript.

Funding: This research was funded by the Augustin Hirschvogel Fellowship at the Leibniz Institute for East and Southeast European Studies in Regensburg, 2020, *Story from the Realm of Maps in Regensburg*.

Institutional Review Board Statement: Not applicable.

Informed Consent Statement: Not applicable.

Data Availability Statement: Not applicable.

Acknowledgments: We would like to share our thought on the research: we had to conduct it in a very short time window between the first and the second wave of the COVID-19 pandemic, which adds unique value to our *Story from the Realm of Maps in Regensburg*. Hence, we would like to thank all IOS employees for their invaluable help with finding the right maps and the selection and provision of materials necessary to the creation of our *Realm of Maps*.

Conflicts of Interest: The authors declare no conflict of interest.

References

1. Tegeler, T.; Bauer, H. Introduction: Maps in Libraries. Trends in Enabling Spatial Information Retrieval. *E-Perimentron* **2019**, *14*, 110–116.
2. Domosh, M.; Heffernan, M.; Withers, C. *The SAGE Handbook of Historical Geography*; SAGE Publications: London, UK, 2021; ISBN 978-1526404558.
3. Kuźma, M.; Bauer, H. Map Metadata: The Basis of the Retrieval System of Digital Collections. *Int. J. Geo-Inf.* **2020**, *9*, 444. [CrossRef]
4. Uhl, J.H.; Leyk, S.; Chiang, Y.-Y.; Duan, W.; Knoblock, C.A. Map Archive Mining: Visual-Analytical Approaches to Explore Large Historical Map Collections. *Int. J. Geo-Inf.* **2018**, *7*, 148. [CrossRef] [PubMed]
5. Medyńska-Gulij, B. How the Black Line, Dash and Dot Created the Rules of Cartographic Design 400 Years Ago. *Cartogr. J.* **2013**, *50*, 356–368. [CrossRef]
6. Medyńska-Gulij, B.; Żuchowski, T.J. *European Topography in Eighteenth-Century Manuscript Maps*; Bogucki Wydawnictwo Naukowe: Poznań, Poland, 2018; ISBN 978-83-7986-204-7.
7. Medyńska-Gulij, B. *Kartografia i Geomedia*; Wydawnictwo Naukowe PWN: Warsaw, Poland, 2021; ISBN 978-83-01-21554-5.
8. Williams, J.P.; Smith, J.H. *The Players' Realm. Studies on the Culture of Video Games and Gaming*; McFarland & Co.: Jefferson, NC, USA, 2007; ISBN 978-0786428328.
9. Kersten, T.P.; Edler, D. Special Issue "Methods and Applications of Virtual and Augmented Reality in Geo-Information Sciences". *PFG—J. Photogramm. Remote. Sens. Geoinf. Sci.* **2020**, *88*, 119–120. [CrossRef]
10. Zagata, K.; Gulij, J.; Halik, Ł.; Medyńska-Gulij, B. Mini-Map for Gamers Who Walk and Teleport in a Virtual Stronghold. *Int. J. Geo-Inf.* **2021**, *10*, 96. [CrossRef]
11. Medyńska-Gulij, B.; Lorek, D.; Hannemann, N.; Cybulski, P.; Wielebski, Ł.; Horbiński, T.; Dickmann, F. Die kartographische Rekonstruktion der Landschaftsentwicklung des Oberschlesischen Industriegebiets (Polen) und des Ruhrgebiets (Deutschland), Cartographic reconstruction of the landscape development of the Upper Silesian industrial area (Poland) and the Ruhr area (Germany). *KN J. Cartogr. Geogr. Inf.* **2019**, *69*, 131–142.
12. Jones, K. David Rumsey Map Collection. *Multimed. Technol. Rev.* **2017**. Available online: https://www.arlisna.org/publications/multimedia-technology-reviews/1200-david-rumsey-map-collection (accessed on 13 October 2021).
13. Maiellaro, N.; Varasano, A. One-Page Multimedia Interactive Map. *Int. J. Geo-Inf.* **2017**, *6*, 34. [CrossRef]
14. Horbiński, T.; Cybulski, P.; Medyńska-Gulij, B. Web Map Effectiveness in the Responsive Context of the Graphical User Interface. *Int. J. Geo-Inf.* **2021**, *10*, 134. [CrossRef]

15. Roth, R.E. Cartographic Design as Visual Storytelling: Synthesis and Review of Map-Based Narratives, Genres, and Tropes. *Cartogr. J.* **2021**, *58*, 83–114. [CrossRef]
16. Thöny, M.; Schnürer, R.; Sieber, R.; Hurni, L.; Pajarola, R. Storytelling in Interactive 3D Geographic Visualization Systems. *Int. J. Geo-Inf.* **2018**, *7*, 123. [CrossRef]
17. Mocnik, F.-B.; Fairbairn, D. Maps Telling Stories? *Cartogr. J.* **2018**, *55*, 36–57. [CrossRef]
18. Horbiński, T.; Zagata, K. Map Symbols in Video Games: The Example of "Valheim". *KN J. Cartogr. Geogr. Inf.* **2021**, *14*, 13. [CrossRef]
19. Edler, D.; Dickmann, F. The Impact of 1980s and 1990s Video Games on Multimedia Cartography. *Cartographica* **2017**, *52*, 168–177. [CrossRef]
20. Medyńska-Gulij, B.; Forrest, D.; Cybulski, P. Modern Cartographic Forms of Expression: The Renaissance of Multimedia Cartography. *Int. J. Geo-Inf.* **2021**, *10*, 484. [CrossRef]
21. Harrower, M. The Cognitive Limits of Animated Maps. *Cartographica* **2007**, *42*, 349–357. [CrossRef]
22. Clarke, K.C.; Johnson, J.M.; Trainor, T. Contemporary American cartographic research: A review and prospective. *Cartogr. Geogr. Inf. Sci.* **2019**, *46*, 196–209. [CrossRef]
23. Bertin, J. *Semiologie Graphique—Les Diagrammes, les Reseaux, les Cartes*; Gaultier: Paris, France, 1967.
24. Hake, G. *Kartographie*, 2nd ed.; Walter de Gruyter: Berlin, Germany; New York, NY, USA, 1976.
25. Freitag, U. Semiotik und Kartographie. Über die Anwendung kybernetischer Disziplinen in der theoretischen Kartographie. *Kartogr. Nachr.* **1971**, *21*, 171–182.
26. Dent, B. *Principles of Thematic Map Design*; Addison-Wesley Publishing Company: Boston, MA, USA, 1985; ISBN 978-0201113341.
27. Imhof, E. *Thematische Kartographie*; Walter de Gruyter: Berlin, Germany; New York, NY, USA, 1972.
28. Robinson, A.H.; Sale, R.D.; Morrison, J.L. *Elements of Cartography*, 4th ed.; John Wiley & Sons, Inc.: New York, NY, USA, 1978; ISBN 9780471017813.
29. Ratajski, L. The methodical basis of the standardization of sign on economic maps. *Int. Yearb. Cartogr.* **1971**, *11*, 137–159.
30. Keates, J. *Cartographic Design and Production*; Longman: London, UK, 1973; ISBN 978-0582482838.
31. van Dijck, J. *The Culture of Connectivity: A Critical History of Social Media*; Oxford University Press: Oxford, UK, 2013; ISBN 978-0199970780.
32. Nash, P. *Short Films. Writing the Screenplay*; Oldcastle Books: New York, NY, USA, 2012; ISBN 978-1842435014.
33. Cooper, P.; Dancyger, K. *Writing the Short Film*, 3rd ed.; Elsevier/Focal Press: Burlington, MA, USA, 2005; ISBN 978-0240805887.
34. Booker, C. *The Seven Basic Plots. Why We Tell Stories*; Bloomsbury Academic: New York, NY, USA, 2006; ISBN 978-0826480378.
35. Kraak, M.J.; Ormeling, F.J. *Cartography. Visualization of Geospatial Data*, 4th ed.; CRC Press: Abingdon, UK, 2020.
36. Medyńska-Gulij, B.; Wielebski, Ł.; Halik, Ł.; Smaczyński, M. Complexity Level of People Gathering Presentation on an Animated Map—Objective Effectiveness Versus Expert Opinion. *Int. J. Geo-Inf.* **2020**, *9*, 117. [CrossRef]
37. Brunnbauer, U. *Globalizing Southeastern Europe. Emigrants, America, and the State Since the Late Nineteenth Century*; Lexington Books: Lanham, MD/Boulder, CO/New York, NY, USA; London, UK, 2016; ISBN 978-1498519557.
38. Magocsi, P.R. *Historical Atlas of Central Europe, 3rd Revised and Expanded Ed.*; University of Toronto Press: Toronto, ON, Canada, 2018; p. 296.
39. Shuman, A. *Storytelling Rights. The Uses of Oral and Written Texts by Urban Adolescents*; Cambridge Studies in Oral and Literate Culture 11; Cambridge University Press: Cambridge, UK, 1986; ISBN 9780511983252.
40. Kühne, O.; Edler, D.; Jenal, C. A Multi-Perspective View on Immersive Virtual Environments (IVEs). *Int. J. Geo-Inf.* **2021**, *10*, 518. [CrossRef]
41. Caquard, S.; Cartwright, W. Narrative Cartography: From Mapping Stories to the Narrative of Maps and Mapping. *Cartogr. J.* **2014**, *51*, 101–106. [CrossRef]
42. Collins, A.; Neville, P.; Bielaczyc, K. The role of different media in designing learning environments. *Int. J. Artif. Intell. Educ.* **2000**, *11*, 144–162.
43. Pham, H.-H.; Farrell, K.; Vuong, Q.-H. Using YouTube Videos to Promote Universities: A Content Analysis. *Tech. Technol. Educ. Manag.* **2017**, *12*, 58–72.
44. Finkler, W.; León, B. The power of storytelling and video: A visual rhetoric for science communication. *J. Sci. Commun.* **2019**, *18*, A02. [CrossRef]
45. DiBiase, D.; MacEachren, A.M.; Krygier, J.B.; Reeves, C. Animation and the Role of Map Design in Scientific Visualization. *Cartogr. Geogr. Inf. Syst.* **2013**, *19*, 201–214. [CrossRef]
46. Cybulski, P.; Wielebski, Ł. Effectiveness of Dynamic Point Symbols in Quantitative Mapping. *Cartogr. J.* **2019**, *56*, 146–160. [CrossRef]
47. Wielebski, Ł.; Medyńska-Gulij, B. Graphically supported evaluation of mapping techniques used in presenting spatial accessibility. *Cartogr. Geogr. Inf. Sci.* **2019**, *46*, 311–333. [CrossRef]
48. León, B.; Bourk, M. *Communicating Science and Technology Through Online Video. Researching a New Media Phenomenon*; Routledge: Abingdon, UK, 2018.
49. Cooke, S.J.; Gallagher, A.J.; Sopinka, N.M.; Nguyen, V.M.; Skubel, R.A.; Hammerschlag, N.; Boon, S.; Young, N.; Danylchuk, A.J.; Metcalf, V. Considerations for effective science communication. *FACETS* **2017**, *2*, 233–248. [CrossRef]

International Journal of
Geo-Information

Article

Cartographic Design and Processing of Originally Printed Historical Maps for Their Presentation on the Web

Petra Justová * and Jiří Cajthaml

Department of Geomatics, Faculty of Civil Engineering, Czech Technical University in Prague, Thákurova 7, 166 29 Praha, Czech Republic; jiri.cajthaml@fsv.cvut.cz
* Correspondence: petra.justova@fsv.cvut.cz

Abstract: On the example of our project on the creation of the historical web atlas on Czech history, we introduce the process of adapting originally printed historical maps for their presentation in the web environment, which overcomes the shortcomings of standard approaches in similar projects based on printed predecessors published only as zoomable scanned analogues or default GIS maps. To simplify the originally complex map and to increase the information potential of the maps, we propose seven different types of additional map functionality according to the specific characteristics of the original map content. In addition, we present a set of rules, principles, recommendations, and methods for the cartographic design and processing of originally printed historical maps that should be considered when they are prepared for presentation on the web, including the description of the specific visualisation processes for the proposed types of map functionality. The proposed complex methodology can be applied to similar projects focused on the conversion of originally printed maps to the web and may contribute to improving the quality of the visualisation and presentation of historical maps on the web in general.

Keywords: web cartography; cartographic design; map processing; historical map; historical atlas; printed maps; web map application; map functionality

Citation: Justová, P.; Cajthaml, J. Cartographic Design and Processing of Originally Printed Historical Maps for Their Presentation on the Web. *ISPRS Int. J. Geo-Inf.* **2023**, *12*, 230. https://doi.org/10.3390/ijgi12060230

Academic Editors: Wolfgang Kainz, Beata Medynska-Gulij, David Forrest and Thomas P. Kersten

Received: 10 February 2023
Revised: 25 May 2023
Accepted: 30 May 2023
Published: 2 June 2023

1. Introduction

1.1. Visualisation of Historical Data

Historical events and processes are inherently related to a specific time and geographical space. When visualising historical data in maps, it is essential to deal primarily with cartographic methods for depicting development over time, because capturing a change in the spatial determination of a phenomenon or a change in the spatial relationships between the depicted phenomena over time is crucial when studying (and presenting) history. In the creation of historical maps, the need to capture multiple temporal states of a given phenomenon is very common, so the knowledge of methods suitable for visualising time in maps is an essential skill for the historical cartographer. The problem of visualising time as a fourth dimension in the two-dimensional map space is summarised by Monmonier [1], who compares the limits of time representation in static maps with the possibilities of its representation in dynamic maps, or by Vasiliev [2], who defines several categories of time according to the type of temporal information. His work is based on the study of earlier map works and presents a comprehensive overview of the methods of the cartographic representation of spatio-temporal information depending on the type of temporal information and the spatial dimension of the phenomenon. His research builds on previously published works on similar topics [3,4]. The difficulty of representing time within a static medium is also discussed by Kraak [5], who describes several temporal concept models and presents multiple cartographic solutions to depict the spatio-temporal information in a map. Wigen and Winterer [6] provide a historical overview of cartographic efforts to represent time on maps. The effective representation of time in dynamic maps is also addressed by other authors [7,8].

A specific problem in visualising and presenting historical data is the representation of the location or distribution uncertainty of a phenomenon [9]. This problem is caused by the way data are acquired during historical research (positional identification of an object or delineation of the area of occurrence of a phenomenon based on textual information, extraction of data from positionally inaccurate map bases). The issue of displaying the position or attribute uncertainty in a GIS environment has long been addressed by numerous authors [10,11].

1.2. Printed Historical Atlases

Historical atlases represent a very specific and interesting type of historical cartographic work. They reflect not only the theoretical and methodological level of the discipline, or the development of historical discourse but also the applied methods of the cartographic representation of historical events and related phenomena. Via school education historical atlases, these works can penetrate into wider society and form the way history is perceived and evaluated by the public [12]. This fact has been grossly abused by political forces, especially in totalitarian systems [13]. The analysis of the approaches to the cartographic processing of historical phenomena in previously issued atlases with a similar focus can become a great source of inspiration and is strongly recommended when compiling an atlas [14–16].

Historical cartography began to develop rapidly in the 19th century when the modern form of the historical atlas began to take shape and the tradition of the first historical school atlases was established [17,18]. This advance continued also in the 20th century, especially after World War II. During this period, more specialised works appeared, and there was also an increase in the range of topics covered in general historical atlases [19,20].

1.3. Electronic Historical Atlases

At the end of the 20th century, the technological progress and the use of digital methods in cartography radically affected the way of the processing and publication of atlases that began to transform from original paper media to digital media. Initially, electronic atlases were released as stand-alone applications, but in the 1990s, due to the progress of the Internet, data transfer standards, and the development of graphic formats, the first electronic historical web-based atlases began to appear [21]. The earliest web-based historical atlases were often developed in parallel with the creation of their analogue version and were presented as browsable applications with limited functionality. Along with the digital revolution, new visualisations and functionalities have entered the field of atlas cartography, and web-based atlases have started to be enriched with a related multimedia content (diagrams, text, visuals, videos, and animations) linked to the displayed geographic entities and have gradually evolved into multimedia atlas information systems (MAIS) [22]. The process of creating web-based atlases and its development over time has been addressed in the studies by Swiss researchers from the Institute of Cartography in Zurich [23–25], along with the discussion on different approaches used in the development of web atlases, which were based on multimedia content, but without including cartographic aspects. The importance of including the cartographic aspect in the process of creating a web-based interactive atlas was also emphasised by Lechthaler [26], who mentions the term cartographic information system (hereafter CIS) in this context. CISs are primarily focused on a high-quality cartographic visualization of the displayed data [27], which includes the additional processing of the presented information, where basic cartographic rules are applied to the displayed data [28].

Over the last three decades, two basic approaches to the creation of historical web atlases have evolved: (1) electronic versions of previously existing printed atlases and (2) stand-alone electronic atlases. They differentiate in the way and the quality of the cartographic visualisation and presentation of data, in the kinds of map interaction/functionality or in the thematic and geographical scope. In general, historical web atlases that were developed in parallel with or as follow-ups to their printed versions present a broad thematic focus of maps (population, economics, culture, military, etc.). They are focused on national or local history (regions or towns) and provide a more accurate historical content. They are often processed in the form of so-called web map portals presenting either zoomable default GIS maps [29] or georeferenced scanned analogue maps enhanced with simple animation [30] or combining scanned analogue maps or raster images with vector graphics and enabling the user to change the map content via layer selection [31]. The later mentioned map functionality is very common for historical town atlases [32–34]. In contrast, stand-alone electronic historical atlases present topic-specific maps on the territorial and political development of the world or its part over a certain time period. Most of them are distributed in the form of non-zoomable static raster images with temporal animation, map switching, or hyperlinks [35–38], or in the form of animated videos composed of static raster images [39]. Only a few examples of stand-alone electronic atlases follow the specific rules of web cartography, which result from the specific characteristics of the web as a presentation medium [40], and present their content as dynamic multiscale vector maps and offer the user the basic web map functionality (zooming, panning, pop-up display, and hyperlinks) or an interactive timeline [41–43].

Many of the above-mentioned projects were developed more than 15 years ago, which is reflected in the technological aspects of the map applications that directly affect their functionality. Some of them (developed in the late 1990s) are even no longer updated and are technologically obsolete [31,44,45]. The data preparation and cartographic processing in most of the above examples were performed in desktop GIS software using standard data formats for the conversion to the web environment. For the web presentation, the authors mostly used a combination of HTML client programming, CSS, and JavaScript. The dynamics and interactivity of web maps were programmed using the open-source libraries Open Layers or Leaflet, or within commercial APIs (ArcGIS API for JavaScript, MapBox GL JS).

1.4. Aims

In this article, we present a set of rules, principles, recommendations, and methods for the cartographic design and processing of historical maps (originally prepared for the print medium) for their presentation in the web environment. The procedure is demonstrated on the example of our recent project on creating the historical web atlas [46], which presents selected historical maps from two printed historical atlases [47,48]. The main aim of our work was to create a historical web atlas that overcomes the shortcomings of similar currently existing historical web atlases (created as electronic versions of previously existing printed atlases) that mostly display the originally printed maps on the web in the form of raster images without adequately adjusting their cartographic design to the specific characteristics of the web as a presentation medium (see Section 2.2).

In this article, we focus on the description of selected cartographic aspects of the visualisation and presentation of historical maps (originally prepared for the print medium) in the web environment. We do not deal with the map interface design and its effectiveness (form, placement, or visual hierarchy of layout elements) [49,50] or with the overall conception of an atlas [51,52] nor with the data, software, or technological aspects of web maps.

The sub-objectives of our project can be summarised as follows:

- To create a historical web atlas that would differ from standard approaches in similar projects that present only digitized analogue maps;
- To design the type of web map functionality (map dynamics and interactivity) according to the map characteristics to increase the information potential of original maps;
- To propose a methodology for the conversion of originally printed historical maps to the web environment;
- To keep the cartographic quality of the original printed maps with respect to the specific characteristics of the web as a presentation medium.

2. Methods

The process of adapting the originally printed historical maps for their presentation in the web environment was solved in several subsequent stages:

- Map analysis (Section 2.1);
- Cartographic design and map processing (Section 2.2);
- Map publishing (Section 2.3).

2.1. Map Analysis

During the initial stage of the preparation of our historical web atlas, an analysis of the maps was carried out to identify the most suitable type and the level of web map functionality that would simplify the originally complex map while enabling a meaningful interaction with the original map content in the web environment that would increase the information potential of the presented map. In the detailed map analysis (Figure 1), we identified three basic categories of maps with similar characteristics of the data component type and map complexity (visual, intellectual, or both) [53]. According to these characteristics, we assigned the appropriate map functionality to each map. Our approach was inspired by the concept of restrictive flexibility when the users explore the map due to predefined interactive functions to fulfil the intended information task of the presented map and to ensure the legibility of the map in all phases of interaction with its content [54].

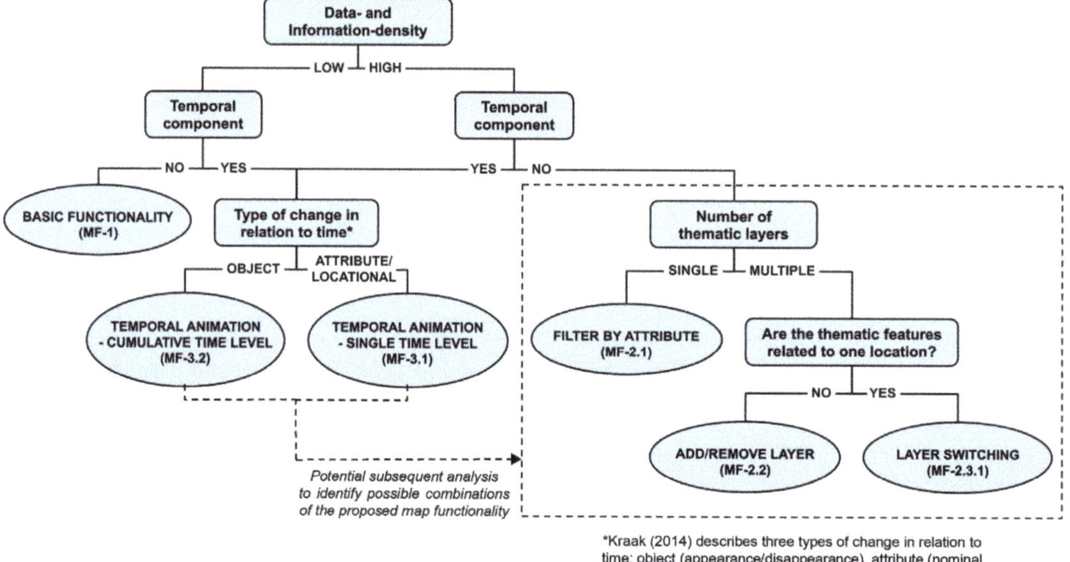

Figure 1. Map analysis process [5].

For the category of simple analytical maps with low data- and information-density, no inset maps, and no temporal data component, we decided to apply only the basic map functionality (MF-1): zooming, panning, feature highlight, and pop-up display, as an additional functionality would not add any more informational value to the user. This set of map interactions was defined as the minimum level of functionality that was applied to all web maps as a default. An additional map functionality was applied to more complex maps with a high data- or information-density.

Based on the similar characteristics of maps, we determined the following types of additional map functionalities:

- Filter by attribute (MF-2.1): The user has the option to filter the features according to the predefined attribute to analyse the spatial pattern/distribution of phenomena and their relations. This functionality is suitable for high data-density maps displaying a single thematic layer of qualitative data with no temporal component.
- Add/Remove layer (MF-2.2): The user is allowed to customise the map content by adding or removing thematic layers while preserving the initial default thematic layer in the map view. This functionality can be applied to maps displaying multiple thematic features with no temporal component that are not related to one location or are of different spatial dimensions.
- Layer/Map switching (MF-2.3): The user can switch the displayed thematic layer (MF-2.3.1) or change the geographical extent of the map (MF-2.3.2). The first mentioned option is suitable for maps displaying multiple thematic layers/features with no temporal component related to one location that are represented by multiple symbols of the same spatial dimension, or for multivariate maps combining different thematic methods. This layer switch option enables the user to reveal the thematic information in a separate visually simplified map while keeping the thematic information of the remaining (not visually represented) layers in a pop-up window. Map switching functionality (MF-2.3.2) is best useable to show more detail of a portion of the main map (originally displayed in an inset map) in a single map view.
- Interactive temporal animation (MF-3): The user can control the position of the slider on the timeline with predefined time sections. Temporal animation enables the user to better perceive the development of depicted phenomena over time as it is presented either as a sequence of maps representing a single time level (MF-3.1) or as a sequence of maps presenting individual time levels cumulatively (MF-3.2). The former option can be applied on maps displaying several states of phenomena in multiple time levels, as the state of the depicted phenomena is valid only for the displayed time interval (e.g., administrative border or military campaign). On the contrary, the cumulative sequence is suitable for maps displaying features with temporal information on their appearance and disappearance, as many features existed over several time intervals (e.g., development of a railway network).

Temporal animation was also applied on a set of simple analytical maps with no temporal data component that capture the state of a phenomenon at one moment in time, but together form a temporal sequence (e.g., political maps).

2.2. Map Processing

The cartographic design and map processing for presentation in a print medium differ greatly from those required for the presentation of maps in the web environment. These differences are mainly due to different characteristics of the web as a presentation medium (hypermedia, interactivity, dynamics, and unlimited map extent) and due to different requirements of the display device (RGB colour model and limited resolution) [55]. Therefore, when maps that were originally prepared for the print medium are to be presented on the web, they must be adequately adjusted to the specific characteristics of this environment, which often requires considerable intervention in the design of the original maps, which were not initially prepared for this purpose.

To meet the requirements for the presentation of maps on the web, the map processing procedure was divided into five consecutive stages that correspond to specific characteristics of the web environment:

- RGB colour mode and symbol simplification (Section 2.2.1);
- Zooming (Section 2.2.2);
- Panning (Section 2.2.3);
- Pop-up window (Section 2.2.4);
- Additional map functionality (Section 2.2.5).

2.2.1. RGB Colour Mode and Symbol Simplification

At the beginning of map processing, we modified the original map key in terms of the colour mode and the complexity of symbols. To preserve the colours of the printed atlas for the web environment, the original CMYK colour model was converted to its RGB equivalent. The complex cartographic symbology was simplified as it could be replaced by the functionality of the web map (Figure 2). The qualitative or quantitative characteristics of the feature were displayed only in the pop-up window. The temporal information about the year of the appearance or disappearance of a feature was displayed in the timeline.

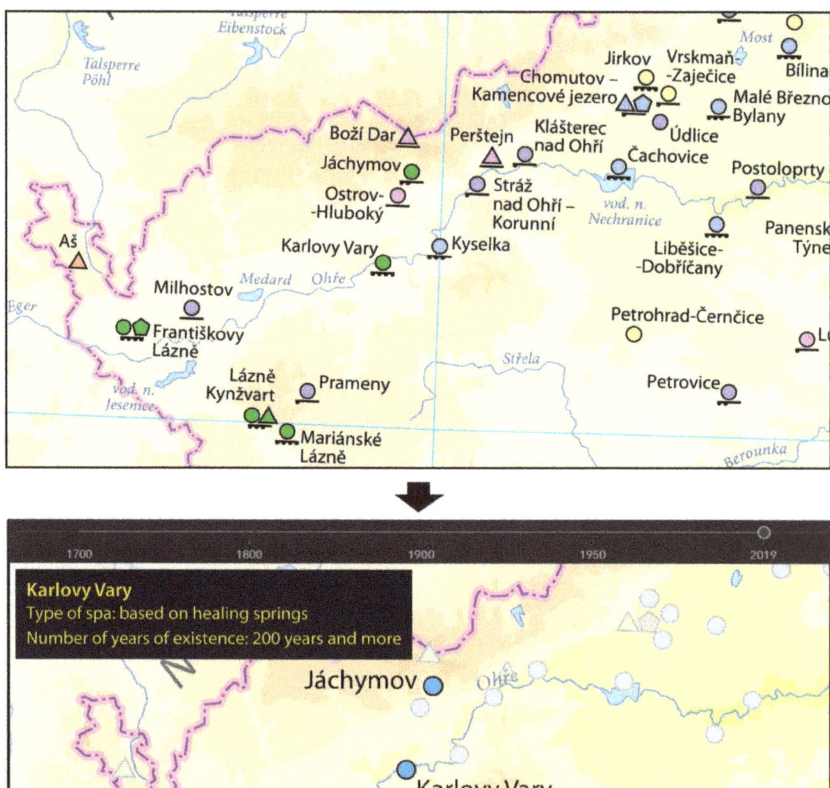

Figure 2. Complex cartographic symbology of the original map (**top**) vs. simplified map symbology replaced by the web map functionality (**bottom**).

2.2.2. Zooming

In the next stage, it was necessary to redesign the initially static maps that were designed at a fixed scale into dynamic multiscale maps. At first, we determined the basic scale to optimally display the whole area of interest in the respective historical time on a common computer screen. Considering the different extents of the territory of the Czech Lands in different historical periods, two basic scales had to be chosen (1:3M for

Czechoslovakia and 1:2M for the Czech Republic or the Czech Lands). This fact affected the way in which other zoom levels were derived (ratio of 1.5) as we aim to unify the scale series to provide a visual comparison of the web maps within a single web map application. For the same reason, we also unified the map projection, since the original maps used the Albers equal area conic projection but with a different position of the central meridian depending on the geographical extent of the historical territory of the Czech Lands. The maximum zoom level was set specifically for each web map with regard to the level of detail and generalisation of the original data. For each zoom level, we set the symbol and label size with respect to the technical limitations of display devices to improve the legibility of the web map.

After setting the symbol and label size for predefined zoom levels, we proceeded to edit the original map projects for a dynamic multiscale display on the web using the modified original map key (RGB colour model, simplified symbols, and symbol and font size for each zoom level). It was necessary to redesign the original map composition by symbol and label replacement to avoid the symbol overlap or the ambiguity of label association that appeared frequently due to the modification (enlargement or reduction) of the original size of cartographic symbols and labels. Automatic dynamic label placement was not applied, as the result did not meet our requirements for a high-quality cartographic design and information value of the maps [56,57]. Along with the symbol and label replacement, we reduced the visual complexity and information density of some maps at smaller scales: (1) by removing the labels showing the name or time attribute of a thematic feature, as these characteristics could be displayed in a pop-up window, (2) by the selection of toponyms according to their thematic importance, or (3) by aggregating the symbols. The latter step was discussed with the subject specialist, historian, or historical geographer. The original labels have been preserved only at the largest viewable scale.

2.2.3. Panning

In addition to processing the maps for zooming (see above), it was necessary to adequately process the maps for the application of the remaining basic map functionality, panning, and pop-up display. As the original maps were designed for presentation in a print medium, the map content was restricted only to the predefined map frame. Although web technologies offer different methods to limit the extent of the web map that the user can explore, none of them completely solved our problem of missing data at the edge of the original map frame. Therefore, we decided to apply a covering polygon layer outside the area of interest (thematic map content) using the feathering effect to cover the area with no data.

2.2.4. Pop-Up Window

The attribute data in the geodatabase of the original maps were often incomplete or in an inappropriate format (abbreviations or numerical codes) as they were primarily used for visualisation. Therefore, it was necessary to edit or complete selected attributes that were intended to be displayed in a pop-up window.

2.2.5. Additional Map Functionality

Besides the above-mentioned general modifications in the cartographic design of maps, it was also necessary to redesign the original visualisation and data structure of some maps according to the type of additional map functionality:

- Filter by attribute (MF-2.1): A copy of the filtered thematic layer with faded symbology was made to visually distinguish inactive features while filtering. This duplicate layer also preserved the original thematic content in the background in all phases of interaction with the map.
- Add/Remove layer (MF-2.2): No additional visualisation process was applied. For the purpose of this functionality, we only differentiated the thematic layers by assigning a unique ID that specified the initial setting of the layer visibility.

- Layer/Map switching (MF-2.3): Originally polythematic or multivariate maps had to be decomposed into several simplified monothematic maps that could be displayed separately in a single map frame (MF-2.3.1). This process often required a complete redesign of the original map composition either by symbol and label replacement (Figure 3), or by modifying the original thematic cartographic method. Inset maps that were going to be displayed in a single map frame (MF-2.3.2) required a more radical intervention in the original map composition as the final scale significantly differs from the original one.

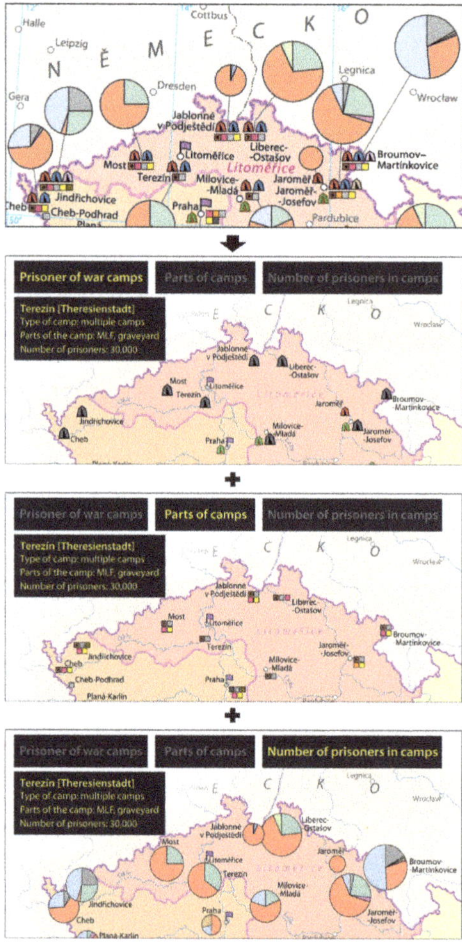

Figure 3. Decomposition of the original polythematic and multivariate map with the detail of the pop-up window showing the thematic information of all original layers.

- Interactive temporal animation—single time level (MF-3.1): Single maps that displayed several states of phenomena in multiple time levels were decomposed into a series of maps, each displaying the state of the phenomena only in a given time interval. If the temporal component of the data was expressed by means of visual variables (e.g., change of administrative borders), it was necessary to redesign the original cartographic visualisation and create new visualisations for each time interval (Figure 4). If the temporal component of the feature was expressed only through the exact time annotation (movement of the army or a battle on military maps), the visualisation for each map view was created by filtering the features according to the time or another predefined attribute. The predefined attribute represented the order of the time interval on the timeline and was derived from the time attribute in the geodatabase. Very often it was necessary to complete the original database of the map, as temporal information for some features was missing. The time intervals (sections of the timeline), their number, and range were set according to important milestones in the development of the depicted phenomena. This process required a discussion with the subject specialist, historian, or historical geographer. In some cases, this information could be easily retrieved from the original map (exact time determination of the change of the borders or the movement of the war front). In the temporal animation of military maps, we decided to visually preserve the previous state of phenomena in the map view to facilitate the comprehension of the spatial context of the movement. For this purpose, the duplicate of the thematic layer with faded symbology was created, to visually distinguish the current and the past time interval (Figure 5).

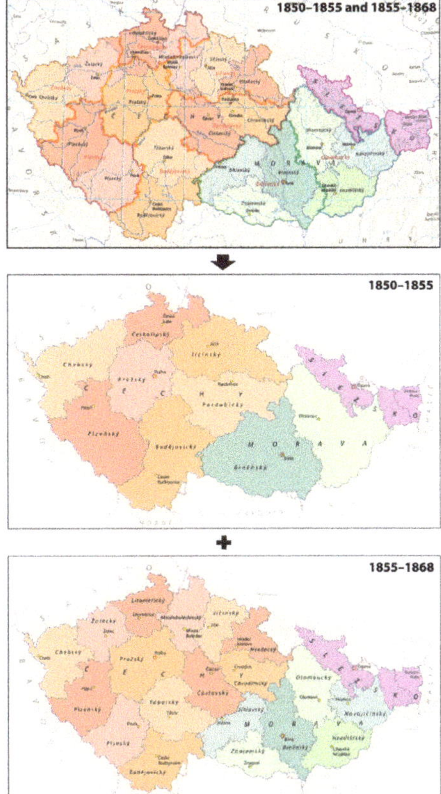

Figure 4. Decomposition of the original map displaying several states of phenomena in multiple time levels by means of visual variables.

- Interactive temporal animation—cumulative time levels (MF-3.2): The complex cartographic symbology of maps that originally displayed features with temporal information on their appearance and disappearance (often along with many other attributes) had already been simplified within the general modifications of the cartographic design (see above). Therefore, the original map composition could be decomposed into multiple temporal views only by filtering features according to the time or another predefined attribute (see above). In contrast to the afore-mentioned functionality (MF-3.1), each temporal view included the content of the previous one, as many displayed features existed over several time intervals (Figure 6). The faded symbology was applied only for disappeared features.

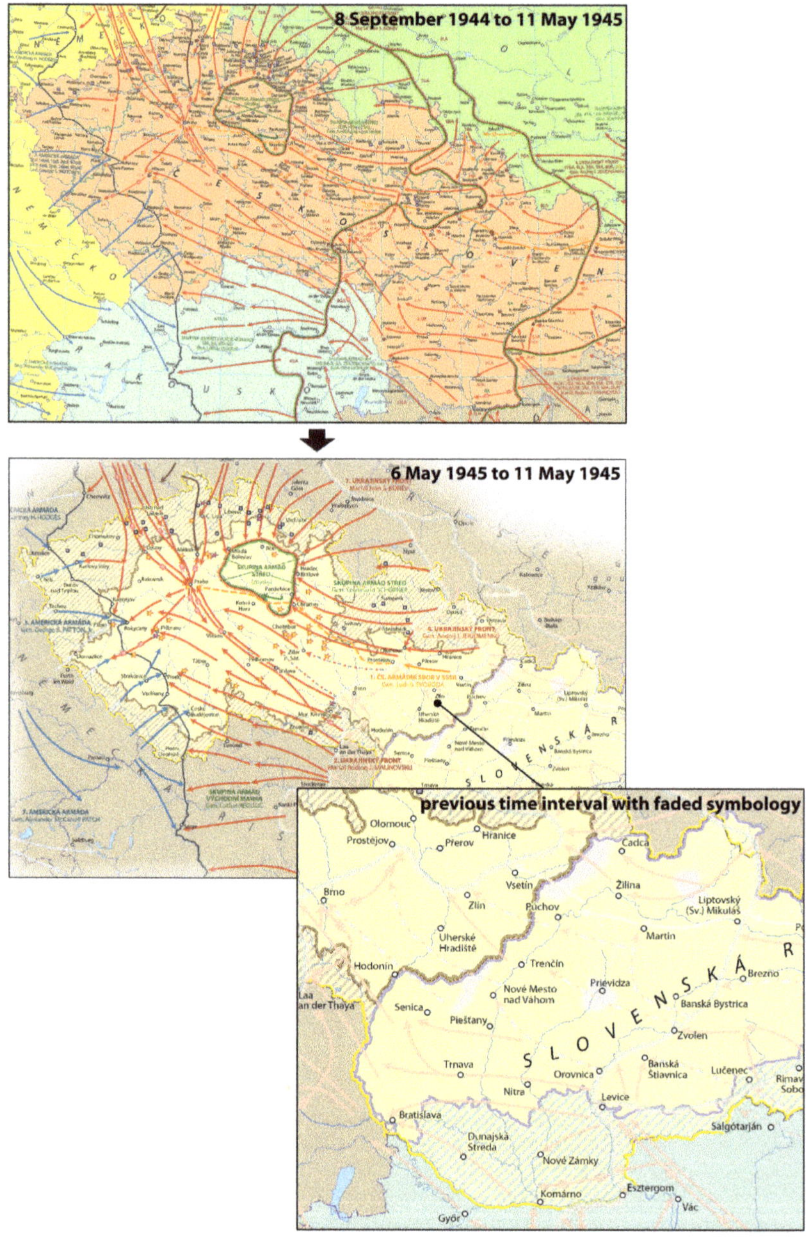

Figure 5. Faded symbology used in a military map to visually preserve the previous state of phenomena in the map view to facilitate the comprehension of the spatial context of the movement.

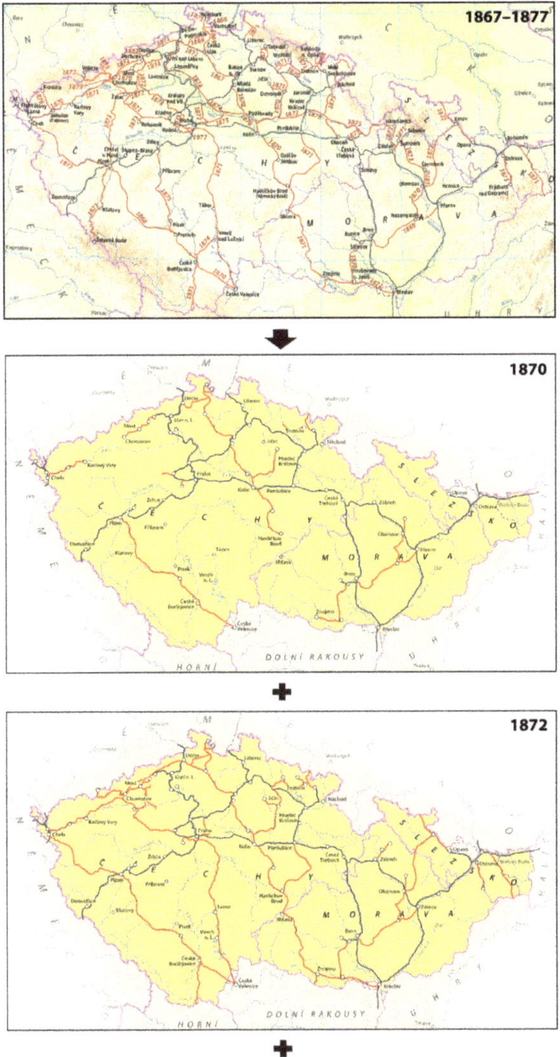

Figure 6. Decomposition of the original map displaying features with temporal information on their appearance into multiple temporal views that present individual time levels cumulatively.

2.3. Map Publication

To publish maps on the web, the map layers were divided into two groups (Figure 7):

- Thematic layers—an active content of the map over which a certain functionality is applied;
- Background layers—an inactive content of the map.

All map layers have been assigned a unique ID that specified the layer behaviour in the web map, such as which layers are jointly controlled by the elements of the additional map functionality (e.g., display in the same section of the timeline or joint switching on/off) or the initial setting of the layer visibility (MF-2.2).

Due to the limitations of the technology used for publishing the maps, a compromise solution in the cartographic design had to be adopted in some cases, as certain layers or drawing techniques could not be published in the web environment properly (e.g., annotation groups or masks).

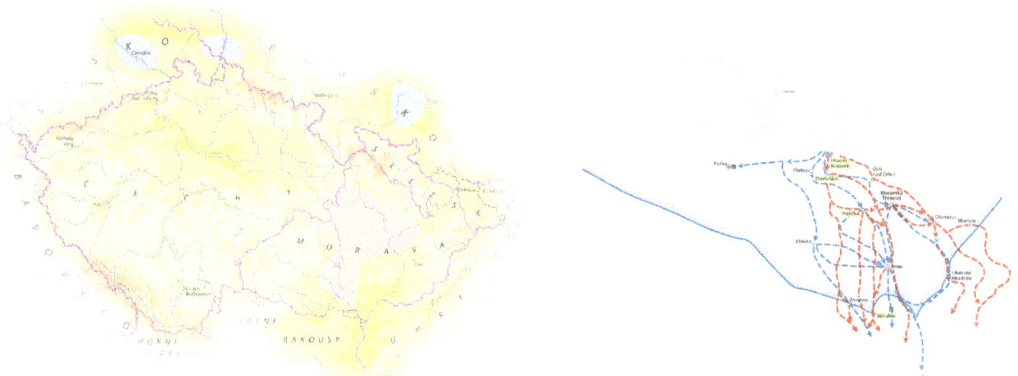

Figure 7. Background (inactive) layers of the map (**left**) vs. thematic (active) layers over which a certain functionality is applied (**right**).

3. Results

On the example of our recent project on creating the historical web atlas on Czech history [46], we introduced the process of adapting originally printed historical maps for their presentation in the web environment. During the initial stage, we carried out a detailed analysis of more than 160 maps to design the appropriate type of map functionality for each map that would simplify the original complex map (both visually and intellectually) and increase the information potential of the presented map (in comparison to its printed version). Based on the analysis (Figure 1), we identified seven categories of maps with similar characteristics of the data component type and map complexity (Figure 8). According to these map characteristics, we designed a web map application template with a predefined type of map functionality for each map category. The types of map functionalities were determined as follows:

- Basic functionality (MF-1);
- Filter by attribute (MF-2.1);
- Add/Remove layer (MF-2.2);
- Layer switching (MF-2.3.1);
- Map switching (MF-2.3.2);
- Interactive temporal animation—single time level (MF-3.1);
- Interactive temporal animation—cumulative time levels (MF-3.2).

In the next stage, we focused on the cartographic design and processing of the maps for their publication on the web. At first, we have described the general modification of maps that should be considered when preparing the map for its presentation on the web and that corresponds with the basic characteristics of the web environment (interactivity, dynamics, and unlimited map extent) and technical requirements of the display device (RGB colour model and limited resolution).

The general steps of map processing with basic recommendations can be summarised as follows:

- Conversion of the original CMYK colour model to its RGB equivalent;
- Simplification of complex cartographic symbology (replacement by map functionality);
- Redesign of the originally static map to a dynamic multiscale map: (1) setting the maximum zoom level according to the level of detail of the original data; (2) reducing the visual complexity and information density of some maps at smaller scales;

— Solving the problem with missing data at the edge of the original map frame: (1) applying a method to limit the extent of the web map; (2) applying a covering polygon layer;
— Editing or enrichment of attribute data of the original maps for displaying additional information in pop-up.

map functionality	component type	map characteristics	description	additional visualization process
Basic functionality (MF-1)	locational attribute	• simple cartographic symbology • low data-density maps	• zooming • panning • pop-up display	none
Filter by attribute (MF-2.1)	locational attribute	• qualitative data (two or more thematic categories) • high data-density maps	• filtering the features according to the predefined attribute • user can analyse the spatial distribution of phenomena and their relations	• duplicate layer with faded symbology for inactive features while filtering
Add/Remove layer (MF-2.2)	locational attribute	• multiple thematic features not related to one location • multiple thematic features of different spatial dimension	• customisation of the map content by adding or removing thematic layers • preserving the initial default thematic layer in the map view	none
Layer switching (MF-2.3.1)	locational attribute	• multivariate map combining different thematic methods • multiple thematic layers/features related to one location represented by multiple symbols of the same spatial dimension	• switching the displayed thematic layer • user can reveal the thematic information in a separate visually simplified map while keeping the thematic information of the remaining (not visually represented) layers in pop-up	• decomposition of the original polythematic (multivariate) map into single monothematic maps • redesign of the original map composition: (1) by symbol and label replacement, or (2) by modifying the original thematic cartographic method
Map switching (MF-2.3.2)	locational attribute	• two or more inset maps	• change the geographical extent of the map • user can show original inset maps in a single map view	• may require more radical intervention in original map composition (symbol and label replacement) as the final scale significantly differs from the original one
Interactive temporal animation– –single time level (MF-3.1a)	locational attribute temporal	• maps displaying several states of phenomena in multiple time levels • temporal component expressed by means of visual variables	• timeline with predefined time sections • the development of depicted phenomena over time is presented as a sequence of a maps representing single time level	• decomposition of the original map into multiple temporal views by modifying the original thematic cartographic method
Interactive temporal animation– –single time level (MF-3.1b)	locational attribute temporal	• maps displaying several states of phenomena in multiple time levels • temporal component expressed via exact time annotation	• timeline with predefined time sections • the development of depicted phenomena over time is presented as a sequence of a maps representing single time level	• decomposition of the original map into multiple temporal views by filtering features according to the time or other predefined attribute • duplicate layer with faded symbology for past time interval
Interactive temporal animation– –cumulative time levels (MF-3.2)	locational attribute temporal	• maps displaying features with the temporal information on their appearance and disappearance	• timeline with predefined time sections • the development of depicted phenomena over time is presented as a sequence of maps presenting the individual time levels cumulatively	• decomposition of the original map into multiple temporal views by filtering features according to the time or other predefined attribute • duplicate layer with faded symbology for disappeared features

Figure 8. Typology of map functionality according to the specific characteristics of the original map content, including the description of the specific visualisation process.

Besides the general modifications, we have also outlined specific visualization processes for each type of map functionality (Figure 8). Additional visualization processes were suggested as follows:

— Using faded symbology to visually distinguish active and inactive features while filtering (MF-2.1) for existing and disappeared features at a given time interval (MF-3.2);
— Redesign of the original map composition either by symbol and label replacement (MF-2.3.1) or by the modification of the original thematic cartographic method (MF-2.3.1, MF-3.1a);
— Creating a visualisation for temporal views by filtering features according to the time or another predefined attribute (MF-3.1b, MF-3.2).

4. Discussion and Conclusions

In comparison with national web-based atlases [24,28], which are created mainly as stand-alone web products based on originally unvisualised GIS data, historical web atlases are very often created as follow-ups to their printed version. In addition, historical atlases present more visually and intellectually complex maps that require a specific approach in terms of their cartographic design and map processing for their presentation on the web (compared to simple thematic maps in national atlases).

Currently, there exists no comprehensive study or research that addresses the issue of the cartographic design and processing of historical maps (originally prepared for the print medium) for their presentation in the web environment. Within our recent project on the creation of the historical web atlas on Czech history [46], we aimed to create an interactive version of the historical web atlas that would differ from standard approaches in similar projects based on printed predecessors that publish scanned analogues or default GIS maps in the form of zoomable raster images [29–31] and that would offer the user an additional map functionality that would increase the information potential of the presented maps. All of the abovementioned examples use the simplest way of displaying maps on the web and do not respect the different characteristics of the web as a presentation medium that require different approach to cartographic design and map processing. If the additional map functionality is applied, it is applied universally to all maps and does not respect the specific characteristics of presented maps.

Therefore, we decided to design the appropriate type of map functionality according to the specific characteristics of the original map content to increase the information potential of the maps. Based on the defined categories of maps with similar characteristics of the data component type and map complexity, we proposed seven different types of additional map functionality suitable for the presentation of historical maps on the web. In addition, we presented a set of rules, principles, recommendations, and methods for the cartographic design and processing of originally printed historical maps that should be considered when preparing the map for presentation on the web, including the description of the specific visualisation processes for the proposed types of map functionality.

Although the map functionality typology and the proposed methodology were derived on the basis of a sample of more than 160 historical maps on Czech history, considering the similar thematic structure of historical atlases [58], we assume that this sample is sufficiently broad and representative that our proposed principles, methods, and recommendations can be applied to other similar projects focused on the conversion of originally printed (digitally created) maps to the web.

In our project, a single map functionality was assigned to each map. We are aware that the application of single map functionality may not be sufficient for more complex maps with temporal data component. Therefore, we suggest a further analysis of maps with temporal data components in terms of the number of thematic layers and the characteristics of the thematic features (Figure 1), that was originally performed only on maps with no temporal component. Based on this further analysis, possible combinations of the proposed types of map functionality can be identified.

By publishing our experience in creating the historical web atlas, we aim to encourage authors from the field of historical cartography to publish originally printed maps/atlases on the web to make them available to a wider audience and to take advantage of the web map functionality for presenting complex historical topics by means of tools that significantly increase the information potential of the displayed content, which a printed or scanned analogue map (or a default GIS map) is unable to provide. Nonetheless, for some projects based on printed predecessors that exist only in analogue form, this may require a very time-consuming vectorisation process of the original maps.

The proposed map functionality typology and the complex methodology on the cartographic design and processing of originally printed historical maps for publication on the web can contribute to the improvement of the quality of the visualisation and presentation of historical maps in the web environment. Some general principles may be applicable even for the creation of stand-alone (historical or non-historical) web projects and provide a guidance for the presentation of complex maps on the web.

In further research, the proposed map functionality and the additional visualisations applied are planned to be evaluated in terms of information transfer efficiency to the target user. For this purpose, a series of usability tests will be performed using the knowledge testing or eye-tracking methods [59,60] to better understand the users of the atlas by researching who they actually are, how they interacted with maps, and for what purpose they used the maps [61,62].

Author Contributions: Conceptualisation, Petra Justová and Jiří Cajthaml; methodology, Petra Justová; writing—original draft preparation, Petra Justová and Jiří Cajthaml; writing—review and editing, Petra Justová and Jiří Cajthaml; visualisation, Petra Justová; supervision, Jiří Cajthaml; project administration, Jiří Cajthaml. All authors have read and agreed to the published version of the manuscript.

Funding: This work was supported by the Grant Agency of the Czech Technical University in Prague, Grant SGS23/051/OHK1/1T/11 "Analysis, visualization and presentation of 2D and 3D geospatial data using modern methods of cartography and GIS".

Data Availability Statement: Not applicable.

References

1. Monmonier, M. Strategies for the Visualisation of Geographic Time-Series Data. *Cartographica* **1990**, *27*, 30–45. [CrossRef]
2. Vasiliev, I. Mapping Time. *Cartographica* **1997**, *34*, 1–51. [CrossRef]
3. Wright, J.K. Appendix 1—Cartographic Considerations. A Proposed Atlas of Diseases. *Geogr. Rev.* **1944**, *34*, 649–651.
4. Hsu, M.-L. The Cartographer's Conceptual Process and Thematic Symbolization. *Am. Cartogr.* **1979**, *6*, 117–127. [CrossRef]
5. Kraak, M.-J. *Mapping Time. Illustrated by Minard's Map of Napoleon's Russian Campaign of 1812*; ESRI Press: Redlands, CA, USA, 2014.
6. Wigen, K.; Winterer, C. (Eds.) *Time in Maps. From the Age of Discovery to Our Digital Era*; University of Chicago Press: Chicago, IL, USA, 2020.
7. Buckley, A. Guidelines for the effective design of spatio-temporal maps. In Proceedings of the 26th International Cartographic Conference ICC2013, Dresden, Germany, 25–30 August 2013.
8. Andrienko, N.; Andrienko, G.; Gatalsky, P. Exploratory spatio-temporal visualization: An analytical review. *J. Vis. Lang. Comput.* **2003**, *14*, 503–541.
9. Knowles, A.K.; Hillier, A. *Placing History. How Maps, Spatial Data, and GIS Are Changing Historical Scholarship*; ESRI Press: Redlands, CA, USA, 2008.
10. MacEachren, A.M. Visualizing Uncertain Information. *Cartogr. Perspect.* **1992**, 10–19.
11. Kubíček, P.; Šašinka, Č. Thematic uncertainty visualization usability—Comparison of basic methods. *Ann. GIS* **2011**, *17*, 253–263. [CrossRef]
12. Bláha, J.D.; Močičková, J. The research-analytic part of preparation of a cartographic work: A case study of an analysis of historical atlases as the basis for creating the Czech Historical Atlas. *AUC Geogr.* **2018**, *53*, 58–69. [CrossRef]
13. Močičková, J. Historický atlas revolučního hnutí–historickokartografické dílo jako nástroj politické propagandy v Československu 50. let 20. století. *Hist. Geogr.* **2016**, *42*, 329–368.
14. Keates, J.S. *Cartographic Design and Production*, 2nd ed.; Longman: Harlow, UK, 1989.
15. Leszczycki, S. The Concept of a New Geographical Atlas. *Geogr. Vestn.* **1977**, *49*, 55–58.

16. Vozenilek, V. Aspects of the Thematic Atlas Compilation. In *Modern Trends in Cartography*; Brus, J., Vondráková, A., Voženílek, V., Eds.; Springer: Cham, Switzerland, 2015; pp. 3–12.
17. Dörflinger, J. Geschichtsatlanten vom 16. bis zum Beginn des 20. Jahrhunderts. In *Vierhundert Jahre Mercator, Vierhundert Jahte Atlas: "Die Ganze Welt Zwischen Zwei Buchdeckeln": Eine Geschichte der Atlanten*; Wolff, H., Ed.; Ausstellungskataloge/Bayerische Staatsbibliothek 65; Anton H. Konrad: Weißenhorn, Germany, 1995; pp. 179–198.
18. Goffart, W. *Historical Atlases. The First Three Hundred Years, 1570–1870*; University of Chicago Press: Chicago, IL, USA, 2003.
19. Black, J. Historical atlases. *Hist. J.* **1994**, *37*, 643–667. [CrossRef]
20. Black, J. *Maps and History. Constructing Images of the Past*; Yale University Press: New Haven, CT, USA, 2000.
21. Siekierska, E.M.; Taylor, D. Electronic mapping and electronic atlases: New cartographic products for the information era: The electronic atlas of Canada. *CISM J.* **1991**, *45*, 11–21. [CrossRef]
22. Cron, J.; Wiesmann, S.; Hurni, L. Facilitating the Handling of Interactive Atlases by Dynamic Grouping of Functions—The Example of 'Smart Legend'. In *Geospatial Vision*; Moore, A., Drecki, I., Eds.; Lecture Notes in Geoinformation and Cartography; Springer: Berlin/Heidelberg, Germany, 2008; pp. 1–18.
23. Bär, H.R.; Sieber, R. Towards high standard interactive atlases: The GIS and multimedia cartography approach. In Proceedings of the 19th International Cartographic Conference, Ottawa, ON, Canada, 14–21 August 1999.
24. Sieber, R.; Huber, S. Atlas of Switzerland 2—A highly interactive thematic national atlas. In *Multimedia Cartography*; Cartwright, W., Peterson, M.P., Gartner, G., Eds.; Springer: Berlin/Heidelberg, Germany, 2007; pp. 161–182.
25. Sieber, R.; Geisthövel, R.; Hurni, L. Atlas of Switzerland 3—A Decade of Exploring Interactive Atlas Cartography. In Proceedings of the 24th International Cartographic Conference ICC, Santiago de Chile, Chile, 15–21 November 2009.
26. Lechthaler, M. Interactive and Multimedia Atlas Information Systems as a Cartographic Geo-Communication Platform. In *Cartography in Central and Eastern Europe*; Gartner, G., Ortag, F., Eds.; Lecture Notes in Geoinformation and Cartography; Springer: Berlin/Heidelberg, Germany, 2009; pp. 383–402.
27. Kelnhofer, F. Geographische und/oder kartographische Informationssysteme. In *Kartographie im Umbruch-Neue Herausforderungen, neue Technologien. Beiträge zum Kartographiekongreß, Interlaken '96*; Nicoline Thieme: Leipzig, Germany, 1996; pp. 9–26.
28. Kelnhofer, F.; Pammer, A.; Schimon, G. "GeoInfo Austria"—Interaktives Multimediales Kartographisches Informationssystem von Österreich. In *Beiträge zum 5. Symposium zur Rolle der Informationstechnologie in der und für Die Raumplanung, CORP 2000*; Schrenk, M., Ed.; TU Wien: Vienna, Austria, 2000.
29. Atlas Obyvatelstva. Available online: http://www.atlasobyvatelstva.cz/ (accessed on 20 December 2022).
30. Atlas of the Historical Geography of the United States. Available online: https://dsl.richmond.edu/historicalatlas/ (accessed on 10 December 2022).
31. Historical Atlas of Canada. Available online: http://www.historicalatlas.ca/website/hacolp/index.htm (accessed on 15 December 2022).
32. Šedivý, J. Map portals and databases of towns in Central Europe: An appendix or substitute for printed historical town atlases? *Città Stor.* **2015**, *10*, 261–280.
33. Historic Towns Atlas of the Czech Republic Web Map Portal. Available online: http://towns.hiu.cas.cz/en/ (accessed on 22 December 2022).
34. Österreichischer Städteatlas. Available online: https://www.arcanum.com/hu/online-kiadvanyok/OsterreichischerStadtatlas-osterreichischer-stadteatlas-1/ (accessed on 3 December 2022).
35. Atlas of World History. Available online: https://atlasofworldhistory.com/ (accessed on 5 December 2022).
36. Mapping History. Available online: https://mappinghistory.uoregon.edu/english/index.html (accessed on 5 December 2022).
37. The Timemap of World History. Available online: https://timemaps.com/ (accessed on 6 December 2022).
38. Omniatlas—Interactive Atlas of World History. Available online: https://omniatlas.com/ (accessed on 3 December 2022).
39. The Map as History: A Multimedia Atlas. Available online: https://www.the-map-as-history.com/ (accessed on 5 December 2022).
40. Muehlenhaus, I. *Web Cartography*; CRC Press: Boca Raton, FL, USA, 2013.
41. Ostellus—The Future of History. Available online: https://ostellus.com/ (accessed on 3 December 2022).
42. Hruška, L.; Jarošová, L.; Lipovski, R. The Great Historical Atlas of Silesia. Available online: https://mapa.atlas-slezska.cz/en.html (accessed on 10 December 2022).
43. The GeaCron. Available online: http://geacron.com/ (accessed on 15 March 2023).
44. Historical Atlas of the 20th Century. Available online: http://users.erols.com/mwhite28/20centry.htm (accessed on 3 December 2022).
45. World History: HyperHistory. Available online: https://www.hyperhistory.com/online_n2/History_n2/a.html (accessed on 6 December 2022).
46. Czech Historical Atlas Portal. Available online: http://www.czechhistoricalatlas.cz/ (accessed on 8 January 2023).
47. Semotanová, E.; Cajthaml, J. *Akademický Atlas Českých Dějin, 2., Doplněné Vydání*; Academia: Praha, Czech Republic, 2016.
48. Semotanová, E.; Zudová-Lešková, Z. *Český Historický Atlas. Kapitoly z Dějin 20. Století = Czech Historical Atlas: Chapters on the History of the 20th Century*; Práce Historického Ústavu AV ČR, v.v.i. Řada A, Monographia = Opera Instituti Historici Pragae. Series A, Monographia Svazek 94; Historický Ústav: Praha, Czech Republic, 2019.
49. Sieber, R.; Schmid, C.; Wiesmann, S. Smart legend–smart atlas. In Proceedings of the 22th International Conference of the ICA, A Coruña, Spain, 9–16 July 2005.

50. Vít, L.; Bláha, J.D. A Study of the User Friendliness of Temporal Legends in Animated Maps. *AUC Geogr.* **2017**, *47*, 53–61. [CrossRef]
51. Caquard, S.; Cartwright, W. Narrative Cartography: From Mapping Stories to the Narrative of Maps and Mapping. *Cartogr. J.* **2014**, *51*, 101–106. [CrossRef]
52. Landaverde Cortés, N.A. A Conceptual Framework for Interactive Cartographic Storytelling. Master's Thesis, University of Twente, Enschede, The Netherlands, 2018.
53. Fairbairn, D. Measuring Map Complexity. *Cartogr. J.* **2006**, *43*, 224–238. [CrossRef]
54. Gartner, G.; Kriz, K.; Spanring, C.; Pucher, A. The concept of "restrictive flexibility" in the "ÖROK Atlas Online". In *Internet-Based Cartographic Teaching and Learning: Atlases, Map Use and Visual Analytics*; TU Wien: Vienna, Austria, 2005; pp. 105–110.
55. Neudeck, S. Zur Gestaltung topografischer Karten für die Bildschirmvisualisierung. Ph.D. Thesis, Universität der Bundeswehr München, Neubiberg, Germany, 2001.
56. Kern, J.P.; Brewer, C.A. Automation and the Map Label Placement Problem: A Comparison of Two GIS Implementations of Label Placement. *Cartogr. Perspect.* **2008**, *60*, 22–45. [CrossRef]
57. Been, K.; Daiches, E.; Yap, C. Dynamic map labeling. *IEEE Trans. Vis. Comput. Graph.* **2006**, *12*, 773–780. [CrossRef] [PubMed]
58. Jílková, P.; Janata, T. Selected Possibilities of Data Excerption from the Database of Historical Atlases. *Civ. Eng. J.* **2019**, *28*, 589–598. [CrossRef]
59. Çöltekin, A.; Heil, B.; Garlandini, S.; Fabrikant, S.I. Evaluating the Effectiveness of Interactive Map Interface Designs: A Case Study Integrating Usability Metrics with Eye-Movement Analysis. *Cartogr. Geogr. Inf. Sci.* **2009**, *36*, 5–17. [CrossRef]
60. Ooms, K.; de Maeyer, P.; Fack, V. Analyzing eye movement patterns to improve map design. In *AutoCarto 2010, Proceedings of the 18th International Research Symposium on Computer-Based Cartography and GIScience (AutoCarto 2010): Geospatial Data and Geovisualization for the Environments, Security, and Society, Orlando, FL, USA, 15–18 November 2010*; Cartography and Geographic Information Society (CaGIS): Cincinnati, OH, USA, 2010.
61. Kramers, E.R. Interaction with Maps on the Internet—A User Centred Design Approach for The Atlas of Canada. *Cartogr. J.* **2008**, *45*, 98–107. [CrossRef]
62. Van Elzakker, C.P.J.M.; Ooms, K. Understanding map uses and users. In *The Routledge Handbook of Mapping and Cartography*; Routledge: Oxford, UK, 2017; pp. 55–67.

International Journal of
Geo-Information

Article

Digitization, Visualization and Accessibility of Globe Virtual Collection: Case Study Jüttner's Globe

Eva Štefanová [1,*], Eva Novotná [2] and Miroslav Čábelka [1]

[1]   Department of Applied Geoinformatics and Cartography, Faculty of Science, Charles University, Albertov 6, 128 00 Prague, Czech Republic; miroslav.cabelka@natur.cuni.cz
[2]   Map Collection, Faculty of Science, Charles University, Albertov 6, 128 00 Prague, Czech Republic; eva.novotna@natur.cuni.cz
*   Correspondence: eva.stefanova@natur.cuni.cz

Abstract: The aim of the article is to prepare a model for making available metadata and digital objects of the new Globe Virtual Collection for the Map Collection of the Faculty of Science of Charles University. The globes are special cartographic documents; therefore, they are also described in a special way. The article deals with the digitization, visualization and accessibility of an old globe by Josef Jüttner from 1839, which comes from the depository of one of the most important central European collections. A simple model for a new virtual processing of the globe collection at Charles University is presented. SfM-MVS photogrammetry was chosen for digitization of the globe. The basic elements of the copperplate were set as basic parameters for image acquisition. Contrast, density, black line, line, dash and dot patterns and their complex use were observed for a good graphic design of the globe. Other parameters included a closer determination of the users for whom the resulting product is intended, as well as the profile of the users' behavior on the site so far. New metadata were extracted from the bibliographic description. The virtual 3D globe was integrated into the database using the Cesium JavaScript library. Metadata and a 3D model of the globe were linked together and made available to the general public on the Globe page of the Map Collection of the Faculty of Science of Charles University. A comparison of web browsers was performed focusing on the loading time of the 3D model on the website. New graphic elements were identified with the new processing. It was possible to read the factual information written on the globe. Different possibilities and limitations of metadata description, photogrammetric methods and web presentation are described. This good practice can be applied by other virtual map collections.

Keywords: historical cartography; 3D objects; SfM-MVS photogrammetry; digital reconstruction; metadata; globe visualization; virtual globes; Jüttner Josef

Citation: Štefanová, E.; Novotná, E.; Čábelka, M. Digitization, Visualization and Accessibility of Globe Virtual Collection: Case Study Jüttner's Globe. *ISPRS Int. J. Geo-Inf.* **2023**, *12*, 122. https://doi.org/10.3390/ijgi12030122

Academic Editors: Beata Medynska-Gulij, David Forrest, Thomas P. Kersten and Wolfgang Kainz

Received: 24 December 2022
Revised: 1 March 2023
Accepted: 7 March 2023
Published: 9 March 2023

1. Introduction

Thanks to digitization, the last ten years have seen a major breakthrough in the processing and accessibility of cartographic documents of Czech and foreign memory institutions. Cartographic cultural heritage is comprised not only of old maps, atlases and globes published before 1850, but also of unique manuscripts, works on original media or other rarities. Globes represent a specific category because their production in manufactories continued practically until the middle of the 20th century. Therefore, it is safe to assume that so-called unique globes were being created until 1950.

Digitization—composed of all its essential elements, i.e., selection, legal aspects, metadata [1–3], technology [4–6], grid data and repositories [7,8]—appears to be the key tool for protection, archiving and enabling public access to globes. This viewpoint is reflected in documents issued by governments and international organizations. The protection efforts are supported by major international institutions such as UNESCO [9]. The greatest attention is paid to the processing of, and ensuring access to, old cartographic documents, which are among the most used but also the most endangered segments of cartographic cultural heritage [10]. It is due to this reason they need special protection.

Many map collections in Europe, the United States of America and Australia are gradually digitizing and making accessible their unique holdings. Among the most im-portant described and digitized map collections today are the David Rumsey Map Center [11] at Stanford University Libraries, the Leventhal Map and Education Center [12] at the Boston Public Library and the Map Collection at the National Library of Australia [13]. However, the description and digitization of globes is different from the processing of other cartographic materials. It is necessary to accurately prepare the sphere, as well as the description, geographical grid and texture of the continents, countries and other details. This is possible by creating a virtual three-dimensional model.

An example is the digital reconstruction of an old 16th century globe by Gerhard Mercator [14] for visitors to the Mercator Museum in Belgium [15]. One of the largest manuscript globes of László Percze with a diameter of 139 cm from 1862 has been digitized by the team of Mátyás Gede and made available in the Hungarian Virtual Globes Museum [16,17]. Valuable exhibits of Dutch pairs of globes by Willem Janszon Blaeu from the 17th century are also made available here.

The French National Library has also published a collection of digitized globes in the Gallica Digital Library [18], including a valuable terrestrial globe with a diameter of 24 cm by Martin Waldseemüller from 1506. Another unique item is an Arabic celestial globe from the 11th century.

Around a hundred digitized globes from various collections of the Czech Republic are published on the website of the Research Institute of Geodesy, Topography and Cartography [19]. Among old works, remarkable is a pair of globes by Vincenzo Maria Coronelli from the end of the 17th century with a diameter of 110 cm, owned by the National Library of the Czech Republic [20].

The problem of describing the globes and making them available was also addressed by the Map Collection of the Faculty of Science of Charles University, which was founded as early as in 1891. It acquired the status of the State Map Collection of the Czechoslovak Republic in 1920. Although this status was changed in the 1950s, the Map Collection has retained its importance. It contains 130,000 map sheets, 3500 atlases and 122 globes. More than 100,000 maps have been scientifically catalogued and 65,000 map sheets have been made accessible electronically [21].

The first attempts to make this part of the collection accessible were published al-ready in 2013 [22]. Photographs of the globes with a brief description were gradually published on separate websites. They are dedicated to each exhibit, showing bibliographic, globographic information on the given specimen (metadata), such as the title, author, place and year of publication, as well as physical data describing the dimensions, the stand and material from which the globe was made. In the history of cartography, the Greenwich meridian was not always the prime meridian. Therefore, the position of the prime meridian (passing through, for example, Cape Verde or the Canary Islands) is also given for each globe. Old celestial globes depict the starry sky with allegorical representations of constellations in the form of animal or human figures.

Attempts at three-dimensional presentation have been published for the eight oldest globes. Researchers can virtually rotate the globe model, zoom in on it, view unique illustrations and notes, or read the historical names of cities and constellations. Due to the large volume of data, only smaller models have been presented so far. Projecting a globe into two-dimensional maps was also a topic of discussions. In 2013, an experienced team from the Belgian University of Ghent, led by Professor Philippe de Maeyer, who had already digitized the above-mentioned G. Mercator's globe [15] in the past, was recruited to collaborate. The digitized globes are presented in different formats (pdf, wrl and WebGL) and in different qualities, so that each user can view the model according to the capabilities of their technical equipment. It is often the case that researchers do not have the necessary browsers, programs or other equipment needed to quickly view the globes.

Physically, old globes are either not displayed or only rarely displayed so as not to damage them. The spheres were often made of plaster and paper and are therefore very fragile. The globes cannot be exposed to light or to changes in climate and temperature, or to human intervention in order to prolong their life. The high financial costs involved in restoration should also be mentioned, as it was not possible to make accessible the original blackened, damaged, cracked or illegible globes [23]. This has also created new re-paired objects that will need to be newly processed. In addition, through gradual acquisitions, the Map Collection of the Faculty of Science, Charles University, has obtained around 50 new globes in the last ten years. In digital format with verified metadata, re-searchers can zoom in on the globes at any time in great detail and verify a lot of the data.

The aim of the article is to prepare a model for making available metadata and digital objects of the new Globe Virtual Collection for the Map Collection of the Faculty of Science of Charles University. Therefore, the research question arose of how to describe the globes in a modern and effective metadata method and of how to digitally process and make them accessible. Simply put, how to digitally process and visualize the repaired and new globes quickly, easily, without major demands on other auxiliary tools and browsers, in order to make them efficiently accessible to researchers.

This article therefore deals with the digital reconstruction and visualization of globes and how to provide access to them. Good practice is demonstrated through a case study of the processing of a terrestrial globe by Josef Jüttner (1775–1848), originating from the repository of the Map collection of the Faculty of Science of Charles University.

1.1. Research of Literature and Existing Methods

Knowledge of the original materials, the process of creating the globes and their individual components was a prerequisite of successful digitization. Numerous authors have studied this topic in the past [24–28]. Initially, the globes used to be of a smaller diameter. As the diameter became larger, it gradually became customary to create globes comprised of two hemispheres, which were made in molds. Gores were cut from a globe map printed on paper and carrying terrestrial or celestial content, and then carefully pasted onto a gypsum or cardboard sphere. The old globes were often supplemented with further instruments and aids such as the compass, the meridian or the horizontal circle. Degrees were engraved into the meridian circle to aid the reading of the latitude. The horizontal circle was usually made of wood and pasted over with paper with a calendarium printed on it [29].

The old terrestrial globes usually included a geographical grid. The choice of the zero meridian could vary—it would pass through the Azores or El Hierro Island in the Canary Islands. It was not until the late 19th century that the prime Greenwich meridian became the established norm.

In order to digitize the globes and provide access to them, it was first necessary to identify a suitable digitalization method and then apply it accurately and efficiently. The criteria for the selection of this method were to determine the overall geometry and small details such as drawings, grids, or captions on the terrestrial globe. Another important parameter to be taken into account was the size of the globe, which could range from 10 cm to 100 cm in diameter. In particular, the Map Collection of the Faculty of Science of Charles University contains old globes ranging from 8.6 cm to 113 cm in diameter. In the whole Czech Republic, according to Karel Kuchař's research, old globes have a diameter from 7.5 cm to 127 cm [30].

The physical mounting and the possibility of removing the globe from the base was equally important. Four basic approaches to the digitization and 3D reconstruction of old globes are described in the literature.

The first attempts focused on creating a high-quality texture and then displaying it on a spherical model. This method was described by [31]. Taking images of the globe was carried out sequentially, according to the geographical grid at regular intervals to ensure sufficient image overlap. The subsequent conversion of the images into a 3D model was very +labor intensive. Each image had to be georeferenced separately using control points. Overlapping images rectified this way could be assembled into a complete mosaic in the shape of a sphere. This method has also been discussed in detail by [32,33]. This procedure involved some difficult manual handling of the globe. This involved dismounting the object from the base, expertly removing the meridian ring and placing the globe into a special static imaging device, where it was fixed and could be moved around on wheels.

The second approach was the creation of a 3D model on the basis of printed globe maps, as described by [17,34–36]. This method involved a reconstruction (rather than digitization) of the globe and its visualization on the Internet. In these cases, gores of the globe printed on paper were scanned. The digitized data obtained in this way had to be georeferenced using the intersections of the geographic grid to create a uniform (spherical) object with the texture of a globe. It was an efficient, accurate but very labor-intensive method.

A third variant was described by [37]. It involved laser scanning combined with the photogrammetric method. In this case, a real 3D globe was scanned and photographed. Laser scanning was used to define the final shape of the globe (sphere) and a texture was created from the photogrammetric images. The resulting globe models had a visible structure (e.g., continents, oceans, descriptions, etc.). Calculation problems may arise in relation to the resulting point cloud. Visible empty space appears in areas with a uniform texture, such as the oceans. This problem is addressed by [38], who used laser scanning data to fill in the gaps in the point cloud.

The last known method of processing three-dimensional objects is close-range photogrammetry. This method was tested by [15]. The gaps occurring in the point cloud were calculated using triangulation. This procedure of contact-free digitization and reconstruction of old globes was used by [39,40]. The testing of the relationship between varying image overlap sizes, the used focal length and the resulting accuracy of the 3D model was presented by [41].

Laser scanning and photogrammetry are suitable for creating a 3D textured model. Laser scanning is recommended when high accuracy over a large area is required. Photogrammetry is more suitable for documenting smaller objects and capturing realistic textures. A comparison of these methods according to the various criteria of equipment, data processing and accuracy are described, for example [42]. SfM-MVS photogrammetry appears to be a suitable method for 3D digitization of the globe, using the automatic techniques of Structure from Motion (SfM) and MultiView Stereo (MVS) to estimate a 3D model from overlapping 2D images. The creation of the 3D model takes three steps. (i) The automatic keypoints extraction from all images and matching across images using the algorithm Scale Invariant Feature Transform (SIFT). SIFT allows for a comparison of corresponding features under changes in the angle of view, scale, partial occlusion and changing illumination. (ii) SfM is used for the calculation of individual camera positions, orientations, focal lengths and relative positions of corresponding features. This step is known as bundle adjustment. (iii) The dense point and 3D surface (model) are determined based on known camera parameters using MVS. The quality and accuracy of the 3D model depends on the image's quality [43].

In general, camera calibration must be performed for non-metric cameras. This determines the Internal Orientation Parameters (IOP) that are necessary for image processing—i.e., the focal length, the coordinates of the camera principal point, and the radial and tangential distortion coefficients. The radial distortion of the lens significantly affects the geometric accuracy of the result. For non-metric cameras, the distortion at the edges of the images can be as high as 200 μm. In the case of high image overlap, self-calibration can be used [44,45]. The digitization parameters are related to good map design. The use of the black line, black dash and black dot in cartography is described [46]. Web presentations represent an increasingly popular means for the public to access three-dimensional globe models in a virtual environment and work with them. These presentations can be created in three ways.

The first option is to make the digitized globes accessible in the form of georeferenced maps using one of the web services designed for this purpose. The Tile Map Service is an example of such a service [33].

The second option is to view a three-dimensional model in one of the graphic formats directly in the web browser. The graphic format determines how and what data are saved in the file. Each format has its own specifics and is suitable for a particular application and purpose. The most commonly used 3D graphic formats are obj, x3d or glb.

The obj format was developed in 1990 by Wavefront Technologies. It is an open text format that describes the geometry of one or more objects. The color and texture data in-formation is usually stored in a separate mtl file, which is distributed with the original obj file [47].

This format is used in most 3D software such as Maya, Blender and 3ds Max. It is suitable for exporting objects, but not for exporting animations [48].

The glb format is a binary file format used to share 3D data. It contains geometric da-ta such as vertices, surfaces and normals. However, it may also contain additional information about models, lighting, materials and animations. The glb format is developed by the Khronos Group, a non-profit consortium that creates open standards in 3D computer graphics or virtual and augmented reality [49].

Extensible 3D, abbreviated x3d, is an open format from the Web3D consortium. It al-lows you to store 3D objects and scenes. It is used, for example, for the visualization of scientific data, multimedia presentations or websites with 3D content. It replaces and extends the original Virtual Reality Modeling Language (VRML) format, which was de-signed primarily for virtual reality, to include new features [50].

The glb format has the advantage over obj and x3d in the better rendering of object shape. This is because it uses not only triangular meshes, but also points or edges. All the elements of a 3D scene are stored in one compressed file, so it speeds up loading in web browsers.

In terms of technology, HTML5 programming language is used nowadays, which allows for rendering graphical elements directly in the core of the web browser [51]. Open-source libraries based on JavaScript and WebGL (Web Graphics Library) web standard are used to render 3D graphics [52]. JSC3D Viewer, the Google Earth plugin or the JavaScript libraries Cesium and X3DOM appear to be the most widely used tools [51,53].

The last option for presenting a 3D model of a globe is the 3D pdf format. This is a document type created by Adobe in the 1990s that can be opened in the freely available Acrobat Reader. This presentation method allows 3D objects to be displayed, as well as zoomed in on, rotated from any angle and interactively manipulated [54].

One of the most popular options for displaying a virtual globe in a web browser is the Cesium JavaScript library. It is available under the Apache 2.0 open-source license for commercial and non-commercial use. It supports many formats such as obj, glb, 3D tiles and more. It is a solution that supports the modification of globe display parameters—including the combination of map layers with a digital model of the Earth's relief—by calling the CesiumJS functions. Cesium makes full use of web technologies, and, unlike software solutions, it does not require any installation, only a web browser [51].

JSC 3D Viewer is a freely available browser based on JavaScript technology. The program uses spatial models that are stored in an obj file. The model can be rendered using points, triangles, a flat surface, smooth flat surface or may be textured. The web browser loads the entire model, saves it in its cache memory and displays it. JSC Viewer is easy to implement in a web browser without the need for any installation [55].

X3DOM is a program library created in JavaScript. It allows for the easy integration of three-dimensional models in x3d formats directly into the web browser. Using the JavaScript language, it is possible to modify the parameters of the scene view, as well as the appearance of the 3D model and more. Like Cesium and JSC 3D Viewer, X3DOM does not require the installation of plugins. The library can be used free of charge for non-commercial and commercial purposes. Detailed description of the X3DOM library is described [56,57].

It is necessary to specify the end users and their demands for making the globes available on the web. The added value of virtual globes compared to traditional objects can be summarized as follows: interactivity, navigation, object as interface, exploration, customizable to user needs, updatable, dynamic animation, multimedia integration [58]. The authors of the web presentation drew/derived primarily from good practice examples of other virtual globe collections, as well as from communication with users. The necessary financial, human and time resources have not yet been obtained to carry out a detailed survey of user needs. The web interface for viewing the globes contains tools related to the experience of cartographers and programmers. It has to be user-friendly and address their spatio-temporal questions for the answers they need to function in their daily life, employment or education [59]. It is assumed that users of the virtual globe collection have a basic level of technical knowledge and have an idea of how to use the online map or globe and also how to access the information. Most of these users are probably influenced by the availability of Google Maps [60]. In general, globe information offers many topics related to geography, cartography, history, astronomy or toponomastics. The restorers can also use the objects to repair and reconstruct globes. In this case, working with users is based on a partnership and collaboration with experts and is aimed at educational institutions. In previous years, the Map Collection organized a series of competitions for the public on the topic of georeferencing old maps. Therefore, it is known that most people interested in these cartographic materials access them from home in their free time and use the site for their personal use [61]. On the other hand, interviews with curators and teachers during school excursions in the Map Collection of the Faculty of Science of Charles University show that the globe's website is also used in teaching. Considering the demands on the quality of data transfer, it can be assumed that users tend to be well technically equipped, not only with good quality computers, but also with good data transfer speeds, as is common in the academic community.

Slow loading of the 3D model can be a significant limitation for working with web pages. If the model does not load in real time, the user may even leave the site and probably not return. Therefore, website speed optimization performance is an important parameter for making the 3D model available. Most web browsers offer a simple tool in their "Developer Tools" to assess a website's loading speed and weak points. Website loading time can be measured using three criteria: Domcontentloaded, Load and Finished. It is made possible thanks to WebGL's ability to work with so-called DOM (Document Object Model) elements [49,62]. DOM is an API allowing access to or modification of the content, structure or style of a document.

The "Domcontentloaded" event indicates the point in time when the DOM has been completed but the external resource has not yet been loaded. The "Load" event indicates when each individual resource (img, glb, style, script, etc.) has been loaded into the browser. "Finish" refers to the period of time from the start and end of loading the 3D model data after each resource has been downloaded.

1.2. Case Study: The Globe of Josef Jüttner

A relatively large globe created by Josef Jüttner in the first half of the 19th century was chosen for the case study.

Josef Jüttner (1775–1848) was a mathematician and military cartographer of the Austro-Hungarian Army. He later worked in Prague as a globemaker and publisher [63]. He also published maps, for example *Grundriss des . . . Hauptstadt Prag*, on two sheets (1811–1815). He created a 12-inch celestial globe in 1:29,000,000 (Prague, 1822, with Franz Lettany) and a new 12-inch celestial globe (1824). An armillary sphere called *Ringkugel* was published in 1828. The Jüttner-Lettany globe from 1822 was reissued in Prague in 1840. Jüttner continued to publish 24-inch pair globes in the 1836–1839 period. The terrestrial globe was republished in 1846 [64].

The subject of our study, i.e., the terrestrial globe called *Erdkugel*, with a diameter of 63.5 cm, a circumference of 199.5 cm and a height of the stand up to the brass meridian circle (measured perpendicularly from the base) of about 41.8 cm, originates from 1839. The full transcription of the German text of the cartouche containing the globe's title (Figure 1) and author reads as follows: *ERDKUGEL von 2 Wiener Schuh im Durchmesser nach den neuesten geographischen Bestimmungen entworfen und herausgegeben von Joseph Jüttner Oberstlieutenant im Kais. Königl. Bombardier Corps; in Kupfer gestochen von Bernhard Bitter und Johann David.*

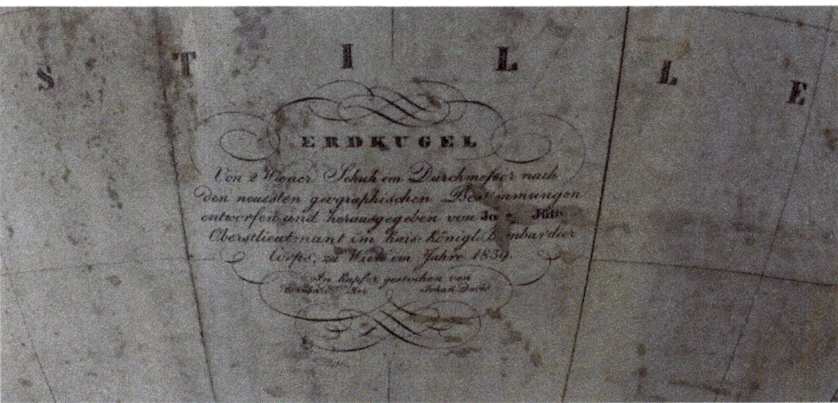

Figure 1. The German text of the cartouche. Source: Map Collection of the Faculty of Science, Charles University.

(Translated: The globe measuring 2 Viennese feet in diameter reflecting the latest geographical knowledge was designed and published by Josef Jüttner, Artillery Lieutenant Colonel; engraved in copper by Bernhard Bitter and Johann David.)

J. Jüttner published the work in Vienna at the scale of 1:20,050,000. Bernhard Bitter and Johann David from Vienna are listed as engravers [64]. The globe is made of gypsum, covered with a copperplate print of 18 paper gores and two polar spherical caps. A wooden lacquered tripod on wheels serves as the base. The globe is encircled by a brass meridian circle with engraved degrees (4 × 0–90°). Furthermore, the instrument is equipped with a horizontal wooden circle illustrated by a calendarium and Zodiac signs.

A detailed metadata description in MARC 21 format was created for the Central Catalogue of Charles University. From it, the basic metadata for the web presentation of globes were extracted (in the Appendix A, Table A1).

The globe was restored in 2016 by Miroslav Široký as it was illegible and damaged in many places [65]. The stand had also seriously deteriorated. The repaired globe was inserted into a triangular base. Currently, the historical teaching aid is in excellent physical condition (Figure 2).

Figure 2. Overall view of the restored terrestrial globe by J. Jüttner, 1839. Source: Map Collection of the Faculty of Science, Charles University.

1.3. Contents of the Globe

Detailed imaging allowed for better identification of the contents of the globe. The zero meridian passes through the east of the island of El Hierro, i.e., through the Canary Islands. The globe features a geographical grid, and the major parallels are marked: the Equator, the Tropics and the polar circles.

The colors of the globe have already faded, especially those of the water bodies where the blue color is preserved only along the coasts. The Earth's surface was rendered in greyish brown. The elevation of the mountains was represented by shading in a brown color and the density of slope shading.

The contemporary content reflects the geopolitical division of the world and the geographical knowledge of the time. The globe contains information about geographical discoveries. For example, the following note appears regarding the Philippines: *Die Filippinen wurden im Jahre 1521 von Ferdinand Magellan entdeckt. Er und mehrere seiner Officire wurden von den Eingebonen gohödtet. Sebastian del Cano setzte die eisen ach Ostindien fort, und war der erste der die Welt umsegelte. (Translation: The Philippines were discovered by Ferdinand Magellan in 1521. He and several of his officers were killed by the natives. Sebastian del Cano continued his voyage to the East Indies and was the first to circumnavigate the world.)*

India is named as *Hindostan*. Tibet, *Nepal*, the Himalayas (*Himalaya Gebirge*), the eastern Indian coast known as *Coromandel*, or the island of Sri Lanka (*Ceylon*) are also shown in the area.

In the east, the description of *Kamtschatka* is emphasized, and present-day Alaska, which was still part of the Russian empire known there as *Russisches America*, is seen.

The continents and the countries within them are marked selectively. Nevertheless, the globe map looks cluttered in some areas and on the coasts. The work also contains a detailed description of the countries in the United States of America (*Vereinigte Staaten*), marked by Roman numerals from I to XXX. Numbers are assigned similarly to the states in Mexico (I.–XIX.) and in Central America (*Mittel America, I.–V.*). In South America, the Amazon River is marked, as well as the following states: Columbia, Guyana and Brazil.

Australia is only presented as an integrated land body without any further geographic structure, just the older name for the territory, *Neu Holland*, is included.

In Africa, the Niger River, Lake Chad (*Der Tschad*) is drawn, but the Congo River is not yet described, which is consistent with contemporary knowledge. An infamous slave trading post is shown on the west coast of the African continent (*Gold Küste Sclaven*, Figure 3). The Pacific Ocean is called *Das nordliche Stille Meer*.

Figure 3. Detail of the researched globe by J. Jüttner. Source: Map Collection of the Faculty of Science, Charles University.

2. Materials and Methods

The high-quality perception and interpretation of the work by the user is related to the appropriate parameters of digitization and three-dimensional visualization of the selected object. The basic rules of graphical map design were established in the middle 16th century in commercial Dutch printing houses. The design aspects (contrast, density, weight of black graphical elements, dot and dash pattern and comprehensive use of black line, dash and dot) are analyzed in [46]. The following parameters were determined for taking the images of the globe for the interpretation of these graphical features on the 3D model (Table 1).

Table 1. Parameters for image acquisition and 3D visualization of black-and-white graphics. Source: Map Collection of the Faculty of Science, Charles University.

Globe Design	Graphic Representation	Geographic Element	Pattern
Contrast	Short lines with horizontal dashes of coast, hatching of coastline, white and yellow land	Land and sea resolution	

Table 1. *Cont.*

Globe Design	Graphic Representation	Geographic Element	Pattern
Density (readability limits)	Varied groups of black elements, different density and various thickness of lines, dashes and dot sizes	Mountains, hypsography, coastline	
Weight of black graphical elements	Number of dots or lines, shades of grey, width of lines, changes in line thickness, size of dots	Edges of desert oases, hypsography, lakes, roads, state borders	
Dot pattern	Regular dots	Edges of desert oases, shallows around islands	
Line pattern	Fine line drawing	Rivers, coastline, outlines of countries, continents, roads, geographical grid	
Complex use of the black line	Elements tying together the graphic of the globe	Hydrographic network, coastline, roads, state border	
Complex use of the black dash	Regular and irregular delicate dashes in direction	Movement of flowing rivers, water surfaces, degrees of zero meridian, hypsography	

Table 1. *Cont.*

Globe Design	Graphic Representation	Geographic Element	Pattern
Complex use of the black dot	Elements tying together the graphic of the globe	Coasts (for islands, e.g., Bank von Neufundland), deserts (for oases), edges of unexplored areas (e.g., Antarctica)	

In accordance with the approach chosen to perform the digitization, a contactless imaging method was used to avoid damaging the surface of the rare and unique globe. Therefore, the globe was placed on a special static base, while the camera could move around it. The aim was to minimize any manual handling of the globe. A non-metric digital camera was chosen to perform the imaging. The camera must fulfil the following parameters: (i) be a high-quality image sensor with excellent detail rendering, (ii) be able to take images of high-contrast subjects, (iii) have a low level of noise, (iv) focus in low light conditions, (v) a high-aperture lens and (vi) greater depth of field.

The same focal length and thus the same distance from the object was set throughout the photographing to maintain the constancy of the IOP. In order to capture all the detail, the globe was not photographed as a single whole but was split into smaller segments (Figure 4).

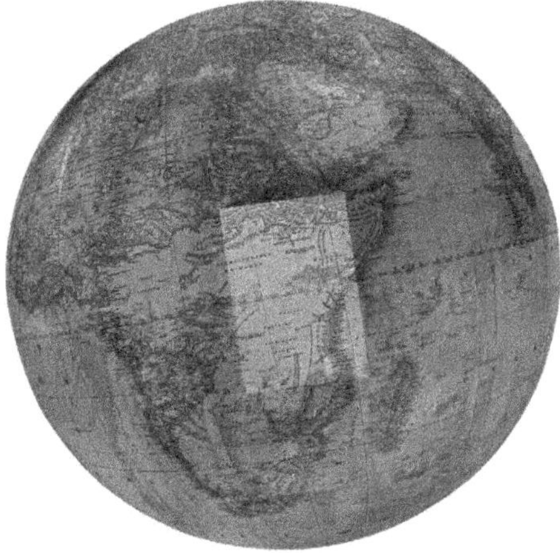

Figure 4. Dividing the whole globe into segments. Source: author, Agisoft.

In the Virtual Globe Collection, the data is cartographically and photogrammetrically modified against geographic information systems (GIS). Functionality will be deliberately limited and focused on users for whom analytical and visualization functions will be customized. The final users are non-expert GIS users who are able to use the interface independently and comprehensively, check functions and data and do not need instructions or a narrative workflow. They focus on the manipulation, analysis and presentation of data. The original data are processed raw and unmodified into the primary model during a short or long processing period. While non-experts use a basic interface, the control of functions and data is performed by the authors of the interface, because users need significant support, tutorials and a narrative workflow. The main focus is the visualization of globes; data must be modified and integrated into the web browser. Users work with a secondary model; the response time must be very short and the purpose of their work is very specific [57].

On the Map Collection globes website there is a counter tracking the number of visits. A "visit" is defined as viewing a page in a browser for a certain period of time, in this case 30 min. This means if a user returns to the site from the same computer after more than 30 min, a new visit is counted. It is possible to track user accesses since 2014 (10,800), weekly (average 17) and daily (average 6). Between 2014 and 2021, the average annual decline in visits was 30%. The highest number of visits was recorded in 2014 (3594) and the lowest in 2021 (249). In 2022, the number of users (329) will increase to approximately the 2019 value after the end of the COVID-19 measures (Figure 5).

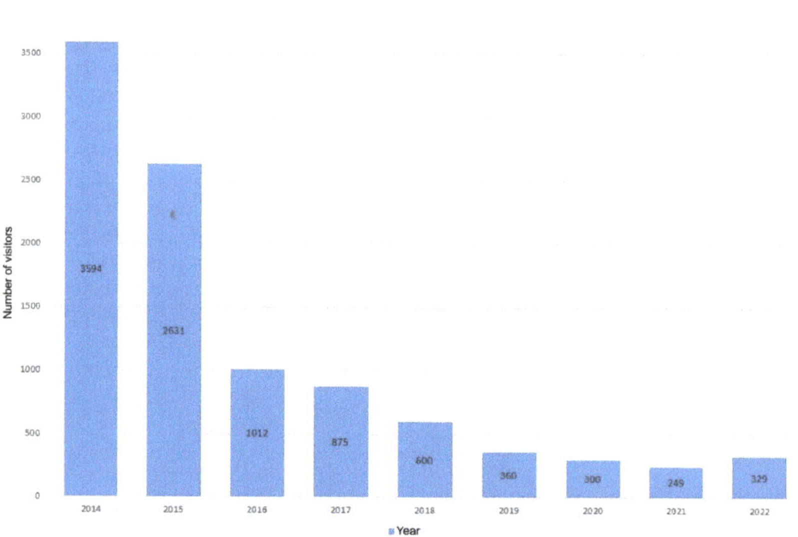

Figure 5. The number of visits for the virtual globe collection in 2014–2022. Source: Map Collection of the Faculty of Science, Charles University.

The Cesium JavaScript library was used to display the spatial model on the web. This library was chosen due to its ability to display 3D file formats (glb, obj) and its easy implementation in a web browser. The ability of this solution to publish source files (the 3D model and the Cesium library) on a selected web server and the support of displaying the files in most web browsers were also important factors. The installation of the Cesium library on the server is not very demanding. The latest version of the installation files was downloaded from http://cesiumjs.org (accessed on 15 May 2022). A folder of approximately 35 MB in size contains all the necessary libraries and documentation. This folder and the 3D model file (glb) were uploaded to the web server and prepared for further editing and basic setup. The detailed process of digitization, visualization and accessibility of the globe is shown in Figure 6.

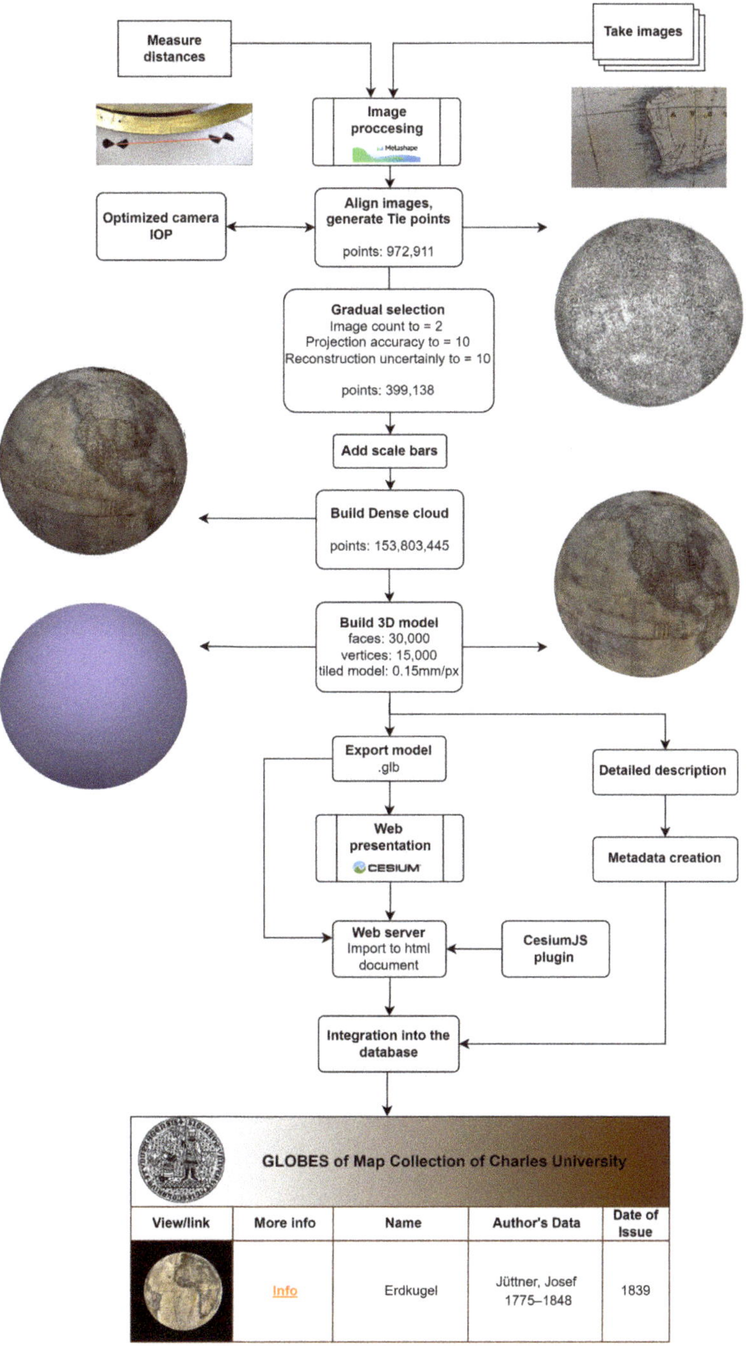

Figure 6. The processing steps of digitization, visualization and accessibility of the globe. Source: author.

3. Results

3.1. Digitization of the Globe

SfM-MVS photogrammetry was used to digitize and visualize the Jüttner's globe due to its size.

Imaging was performed with a Nikon D750 digital non-metric camera (DSLR) according to the principles of SfM-MVS photogrammetry, with a fixed focal length of 85 mm (Nikkor lens AF-S, 85 mm, f/1,4G, Nikon, Tokyo, Japan). This camera fulfils the conditions described in Materials and Methods. The images were taken in the aperture priority mode because the higher the f/stop, the smaller the opening in the lens; the greater the depth of field, the sharper the background. At the same time, the focusing distance was maintained the same, thus IOP did not change at all. The minimum focusing distance for this type of camera is 850 mm. In order to obtain the detail, it was necessary to reduce the focusing distance, so the extension tubes were used. The extension tubes contained no optical elements; thus, they did not affect the image quality or change the focal length (Table 2).

Table 2. Camera parameters. Source: The Nikon D750 manual.

Image resolution	6.016 × 4.016 px
Sensor size	full frame (35.9 × 24.0 mm)
Focal length	fixed 85.0 mm
F-stop	f/10
Pixel size	0.006 mm
Image format	raw/jpeg

Images were taken sequentially along the meridians with an 80% overlap. So a high overlap is necessary for the optimization of the camera (IOP calculation). The configuration of the images can be seen in Figure 7. Control points, i.e., round white-black targets, were placed around the globe and the distances between them were measured. A total of 820 images were taken in raw format. The sharpness and color of the images were additionally checked and edited (sharpening, white balance and color) in Zoner Photo Studio X. Finally, the images were converted into jpeg format for further processing.

Figure 7. Image configuration. Source: author, Agisoft.

The image processing was performed on a PC (Intel Core i7 3.8 GHz, 64 GB RAM, NVIDIA GeForce GTX 1650 SUPER, Win10) in Agisoft Metashape Professional (Agisoft) version 1.8.2.0, which is suitable for photogrammetric image processing and uses the SfM-MVS technique. The first step was the alignment of the images. First, feature points were identified on the images and then linked across all images to tie points. Subsequently, a dense cloud was created. Point cloud cleaning was performed according to the parameters described in Figure 6. The size of the point cloud and the resulting model was determined using measured distances between ground control points. The next step is to generate a 3D mesh and textured it into a 3D photorealistic model (Figure 8). The results of the individual steps are summarized in Table 3. The resulting model was exported into a glb format.

Figure 8. 3D textured model. Source: author, Agisoft.

Table 3. Overview of processing results. Source: Agisoft.

	Number	Quality	File Size	Time of Processing
Tie points	points: 972,911/399,138	Medium	314 MB	1 h 49 min
Dense cloud	points: 153,803,445	Medium	223 MB	1 h 05 min
3D model	faces: 30,000 vertices: 15,000	Medium	84 MB	55 min

3.2. Accessibility of 3D Model

One of the goals was the interconnection of information, the 3D model and the presentation on the web. For the purpose of displaying the 3D model on the web, the glb binary format was chosen. It is a compressed file that was chosen for its faster loading in web browsers. The Cesium JavaScript library was used to display the model on the website.

The file with the 3D model in the glb format and the Cesium library source files have been uploaded to the web server of the Geography Section of the Faculty of Science of Charles University and are accessible to the public (Figure 9).

After entering the web address, the web browser first shows the basic interface of the Cesium application. This resembles a space environment in which a 3D model of the globe is subsequently displayed. The scene is reminiscent of a starry sky, in which the model evokes a planet. Due to the size of the file (14 MB), the user has to wait a few seconds before the model is loaded into the web browser.

The model can be rotated as desired, as well as zoomed in or out. The bottom bar, which is a standard part of the Cesium application, is in the form of a timeline and is used to adjust the lighting of the model (Figure 9).

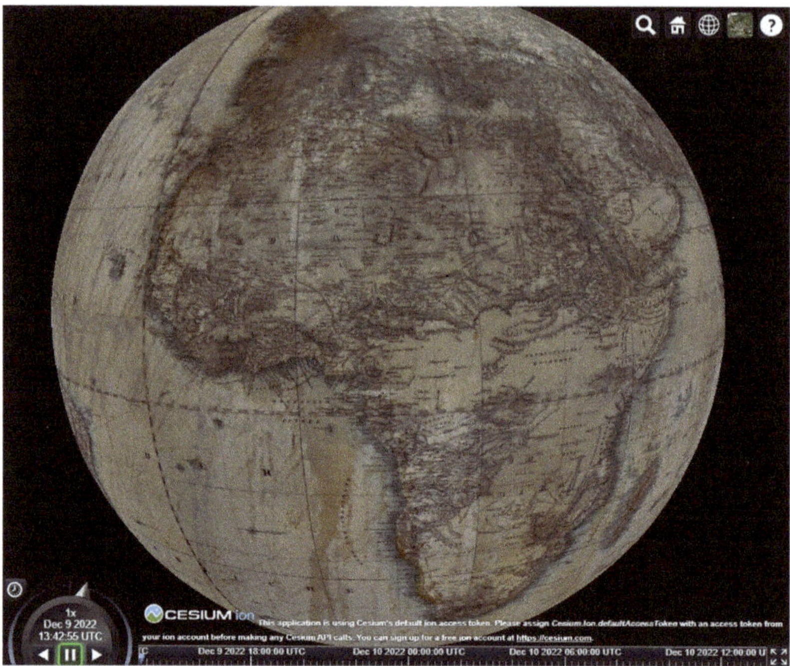

Figure 9. Example of access to the globe in the web browser. Source: Map Collection of the Faculty of Science, Charles University, Cesium.

Different web browsers can affect the speed of loading the 3D model. Therefore, a comparison of web browsers was made focusing on the speed of loading the 3D model on the web. The loading of the 3D model data was tested in Chrome, Firefox, Microsoft Edge and Opera browsers. The loading time of the website with the model in the glb format (14 MB) was measured with Cache disabled based on three criteria: Domcontentloaded, Load and Finished. The tests were conducted on a PC Intel Core i7 3.8 GHz, 64 GB RAM, NVIDIA GeForce GTX 1650 SUPER, Win10 using the internet network of the Faculty of Science. Table 4 shows the average values after the loading of the 3D model data 30 times by each of the listed browsers.

Table 4. Browser-specific loading time. Source: Map Collection of the Faculty of Science, Charles University.

Browser	Domcontentloaded (s)	Load (s)	Finished (s)
Chrome	0.578	0.913	1.057
Firefox	0.213	0.885	2.000
MS Edge	0.589	0.871	1.000
Opera	0.636	2.000	2.000

The Domcontentloaded event indicates when the DOM has been completed but the external resource has not yet been loaded. Load indicates when each individual resource (eg glb) was loaded into the browser. Finished refers to the period of time from the start and end of loading the 3D model data to the download of each resource.

3.3. Metadata and Selected Aspects of the Globe

A unique description of the Jüttner globe has been created, which has not been done to such an extent before. Based on this, user metadata were created (Table 5). These metadata provide information about the globe and its 3D model. They contain the following items: author, publisher information, scale, physical description of the globe, dimensional information, geographic grid, material, printing technique, color, format information, size and web presentation of the 3D model and more. The results have been integrated into the globe database of the Map Collection of the Faculty of Science of Charles University on the website [66]. In this way, the metadata, image information and 3D model of the globe have been interlinked and made accessible to the general public in Czech and English versions (the 3D model and metadata can be accessed at http://www.mapovasbirka.cz/globy/english/index_eng.html; Supplementary Materials).

Table 5. Globe and 3D model metadata. Source: Map Collection of the Faculty of Science, Charles University.

Information about the Globe	
Title	Erdkugel [cartographic document]/von 2 Wiener Schuh im Durchmesser nach den neuesten geographischen Bestimmungen entworfen und herausgegeben von Joseph Jüttner Oberstlieutenant im Kais. Königl. Bombardier Corps; in Kupfer gestochen von Bernhard Bitter und Joh. David.
Author's Data	Jüttner, Josef, 1775–1848
Engraver	Bernhard Bitter and Johann David from Vienna
Publisher	Zu Wien: Joseph Jüttner, 1839
Scale	1:20,050,000
Physical Description	1 globe: in color, cooperlate, on a wooden stand with a tripod, a brass meridian circle with an engraved scale, a wooden horizontal circle with zodiac signs; 63.5 cm in diameter.
General Comments on the Globe Height	Globe diameter: 63.5 cm, Globe circumference: 199.5 cm, Pedestal height: 41.8 cm, Total height of the globe: 105.3 cm.
General Comments on the Prime Meridian Position	The Prime Meridian passes through the Canary Islands.
Geographic grid	The globe has a geographical grid, and the major parallels are marked: the Equator, the Tropics, and the Polar circles.
Color of watercourses	Blue
Color of Earth's surface	Grey-brown
Altitude	The altitude of the mountain range is represented by shading in brown and the shading density of the slopes.
Material	Gypsum
Printing technology	A copperplate print of eighteen paper gores and two polar spherical caps.
Language	German
Legend	No
Restoration year	2016
Restorer	Miroslav Široký
Globe owner	Map Collection of Charles University
2D model	Link to the 2D presentation of the globe.

Table 5. *Cont.*

Information about the 3D Model	
Comments on the 3D model creation	The globe model was created using the photogrammetric method in the Agisoft software.
3D Model Format	glb
3D Model Size	14 MB
Web presentation	Due to the size of the model (14 MB), it is necessary to wait a few seconds for the model to load in the web browser. By using the slider in the bottom horizontal bar in CesiumJS, you can change the direction of the globe model's lighting.
Genre/Form	Globes
Autor of 3D Model	Map Collection of Charles University
3D model	Link to the 3D model of the globe.

The detailed photogrammetric processing by J. Jüttner for the terrestrial globe revealed completely new details, which were previously invisible to the human eye. Patterns for black dots, dashes and lines were identified. The dots were used to represent desert oases and also shallows around smaller islands, for example in the Pacific Ocean. The lines were used to represent elevation, as landscape dashes, to indicate the rivers or to separate administrative areas on the map. The lines were used to represent the outlines of continents, lakes, rivers and roads. The rivers have thinner simple lines at their headwaters and thicken towards the estuary, where they diverge into several rivers in the estuary.

The density of the graphic elements was used to highlight the elevation where the lines thicken or even cross; then, in the case of dense dots, the edge of the oasis is marked. The artist achieved the contrast of the graphic by depicting the light area and the blue edges of the sea along the coast and by using short parallel lines. The complex use of the black line on the globe map is visible in the elaborate river networks. Irregular lines in the direction of flowing rivers refer to the complex use of lines. The complex use of the black dot is particularly visible in desert areas, where small and large, dense and sparse dots are depicted to achieve the effect of grey and dark grey.

A very interesting and new result is the visibility of factual data on the globe, which were previously not visible to the human eye. They open completely new possibilities of using this cartographic artwork. They reflect the contemporary state of scientific geographical knowledge. Geographical discoveries with names and dates are depicted here, e.g., the discoveries of W. Barentze (von Barentz 1596 entdeckt—Figure 10, Neu Seeland von Abel Tasman entdeckt 1642, Neu Georgie. Endeckt von Roche 1675, wieder aufgesucht von Cook 1775 u. Bellingshausen 1819).

The territory of the Unknown Land (Unbekannte Laender) is depicted in the area of the African continent (Unbekannte Laender). Curiously, there is a description of the occurrence of giraffes (Ebene voller Giraffen) in southern Africa (Figure 11).

The previously mentioned dashed lines depicting national borders show the contemporary geopolitical division of the world. It is also possible to recognize detailed descriptions of contemporary German geographical names (Figure 12). This is given in normal type, capitals, block capitals and italics. The font size and fillers are also differentiated according to the meaning. Especially on the coast, the descriptions of ports are so dense that they resemble old portulan maps (Figure 11). Toponyms in Europe are so oversaturated that they are almost unreadable (Figure 13). The author was obviously aiming for completeness, but a bit of generalization would be appreciated by the user.

Figure 10. Barentze' discovery—von Barentz 1596 entdeckt. Source: Map Collection of the Faculty of Science, Charles University.

Figure 11. The description of the occurrence of giraffes. Source: Map Collection of the Faculty of Science, Charles University.

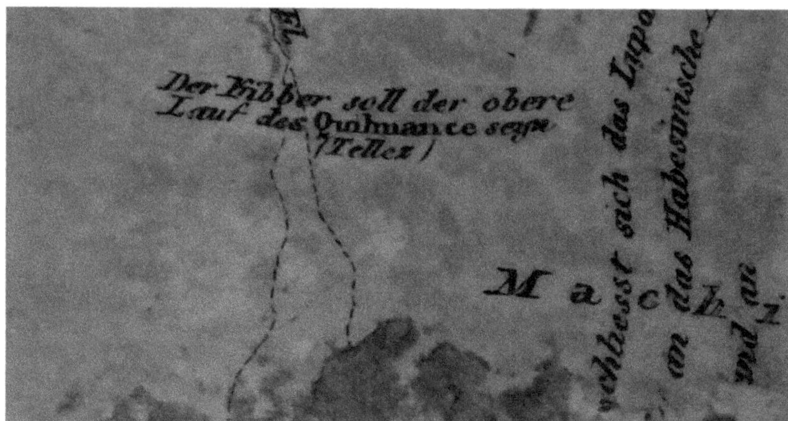

Figure 12. The detailed descriptions of contemporary German geographical names. Source: Map Collection of the Faculty of Science, Charles University.

Figure 13. The example of dense description. Source: Map Collection of the Faculty of Science, Charles University.

These types of old globes should be represented in the models on the web by individual cartographic schools (Arabic, German, Italian, Dutch, French, Austrian, etc.). Furthermore, the globes should be visualized in great detail and in true colors. They should not be artificially colored, lettered or re-edited graphically in order to preserve as faithfully as possible the original intention of the author of the globe. Of course, it also depends on the original, because it may have been artificially devalued by non-expert restoration. Attention should also be paid to the user's perception of the globe. For example, it is not appropriate to zoom in on the globe to unreadable details that do not add new information. Critical comments on the digitalization and content of the globe are ideal, as well as links to relevant information about the globe and its author. These cartographic artworks were produced in manufactories, so minor differences are always to be expected. It can be assumed that wear has occurred, especially in the upper parts, i.e., in the northern hemisphere. Each digitized globe may be important for understanding parts that are already damaged or unreadable on another globe. Therefore, there should be no obstacle to digitizing the same globe by the same author and from the same year. On the other hand, it is desirable to digitize as many of the available old globes as possible for successive comparisons, which we lack so far.

4. Discussion

Many globes contain very little descriptive data of their own. Therefore, good quality visualizations can play a crucial role in the identification of a particular globe, especially in the case of older works. The virtual rotation and zooming of the globe or the celestial sphere make it easier for the researcher to read and verify the title, author and publisher details, as well as other supplementary texts that can be found on old globes. Such texts contain information about the history of the globe, or the state of geographical and cartographic knowledge at the time of the work's creation. Researchers can usually discover a lot of details that map collection staff cannot process when creating general metadata.

The digitization of globes is a significant contribution to the creation of scholarly digital editions (SDEs). They are a valuable historical cartographic source for many specialists. The digital globe edition in the web environment provides and integrates three levels of data and information, namely pictorial (globe as an image), factual (globe as a source of information) and secondary (globe metadata).

A globe, seen as an image or object, also represents a map when turned into a two-dimensional image. It is possible to add additional information or make a link to the original globe map image, if available. Globe maps are also often used to reconstruct unavailable old globes. However, this fact needs to be mentioned in the metadata in order not to confuse researchers. Old globes are interesting art objects illustrated with not only map symbols and compass roses, but also with animals, people, sea monsters and large sailing ships. The question remains of how to describe individual images on the globe or how to search for them automatically.

Old globes are not only cultural artifacts, but also a source of much factual information for historians, librarians, geographers, cartographers, philologists, teachers, artists and others. They can be used to examine information related to the environment, ecology or socio-economic phenomena such as land-use development and changes in or reconstruction of river flows or coastline shapes. They also record geographical names as well as cultural, political and economic facts. They are an important source of information on the stage of development for technology and surveying and for cartographic knowledge at the time of the document creation. Last but not least, they are an essential study material for the history of the development of science and scientific instruments. Here, the instruments can be connected to similar ones or to information on the development of science and technology in a given period (for example [67]).

As a source of factual information, the globe offers specific data, whether provided on a physical, political or celestial map. It shows rivers, ponds (some no longer existent), the shape of the coastline, which was also subject to changes, or sea currents. There are mountains and mountain ranges with the information, names and elevations of the time. There are images of woodlands, often depicted with very specific, unified features. There are historical toponyms. Philologists will appreciate the possibility of studying the development of the national geographical nomenclature over time. The extent to which descriptions and comparisons of historical toponyms on globes should be carried out remains a matter for discussion, and how to directly link the search for unified toponyms with the digital globe is also open for debate.

The map legend explains to the researcher the size and status of settlements, the economy and the political context. It may also contain statistical information. It should be transcribed in detail in all its original language versions and ideally also translated into the current language. All this can serve to annotate the globe in detail.

Celestial globes are important for the history of astronomy. Astronomical globes depict ancient constellations, allegories, ideas about the size of stars and so on [68]. For celestial globes, it will be necessary to address the issue of the unification of nomenclature so as to avoid any confusion of terms. The history of globes is also connected with decorations in the ceremonial halls of aristocratic mansions, in monastic libraries as a symbol of education, etc. Historians or archivists can research the history of ownership, or the actual paths taken by these art objects and the reasons behind this movement. The projected globes could theoretically be linked to information on original mansions, castles, chateaus or presentation libraries, with their history and interiors.

Globes have been used as teaching aids since the 19th century and were usually supplemented with instructions for use and examples for teaching geography. They are valuable for research into the history of the didactics of geography and mathematical cartography. They acquaint researchers with the state of knowledge in the field and the quality of teaching at various types of schools where these aids were compulsorily introduced. In this connection, it is also possible to examine how these aids are promoted by the ministries of education, and how well the schools and school offices are equipped. Comparisons can also be made within different regions. In this case, the digital globes could be supplemented with full texts of scanned manuals, textbooks, ministerial decrees or links to archival material, all in compliance with copyright and proprietary rights. Examples of period school offices can also shed light on the history of the use of globes.

Globes provide information on cartographers, engravers, translators, publishers and makers. Cartographers who produced globe maps were usually involved in the production of old maps, and this production complemented their wider activity in publishing maps, atlases and globes. This information offers a wide range of options for making links to digitized biographical dictionaries, bibliographies of authors, information about their lives, and their images.

Additional factual information can be obtained in the fields of book culture, the history of publishing houses, their production and the production of school geographical aids in the region concerned. Of equal interest is the development of trades, the origins of a company and the beginning of and the reason behind the production or sale of a globe. This information retrospectively helps to identify the missing data on the date of the publication of the document. The researcher can also examine the printing techniques used to make the globe map. If in some cases the printing plates have survived, then valuable hands-on experiments can be made with reprints of the work. Even without an imprint, it is possible to reconstruct no longer existing globes that have survived only on printing plates and to use them as a source of new information. This is why printing plates and stones also become items of map collections. Here, it is quite appropriate to make a link to dictionaries of printers, publishers and companies involved in the publication of globes or teaching aids, as the case may be.

Information about the manufactories producing globes is essential for the history of the industry. Of interest to humanities studies is the impact of the war on small manufacturers and education systems, and the interventions of totalitarian regimes and censorship in the production and creation of these teaching aids. Some patented inventions, such as composing globes, are an interesting part of the history of patents [69].

Furthermore, it is possible to study the development of geographical instruments in their entirety; that is, globes, both terrestrial and celestial, planetariums, armillary spheres and sololunaria. The question is how to digitize and present these complex devices. The digitized objects can be used to create further reconstructions of possible usage, not only to enable rotating and zooming in on them, for example [70].

Like the bibliographic description of an old map, the metadata description of a globe should be improved from the simplest to the most complex form of description. Metadata not only enable subsequent retrieval by the user, but in the event of damage or even loss, they can provide very accurate information that can be used to reconstruct or trace the given collection item.

Globes are usually described individually, but there are cases where they are described as a set or pair. Indeed, old globes were usually made in pairs, that is, as terrestrial and celestial globes [71]. Therefore, it is advisable to try to track down information about the second, paired globe in information sources and mention it at least in the metadata description, if there is no online availability of the paired globe. According to the available information, J. Jüttner did not make a star globe as the second in the pair. Therefore, this matter need not be given further attention.

In addition to basic descriptive information about the globe, such as the title, subtitle, author, secondary authors, scale (star globes do not have scales), place, publisher and the year of publication, the physical description fields (diameter, circumference, base) are much richer for globes than for conventional maps. This is due to the different materials the globes are made of. The accuracy and quality of metadata are very important, especially in the physical description, because the researcher of a digitized product does not have access to the original physical product. Important is the exact size, and the globe circumference from which both the diameter and the scale are derived.

From the 15th to the early 19th century, intaglio copperplate engraving techniques were used in cartographic production. These techniques involve making indents or incisions into a metal plate surface (most often copper) which hold the ink when ink is applied to handmade paper. The quality of the printing plate determined the quality and even distribution of the printed ink. The quality of the print was also influenced by the intensity of the paper wetting, the ink density and smear, as well as the smooth passing of the print through the intaglio press. The scale and dimensions of the original globe map may therefore vary from source to source, as may the dimensions of the gores, depending on the time and method of obtaining the relevant data and the shrinkage of the handmade paper for a particular map. Similarly, the plaster base, the hemispheres, were handmade, and this could also cause differences in the diameter of the base sphere. Therefore, it will be very useful in the future to compare similar specimens of the same globe. This could shed light on what variations can occur in handmade globes. Further interventions by restorers, especially wet process cleaning, can lead to changes in the paper or in individual gore joints. It is therefore a good idea to take these into account.

As far as further processing is concerned, the globe gores were cut out by hand, so they could also have been different sizes before gluing. They were glued on a plaster or paper base material by hand. A high-quality globe product was characterized by the precise connection of the individual globe gores in the map image as well as in the description of geographical names. In inaccurate products, the globe gores were glued sloppily or did not connect in some parts. The question is whether to make (or were made) corrections of these manufacturing errors during digitization, when the segments can be accurately connected.

Information on the size of the base, and its height below the globe, should also be accurate. It is equally important to know whether the globe is furnished with a meridian semicircle or circle, how it is divided into gores and whether it has an hour circle, compass or calendarium. When digitizing, the meridian circle as well as the hour circle are often removed from the sphere or digitally eliminated. It is also appropriate to ask how to incorporate these parts into the digital presentation of the globe. Whether to settle for a photograph, a description or whether to try to combine them into one object remains an issue.

If possible, the map legend should be detailed in the digitized object's metadata. We can also expand and add information to the note fields. These include characteristic features of decoration, such as portraits, scenes, allegories, parerga and cartouches. It is possible to add literature references that cite the source of studies on the globe or the existence of the second globe to the pair. To describe in detail the method of manufacturing and the material, you may pay detailed attention to the content and write down interesting toponyms or factual information that is described on the globe. The note fields also include information on restoration efforts, repairs and changes to the globe in terms of connections of paper gores, color, material or replica stands. It is also a good idea to include the history of ownership and provenance of the globe, if known. Additional notes should be made regarding language, edition, version, reprint, etc. A lot of other information can be drawn from supplementary documents such as the cover, information manuals, etc. A note on a defective specimen informs about the physical condition of the globe specimen, especially its damage.

The user should be knowledgeable of what type of globe they have access to. The objects are classified by: displayed object (terrestrial globes, planet globes, star globes), purpose (school, decorative, etc.), content (geographic and thematic), method of execution (smooth, embossed, luminous, duplex globes), induction globes (teaching aid without map content), free globes (have no axis and sit in a bowl), pneumatic (inflatable), processing technology (glued, injection-molded) or scale. So far, only celestial, terrestrial and embossed globes have been used for digitization purposes.

It is essential to always describe changes in the document history caused by restoration, preferably with pictorial documentation, original damage, various crooked paper gluing, inaccurate connections of the geographical names descriptions that have been corrected, different sizes or changes in the printing paper or other printing and manufacturing errors.

The physical condition or unique nature of a work may be crucial factors to be considered, as it may not be feasible to regularly lend out such works to the public due to security reasons or other restrictions imposed by restorers.

Many map collections are unable to describe their holdings for financial reasons. This is especially true about large archival or museum collections. Therefore, attempts have been made to digitize and display the globes owned by such institutions with minimum metadata, as has been performed, for example, in the Virtual Map Collection [19]. Although metadata are clearly highly beneficial for all digital representations as described above, in these special cases, even displaying a globe on the Internet with minimum descriptive metadata seems to be highly beneficial as well.

By virtue of such accessibility, the work has been recorded at a certain point of time, and thus documented. According to the depth of their interest and knowledge, the researchers can verify and complete the description of the work. If the work was to be lost or damaged in the future, there is a real chance of making a copy or at least creating a detailed description of it. Then, this can make it much easier for the police or the original owner to locate the work.

The method of digitization and visualization described above was tested on the antique Jüttner globe. There are other specimens of this globe in the Czech Republic. In the future, it would therefore be useful to find out about their physical condition and compare them physically. Furthermore, an attempt should be made to digitally document and compare them.

The suitability of SfM-MVS photogrammetry for the digitization and visualization of old globes was demonstrated. The method is simple and there is no need for expensive specialized equipment or software such as laser scanners or expensive metric cameras.

The globes can be digitized on site and do not have to be transported to a specially designed facility. The method is contactless, posing no threat of damage to rare old globes. Photographing can be performed with digital cameras now commonly available and boasting sufficient resolution to capture even the smallest details on the surface. Any number of images can be recorded and processed. Image acquisition is relatively quick compared to the computer conversion of data into a 3D model.

The relatively high processing power required of the computer used for processing the images appears to be a disadvantage of this method. The method cannot be applied in a universal way with the same setup and implementation. It always depends on the size of the object to be digitized, the possibility of handling the object and the site of the imaging. The lighting of the photographed scene was a serious problem. In the case of smaller globes, the use of a diffusion tent to improve the properties of the light can be useful. This is one of the reasons why all images had to undergo additional editing. The quality of the final textured model depends on the quality of the images. The quality of the camera used is a closely related factor. A significant problem with cheaper cameras is lens distortion, which affects the geometric accuracy of the model. Every camera must be calibrated before use.

The digital 3D model of the Jüttner's globe was made available on the web using the Cesium JavaScript library. The advantage of this solution is the possibility to alter the source code and modify the visualization parameters of the globe. The solution makes full use of web technologies, and unlike purely software-based solutions, it does not require any software installation, only a web browser. Another advantage is the easy implementation of the model of a web browser. A common disadvantage of the internet presentations discussed here is the user's dependence on an internet connection.

The problem with making this globe available online was the considerable size of the resulting 14 MB glb format model. This resulted in the lengthy loading time of the model. This problem could have been eliminated by using image pyramids, which would have increased the speed of image data processing.

The tests performed showed the effect of using different web browsers on the speed of loading the 3D model and the entire website. They showed how quickly users can view content and interact. The resulting values are dependent on the speed of the internet connection and the performance of the PC, and are therefore not absolute. However, the speed of different browsers under the same conditions can be compared with each other. From the point of view of loading the model itself (Domcontentloaded and Load), Firefox emerged as the fastest browser. In terms of loading the entire website, the test identified MS Edge (1.000 s) and Google Chrome (1.057 s) as the fastest browsers.

When making globes available on the Internet, it is always important to respect the copyright and property rights associated with the use of the data in accordance with the law and licensing terms. However, old globes, published before 1850, are not affected by copyright restrictions.

5. Conclusions

The chosen method is universally suitable, but it is essential to adapt it to the selected globe. The size of the globe and the ease of manual handling of the object must be considered and the right camera chosen accordingly. However, creating the images is the most important step, as their quality determines the quality of the final model.

The proposed methodology of digitization and accessibility was verified by practice on the globe by J. Jüttner and originated from the first half of the 19th century. Another valuable collection item is now available to researchers 24 h a day. It can be distributed in a virtual environment to users who do not have any specialized software installed on their computers or mobile devices.

This piece of cartographic heritage is no longer exposed to changes in light, climate or temperature. Therefore, its lifetime can be substantially extended and the likelihood of preserving such a precious exhibit for future generations is increased.

The employees of the Map Collection of the Faculty of Science of Charles University will be inspired by this good practice when creating the new Virtual Globe Collection. Further research into user requirements is planned, taking into account the excellent results achieved via the User-Centered Design method; see, for example, the team creating the Virtual Atlas of Canada [61]. The authors plan to examine how this new model is used by both experts and non-experts.

This method is particularly suitable for raising awareness about collection exhibits and aiding the work of curators and restorers. Virtual presentation is also widely used in the field of cartography. In the future, the focus will be on further developing annotations with spatial information, i.e., the globes will be georeferenced. The user will be able to easily compare the quality of the old and new globe maps.

Supplementary Materials: Virtual Jüttner's globe is available online on the website http://www.mapovasbirka.cz/globy/english/index_eng.html.

Author Contributions: Conceptualization, Eva Štefanová, Eva Novotná and Miroslav Čábelka; Methodology, Eva Novotná, Eva Štefanová and Miroslav Čábelka; software, Miroslav Čábelka and Eva Štefanová; Validation, Eva Novotná; formal analysis, Eva Štefanová, Miroslav Čábelka and Eva Novotná; investigation, Eva Novotná; Resources, Eva Novotná; data curation, Eva Novotná; writing—original draft preparation, Eva Novotná, Eva Štefanová and Miroslav Čábelka; writing—review and editing, Eva Štefanová; visualization, Miroslav Čábelka; supervision, Eva Novotná; project administration, Eva Štefanová; funding acquisition, Eva Štefanová, Eva Novotná and Miroslav Čábelka. All authors have read and agreed to the published version of the manuscript.

Funding: This research received no external funding.

Data Availability Statement: The data are not publicly available. They are the property of the Map Collection of the Faculty of Sciences, Charles University.

Acknowledgments: We would like to thank the Map Collection of the Faculty of Science of Charles University for making the rare Jüttner's globe available for digitization.

Conflicts of Interest: The authors declare no conflict of interest.

Appendix A

Table A1. Example of a globe metadata record in MARC21 format. Source: Map Collection of the Faculty of Science, Charles University.

Tag	Value
LDR	02546nem a2200433 i 4500
001	9900015521830106986
003	CZ-PrCU
005	20230130172719.0
007	dc#cpz
008	130218s1839 au d 0 ger d
034	1_ ⏐a a ⏐b 21000000 ⏐d W1800000 ⏐e E1800000 ⏐f N0900000 ⏐g S0900000
035__ ⏐	a (CZ-PrCU)001552183
035__ ⏐	a (CZ-PrCU)001552183CKS01-Aleph
040__ ⏐	a ABD065 ⏐b cze ⏐c ABD065 ⏐e rda
072_7 ⏐	a 913 ⏐x Regionální geografie ⏐2 Konspekt ⏐9 7
072_7 ⏐	a 912 ⏐x Mapy. Atlasy. Glóby ⏐2 Konspekt ⏐9 7
080__ ⏐	a 913:912.643 ⏐x (100) ⏐2 MRF
1001_ ⏐	a Jüttner, Josef, ⏐d 1775-1848 ⏐7 jk01052284 ⏐4 ctg ⏐4 pbl
245_10 ⏐	a Erdkugel / ⏐c von 2 Wiener Schuh im Durchmesser nach den neuesten geographischen Bestimmungen entworfen und herausgegeben von Joseph Jüttner Oberstlieutenant im Kais. Königl. Bombardier Corps; in Kupfer gestochen von Bernhard Bitter und Joh. David
255__ ⏐	a Scale 1:20,050,000 ⏐c (W 180°00′00″–E 180°00′00″/N 90°00′00″–S 90°00′00″)
264_1 ⏐	a Zu Wien: ⏐b Joseph Jüttner, ⏐c 1839
300__ ⏐	a 1 globe: ⏐b colored copper engraving, plaster covered with paper, on a wooden tripod, a metal horizontal circle with an engraved scale, a vertical wooden circle with a calendar and zodiac signs; ⏐c diameter 63.5 cm
336__ ⏐	a cartographic image ⏐b cri ⏐2 rdacontent
337__ ⏐	a unmediated ⏐b n ⏐2 rdamedia
338__ ⏐	a object ⏐b nr ⏐2 rdacarrier
500__ ⏐	a The Prime Meridian passes through the Canary Islands
500__ ⏐	a Globe circumference 199.5 cm, height from the original base pad to the pad of the copper meridian circle 41.8 cm
583__ ⏐	a Restored in 2016; ⏐h RNDr. Miroslav Široký, CSc.; ⏐5 CZ-PrUPM
590__ ⏐	a Before restoration, the wooden circle was cracked, the coloring scraped off in many places, the globe tilted to one side ⏐5 CZ-PrUPM
650 07 ⏐	a world geography ⏐7 ph120523 ⏐2 czenas
651 _7 ⏐	a countries of the world ⏐7 ge131523 ⏐2 czenas
655 _7 ⏐	a terrestrial globes ⏐7 fd796644 ⏐2 czenas
700 1_ ⏐	a Bitter, Bernhard ⏐4 egr
700 1_ ⏐	a David, Johann, ⏐d active in the 19th century ⏐7 jx20071127032 ⏐4 egr
856 41 ⏐	u http://www.mapovasbirka.cz/globy/gallery/g79.html ⏐ Digitized document
910__ ⏐	a ABD065
988__ ⏐	a GEOUK
997__ ⏐	a MP
998__ ⏐	a 20140407

References

1. Andrew, P.G.; Larsaard, M.L.; Moore, S.M. *RDA, Resource Description & Access and Cartographic Resources*; ALA, Ed.; American Library Association: Chicago, IL, USA, 2015; ISBN 978-0-8389-1131-0.
2. Gede, M. The possibilities of globe publishing on the web. In *Online Maps with APIs and WebServices*; Peterson, M.P., Ed.; Springer: Berlin, Germany, 2012; pp. 219–238; ISBN 978-3-642-27484-8.
3. Novotná, E. Catalogación de documentos cartográficos en RDA. *Prof. Inf.* **2014**, *23*, 195–203. [CrossRef]
4. Přidal, P. Bounding Box. Unterageri: Klokan Technologies. 2017. Available online: https://boundingbox.klokantech.com (accessed on 23 January 2023).
5. Bayer, T. Advanced methods for the estimation of an unknown projection from a map. *GeoInformatica* **2016**, *20*, 241–284. [CrossRef]
6. Roset, R.; Pacsual, V.; Montaner, C. From gazetteer to bounding box: Using SDI standards to build a geoportal for ancient maps in Catalonia. *E-Perimetron* **2015**, *10*, 11–20. Available online: http://www.e-perimetron.org/Vol_10_1/Roset_Pascual_Montaner.pdf (accessed on 15 May 2022).
7. Oliver, G.; Harvey, R. *Digital Curation*, 2nd ed.; Facet Publishing: London, UK, 2016; p. 240; ISBN 978-1-78330-097-6.
8. Jobst, M. (Ed.) Lecture notes in geoinformation and cartography. In *Preservation in Digital Cartography: Archiving Aspects*; Springer: Berlin, Germany, 2011; ISBN 978-3-642-12732-8.
9. Novotná, E. Cartographic Culture Heritage Belongs to UNESCO. *E-Perimetron* **2016**, *11*, 150–159. Available online: http://www.e-perimetron.org/vol_11_4/novotna.pdf (accessed on 5 October 2022).
10. Southall, H.; Přidal, P. Old Maps Online: Enabling Global Access to Historical Mapping. *E-Perimetron* **2012**, *7*, 73–81. Available online: http://www.e-perimetron.org/Vol_7_2/Southall_Pridal.pdf (accessed on 24 January 2023).
11. David Rumsey Map Center. Available online: http://www.davidrumsey.com (accessed on 24 January 2023).
12. Leventhal Map and Education Center. Available online: https://www.leventhalmap.org (accessed on 24 January 2023).
13. National Library of Australia. Available online: https://www.nla.gov.au/collections/what-we-collect/maps (accessed on 23 January 2023).
14. Mercator Museum in Belgium. Available online: http://www.kokw.be/m4FAB.html (accessed on 23 January 2023).
15. Stal, C.; De Wulf, A.; De Coene, K.; De Maeyer, P.; Nuttens, T.; Ongena, T. Digital Representation of Historical Globes: Methods to Make 3D and Pseudo-3D Models of Sixteenth Century Mercator Globes. *Cartogr. J.* **2012**, *49*, 107–117. [CrossRef]
16. Hungarian Virtual Globes Museum. Available online: http://terkeptar.elte.hu/vgm (accessed on 23 January 2023).
17. Gede, M.; Márton, M.; Ungvári, Z. Digital reconstruction of Perzel's globe. *E-Perimetron* **2011**, *6*, 68–76. Available online: https://edit.elte.hu/xmlui/bitstream/handle/10831/50961/Gede_Marton_Ungvari_e_perimetron_6_evf_68_76.pdf (accessed on 7 February 2022).
18. Gallica Digital Library. Available online: https://gallica.bnf.fr (accessed on 23 January 2023).
19. Virtual Map Collection. Available online: http://www.chartae-antiquae.cz/en/globes (accessed on 15 May 2022).
20. National Library of the Czech Republic. Available online: http://www.chartae-antiquae.cz/en/globes/76912 (accessed on 2 October 2022).
21. Novotná, E. Database of digitized map collections of the Czech Republic. *E-Perimetron* **2021**, *16*, 72–77. Available online: http://www.e-perimetron.org/Vol_16_2/Novotna.pdf (accessed on 13 October 2022).
22. Hyndráková, M. Virtual exhibition of Globes. In Proceedings of the 9th International Workshop on Digital Approaches to Cartographic Heritage, Budapest, Hungary, 4–5 September 2014. Commision on Digital Technologies in Cartographic Heritage.
23. Široký, M. *The Restoration Report: J. Jüttner*; 77 Sheets, Colour Photos and 1 CD; Karolinum Press: Prague, Czech Republic, 2015. (In Czech)
24. Stevenson, E.L. *Terrestrial and Celestial Globes: Their History and Construction, Including a Consideration of Their Value as Aids in the Study of Geography and Astronomy*; Yale University Press: New Haven, CT, USA, 1921.
25. Mucha, L. The Czech globe-maker Jan Felkl. *Sborník Ceskoslov. Společnosti Zeměpisné* **1960**, *65*, 241–245.
26. Muris, O.; Saarman, G. *Der Globus im Wandel der Zeiten: Eine Geschichte der Globen*; Columbus Verlag: Berlin, Germany, 1961; p. 287.
27. Mokre, J.; Allmayer-Beck, P.E. (Eds.) *Rund um den Globus: Über Erd-und Himmelsgloben und ihre Darstellungen*; Bibliophile Edition: Wien, Austria, 2008; p. 224; ISBN 978-3-9502052-3-7.
28. Sumira, S. *The Art and History of Globes*; The British Library: London, UK, 2014; p. 224; ISBN 978-0-7123-5868-2.
29. Lister, R. *How to Identify Old Maps and Globes: With a List of Cartographers, Engravers, Publishers and Printers Concerned with Printed Maps and Globes From c. 1500 to c. 1850*; Bell: London, UK, 1965; p. 256; ISBN 0-7135-0647-4.
30. Kuchař, K. Register of old globes in Czechoslovakia. *Zprávy Geogr. Ust. Čsav* **1964**, *5*, 7–12. (In Czech)
31. Hruby, F.; Plank, I.; Riedl, A. Cartographic heritage as Shared Experience in Virtual Space: A Digital Representation of the Earth Globe of Gerard Mercator (1541). *E-Perimetron* **2006**, *1*, 88–98. Available online: www.e-perimetron.org/Vol_1_2/Hruby_etal/Hruby_et_al.pdf (accessed on 29 March 2021).
32. Talich, M.; Ambrožová, K.; Havrlant, J.; Böhm, O. Digitization of Old Globes by a Photogrammetric Method. In *Cartography—Maps Connecting the World. Lecture Notes in Geoinformation and Cartography*; Robbi Sluter, C., Madureira Cruz, C., Leal de Menezes, P., Eds.; Springer: Cham, Switzerland, 2015; p. 15. [CrossRef]
33. Ambrožová, K.; Havrlant, J.; Talich, M.; Böhm, O. The process of digitizing of old globe. *Int. Arch. Photogramm. Remote Sens. Spatial Inf. Sci.* **2016**, *XLI-B5*, 169–173. [CrossRef]

34. Adami, A.; Guerra, F. Coronelli's Virtual Globe. *E-Perimetron* **2008**, *3*, 243–250. Available online: http://www.e-perimetron.org/Vol_3_4/Adami_Guerra.pdf (accessed on 31 January 2022).

35. Márton, M.; Gercsák, G. Virtual Globes Museum. In Proceedings of the XXIV International Cartographic Conference, Santiago, Chile, 15–21 November 2009; International Cartographic Association: Bern, Switzerland, 2009; pp. 1–9. Available online: http://real.mtak.hu/13082/1/1282423.pdf (accessed on 5 February 2022).

36. Gede, M. Publishing globes on the Internet. *Acta Geod. Geophys. Hung.* **2009**, *4*, 141–148. [CrossRef]

37. Adami, A. From real to virtual globe: New technologies for digital cartographic representation. *E-Perimetron* **2009**, *4*, 144–160. Available online: http://www.e-perimetron.org/Vol_4_3/Adami.pdf (accessed on 31 January 2022).

38. Menna, F.; Rizzi, A.; Nocerino, E.; Remondino, F.; Gruen, A. High resolution 3d modeling of the Behaim Globe. *Int. Arch. Photogramm. Remote Sens. Spatial Inf. Sci.* **2012**, *XXXIX-B5*, 115–120. [CrossRef]

39. Bílá, Z.; Řezníček, J.; Pavelka, K. Non-contact digitisation and visualisation of historical globes using photogrammetry. In Proceedings of the International Multidisciplinary Scientific GeoConference: SGEM, Albena, Bulgaria, 16–22 June 2013; Surveying Geology & Mining Ecology Management (SGEM): Sofia, Bulgaria, 2013; pp. 805–812. [CrossRef]

40. Gede, M. Automatic Reconstruction of Old Globes by Photogrammetry and Its Accuracy. *E-Perimetron* **2017**, *12*, 132–141. Available online: http://www.e-perimetron.org/Vol_12_3/Gede.pdf (accessed on 15 February 2022).

41. Marshall, M.E.; Johnson, A.A.; Summerskill, S.J.; Baird, Q.; Esteban, E. Automating photogrammetry for the 3D digitisation of small artefact collections. *Int. Arch. Photogramm. Remote Sens. Spatial Inf. Sci.* **2019**, *XLII-2/W15*, 751–757. [CrossRef]

42. Al Khalil, O. Structure from Motion (SfM) Photogrammetry as Alternative to Laser Scanning for 3D Modelling of Historical Monuments. *Open Sci. J.* **2020**, *5*, 1–17. [CrossRef]

43. Iglhaut, J.; Cabo, C.; Puliti, S.; Piermattei, L.; O'Connor, J.; Rosette, J. Structure from Motion Photogrammetry in Forestry: A Review. *Remote Sens.* **2019**, *5*, 155–168. [CrossRef]

44. Kraus, K. *Photogrammetry: Geometry from Images and Laser Scans*, 2nd ed.; Walter de Gruyter: Berlin, Germany, 2007; p. 459; ISBN 13 978-3110190076. [CrossRef]

45. Luhman, T.; Robson, S.; Kyle, S.; Harley, I. *Close Range Photogrammetry: Principles, Techniques and Applications*; Whittles Publishing: Caithness, UK.

46. Medyńska-Gulij, B. How the Black Line, Dash and Dot Created the Rules of Cartographic Design 400 Years Ago. *Cartogr. J.* **2013**, *50*, 356–368. [CrossRef]

47. Wavefront OBJ File Format. Available online: https://www.loc.gov/preservation/digital/formats/fdd/fdd000507.shtml (accessed on 22 April 2022).

48. Wavefront 3D Object File. Available online: https://fileinfo.com/extension/obj (accessed on 25 January 2023).

49. Khronos Group. Available online: https://www.khronos.org/about (accessed on 25 January 2023).

50. Extensible 3D (X3D): Part 1: Architecture and Base Components. Available online: https://www.web3d.org/documents/specifications/19775-1/V3.3/Part01/introduction.html (accessed on 25 January 2023).

51. Gutbell, R.; Pandikow, L.; Kuijper, A. Web-Based Visualization Component for Geo-Information. In *Human Interface and the Management of Information. Interaction, Visualization, and Analytics*; Yamamoto, S., Mori, H., Eds.; HIMI 2018; Lecture Notes in Computer Science, 10904; Springer: Cham, Switzerland, 2018. [CrossRef]

52. Evans, A.; Romeo, M.; Bahrehmand, A.; Agenjo, J.; Blat, J. 3D Graphics on the Web: A Survey. *Comput. Graph.* **2014**, *41*, 43–61. [CrossRef]

53. Loesch, B.; Christen, M.; Nebiker, S. OpenWebGlobe—An open source SDK for creating large-scale virtual globes on a webgl basis. *Int. Arch. Photogramm. Remote Sens. Spat. Inf. Sci.* **2012**, *XXXIX-B4*, 195–200. [CrossRef]

54. Newe, A. Towards an easier creation of three-dimensional data for embedding in-to scholarly 3D PDF (Portable Document Format) files. *PeerJ* **2015**, *3*, e794. [CrossRef] [PubMed]

55. JSC Viewer. Available online: https://code.google.com/archive/p/jsc3d (accessed on 18 June 2022).

56. Behr, J.; Jung, Y.; Drevensek, T.; Aderhold, A. Dynamic and Interactive Aspects of X3DOM. In Proceedings of the 16th International Conference on 3D Web Technology, Paris, France, 20–22 June 2011; pp. 81–87. [CrossRef]

57. Herman, L.; Řezník, T. Web 3D Visualization of Noise Mapping for Extended INSPIRE Buildings Model. In *Environmental Software Systems. Fostering Information Sharing*; Hřebíček, J., Schimak, G., Kubásek, M., Rizzoli, A.E., Eds.; ISESS 2013; IFIP Advances in Information and Communication Technology, 413; Springer: Berlin/Heidelberg, Germany, 2013. [CrossRef]

58. Hurni, L. *Multimedia Atlas Information Systems*; Springer Science and Business Media LLC: Berlin/Heidelberg, Germany, 2008; pp. 759–763.

59. Van Elzakker, C.P.J.M.; Ooms, K. Undestanding map uses and users. In *The Routledge Handbook of Mapping and Cartography*; Kent, A., Vujakovic, P., Eds.; Routledge: London, UK, 2017; pp. 55–67; ISBN 9781315736822.

60. Kramers, E.R. Interaction with Maps on the Internet: A User Centred Design Approach for The Atlas of Canada. *Cartogr. J.* **2008**, *45*, 98–107. [CrossRef]

61. Old Mas Portal. Available online: https://www.natur.cuni.cz/geography/map-collection/science-and-research/old-maps-portal-staremapy.cz (accessed on 24 February 2023).

62. Parisi, T.; Treseler, M. *WebGL: Up and Running*; O'Reilly Media: Sebastopol, CA, USA, 2012; ISBN 978-1-449-32357-8.

63. Regele, O. Die Globen des Josef Jüttner (1775–1848) und des Franz Ritter von Hauslab (1798–1883). *Der Glob.* **1953**, *2*, 16–22.

64. Tooley, R.V. *Tooley's Dictionary of Mapmakers*; E–J. Revised Edition; Connecticut Early World Press: Riverside, CT, USA, 2001; ISBN 0-906430-14-4.
65. Restored Globes of Map Collection of Charles University. Available online: https://www.natur.cuni.cz/geography/map-collection/the-map-collection/restoration/restored-globes#section-0 (accessed on 15 January 2022).
66. GLOBES of Map Collection of Charles University. Available online: http://www.mapovasbirka.cz/globy/english/index_eng.html (accessed on 15 January 2022).
67. Surveying Instruments. Available online: https://www.surveyinginstruments.org/col/en (accessed on 20 January 2022).
68. Kanas, N. *Star Maps: History, Artistry, and Cartography*, 3rd ed.; Springer: Chicester, UK, 2019; p. 599; ISBN 9783030136123.
69. Novotná, E. *Jan Felkl & Syn: Továrna na Glóby = Jan Felkl & Son: A Globe-Making Factory*; Faculty of Science, Charles University: Prague, Czech Republic, 2017; ISBN 978-80-7444-053-3. (In Czech)
70. Planetary Machines Exhibition Animation—Tellurium Felkl. Available online: https://www.youtube.com/watch?v=ghiwUPWxHYM (accessed on 20 January 2022).
71. McEathron, S.R. The cataloging of globes. *Cat. Classif. Q.* **1999**, *27*, 103–112. [CrossRef]

International Journal of
Geo-Information

Article

Geomedia Attributes for Perspective Visualization of Relief for Historical Non-Cartometric Water-Colored Topographic Maps

Beata Medyńska-Gulij

Department of Cartography and Geomatics, Adam Mickiewicz University Poznan, 61-680 Poznan, Poland;
bmg@amu.edu.pl or beata.medynska-gulij@amu.edu.pl

Abstract: The selection of appropriative geomedia attributes for constructing natural and suggestive perspective visualizations of historical non-cartometric manuscript topographic works is investigated, to enable an intuitive perception of relief landforms. The main objective of the study is to demonstrate geomedia parameters for representing the third dimension in topographic watercolor maps from the eighteenth century, using cartographic rules and geoinformation operations for transforming graphic means of expression. The following methods were used: the choice of representative map fragments with specific painterly means of expression; the analysis of main relief forms on historical and modern maps; the rectification; vectorization of contour lines, and the transformation to a GRID model; the use of parameter variations: elevation rise, azimuth and altitude, contrast of illumination; and the creation of the final bird's-eye-view visualization, with appropriate parameters. It is found that the parameters for the visualization of the non-cartometric water-colored topographic image on a 3D model can be selected in turn. However, what matters is maintaining their complementarity. The proposed parameters for the three maps work well for creating the general static bird's-eye-view visualization, with the natural and suggestive perception of the landscapes' relief.

Keywords: geomedia attributes; topographic map; non-cartometric map; historical manuscript map; perspective visualization; land relief; watercolor; geoinformation operation; graphical means of expression

Citation: Medyńska-Gulij, B. Geomedia Attributes for Perspective Visualization of Relief for Historical Non-Cartometric Water-Colored Topographic Maps. *ISPRS Int. J. Geo-Inf.* **2022**, *11*, 554. https://doi.org/10.3390/ijgi11110554

Academic Editors: Florian Hruby and Wolfgang Kainz

Received: 14 September 2022
Accepted: 5 November 2022
Published: 8 November 2022

Publisher's Note: MDPI stays neutral with regard to jurisdictional claims in published maps and institutional affiliations.

1. Introduction

By definition, a map has no perspective, because the observer is located above all parts of the graphic image at the same time [1]. However, for centuries, techniques of adapting perspective–related parameters for cartographic representations of the terrain are associated with the use of geometrical principles and painting techniques [2]. Contour lines have represented relief on topographic maps since the early nineteenth century, i.e., since the first rigorous national triangulation and levelling surveys based on an agreed datum, typically sea level [3]. The arranging of contour lines shows mainly the measurable height of features, whereas supplementary shading helps one notice the plasticity of landform relief [4]. The representation of the third dimension in the context of the topographic content on the map has been an object of research on measurability, imageability, and the esthetics of cartographic images for a long time [5].

At the beginning of professional topography in the eighteenth century, monarchs would order their military engineers to work out multi-sheet topographic works for military and/or administrative purposes [6]. The field survey of that time included reconnaissance by eye, acute observation, and correct distance evaluation and sketching, both in rough sketchbooks and using the plane table [7]. Such secret maps were usually created for the ruler as manuscript works that used water-based media and, as such, few copies were made [8]. At that time, no other technique offered map makers the opportunity to reflect landscape so realistically. This artistic painterly manner, which employed sophisticated graphical means of expression, with harmonious colors and imitation of natural features, is highly predisposed to transformations from the 2D cartographic image to the 3D perspective visualization [6].

The significance of historical topographic maps as a unique window into the state of topography in the eighteenth century is proven by dedicated geoportals and multimedia websites [9,10]. Sheets are subject to georeferencing to current spatial systems, in which the transparency function allows one to manipulate the view of several overlapping maps from different historical periods [11,12]. Unfortunately, such geometric georeferencing to calibration points distorts the graphics of a historical map and, as a consequence, causes the deformation of the means of graphical expression created in such artistic water-based media.

Most eighteenth-century topographic maps are non-cartometric, because they betray highly irregular distortion within single sheets [6]. A non-cartometric map means that the values of angles and distances measured on that map are approximate values to the real azimuths and real distances derived from the map scale. This is why in catalogs and cartobibliographies, the scale figures relating to the eighteenth-century topographic maps are often preceded by the word 'ca' (circa).

Therefore, the issue of the appropriate transformation of such artistic graphics from maps to the third dimension in relation to creating perspective images for areas of the landscape, will be discussed in this study. Such panoramas, with highly individualized images of mountains and valleys, have been created by manipulating parameters and painting and drawing effects [13,14].

On the other hand, to be able to use historical maps more widely, it is important to prepare digital terrain TINs (triangulated irregular networks) and GRID models that make it possible to select measurable and visual parameters in 3D in cartographic-geoinformation applications [15,16]. The creation of realistic visualizations from such models to help explain landscape processes [17], or create a real-time interactive visualization [18], has become the focus for research. The freedom to render topographic images on screen allows more than appropriately shaded slopes [19].

The minimization of the interference of the original graphic elements on the historical map, and allowing one to interpret relationships between relief and topographic objects correctly, remains a great challenge [20]. Hence, the focus of this study is related to a complementary selection of geomedia attributes for constructing the most beneficial 3D visualization from manuscript topographic maps for the simple intuitive perception of relief landforms and a good interpretation of their mutual relations, while maintaining a high level of esthetics.

2. Aims and Questions

The main objective of the study was to demonstrate geomedia parameters for representing the third dimension on non-cartometric topographic water-colored maps, using cartographic rules and geoinformation operations for transforming the graphic means of expression. To meet this objective, the following questions were raised:

— Which cartographic rules are advantageous, and which of them limit the opportunity to create 3D views from topographic maps in the painterly watercolor style?
— Which measurable parameters can be applied to extract visual suggestiveness of relief and landforms via a graphic means of expression in water-color?
— How can one achieve naturalness in terms of color adequacy of a specific existing landscape?
— What are the separate and complementary issues of using measurable parameters and the manual modelling of bird-eye perspective in the construction of a 3D visualization?

3. Materials and Methods

The fundamental methods of research were: the visual analysis of cartographic content; the classification and selection of cartographic material; the comparison of the representations of relief forms on historical and modern topographic maps; the selection of parameters in geoinformation software; the employment of cartographic design rules; the adaptation of the graphical means of expression in water-based media to promote the suggestiveness of relief-form representation.

To meet the objective and answer the questions raised, the stages of research were:

— Formulation of the concept of the research process (Section 3.1);
— The choice of representative non-cartometric sheets/map fragments that demonstrate specific painterly means of expression: cartographic materials (Section 3.2, Figures 1 and 2);
— Comparative analysis of main relief forms on historical and modern maps, map rectification, and vectorization of contour lines (Section 3.3., Figures 3 and 4);
— Transformation to GRID model and parameter variations: vertical exaggeration (vertical scaling factor) and illumination features (Section 3.4., Figures 5–7);
— Creation of the final effect: 3D visualization using the favorable parameters to achieve a suggestive and close natural landscape of the terrain (Section 4., Figures 8 and 9).

3.1. Conception

The following points were included in the concept of the research:

- Cartographic material: representative topographic works from the eighteenth century with painterly graphic means of expression in water-based media;
- Digital interference in cartographic image: rasterized image, no map georeferenced to the coordinate system, transformation to GRID adjusted to graphical means of expression in water-based media;
- Cartographic rules: plasticity of the relief, shape of contour lines, excessive 'meandering' of rivers, vertical-scale exaggeration, bird's-eye view;
- Measurable geoinformational parameters: elevation rise (vertical scale exaggeration); illumination: azimuth–altitude contrast;
- Artistic and esthetic context: maintaining high expression of the painterly style, landscape naturalness of colors, painterly perspective in the artistic image, rendering and perspective changes;
- Technology: rasterized images, raster graphics (Photoshop), manual vectorization: modelling of line arches, Bézier curves, GRID model, variants of parameters (ESRI ArcMap and ArcScene 10.6.1.9270, Environmental Systems Research Institute, Inc. (ESRI), Redlands, CA, USA);
- The expected effect: visualization of high suggestiveness of plasticity (imageability) representation from non-cartometric topographical maps with painterly style, recommendations for favorable arrangement of the perception point and image plane, bird's-eye view.

3.2. Cartographic Materials

Figure 1 presents three map fragments selected from manuscript topographic works made in watercolor for European monarchs from the eighteenth century. The map of Sicily [21] is distinguished by a depiction of woodland in graduated varieties of green and a perfectly toned grey to render the relief forms (Figure 1A). The map of Scotland (Figure 1B) [22] and map of Susa (Figure 1C) [23], despite the richness of the pigments, seem considerably toned and muted. On these three maps, the landscape effect was achieved by employing subtle painterly techniques, predominantly watercolor and washes. Taking a look at originals directly on-site in cartographic archives was indispensable in order to select representative sheets of topographic maps with a favorable layout of the main relief forms.

A painterly means of expression, with blots of color and brushstrokes, prevails in Roy and Sandby's maps of Scotland (Figure 2B) [24] and in Avico and Carelli's map of the Susa Valley (Figure 2C) [25]. These maps seem considerably artistic, in contrast to other maps from the the eighteenth century, with more drawing (linear) means of graphical expression [6]. The three-dimensionality of landforms was rendered most realistically in these maps, due to the use of painterly means of graphical expression. Figure 2 shows the sophisticated painterly means of expression, with harmonious colors and imitation of natural features: stippling, spotting or blotting in grey (relief grading; brush; Indian ink and wash); and blotting, banding with multi-color and multi-shade spotting (landform hatching, land-relief shading on woods or on fields; varying use of landform; brush; Indian ink, bistre, and wash) [26].

3.3. Comparative Analysis of Main Relief Forms, Rectification, and Vectorization of Contour Lines

The first step of the analysis of the main relief landforms was to compare the location of mountain peaks, and the course of mountain chains and rivers on historical and modern maps. Figure 3 presents two maps of the middle part of Sicily (Figure 3A,B) with the location of four peaks, three towns, and the main rivers, to allow researchers to establish the total direction of the fall of the land and the volume of the main relief forms. Irregular distortions of the topographic objects' location on the old map at the scale ca. 1:80,000 do not facilitate the evaluation of the main directions of the fall of the land, even when compared with the drawing of contour lines on the modern topographic map at the scale 1:100,000. The next step was to prepare for the vectorization of contour lines through the registration of the cartographic image in the geoinformation workspace. In line with the concept, a layer with a raster image without georeferences was registered with the modern spatial coordinate system.

A routine transformation to the coordinate system was performed, to check the degree of distortion of the graphical means of expression in the ESRI ArcMap 10.6.1 software (Figure 3C). The UTM 33 North coordinate system according to the UTM grid was specified for the modern topographic map. The next step was the rectification of the historical map according to common points on both maps. Seventeen points were identified, and a 3rd order polynomial transformation was used. The total RMS (root-mean-square) error was 522.342 m. Figure 3C shows a very irregular distortion of the graphics of the old map. In the case of these studies, the most important thing is to clearly indicate the distortion of the painterly style of the original map (Figure 3E) in the image, after rectification to the modern coordinate system (Figure 3D).

The next step was to prepare for the vectorization of contour lines through the registration of the cartographic image in the geoinformation workspace. A layer with the raster image without georeferences to the modern spatial coordinate system was registered (Figure 4B). Then, the vectorization for nine contour lines with agreed-upon altitude values took place, starting with the lowest value, 105, in the bed of the main valley (Figure 4A,D). The vectorization of contour lines was based on the polyline geometrical shapefiles with the use of Bezier curves in a freehand mode. Figure 4C in the bottom right corner shows a manual way of drawing contour lines by means of Bézier curves, which makes it possible to model smooth curves. The isochromatic map in the left part of Figure 4 presents the system of contour lines according to cartographic rules, whereas in the right part it shows the adjustment to a painterly means of expression on the map from the eighteenth century. All of these actions were performed using the ESRI ArcMap 10.6.1 software.

Figure 1. Map fragments with reduction from the originals by approximately 70%: (**A**): Nova et accurata Siciliae Regionum, Schmettausche Karte von Sizilien (1722), ca. 1:80.000; Vienna, Österreichische Nationalbibliothek; Schmettau: Sizilien, ÖNB/KAR: AB141, E19.585-D: 12; (**B**): Military Survey of Scotland—Highlands, ca. 1:36.000; 1747–1752. London, British Library. K.Top: Maps. CC.5a441/16-3f BL; (**C**): Valle di Susa: CARTA TOPOGRAFICA in misura, delle Valli di Cezana, e Bardoneche . . . , (1764), ca. 1:19.000; Turin, Archivio di Stato; Cartella 7, Susa.

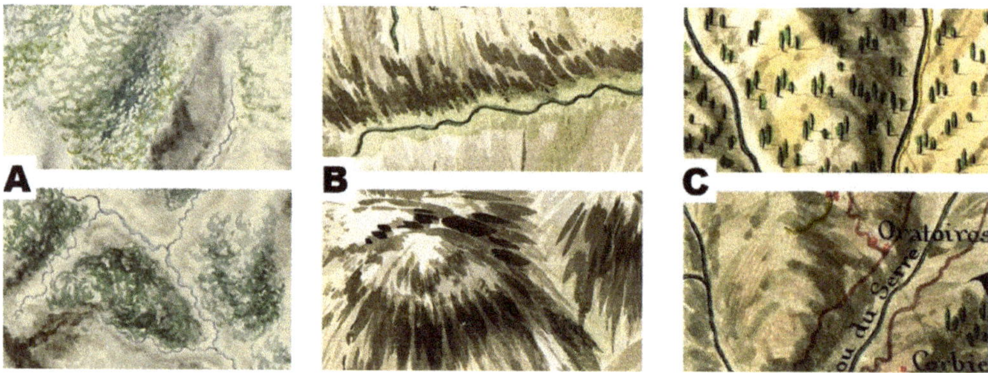

Figure 2. The painterly graphical means of expression in map details from Figure 1. (**A**): map of Sicily, (**B**): map of Scotland, and (**C**): map of Susa.

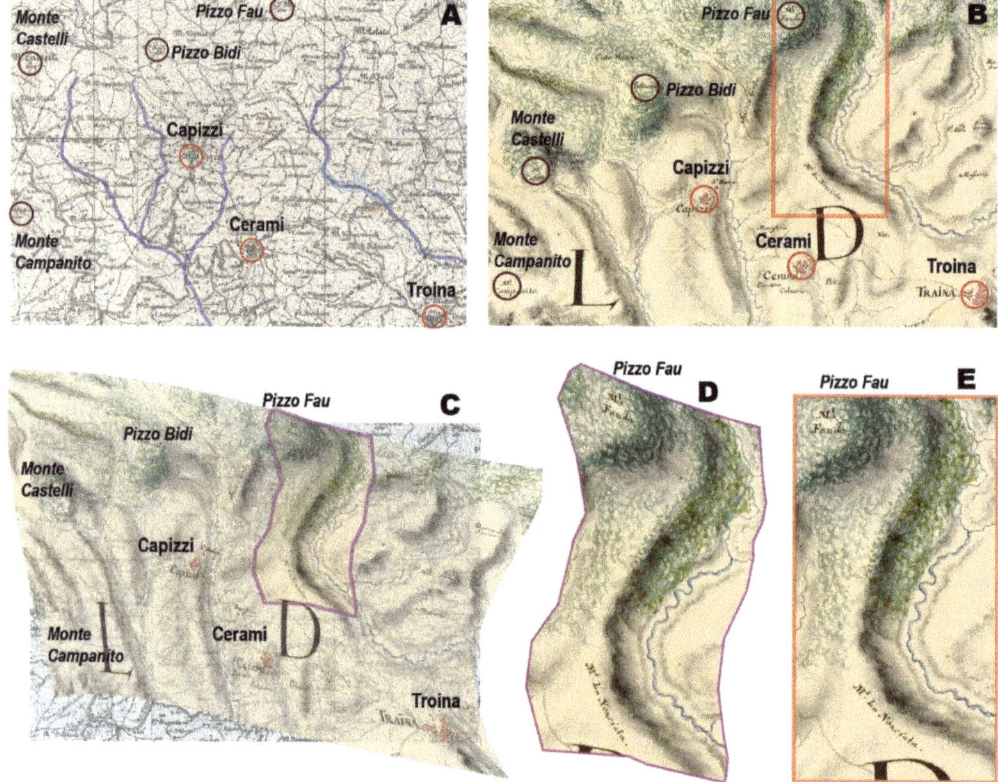

Figure 3. Comparative analysis of main relief forms on map of Sicily, comparing the location of mountain peaks, the course of mountain chains and the rivers (with the location of four peaks, three towns, and main rivers) on the map of Sicily, ca.1:80,000, 1722 (**B**), and topographic map of Italy, sheet: Bronte, 1955, UTM, 1:100,000 (**A**): irregular distortion of old map frame (**C**): distortion of the center of the graphic expression of the valley/ridge south of Pizza Mountain (**D**): original painting style before rectification (**E**).

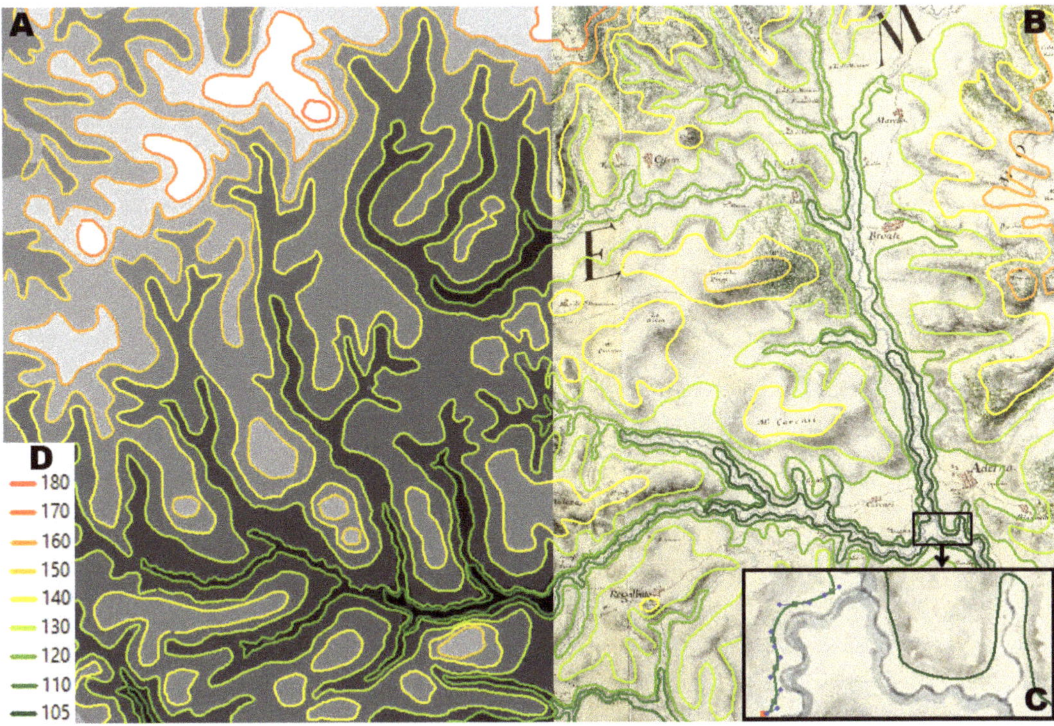

Figure 4. Vectorization of contour lines on the map of Sicily: (**A**): isochromatic map after vectorization; (**B**): contour lines on the layer with raster image; (**C**): manual way of drawing contour lines via Bézier curves; (**D**): nine contour lines with contrasting color.

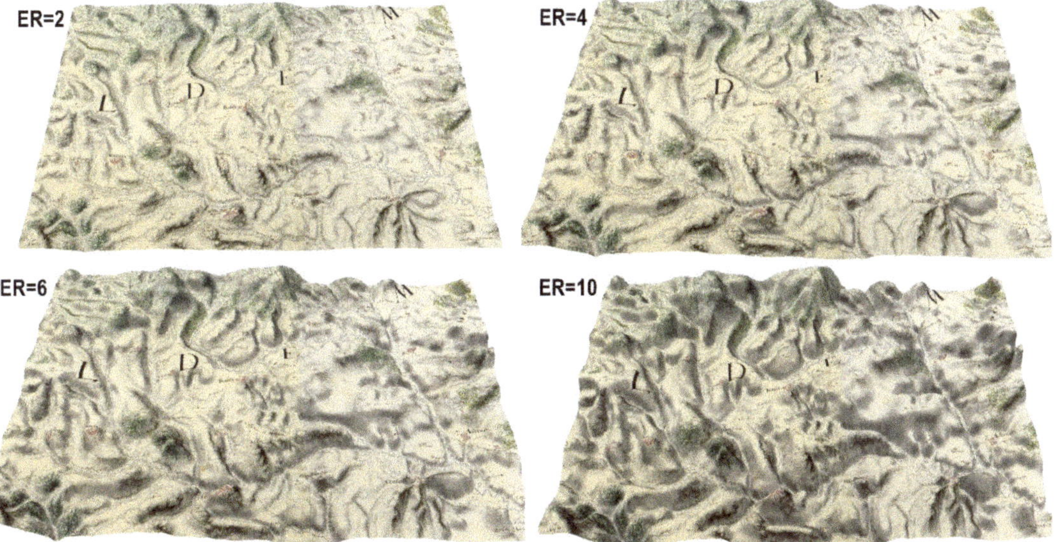

Figure 5. Four values for vertical scaling factor (elevation rise): 2, 4, 6, and 10.

3.4. Transformation to GRID Model and Versioning Parameters

The contours were then used to interpolate a DRID Digital Terrain Model. The cells of the GRID have been calculated for 5-m sides and the height values in adjacent points have clearly been averaged. This was carried out in the ESRI ArcMap 10.6.1 software using the Topo-to-Raster tool which is based on the ANUDEM program [27]. The DTM output was transferred into ESRI ArcScene 10.6.1 software, which enables 3D representation. Modifying the height scale can be done by using elevation values in the raster layer and changing the view settings of the main viewer, (e.g., changing view field and roll angle).

To achieve a compromise between the distortions of the cartographic image and the suggestiveness of the land-relief visualizations, the variation in parameters were tested. Four values for the vertical scaling factor (elevation rise) show results according to the vertical exaggeration rule on the maps of smaller scales (Figure 5). As far as the area in Sicily is concerned, vertical scaling factors of 2–4 seem the most appropriate, and values of 6–10 should be rejected [28,29].

Setting the azimuth and the altitude of illumination parameters is a simple activity, because a cartographic rule says that azimuth NW = 315° and altitude 45° are favorable for intuitive interpretation [29,30]. A juxtaposition of two opposite azimuths of illumination in Figure 6 confirms that rule, as with azimuth SE = 135°, which is more natural for Europe. Rivers seem to be located on mountain ridges, whereas mountain ridges seem to be located in valleys.

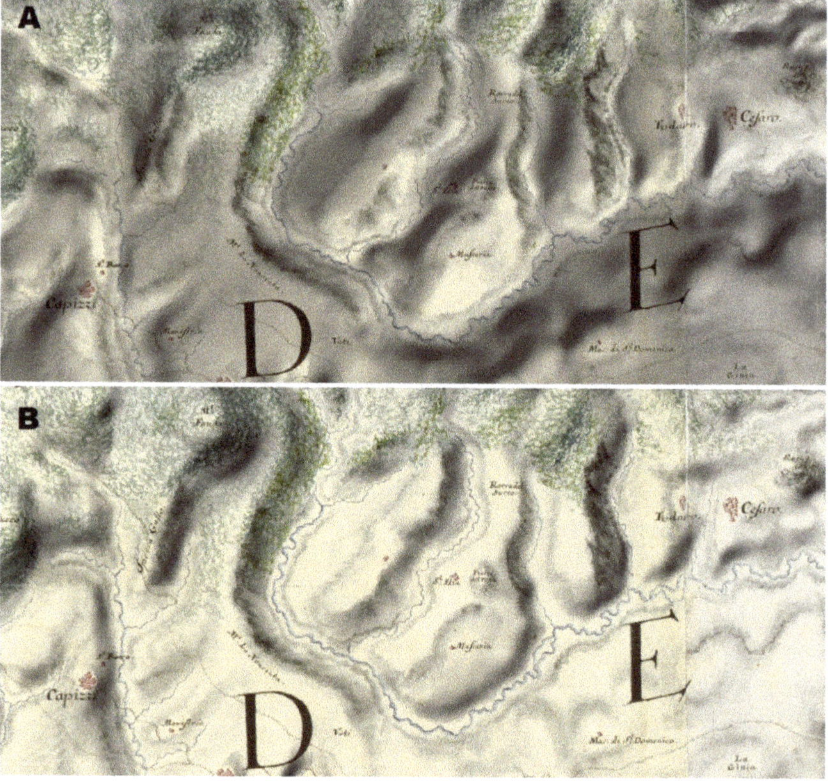

Figure 6. Juxtaposition of two opposite azimuths of illumination (Map of Sicily: (**A**): SE = 135° and (**B**): NW = 315°.

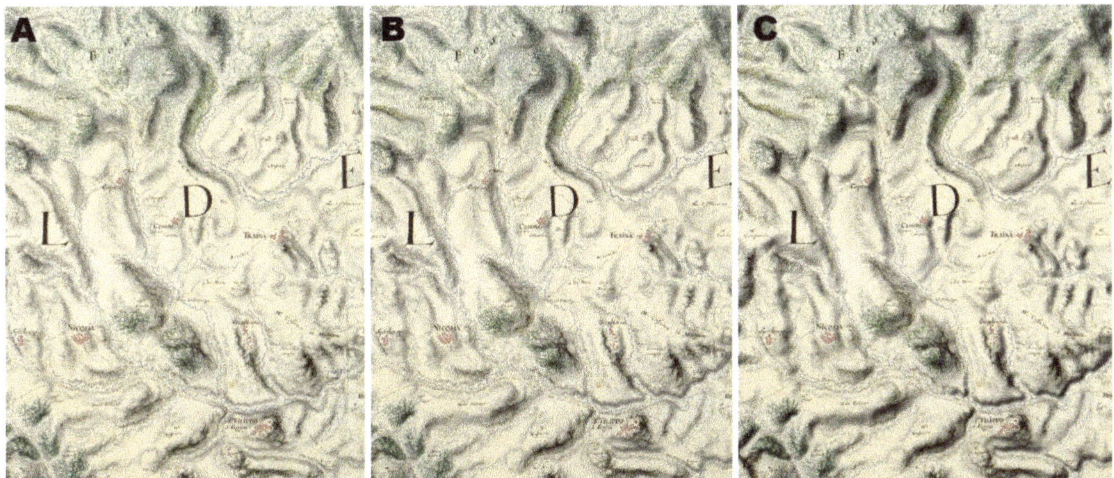

Figure 7. Three contrast values (Map of Sicily): (**A**): 20, (**B**): 40, and (**C**): 100.

Figure 8. Natural landscape colors: high similarity of the map of Sicily (**A**) and the map of Scotland (**B**), (compare the photographs with Figures 1 and 9).

Figure 9. Final bird's-eye view with the following values adopted: vertical scaling = 4, contrast 40%; illumination: azimuth 315° = NW, altitude 45°, slightly inclined towards the viewer (to the right) by approximately 20°; (**A**): from a detail of the Sicily map; (**B**): from a detail of the Scotland map; (**C**): from a detail of the Susa map.

Contrast is a more subjective parameter; therefore, it is necessary to consider the quality of the raster image in order to achieve an effect similar the original paper in terms of the three most important features of color: hue, saturation and value. If an adequate 3D effect for relief-form visualizations is to be achieved, the level of contrast is related to enhancing grayness for the shaded mountain slopes and valley sides. Figure 7 presents three contrast values, with value 20 and 40 appearing to be more favorable than the maximum value of 100. A similar process using appropriate data, map projections, and control points was then carried out for the other two landscape fragments.

4. Results

In the process of creating the final perspective visualization, the author focused on a complementary selection of appropriate perspective parameters: the variable of vertical scaling factor (vertical exaggeration) and slope exposition, and exposure of the cartographic image surface for the suggestive whole representation of land relief. Due to the high similarity of the map of Sicily and the map of Scotland in natural landscape colors, the author used observations and photographs taken from observation points on-site to aid the process (Figure 8).

Considering the variations of parameters demonstrated above, in the context of both naturalness and suggestiveness of relief-form representation on the water-colored topographic maps, to create a 3D model the following values were adopted: vertical scaling = 4; illumination: azimuth 315° = NW, altitude 45°, and contrast 40%. For a correct comprehensive interpretation of the relief for the entire area in 3D view, three visualizations were prepared, using the same parameters (Figure 9). The author used the possibility of smooth rendering with a computer mouse button for setting the bird's-eye view: the height of the observation point, range of the field of vision, and the slope and rotation of the cartographic image plane [31]. In general, each of the three cartographic images (along with previously selected versions of parameters presented above for the map of Sicily) was slightly inclined towards the viewer (to the right) by approximately 20°.

5. Discussion

The research path suggested here was based on three non-cartometric cartographic images with favorable layouts of main relief forms. The fragment of the map of Sicily includes high mountains in the north, and valleys lowering towards the south east. In this case, watercolor banding with multi-color and multi-shade spotting and stippling, and blotting in gray make an excellent match with the parameters selected to highlight the 3D effect. The relief of the map of Scotland is generally lowering southwards with wide valleys; however, massive elevations at the bottom of the sheet near longer multi-color and multi-shade spotting may evoke slight visual discomfort. On the map of Susa, neither high mountain ridges in the north and west, nor a broad valley (along with the tree pattern on mountain slopes) distorts the image perception. The peak with mountain ridges in the south east with slopes directed northwards may cause difficulties with visual interpretation. Thus, when dealing with historical maps in water-based media, one needs to consider limitations that result from digital transformation of the paper map to the 3D model [15]. With the modern panoramas created at present, the problem is minimized by rigorous geometric distortions in favor of readability and the illustrative aspect.

A remaining issue is the establishment of the lowest possible number of contour lines required to create an adequate GRID DTM to match the non-cartometric topographic map. This will have a direct impact on the parameter variants in the 3D model to best represent the artistic water-color relief representation. If one compares the number and density of contour lines on the modern map, one can observe a great reduction in the number of contour lines used for creating the DRID model. The smaller contour interval has been replaced by an approach using river-valley framing and mountain ridge/domineering mountain-peaks framing. When it comes to drawing (digitizing) contour lines along rivers, based on maps from the sixteenth to the eighteenth century, one needs to consider the graphical manner of showing excessive 'meandering' of the river [32]. Those small meanders on the map should not be taken into account in the process of vectorization of contour lines in river valleys (Figure 4C). Thus, in the contour-line vectorization stage, the initial comparative analysis of the river course on the historical non-cartometric map and the modern map, becomes significant.

The order of variations of parameters for the historical water-colored map may also be subject to discussion, because in the line of study suggested in this research, cartographic rules were prioritized. Therefore, the author focused on subjective esthetic feelings and sensations, which are a lot more difficult to verify and are more individualized. The fact that users of interactive atlases have tools for any changes in parameters of 2D and 3D views on the monitor allows users to select the visualization that suits them best [33], and it is useful to give them an appropriate starting point and some guidance.

What is becoming more and more significant is the subjective evaluation of the esthetics of the topographic image in relation to the natural landscape [34], which does not need to go hand in hand with a high effectiveness of interpreting mutual volume relationships of relief forms. For instance, on the one hand, high contrast highlights the most significant features for larger valleys and mountain ridges. On the other hand, such contrast does not allow one to notice the morphology of small forms [35,36]. High contrast is, however, appreciated in designing map symbols that are supposed to be seen in the context of the entire map content.

6. Conclusions

To summarize this analysis, one may conclude that water-based media offered an opportunity to represent relief suggestively and landscape realistically. Focusing on the discussed sophisticated painterly means of expression, the author concludes that the appropriate application of cartographic rules and the complementary selection of measurable geomedia parameters becomes a decisive factor when it comes to the creation of 3D visualizations from historical non-cartometric topographic maps. The rules for drawing contour lines should be applied, but along with the generalization of a necessary number of them, adequate for banding with multi-color and multi-shade spotting and stippling, and spotting or blotting in gray by brush, using water media (Indian ink, bistre, and wash). The vectorization of contour lines suggested here was based on the analysis of topography on the historical and modern maps. The image of the historical map was registered in the graphical layout with no georeference to the standard modern spatial system, because such georeference would distort the graphical means of expression. On the other hand, the adequate transformation of the image to the 3D model, based on vectorized contour lines, adjusts any distortions of the appropriate perception of the artistry in the water-colored image [37].

Generally speaking, the parameters of the non-cartometric topographic image in a 3D model can be selected in turn; however, what matters is to maintain their complementarity, as their effects will interact. On the basis of several parameter variants, the following parameter recommendations are as follows: vertical scaling = 4, illumination: azimuth 315° = NW, altitude 45°, contrast 40% (Figures 4–7). Those parameters work well for creating the general static 3D visualization for the perception of relief features, particularly with a bird's-eye view. On the other hand, the location of the observation point for the bird's-eye view and the rotation of the map surface, are subject to the individual manual action of the designer and an intention to balance the artistry and the measurability of the land-relief representation. In the case of these three watercolor maps, for a favorable overall perception of the relief (suggestive and close to landscape naturalness), the author proposes a gentle tilt and clockwise rotation of the topographic image surface, and an appropriate distance for the observation point (Figure 9).

The limitations on creating the fully renderable 3D models for topographic images in the painterly style result from the existing shading of mountain slopes and valley sides on the north-western side. Hence, the best effect in terms of suggestiveness in perceiving the landscape naturalness, characterizes a static 3D-visualization that resembles professional panoramas of mountain landscapes, created in artificial perspectives [38]. In the research, it is also recommended that a cartographic rule is applied that says that the mountains should be located further toward the top part of the view, whereas valleys and the entire area should be lowered in the direction of the viewer, who is located in the lower part.

Considering the high artistic and documenting value of topographic works from the eighteenth century, the establishment of parameters for interactive rendering of 3D models remains the issue for further research, as each work has its own individual use of artistic application of water-based media. The features of water-colored visualizations for the construction of suggestive and natural 3D visualizations presented here may be of great importance for the construction of scientific visualizations explaining the variability of phenomena on topographic scales since the eighteenth century.

Funding: This research received no external funding.

Data Availability Statement: Not applicable.

Conflicts of Interest: The author declares no conflict of interest.

References

1. Robinson, A.H.; Sale, R.D.; Morrison, J.L. *Elements of Cartography*, 4th ed.; John Wiley & Sons, Inc.: New York, NY, USA, 1978; ISBN 9780471017813.
2. Medyńska-Gulij, B. *Kartografia i Geomedia*; Wydawnictwo Naukowe PWN: Warsaw, Poland, 2021; ISBN 978-83-01-21554-5.
3. Edney, M. Mapping, Survey and Science. In *The Routledge Handbook of Mapping and Cartography*; Kent, A., Vujakovic, P., Eds.; Routledge: London, UK, 2018; pp. 292–325. ISBN 9781138831025.
4. Imhof, E. *Cartographic Relief Presentation*; Walter de Gruyter: Berlin, Germany, 1982; ISBN 9783110067118.
5. Kent, A.; Vujakovic, P. Cartographic Language: Towards a New Paradigm for Understanding Stylistic Diversity in Topographic Maps. *Cartogr. J.* **2011**, *48*, 21–40. [CrossRef]
6. Medyńska-Gulij, B.; Żuchowski, T.J. *European Topography in Eighteenth-Century Manuscript Maps*; Bogucki Wydawnictwo Naukowe: Poznań, Poland, 2018; ISBN 9788379862047.
7. Withers, C.W. *Placing the Enlightenment: Thinking Geographically about the Age of Reason*; The University of Chicago Press: Chicago, IL, USA, 2008; ISBN 9780226904054.
8. Medyńska-Gulij, B.; Żuchowski, T.J. An Analysis of Drawing Techniques used on European Topographic Maps in the Eighteenth Century. *Cartogr. J.* **2018**, *55*, 309–325. [CrossRef]
9. Mapire.eu. Available online: https://maps.arcanum.com/en/ (accessed on 6 September 2022).
10. Roy Military Survey of Scotland, 1747–1755. Available online: https://maps.nls.uk/geo/roy/#zoom=7&lat=56.8860&lon=-4.0709&layers=0 (accessed on 6 September 2022).
11. Fleet, C.; Kowal, K. Roy Military Survey map of Scotland (1747–1755): Mosaicing, geo-referencing, and web delivery. *E-Perimetron* **2007**, *2*, 194–208.
12. Timár, G.; Biszak, S.; Székely, B.; Molnár, G. Digitized Maps of the Habsburg Military Surveys: Overview of the Project of ARCANUM Ltd. (Hungary). In *Preservation in Digital Cartography*; Jobst, M., Ed.; Springer: Berlin-Heidelberg, Germany, 2011; pp. 73–283. ISBN 9783642127335.
13. Gartner, G. Seeing the "perfect world" through Heinrich Berann's Panorama Map of the Alps. *Int. J. Cartogr.* **2021**, *7*, 240–244. [CrossRef]
14. Veronesi, F.; Hurni, L. Changing the light azimuth in shaded relief representation by clustering aspect. *Cartogr. J.* **2014**, *51*, 291–300. [CrossRef]
15. Svobodová, J.; Voženílek, V. Relief for Models of Natural Phenomena. In *Landscape Modelling: Geographical Space, Transformation and Future Scenarios*; Anděl, J., Bičík, I., Dostál, P., Shasneshin, S., Eds.; Springer: Dordrecht, The Netherlands, 2009; Volume 8, pp. 183–196. ISBN 9789048130528.
16. Kettunen, P.; Koski, C.; Oksanen, J. A design of contour generation for topographic maps with adaptive DEM smoothing. *Int. J. Cartogr.* **2017**, *3*, 19–30. [CrossRef]
17. Rink, K.; Chen, C.; Bilke, L.; Liao, Z.; Rinke, K.; Frassl, M.; Yue, T.; Kolditz, O. Virtual geographic environments for water pollution control. *Int. J. Digit. Earth* **2018**, *11*, 397–407. [CrossRef]
18. Cornel, D.; Buttinger-Kreuzhuber, A.; Konev, A.; Horvath, Z.; Wimmer, M.; Heidrich, R.; Waser, J. Interactive Visualization of Flood and Heavy Rain Simulations. *Comput. Graph. Forum* **2019**, *38*, 25–39. [CrossRef]
19. Pingel, T.; Clarke, K. Perceptually Shaded Slope Maps for the Visualization of Digital Surface Models. *Cartographica* **2014**, *49*, 225–240. [CrossRef]
20. Collier, P.; Forrest, D.; Pearson, A. The Representation of Topographic Information on Maps: The Depiction of Relief. *Cartogr. J.* **2003**, *40*, 17–26. [CrossRef]

21. Dufour, L. *La Sicilia Disegnata. La Carta di Samuel Von Schmettau 1720–1721*; Società Siciliana per la Storia Patria: Palermo, Italy, 1995; ISBN 9788874010660.

22. Anderson, C. Constructing the Military Landscape: The Board of Ordnance Maps and Plans of Scotland, 1689–1815. Ph.D. Thesis, University of Edinburgh: Edinburgh, Scotland, 2009.

23. Garis, E. La Carta in Nove Parti Della Valle di Susa. In *Il Teatro Delle Terre: Cartografia Sabauda tra Alpi e Pianura*; Ricci, I., Gentile, G., Raviola, B.A., Eds.; L'Artistica Savigliano: Turin, Italy, 2006; pp. 212–240.

24. Hodson, Y. The Highland Survey 1747–1755 and the Scottish School of Cartography. Opening up the Highlands Highland History and Archives. *Scott. Rec. Assoc. Conf. Rep.* **1991**, *17*, 1–4.

25. Sereno, P. Li Ingegneri Topograffici di Sua Maesta. La formazione del cartografo militarenegli stati sabaudi e l'istituzione dell'Ufficio di Topografia Reale. In *Rappresentare uno Stato. Carte e Cartografi degi Stati Sabaudi dal XVI al XVIII secolo*; Comba, R., Sereno, P., Eds.; Umberto Allemandi & C.: Turin, Italy, 2002; Volume 1, pp. 61–102. ISBN 9788842207177.

26. Teissig, K. *Drawing Techniques*; Octopus Books Ltd: London, UK, 1983; ISBN 9780706417395.

27. Hutchinson, M.F.; Xu, T.; Stein, J.A. Recent Progress in the ANUDEM Elevation Gridding Procedure. In *Geomor-Phometry*; Hengel, T., Evans, I.S., Wilson, J.P., Gould, M., Eds.; Elsevier: Amsterdam, the Netherlands, 2011; pp. 19–22. Available online: http://geomorphometry.org/HutchinsonXu2011 (accessed on 6 August 2022).

28. Jenks, G.; Caspall, F. *Vertical Exaggeration in Three- Dimensional Mapping*; Technical Report No. 2.; Office of Naval Research: Washington, DC, USA, 1967.

29. Castner, H.; Roger Wheate, R. Re-assessing the Role Played by Shaded Relief in Topographic Scale Maps. *Cartogr. J.* **1979**, *16*, 77–85. [CrossRef]

30. Imhof, E. *Gelände und Karte*, 2nd ed.; Eugen Rentsch: Zürich, Switzerland, 1958.

31. *Lexikon der Kartographie und Geomatik*; Bollman, J.; Koch, W.-G. (Eds.) Spektrum Akademischer: Berlin/Heidelberg, Germany, 2001; Volume 1, p. 451. ISBN 978-3827410559.

32. Medyńska-Gulij, B. How the Black Line, Dash and Dot Created the Rules of Cartographic Design 400 Years Ago. *Cartogr. J.* **2013**, *50/4*, 56–368. [CrossRef]

33. Sieber, R.; Serebryakova, M.; Schnürer, R.; Hurni, L. Atlas of Switzerland Goes Online and3D—Concept, Architecture and Visualization Methods. In *Progress in Cartography*; Gartner, G., Jobst, M., Huang, H., Eds.; Springer: Cham, Switzerland, 2016; ISBN 9783319196022.

34. Verdier, N.; Besse, J.-M. Color and Cartography. In *The History of Cartography*; Edney, M.H., Pedley, M.S., Eds.; The Chicago University Press: Chicago, IL, USA, 2020; Volume 4, pp. 294–302. ISBN 9780226339221.

35. Leonowicz, A.M.; Jenny, B.; Hurni, L. Terrain sculptor: Generalizing terrain models for relief shading. *Cartogr. Perspect.* **2010**, *67*, 51–67. [CrossRef]

36. Patterson, T. A desktop approach to shaded relief production. *Cartogr. Perspect.* **1997**, *28*, 38–39. [CrossRef]

37. Medyńska-Gulij, B. Linear and painterly expression in topographic works of art during the Enlightenment. *Int. J. Cartogr.* **2021**, *7*, 158–163. [CrossRef]

38. Patterson, T. A View From On High: Heinrich Berann's Panoramas and Landscape Visualization Techniques for the U.S. National Park Service. *Cartogr. Perspect.* **2000**, *36*, 38–65. [CrossRef]

International Journal of
Geo-Information

Article

Geographic Approach: Identifying Relatively Stable Tibetan Dialect and Subdialect Area Boundaries

Mingyuan Duan and Shangyi Zhou *

Faculty of Geographical Science, Beijing Normal University, Beijing 100875, China;
202031051023@mail.bnu.edu.cn
* Correspondence: shangyizhou@bnu.edu.cn

Abstract: Updating dialect maps requires extensive language surveys. Geographic methods can be applied to identify relatively stable boundaries of dialect and subdialect areas, allowing language surveys to focus on boundaries that may change and thereby reduce survey costs. Certain scholars have pointed out that the watershed boundary can be employed as the boundary of Tibetan dialect areas. This paper adds that the lowest-grade road breakpoint line and no-man's-land boundary can also be used as essential indicators for determining stable (sub)dialect area boundaries. Combined with the revised First Law of Geography and the method of superposition analysis of geographic elements, this study identifies indicators that affect the stability of the Tibetan (sub)dialect area boundaries and evaluates the stability of each boundary segment. Due to the particularity of the study area, most Qinghai–Tibet Plateau (Chinese part) (sub)dialect area boundaries are stable. In addition, boundary inaccuracies caused by defects in the distribution of language survey samples can be identified by geographic approaches.

Keywords: dialect map; boundary; Qinghai–Tibet plateau; linguistic geography; cultural mapping

Citation: Duan, M.; Zhou, S. Geographic Approach: Identifying Relatively Stable Tibetan Dialect and Subdialect Area Boundaries. *ISPRS Int. J. Geo-Inf.* **2022**, *11*, 280. https://doi.org/10.3390/ijgi11050280

Academic Editors: Beata Medynska-Gulij, David Forrest, Thomas P. Kersten and Wolfgang Kainz

Received: 6 March 2022
Accepted: 26 April 2022
Published: 27 April 2022

Publisher's Note: MDPI stays neutral with regard to jurisdictional claims in published maps and institutional affiliations.

1. Introduction

Language or dialect maps are important references for cross-regional communication. Currently, the frequency of cross-regional movement is increasing [1], which makes it imperative to determine whether language or dialect area boundaries are crossed. In the past three years, our group members have conducted four field surveys on the Qinghai–Tibet Plateau, covering three Tibetan dialect areas. People from different dialect areas were discovered to have little understanding of each other's dialect. These survey experiences prompt us to consider why dialects differ so much in today's age of increased mobility. Human geographers tend to analyze boundary issues from the perspective of man–land relationships, believing the stable boundaries of dialect areas can be seen as the communication boundaries that are constantly maintained under the dynamic interactions between human beings and the earth. However, the boundaries have been further questioned regarding their either-or dualistic nature since they are considered to imply more hybridity and permeability [2]. Therefore, in addition to discussing how to accurately determine the boundaries of Tibetan dialect and subdialect areas by geographic approaches, the more significant contribution of this paper is to further study the stability of each boundary segment. In the context of accelerating mobility, this study tells people which segments of the boundaries would not change if the Language Atlas of China could not be updated annually.

Understanding whether the boundaries of language or dialect areas are crossed can help predict and circumvent the linguistic challenges associated with immigration, investment, travel, and other spatial behaviors. Mapping the dialect area boundary is necessary to help multiple subjects establish the structure of regions. The distribution of dialect areas is vague in most people's minds, but it is human instinct to seek lucidity in the world [3] since it helps make more favorable decisions. In addition, maps are always context-related [4], leading to different understandings of their meanings by different subjects. For individuals, clarifying the location of dialect area boundaries has important implications for interregional communication and the establishment of a sense of place. Language or dialect area boundaries comprise the medium for judging the depth of mutual understanding [5]. The less permeable the dialect area boundaries are, the greater the linguistic challenges people encounter when crossing them. As a result, individuals tend to connect with others in places where the language or dialect is closer to their own to improve satisfaction and efficiency of communication [6]. For firms, identifying language or dialect area boundaries can help determine whether cross-regional investment and business can bring benefits. Language or dialect similarity can facilitate the cross-regional blending of knowledge for spillover benefits [7]; cultural differences in language and dialect are the source of cultural innovation and new economic growth [8]. For the government, dialect area boundaries can strengthen the sense of place and identity of people within the boundaries [9,10], which also contribute to promoting social cohesion [11].

Language surveys are costly and time-consuming [12] since an insufficient number of survey samples can yield inaccuracies in the boundaries of dialect areas [13]. The linguistic method is based on vocabulary, pronunciation, and grammar surveys and is utilized to determine the boundaries of language and dialect areas [14,15]. There are currently only two atlases of the Chinese language: the 1987 edition and the 2012 edition. The Survey Operation Technical Specification (2012) stipulates that in minority areas, at least one survey site should be set for each language area and that for languages with large dialect differences, one survey site should be established for each dialect area [16]. However, the spatial units in the Tibet Autonomous Region and Qinghai Province are relatively large, and the internal geographic conditions vary greatly. The limitation of survey sites cannot fully reflect the differences in dialect areas in the region, causing a decrease in the accuracy of the boundaries. Additionally, the dialect area boundaries determined by linguistics methods cannot reflect the stability of every segment.

Analyzing the superposition of multiple geographic elements through maps can help quickly locate the boundaries of dialect areas and subdialect areas and assess the stability of the boundaries. A map is not only a medium of representation but also a means of analysis. Zelinsky claimed that maps are an important way to understand the meaning of language, but only a few studies have combined maps (such as topographic maps) with linguistics [17]. Jagessar believes that linguistics needs to be analyzed in conjunction with the spatial knowledge and geographic methods [18]. Moreover, language diffusion is a process that is constantly changing in time and space [19,20]. It is necessary to identify ways to optimize the survey plan to update the language map continuously. This paper utilizes a geographic approach to identify relatively stable (sub)dialect area boundaries, allowing language surveys to be concentrated in places more prone to change and reducing survey costs.

2. Literature Review

2.1. Geographic Approach to Determining Boundaries of Dialect Areas

Stable physical geography elements can be employed to accurately determine the boundaries of (sub)dialect areas. The Annales School suggests that physical geographic elements, such as the environment, ecology, and climate, maintain the basic stability of the regional structure over a long period [21]. The stability of this physical condition is the basis of the stability of human activities, thus affecting the formation of dialect area boundaries. Certain studies have discussed the influence of geographic elements on dialect area boundaries [22,23], for example, explaining the reasons for different vocabularies and grammars from a topographic perspective [24]. Mountains, dense forests, swamps [25], and beaches [26] can also serve as dividing lines between language and dialect areas., For example, some German and Italian speakers live on either side of a major ridge of the Alps [27]. The Qinling–Huaihe Line divides North China and South China and is an important Chinese dialect area boundary [28].

However, the boundaries formed by some physical elements are no longer insurmountable dialect area boundaries [1]; thus, it is more important to determine the stability of dialect area boundaries through interregional communication. In a unified communication space, people interact more frequently, and their dialects are more consistent. Geographers have agreed that watershed boundaries can serve as stable boundaries for dialect areas [29,30] since valleys can become centers for the locals to organize their activities, thus forming similar dialects [31]. The Basque language in northern Spain is centered in the barren limestone mountain area, and the lack of resources has kept the language boundary stable for a long time without attracting attention and entry from outsiders [32]. There is an Indo-European and Dravidian boundary between the water-retaining black soil and the thin red dry soil [33]; this soil-type boundary determines a human's life boundary. Another example is Bhutan, where the farmers in Mangde Valley and pastoralists in nearby mountains communicate by exchanging agricultural and pastoral products, thus forming a unified dialect area [34].

Researchers have discussed geographic methods to identify language or dialect area boundaries, but improvement is still needed. First, researchers have focused on geographic elements without judging the stability of these elements as dialect area boundaries. The types and levels of geographic elements reflect the difficulty of language transmission to a certain extent and affect the stability of language or dialect area boundaries. For instance, a higher mountain represents a great barrier to communication on both sides of the mountain. Second, how to identify the communication boundaries through geographic elements is a problem that needs to be further explored. The layers of the analysis are continuously enriched by the method of overlaying geographic elements. An additional geographic element layer means one more indicator for correcting the dialect area boundary.

2.2. Existing Methods for Determining the Tibetan Dialect Area Boundary

The Tibetan language in China has three major dialect regions (refer to Figure 1). Hermanns [35] suggested that the boundary between the Amdo dialect and the Kham dialect area is the watershed between the Yellow River Basin and the Yangtze River Basin. Chamberlain [36] comprehensively plotted the basin boundaries in the Qinghai–Tibet Plateau and revealed a high degree of coincidence with the Tibetan language-type distribution classified by Tournadre [37], a well-known expert in Tibetan studies. Roche [38] also drew the boundaries of the three Tibetan dialect areas according to watershed boundaries. Rongze Chen [39] proposed that the distribution boundary of the three Tibetan dialect areas basically overlapped with the watershed boundaries of the main rivers. Scholars in China and abroad have agreed that the Bayan Har Mountains and the Gangdise-Nyenchen Tanglha Mountains are the boundaries of the three major Tibetan dialect areas. However, researchers have not analyzed the level of each watershed. The influence of different watersheds on the Qinghai–Tibet Plateau in forming dialect area boundaries should differ.

Certain researchers have investigated the relationship between the socioeconomic region boundaries and the Tibetan subdialect area boundaries. The 1987 edition of the Tibetan dialect map was based on the classification of Aitang Qu [40]. According to the living forms of the locals, Qu divided the Amdo dialect into agricultural, pastoral, semiagricultural and semipastoral subdialect areas. According to the historical administrative region, the Dbus–Gtsang dialect area was divided into subdialect areas of Dbus, Gtsang, and Ngari. The Kham dialect area was divided into core, northern, western, and southern subdialect areas. Jumian Kelsang and Yangjing Kelsang divided the Kham dialect area into northern, southern, and pastoral subdialect areas according to the distribution of two highways and people's living forms [41]. Di Jiang further integrated the Tibetan dialect and subdialect areas into the Language Atlas of China published in 2012 [42] (refer to Figure 1).

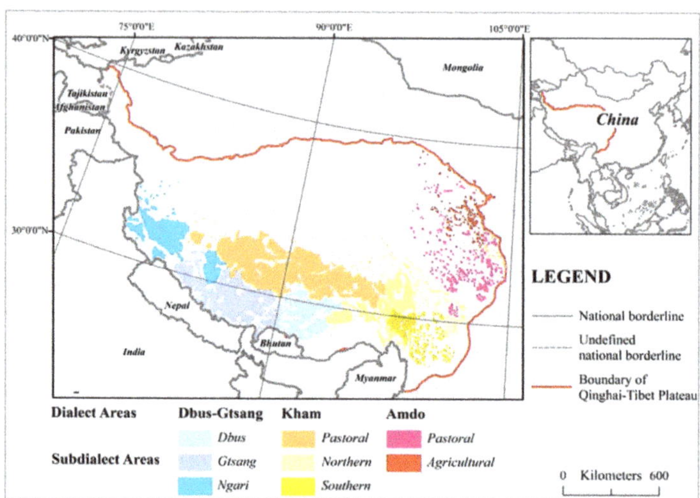

Figure 1. Location of the study area. (The Qinghai–Tibet Plateau of the Chinese part is located in southwestern China. The distribution map of the Tibetan dialect areas and subdialect areas was drawn by the Chinese Academy of Social Sciences in 2012).

In addition, there are some studies on the relationship between roads and (sub)dialects, but most of them are about the similarity of (sub)dialects caused by road connectivity, while studies on (sub)dialect area boundaries are insufficient. The roads are the carrier of the flow of people so that the spread of dialects can be continuous [43]. Some scholars have found that the dialects of villages situated near the same road are more similar to each other [44]. However, the boundaries of (sub)dialect areas still consist of the counties' boundaries, reducing the accuracy to some extent. Further exploration is needed to find a way to more accurately determine the boundaries of the (sub)dialect areas by road.

2.3. Combination of Maps and Language

Mapping is an important method of visual representation. Rich cultural meanings contained in language can be recorded and presented through maps. Geomedia is an intersection field of communication and geography, playing an essential role in helping spread knowledge [45]. Language and dialect act as carriers for promoting knowledge and thought transfer [46] and cultural sharing [47]. Place names are the most obvious linguistic marks and cultural landscapes on maps. Griebel gives an example to show that the place names on the map reflect Inuinnait's understanding of local natural conditions and provide valuable local knowledge [48]. Aporta believes adding the stories narrated by the indigenous Gwich'in to each place's name on the map is beneficial for recording the cultural significance of their living places [49].

Compared with the traditional language or dialect map, the multimedia sound map has more advantages in conveying embodied local feelings. The auditory dimension is the most prominent nonvisual dimensions, whose stimuli are highly correlated with the impressions and meanings generated when people experience the place [50]. Krygier first proposed that sound can expand the scope of map visualization [51]. The embedding of real soundscapes into multimedia maps can evoke people's feelings about places and maintain more lasting memories [50]. For instance, the background sound of the howling of the wind in a multimedia map is a way to enhance the sense of the Antarctic continent [52]. In addition, the development of sound in 3D and virtual reality technology allows people to be more immersive and feel an unprecedented sense of realism [53]. More real feelings of the first person can be obtained in the sound environment of the virtual map, which cannot be formed from God's perspective of the flat map. It is worth noting that sound maps can be applied not only to tourism, leisure, and entertainment but also to the medical and healthcare fields. Auditory maps help people better identify terrain [54], and users can generate a significantly stronger perception of spatial presence under immersive sound conditions [53]. All these practical studies are beneficial for improving the ability of visually impaired groups to perceive and acquire spatial information more accurately.

3. Study Area

The Qinghai–Tibet Plateau is an inland plateau in the southwestern part of China known as the "roof of the world". This plateau includes the entire territory of the Tibet Autonomous Region and Qinghai Province, as well as parts of the Xinjiang Uygur Autonomous Region, Gansu Province, Sichuan Province, and Yunnan Province. Following the Language Atlas of China (the 2012 edition): Minority Languages Volume (C1–25 Tibetan), this paper divides the Tibetan dialect area into three dialect areas and several subdialect areas (refer to Figure 1).

There are two reasons for choosing the Qinghai–Tibet Plateau (Chinese part) as the study area. First, the three major Tibetan dialect areas have a low degree of interoperability and relatively clear dialect area boundaries. Second, the density of the road network in the Qinghai–Tibet Plateau is low, and the breakpoints of the road network can be clearly identified on the map. The terrain of the Qinghai–Tibet Plateau fluctuates sharply, and the weather is harsh, making it difficult to build roads. In addition, the Qinghai–Tibet Plateau has a low population density, low total demand for road traffic, and low road network density in sparsely populated areas.

4. Research Methods

The main method of studying the (sub)dialect area boundary was to define the more accurate human communication boundary by road breakpoint lines and no-man's-land boundaries on the basis of the watershed boundary and then judge the stability of every segment of the (sub)dialect area boundary. The logical basis of this research was the revised First Law of Geography (Figure 2). The First Law of Geography states that everything is related to other things, but things that are spatially close are more closely related [55]. However, the law does not clearly define spatial proximity. This work defined spatial proximity as the distance that people in the Qinghai–Tibet Plateau interact with each other by road.

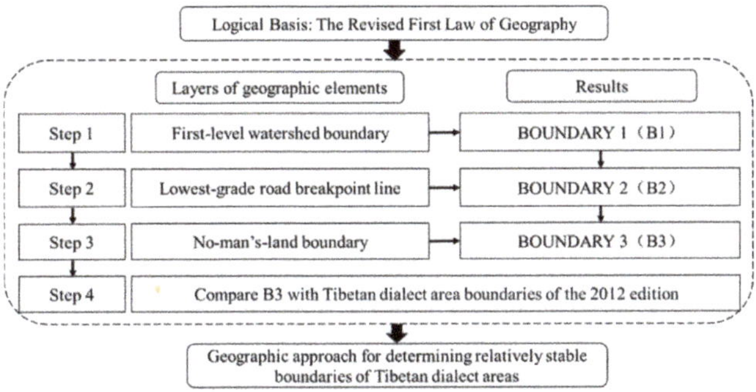

Figure 2. Method framework. (The logical basis of this study is the revised First Law of Geography. Three superimposed geographic elements are selected to determine the (sub)dialect area boundary: the boundary of the first-level watershed, lowest-grade road breakpoint line, and no-man's-land boundary).

The method employed in this paper was the layer overlay Geographic Information System (GIS) analysis method, which was divided into four steps. The first step was to select natural elements with high stability as basic dialect area boundaries. The second step was to determine the (sub)dialect area boundaries by connecting the breakpoints of the lowest-level roads. The third step was to refine the boundaries of (sub)dialect areas by identifying the boundaries of no-man's-land in the Qinghai–Tibet Plateau. The fourth step was to compare the boundaries of Tibetan (sub)dialect areas drawn by the above steps with the Tibetan dialect map of the 2012 edition and determine the relatively stable boundary segments. This paper aimed to develop a set of geographic methods for identifying the boundaries of (sub)dialect areas.

5. Analysis

5.1. Determine First-Level Watershed Boundaries (B1)

The high-grade mountains that constitute the boundaries of the first-level watershed in the Qinghai–Tibet Plateau are selected as the basis for the boundaries of the Tibetan dialect areas. There are many first-class rivers on the Qinghai–Tibet Plateau, including the Yangtze River, the Yellow River, the Lantsang River (also referred to as the Mekong River in Southeast Asia), the Nujiang River (also referred to as the Salween River in Southeast Asia), the Yarlung Zangbo River (also referred to as Brahmaputra River in Southeast Asia), and the Seng-ge Kambab River (which downstream becomes the Indus River in Pakistan). Not every mountain range that forms the boundary of a first-level watershed can be a boundary for dividing dialect areas. Referring to the hierarchical structure of the mountain ranges, we selected the highest-level mountain ranges as the basis for the boundaries of dialect areas. The Geographic Atlas of China [56] classified the mountain system in China. Based on the absolute and relative altitudes, the authors divided the mountains into extremely high mountains, moderately high mountains, and low mountains. However, most mountains on the Qinghai–Tibet Plateau are extremely high mountains (more than 3500 m above sea level), and differences in height cannot be identified on the map. Therefore, digital elevation model data of the Qinghai–Tibet Plateau are utilized to extract the ridgelines and divide the mountains into three grades according to their height and length.

The first-level mountain ranges have established the basic pattern of the three Tibetan dialect areas, which is consistent with the judgment of other scholars. As shown in Figure 3, the Kunlun-Bayan Har Mountains and the Gangdise-Nyenchen Tanglha Mountains form two important dividing lines between the Amdo and Kham and the Kham and Dbus-Gtsang dialect areas, respectively. This result is basically consistent with the boundaries of the three major Tibetan dialect regions presently agreed upon by related researchers. Therefore, the watershed boundaries formed by the first-class rivers are selected as the basic boundaries of the three Tibetan dialect areas, as shown in Figure 3, B1.

The hierarchy of mountain ranges affects the stability of (sub)dialect area boundaries, and the boundaries of (sub)dialect regions formed by high mountain ranges are more stable. In general, the stability of the boundaries of the three Tibetan dialect areas is higher than that of the subdialect areas, with exceptions. The mountain pass is an area with strong population flow, reducing the stability of the dialect area boundary. The exact location of the dialect area boundary here cannot be determined by the mountain range element but needs the assistance of other elements. For instance, the southeastern end of the Bayan Har Mountains has a wide mountain pass, which has become a communication channel between the Amdo dialect area and the Kham dialect area. Another example is the mountain pass between the Hengduan mountains and the Nyenchen Tanglha Mountains, which is also the intersection of the Kham and Dbus-Gtsang dialect areas.

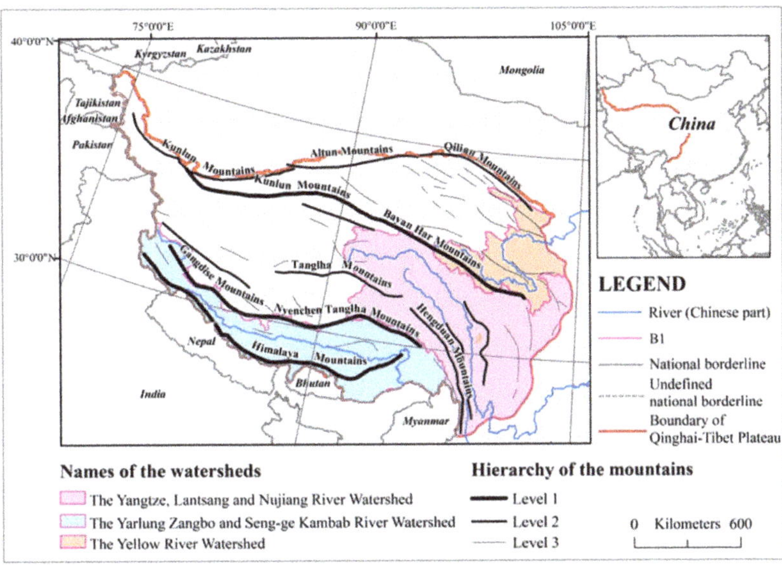

Figure 3. First-level watershed boundaries (B1). (The first-class rivers and high-grade mountains define the basic boundaries of the three dialect areas on the Qinghai–Tibet Plateau of the Chinese part).

5.2. Modify B1 with the Lowest-Grade Road Breakpoint Line (B2)

The breakpoint lines of the lowest-grade road (B2) form the edge of the road network. Among places where even the lowest-level of road connections are not available, dialects and subdialects can vary widely. The first law of geography points out the relationship between the closeness of connection and spatial proximity. People's communication and other activities are basically carried out along roads, so the similarity of (sub)dialects among areas with a short communication distance through roads is higher. Therefore, the road breakpoint line defines where the (sub)dialect area boundary is easily formed. Road grades in China include national highways, provincial highways, county roads, township roads, and other roads from high to low. This paper designates "other roads" as the lowest-level roads and uses GIS to identify the breakpoint line (B2) (Figure 4). Table 1 shows two methods for determining B2.

Table 1. Two ways to determine B2.

Situation	Schematic	Drawing Method
Lowest-grade road breakpoints appear on both sides		Join the midpoint of the two endpoints in turn
Only one side has the lowest-grade road breakpoints		Connect the road breakpoints directly in turn

[1] Note: Red lines represent lowest-grade road breakpoint lines; black lines represent roads.

Generally, the lowest-level road breakpoint lines between two dialect regions are relatively clear, while those between two subdialect regions are blurred. In areas with high connectivity of roads, it is impossible to identify the boundaries of subdialect areas based on road breakpoints, such as the boundaries of the Dbus and Gtsang subdialect areas. Therefore, other potential factors need to be included in the superposition analysis to identify the boundaries of the (sub)dialect areas.

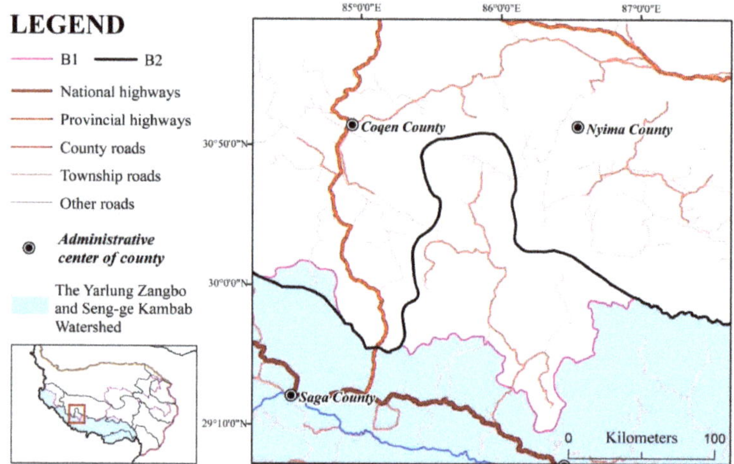

Figure 4. Example Area with B1 and B2. (At the junction of the three counties of Saga, Coqen, and Nyima, the basic boundary of the Yarlung Zangbo and Seng-ge Kambab River Basins (B1) is modified by the lowest-grade road breakpoint line (B2)).

5.3. Amend B2 with No-Man's-Land Boundaries (B3)

This paper modifies B2 with the no-man's-land boundary (B3) in the land cover type data. GlobeLand30 data include 10 main land cover types in the Qinghai–Tibet Plateau: cropland, forest, grassland, shrub, wetland, water, tundra, man-made surface, bare land, glacier, and permanent snow. Wetlands, water bodies, glaciers, and permanent snow are not suitable for human habitation; these three categories are defined here as depopulated areas. The boundaries of dialect or subdialect area defined by the lowest-grade road breakpoint lines may continue to extend, but no-man's-land boundaries are the limits of future road extension and can form more stable (sub)dialect area boundaries. Figure 5 shows the correction of B3 to B2 in Cuona and Lhunze County in southeastern part of Shannan city.

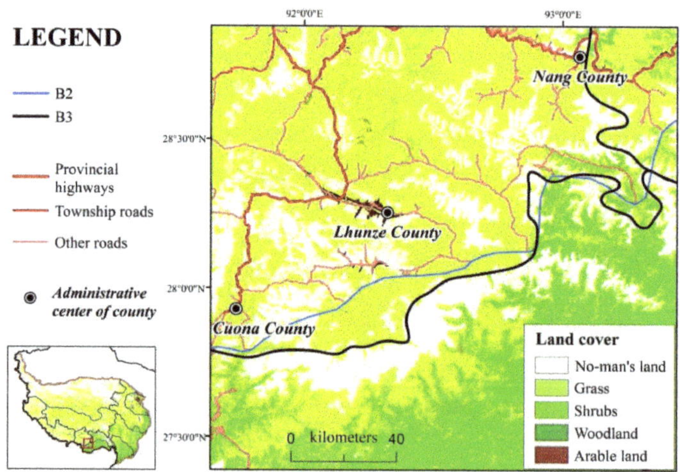

Figure 5. Example Area with B2 and B3. (In the southern part of Lhunze, Cuona, and Nang Counties, the breakpoint line of the lowest-grade roads (B2) is amended by the boundary of no-man's land (B3)).

5.4. Correct Seven Segments of the (Sub)Dialect Area Boundaries

Comparing the revised Tibetan (sub)dialect area boundaries with the 2012 version of the Tibetan dialect map, it reveals that 82.67% of the boundaries are consistent and that only seven boundary segments have significant differences, as shown in Table 2. The boundaries of the (sub)dialect areas, where the differences appear, are mainly composed of the lowest-grade road breakpoint lines, whose stability is lower than that of the physical elements.

Table 2. Seven boundary segments of (sub)dialect areas with significant differences and reasons.

Location	Type in the 2012 Map	Type after Modification	Possible Reasons for the Difference
Northeast of Coqen and Zhongba County (Ngari and Shigatse, Tibet)	Ngari and Gtsang (Dbus-Gtsang)	Pastoral (Kham)	First-level watershed boundary and road breakpoint line
Northern Baxoi County (Qamdo city, Tibet)	Northern (Kham)	Pastoral (Kham)	Road breakpoint line extends south to Bowo county and northern Baxoi County
Most of Gongbo'gyamda County (Nyingchi city, Tibet)	Gtsang (Dbus-Gtsang)	Northern (Kham)	No-man's-land boundary on the west
Northeastern Batang County and Litang County (Garze Prefecture, Sichuan)	Southern (Kham)	Northern (Kham)	Clear road breakpoint line
Nangchen County and Yushu city (Yushu city, Qinghai)	Northern (Kham)	Pastoral (Kham)	Road breakpoint line
Xinghai County and Zeku County (Hainan Tibetan Autonomous Prefecture, Huangnan Tibetan Autonomous Prefecture, Qinghai)	Pastoral (Amdo)	Agricultural (Amdo)	Road breakpoint line
Eastern Gangca County, eastern Menyuan Hui Autonomous County (Haibei Tibetan Autonomous Prefecture, Qinghai)	Pastoral (Amdo)	Agricultural (Amdo)	Road breakpoint line

5.5. Evaluate the Stability of (Sub)Dialect Area Boundaries

Fragments of (sub)dialect area boundaries are graded for stability by evaluating the stability of forming elements. This work classifies the level of mountains, clarity of the lowest-grade road breakpoint line and no-man's-land boundaries into two or three grades. The specific method for evaluating the stability level is shown in Table 3. The (sub)dialect area boundaries are divided into 27 segments according to the different combinations of geographic elements. The stability ratings for each segment are shown in Table 4.

Table 3. Method to determine the stability of the (sub)dialect area boundaries.

Indicators	Level of Stability	Grade	Index
First-level watershed boundary (mountains)	High	Level 1	++++ [1]
		Level 2	+++
		Level 3	++
No-man's-land boundary	Medium	Clear	+++
		Generally clear	++
Lowest-grade road breakpoint line	Low	Clear	++
		Generally clear	+

[1] Note: The symbol "+" represents the degree of stability of indicators of different types and grades, and the more "+", the higher the degree of stability.

Table 4. The influencing indicators and level of stability of each (sub)dialect area boundary segment.

Segment Number	Boundary	Stability Influencing Indicators	Level of Stability
1	Dbus-Gtsang and Kham	Clear road breakpoint line, clear no-man's-land boundary, level 1 mountains	+++++++++ [1]
2	Dbus-Gtsang and Kham	Generally clear road breakpoint line, generally clear no-man's-land boundary	+++
3	Dbus-Gtsang and Kham	Clear road breakpoint line, clear no-man's-land boundary	+++++
4	Dbus-Gtsang and Kham	Clear no-man's-land boundary, level 1 mountains, generally clear road breakpoint line	++++++++
5	Dbus-Gtsang and Kham	Clear road breakpoint line	++
6	Dbus-Gtsang and Kham	Generally clear road breakpoint line	+
7	Dbus-Gtsang and Kham	Generally clear road breakpoint line, level 1 mountains, generally clear no-man's-land boundary	+++++++
8	Dbus-Gtsang and Kham	Generally clear road breakpoint lines, generally clear no-man's-land boundary, level 2 mountains	++++++
9	Northern and Southern subdialect area	Clear road breakpoint lines, clear no-man's-land boundary, level 3 mountains	+++++++
10	Dbus-Gtsang/Kham dialect area and no-man's land	Clear road breakpoint lines, clear no-man's-land boundary, level 1 mountains	++++++++++
11	Dbus-Gtsang and Kham	Generally clear no-man's-land boundary, level 3 mountain range, generally clear road breakpoint line	+++++
12	Dbus and Gtsang	Generally clear no-man's-land boundary, generally clear road breakpoint line	+++
13	Ngari and Gtsang	Clear road breakpoint line, generally clear no-man's-land boundary	++++
14	Kham dialect area and no-man's land	Clear road breakpoint line	++
15	Kham dialect area and no-man's land	Clear road breakpoint line, generally clear no-man's-land boundary, level 2 mountains	+++++++
16	Northern and pastoral subdialect area	Clear road breakpoint line	++
17	Northern and pastoral subdialect area	Generally clear road breakpoint line, clear no-man's-land boundary	++++
18	Northern and pastoral subdialect area	Clear road breakpoint line, generally clear no-man's-land boundary	++++
19	Southern and Northern subdialect area	Clear road breakpoint line	++
20	Amdo dialect area and no-man's land	Clear road breakpoint line, generally clear no-man's-land boundary	++++
21	Amdo dialect area and no-man's land	Clear road breakpoint line, generally clear no-man's-land boundary	++++
22	Amdo and Kham	Clear road breakpoint line, level 1 mountains	++++++
23	Amdo and Kham	Generally clear road breakpoint line, level 1 mountains	+++++
24	Amdo and Kham	Generally clear road breakpoint line, level 3 mountains	+++
25	Agricultural and pastoral subdialect area	Generally clear road breakpoint line, level 2 mountains	++++
26	Agricultural and pastoral subdialect area	Clear road breakpoint line	++
27	Agricultural and pastoral subdialect area	Clear road breakpoint line, level 3 mountains	++++

[1] Note: The symbol "+" represents the level of stability of different segments of (sub)dialect area boundaries. The more "+", the higher the level of stability.

Judging the stability of dialect area boundaries is a refinement of and supplement to the Tibetan Dialect Map of the 2012 edition. The results show that the middle section of the boundary between the Dbus-Gtsang dialect area and the Kham dialect area is affected by the road passing through the Nyenchen Tanglha Mountains, where the stability decreases. The stability of the dialect area boundary in the southern Dbus-Gtsang area is high, mainly due to the blocking effect of the Himalaya Mountains. However, the boundaries of the Dbus and Gtsang subdialect areas have low stability, which is mainly affected by the dense road networks. The northwestern boundary of the Kham dialect area is relatively stable and largely blocked by the Tanglha Mountains, but the central part continues to expand to the interior of the Qinghai–Tibet Plateau. The boundaries of the subdialect areas within the Kham dialect area are mainly affected by roads and have low stability. The western part of the Amdo dialect area is the uninhabited area, and the boundary is relatively stable, while on its southeast side, the blocking effect of the lowest-level mountain is reduced by the development of roads. The internal road networks in the Amdo dialect area form gradually, and the stability of this subdialect area boundary decreases.

6. Conclusions and Discussion

6.1. Conclusions

Three conclusions are drawn based on the above analysis.

First, the road breakpoint line can be a reliable and effective indicator for determining the boundaries of (sub)dialect areas. This paper shows that the stability of each (sub)dialect area boundary segment is not the same everywhere in the Qinghai–Tibet Plateau. The huge mountains determine the breakpoint lines of the road systems and the frequency of people's interaction, thereby affecting the consistency of the dialect or subdialect.

Second, due to the particularity of the study area, most of the (sub)dialect area boundaries are stable. Owing to the superposition analysis of the first-level watershed boundaries, the lowest-level road breakpoint lines and no-man's-land boundaries, the stability of the dialect area boundaries is basically higher than those of the subdialect area boundaries, with some exceptions in the Dbus-Gtsang and the Kham dialect areas.

Third, the geographic method can be applied to identify inaccurate locations of (sub)dialect area boundaries that lack survey samples, and it also comprises a good alternative to linguistic methods for mapping (sub)dialect area boundaries. There are only a few inconsistencies between the (sub)dialect map drawn by geographic methods and the 2012 edition of the Tibetan dialect map. The main reason for the inconsistency is the instability caused by road expansion. Geographic methods can update (sub)dialect area boundaries according to physical and human geographic data, saving considerable manpower and material resources. With the continuous development of geographic technologies such as remote sensing and geographic information system, various methods of obtaining geographic data have become increasingly convenient. Therefore, exploring the (sub)dialect area boundaries through geographic methods before linguistic fieldwork would be more feasible.

6.2. Discussion

The boundaries between the three Tibetan dialect areas remain remarkably stable, predicting that it will continue to maintain the basic pattern of the three dialect areas in the Qinghai–Tibet Plateau in the future. High-grade mountains have been confirmed to have the greatest influence on the stability of dialect area boundaries. The boundaries of dialect areas with high stability are almost always east–west, which is consistent with the direction of the main mountain ranges. In addition, the mountain passes are the locations where roads are most likely to extend. Almost all the road breakpoint lines have changed at mountain passes, thus changing the stability of the boundaries of the dialect area. From the perspective of time geography, people's travel behavior will be constrained to a certain space-time range [57]. However, the construction of roads saves people's travel time, thus greatly expanding the scope of people's movement. Therefore, with the construction of roads, the boundaries of the three dialect areas not only expanded to the interior part of the Qinghai–Tibet Plateau but also overlapped and intersected with each other, thus decreasing the stability of certain boundary segments of the three dialect areas.

This study has some limitations. First, most of the seven differences between the results of this study and the Tibetan dialect map of the 2012 edition are related to road breakpoint lines, which create some uncertainty. We speculate that there are two possible reasons for the difference. One is that there are not enough superimposed layers, leading to a lack of accurate judgment of the results. More elements need to be explored for overlay analysis to further determine the location and stability of the (sub)dialect area boundary. The other reason would be that the boundary of the (sub)dialect area has already changed to some extent due to road network expansion, especially in some places without any impact of insuperable natural factors. Hence, we strongly recommend that linguists conduct language surveys for further confirmation in areas where the results drawn by geographic methods are inconsistent with those of linguistic studies.

Second, due to data availability, this study utilized only data on watersheds, roads, and land cover types to delineate the boundaries of dialect and subdialect areas. More overlapping influencing factors should be explored in the future. The habit of following historical administrative boundaries is a factor that affects the stability of the boundaries of (sub)dialect areas. Certain scholars suggest that the boundaries of administrative regions are representative of other boundaries formed by various elements over a long period and that they are better suited to become the boundaries of dialect areas than the boundaries of religious or economic regions [58]. Geographic boundaries are also constantly undergoing the process of reification–naturalization–fetishization [59]; that is, once an administrative boundary is formed, people will continue to strengthen their awareness of it as a boundary, and this habitual consciousness will become an important factor affecting the stability of the boundaries of (sub)dialect areas. For example, a vast swamp between Upper Carniola and Lower Carniola in Slovenia has long formed a boundary between dialect areas in the two regions, but a few years after the swamp dried up, the dialect area boundary did not disappear [25]. In addition to determining boundaries through tangible entities, the influence of historical, political, emotional, and other factors must be considered, and the influence of these factors will be more difficult to identify.

The administrative boundaries of the Yuan Dynasty had a great influence on the division of dialect areas and showed a certain continuity and heritability. Xuan Zheng Yuan was established during the Yuan Dynasty in China. It was an administrative department directly under the jurisdiction of the central government, responsible for national Buddhist affairs and managing military and political affairs in Tubo (roughly corresponds to the Qinghai–Tibet Plateau). It governed three administrative regions that were consistent with the three Tibetan dialects areas of Amdo, Kham, and Dbus-Gtsang. Future studies will continue to explore the historical impact on the stability of (sub)dialect area boundaries.

Third, our research still needs to be improved in terms of the combination of Tibetan dialects and mapping, which will be a vast and promising field. In terms of the representation forms, our future research will try to demonstrate more audio-visual materials of Tibetan dialect culture to help map readers have more embodied experiences. Tibetan opera, songs, and legends are all spread through different (sub)dialects. We believe that attaching these audio-visual materials to maps could greatly enhance users' perception dimensions of the Qinghai–Tibet Plateau, thereby expanding the usability of the dialect map. As for the application fields, the (sub)dialect difference reflected by the Tibetan dialect can be seen as a cultural parameter to explain social and economic relationships between different regions. Jackson proposed that cultural geographers should pay more attention to the social meanings mapped by language [60]. Therefore, Figure 6 shows not only the boundaries of dialect areas or subdialect areas but also the boundaries of people's social interactions on the Qinghai–Tibet Plateau.

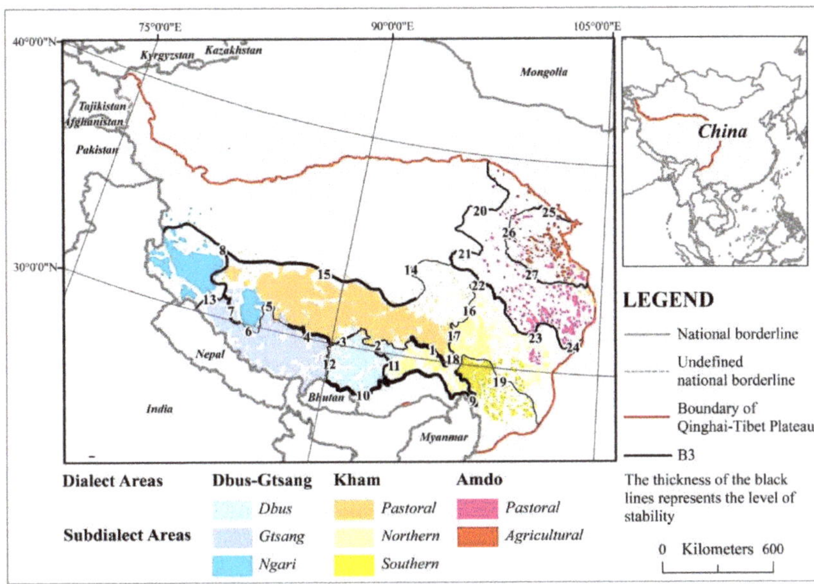

Figure 6. Segmentation of Tibetan (sub)dialect area boundaries. (Compare B3 with boundaries drawn by the Chinese Academy of Social Sciences in 2012 formed by language surveys. The positions of the two boundaries are similar, but the stability of the boundaries of each segment differs. The numbers in Figure 6 represent the serial numbers of the (sub)dialect area boundaries of each segment, corresponding to those in Table 4).

Author Contributions: Mingyuan Duan wrote the major parts of the article. Shangyi Zhou provided important suggestions and modified the content of the paper. All authors have read and agreed to the published version of the manuscript.

Funding: This research was supported by the Second Tibetan Plateau Scientific Expedition and Research Program (Grant No. 2019QZKK0608).

Institutional Review Board Statement: Not applicable.

Informed Consent Statement: Not applicable.

Data Availability Statement: Four types of data were used in this work. The digital elevation data were obtained from the National Catalogue Service for Geographic Information (https://www.webmap.cn/commres.do?method=result100W, (accessed on 8 May 2021)); the watershed boundary data were obtained from the Geographic Data Sharing Infrastructure, College of Urban and Environmental Science, Peking University (http://geodata.pku.edu.cn, (accessed on 5 November 2020)); the road data were obtained from OpenStreetMap (http://download.geofabrik.de/, (accessed on 20 June 2021)); and the land cover type data were obtained from the Globeland30 Database (http://globeland30.org/, (accessed on 8 May 2021)). To match the Tibetan dialect map published in 2012, the production year of all data applied in this paper is approximately 2012.

Acknowledgments: We would like to thank the anonymous reviewers for helping us improve the manuscript and thank for the data support from Geographic Data Sharing Infrastructure, College of Urban and Environmental Science, Peking University (http://geodata.pku.edu.cn, (accessed on 5 November 2020)).

Conflicts of Interest: The authors declare no conflict of interest.

References

1. Sheller, M.; Urry, J. The new mobilities paradigm. *Environ. Plan. A* **2006**, *38*, 207–226. [CrossRef]
2. Atkinson, D.; Jackson, P.; Sibley, D.; Washbourne, N. *Cultural Geography. A Critical Dictionary of Key Concepts*; IB Tauris: London, UK, 2005; pp. 153–154.
3. Tuan, Y.F. *Escapism*; JHU Press: Baltimore, MD, USA, 1998; pp. 22–24.
4. Kitchin, R.; Dodge, M. Rethinking maps. *Prog. Hum. Geogr.* **2007**, *31*, 331–344. [CrossRef]
5. Wardhaugh, R.; Fuller, J.M. *An Introduction to Sociolinguistics*; John Wiley & Sons: Hoboken, NJ, USA, 2021; p. 27.
6. Dooly, M.; Vallejo Rubinstein, C. Bridging across languages and cultures in everyday lives: An expanding role for critical intercultural communication. *Lang. Intercult. Commun.* **2018**, *18*, 1–8. [CrossRef]
7. Basile, R.; Capello, R.; Caragliu, A. Interregional knowledge spillovers and economic growth: The role of relational proximity. In *Drivers of Innovation, Entrepreneurship and Regional Dynamics*; Springer: Berlin/Heidelberg, Germany, 2011; pp. 21–43.
8. Weidenfeld, A.; Björk, P.; Williams, A.M. Cognitive and cultural proximity between service managers and customers in cross-border regions: Knowledge transfer implications. *Scand. J. Hosp. Tour.* **2016**, *16* (Suppl. 1), 66–86. [CrossRef]
9. Segrott, J. Language, geography and identity: The case of the Welsh in London. *Soc. Cult. Geogr.* **2001**, *2*, 281–296. [CrossRef]
10. Stafecka, A. Latvian dialects in the 21 century: Old and new borders. *Acta Balt. Slav.* **2015**, *39*, 1–13. [CrossRef]
11. Haugen, E. Dialect, Language, Nation. *Am. Anthropol.* **1966**, *68*, 922–935. [CrossRef]
12. Fernández-Villanueva, M.; Jungbluth, K. *Beyond Language Boundaries: Multimodal Use in Multilingual Contexts*; Walter de Gruyter GmbH & Co KG: Berlin, Germany, 2016.
13. Kessler, B. Computational dialectology in Irish Gaelic. In Proceedings of the Seventh Conference on European Chapter of the Association for Computational Linguistics, Dublin, Ireland, 27–31 March 1995.
14. Auer, P. The geography of language: Steps toward a new approach. *FRAGL Freibg. Arb. Zur Ger. Linguist.* **2013**, *16*, 11.
15. Gregory, D.; Johnston, R.; Pratt, G.; Watts, M.; Whatmore, S. *The Dictionary of Human Geography*; Basil Blackwell: Oxford, UK, 1981; p. 441.
16. Survey Manual of Chinese Language Resources • Ethnic Languages (Tibeto-Burman Group). Available online: http://www.chinalanguages.cn/gongjuyangben.html (accessed on 6 March 2022).
17. Zelinsky, W.; Williams, C.H. The mapping of language in North America and the British Isles. *Prog. Hum. Geogr.* **1988**, *12*, 337–368. [CrossRef]
18. Jagessar, P. Geography and linguistics: Histories, entanglements and departures. *Geogr. Compass* **2020**, *14*, 1–10. [CrossRef]
19. Heeringa, W.; Nerbonne, J. Dialect areas and dialect continua. *Lang. Var. Change* **2001**, *13*, 375–400. [CrossRef]
20. Britain, D. Space, Diffusion and Mobility. *Handb. Lang. Var. Change* **2013**, *129*, 471.
21. Braudel, F. History and the social sciences: The long term. *Soc. Sci. Inf.* **1970**, *9*, 144–174. [CrossRef]
22. De Busser, R.; LaPolla, R.J. *Language Structure and Environment: Social, Cultural, and Natural Factors*; John Benjamins Publishing Company: Amsterdam, The Netherlands, 2015; Volume 6.
23. Mullonen, I.; Zaiceva, N. Areal distribution of Veps topographical vocabulary. *Linguist. Ural.* **2017**, *53*, 106.
24. Bisang, W. Areal typology and grammaticalization: Processes of grammaticalization based on nouns and verbs in East and mainland South East Asian languages. *Stud. Lang.* **1996**, *20*, 519–597. [CrossRef]
25. Podobnikar, T.; Škofic, J.; Horvat, M. Mapping and analysing the local language areas for Slovenian linguistic atlas. In *Cartography in Central and Eastern Europe*; Springer: Berlin/Heidelberg, Germany, 2009; pp. 361–382.
26. Tönnies, M. Foregrounding Boundary Zones: Martin Parr's Photographic (De-)Constructions of Englishness. In *Landscape and Englishness*; Brill Academic Publishers: New York, NY, USA, 2006; pp. 225–240.
27. Terry, G.J.; Lester, R.; Hsu, M.L. *The Human Mosaic: A Thematic Introduction to Cultural Geography*; Harpercollins College Div: New York, NJ, USA, 1976; pp. 141–168.
28. Zhang, W.J. Three important north-south lines in Chinese dialects. *Lang. Res.* **2018**, *38*, 1–14.
29. Shafer, R. Classification of the Sino-Tibetan languages. *Word* **1955**, *11*, 94–111. [CrossRef]
30. Malmberg, B. *Structural Linguistics and Human Communication*; Springer: Berlin/Heidelberg, Germany, 1963.
31. Rohlf, G. A preliminary investigation of the urban morphology of towns of the Qinghai-Tibet plateau. *Chin. Hist. Geogr. Perspect.* **2015**, 159–178.
32. Urban, M. The geography and development of language isolates. *R. Soc. Open Sci.* **2021**, *8*, 202–232. [CrossRef]
33. Bennett, C.J. The morphology of language boundaries: Indo-Aryan and Dravidian in peninsular India. *Geogr. Perspect. Soc. Cult.* **1980**, 234–251.
34. Dorji, J. Hen Kha: A Dialect of Mangde Valley in Bhutan. *J. Bhutan Stud.* **2011**, *24*, 69–86.
35. Hermanns, M. Tibetische Dialekte von Amdo. *Anthropos* **1952**, *H. 1/2*, 193–202.
36. Chamberlain, B. Linguistic watersheds: A model for understanding variation among the Tibetic languages. *J. Southeast Asian Linguist. Soc.* **2015**, *8*, 71–96.
37. Tournadre, N. The Tibetic languages and their classification. In *Trans-Himalayan Linguistics*; De Gruyter Mouton: Berlin, Germany, 2013; pp. 105–130.
38. Roche, G. Introduction: The transformation of Tibet's language ecology in the twenty-first century. *Int. J. Sociol. Lang.* **2017**, *245*, 1–35. [CrossRef]

39. Chen, R.Z. The distribution pattern of Tibetan dialects and its historical, geographical and humanistic background. *J. Minzu Univ. China Philos. Soc. Sci. Ed.* **2016**, *43*, 128–134.
40. Qu, A.T.; Jin, X.J. Research methods of Tibetan dialects. *J. Southwest Inst. Natl. Philos. Soc. Sci.* **1981**, *3*, 79–87.
41. Kelsang, J.M.; Kelsang, Y.J. *An Introduction to Tibetan Dialects*; Nationalities Publishing House: Beijing, China, 2002; pp. 1–3.
42. Chinese Academy of Social Sciences. *Language Atlas of China*; Commercial Press: Beijing, China, 2012.
43. Britain, D. The role of mundane mobility and contact in dialect death and dialect birth. In *English as a Contact Language*; Hundt, M., Schreier, D., Eds.; Cambridge University Press: Cambridge, UK, 2013; pp. 165–181.
44. Hildebrandt, K.A.; Hu, S. Areal analysis of language attitudes and practices: A case study from Nepal. In *Documenting Variation in Endangered Languages*; University of Hawai'i Press: Honolulu, HI, USA, 2017; pp. 1–5.
45. Fast, K.; Ljungberg, E.; Braunerhielm, L. On the social construction of geomedia technologies. *Commun. Public* **2019**, *4*, 89–99. [CrossRef]
46. Akmajian, A.; Farmer, A.K.; Bickmore, L. *Linguistics: An Introduction to Language and Communication*; MIT Press: Cambridge, MA, USA, 2017.
47. Keesing, R.M. Theories of culture. *Annu. Rev. Anthropol.* **1974**, *3*, 73–97. [CrossRef]
48. Griebel, B.; Keith, D. Mapping Inuinnaqtun: The Role of Digital Technology in the Revival of Traditional Inuit Knowledge Ecosystems. *ISPRS Int. J. Geo-Inf.* **2021**, *10*, 749. [CrossRef]
49. Aporta, C.; Kritsch, I.; Andre, A. The Gwich'in Atlas: Place names, maps, and narratives. In *Modern Cartography Series*; Academic Press: New York, NJ, USA, 2014; Volume 5, pp. 229–244.
50. Edler, D.; Kühne, O.; Keil, J. Audiovisual cartography: Established and new multimedia approaches to represent soundscapes. *KN-J. Cartogr. Geogr. Inf.* **2019**, *69*, 5–17. [CrossRef]
51. Krygier, J.B. Sound and geographic visualization. In *Visualization in Modern Cartography*; MacEachren, A.M., Taylor, D.R.F., Eds.; Elsevier: Oxford, UK, 1994; pp. 149–166.
52. Pulsifer, P.L.; Caquard, S.; Taylor, D.R. Toward a new generation of community atlases—The cybercartographic atlas of Antarctica. In *Multimedia Cartography*; Springer: Berlin/Heidelberg, Germany, 2007; pp. 195–216.
53. Hruby, F. The sound of being there: Audiovisual cartography with immersive virtual environments. *KN-J. Cartogr. Geogr. Inf.* **2019**, *69*, 19–28. [CrossRef]
54. Schito, J.; Fabrikant, S.I. Exploring maps by sounds: Using parameter mapping sonification to make digital elevation models audible. *Int. J. Geogr. Inf. Sci.* **2018**, *32*, 874–906. [CrossRef]
55. Tobler, W.R. A Computer Movie Simulating Urban Growth in the Detroit Region. *Econ. Geogr.* **1970**, *46* (Suppl. 1), 234–240. [CrossRef]
56. Wang, J.; Zuo, W. *The Geographical Atlas of China*; SinoMaps Press: Beijing, China, 2010.
57. Golledge, R.G. *Spatial Behavior: A Geographic Perspective*; Guilford Press: New York, NJ, USA, 1997.
58. Derungs, C.; Sieber, C.; Glaser, E.; Weibel, R. Dialect borders—Political regions are better predictors than economy or religion. *Digit. Scholarsh. Humanit.* **2020**, *35*, 276–295. [CrossRef]
59. Schaffter, M.; Fall, J.J.; Debarbieux, B. Unbounded boundary studies and collapsed categories: Rethinking spatial objects. *Prog. Hum. Geogr.* **2010**, *34*, 254–262. [CrossRef]
60. Jackson, P. *Maps of Meaning*; Taylor & Francis: London, UK, 2012; pp. 155–170.

 International Journal of
Geo-Information

Article

Web Mapping and Real–Virtual Itineraries to Promote Feasible Archaeological and Environmental Tourism in Versilia (Italy)

Marco Luppichini [1,2], Valerio Noti [3], Danilo Pavone [4], Marzia Bonato [5], Francesco Ghizzani Marcìa [6], Stefano Genovesi [6], Francesca Lemmi [5], Lisa Rosselli [6], Neva Chiarenza [7], Marta Colombo [7], Giulia Picchi [7], Andrea Fontanelli [8] and Monica Bini [1,9,10,*]

[1] Department of Earth Sciences, University of Pisa, Via S. Maria, 52, 56126 Pisa, Italy
[2] Department of Earth Sciences, University of Study of Florence, Via La Pira 4, 50121 Florence, Italy
[3] TerreLogiche S.r.l., Via G. Verdi 3, 57021 Venturina Terme, Livorno, Italy
[4] National Research Council (CNR), ISPC—Isituto di Scienze del Patrimonio Culturale, 56126 Catania, Italy
[5] Civico Museo Archeologico di Camaiore, 55041 Camaiore, Italy
[6] Department of Civilisation and Forms of Knowledge, University of Pisa, Via dei Mille 19, 56126 Pisa, Italy
[7] Soprintendenza Archeologica Belle Arti e Paesaggio per le province di Lucca e Massa Carrara, Piazza della Magione, 55100 Lucca, Italy
[8] Oasi LiPU del Chiarone Parco Migliarino San Rossore Massaciuccoli, Via del Porto in Massaciuccoli 154, 55054 Massarosa, Italy
[9] Centro Interdipartimentale di Ricerca per lo Studio degli Effetti del Cambiamento Climatico (CIRSEC), University of Pisa, Via del Borghetto 80, 56124 Pisa, Italy
[10] Istituto Nazionale di Geofisica e Vulcanologia (INGV), sez. Pisa, via Cesare Battisti 53, 56125 Pisa, Italy
* Correspondence: monica.bini@unipi.it

Citation: Luppichini, M.; Noti, V.; Pavone, D.; Bonato, M.; Ghizzani Marcìa, F.; Genovesi, S.; Lemmi, F.; Rosselli, L.; Chiarenza, N.; Colombo, M.; et al. Web Mapping and Real–Virtual Itineraries to Promote Feasible Archaeological and Environmental Tourism in Versilia (Italy). *ISPRS Int. J. Geo-Inf.* **2022**, *11*, 460. https://doi.org/10.3390/ijgi11090460

Academic Editors: Wolfgang Kainz, Beata Medynska-Gulij, David Forrest and Thomas P. Kersten

Received: 20 June 2022
Accepted: 18 August 2022
Published: 28 August 2022

Publisher's Note: MDPI stays neutral with regard to jurisdictional claims in published maps and institutional affiliations.

Abstract: The Versilia plain (NW Italy) experiences forms of tourism that are mainly limited to the beach area and concentrated in the summer season. The area is rich in cultural and natural heritage, not yet adequately enhanced. The presence of four local archaeological museums and a natural park offers a great opportunity to favour feasible archaeological and environmental tourism. The aim of this study is to use a holistic methodology to improve a different type of tourism in the study area. We propose a consilient multidisciplinary approach based on geological, biological and archaeological data in order to enhance the cultural and natural heritage of the Versilia plain. We have based our study on the reconstruction of palaeoenvironment maps showing the evolution of the territory and used them as a leitmotiv to link the archaeological museums and the natural park. We define real and virtual itineraries to create a synergy between the most important archaeological and natural sites and museums. It is possible to promote a different type of tourism in the study area by decreasing human impact and creating a relationship between the fragmented natural and archaeological heritage. Palaeoenvironment maps and real and virtual itineraries can be consulted with the aid of a web application, more specifically web mapping, developed with free and open-source libraries. The web mapping also contains other geological, geomorphological and archaeological datasets, which allow to understand the evolution of the environment and the cultural and natural heritage of the study area. The dataset available on the web mapping is also downloadable.

Keywords: consilient approach; real itinerary; virtual tour; web application; sustainable tourism; web mapping; palaeoenvironment map

1. Introduction

Tourism on a global scale is an economic sector that has undergone tremendous growth over the last 50 years [1,2]. Unfortunately, this growth has not always had a positive effect on the destination locations that have sometimes experienced overtourism [3]. Fortunately, in recent years, this marked growth in the tourism sector has also seen a shift towards sustainable forms of tourism [4]. Sustainable tourism can be defined as tourism that considers its current and future impact on the environment, society and economy whilst meeting the needs of customers, industry, the environment and host communities [1,2]. In sustainable tourism, the negative impacts on the destination (e.g., economic leakage, damage to the natural environment and overcrowding) need to be compensated by positive impacts (e.g., job creation, cultural heritage preservation and interpretation, wildlife preservation and landscape restoration) [5]. After 2017 (the year of sustainable tourism), public awareness of the concept has increased, consolidating a trend that is starting to make inroads in terms of expectations and conduct of both those who travel and those who operate in the industry [6]. In this global trend, Italy—for which the tourism sector accounts for more than 5% of GDP and more than 6% of employment—is not an exception [7]. Particularly relevant are the safeguarding and promotion of its cultural heritage, one of the country's indisputable assets, as well as the protection of its environment [7]. However, tourism potential still needs to be better exploited by the coordination of tourism policies [7]. Indeed, some areas still experience undersized tourism, while other areas risk overcrowded tourism [7]. Versilia is mainly characterised by seaside tourism focused on the beach, where there are overcrowded phenomena, especially during the summer season; on the other hand, the numerous cultural and environmental assets of the immediate hinterland are hardly known and visited. In this territory, there are four archaeological museums, several archaeological sites and important environmental assets, among which the Lake of Massaciuccoli (Site of Community Importance and hot spot of biodiversity—one of the most important wetlands of the Italian peninsula) stands out [8]. However, as a result of the lack of synergy and of a shared strategy among the managing bodies, tourism linked to these cultural/natural sites is still undersized.

Web mapping [9], which is the process of implementing and displaying maps on the World Wide Web, allows the visualisation and updating of geospatial data. Web mapping facilities are widely used to store and display geological and geomorphological data from national geological and cartographic surveys. The aim of web mapping is to create platform-independent software systems open to every computer connected to the internet and running on an appropriate web server [10].

Web mapping integrates two powerful technologies: GIS and the internet, by providing global connectivity [11]; the result of this synergy translates into greater ease in finding data, sharing analytical tools and reaching a greater number of users [12,13].

Several studies have developed WebGIS and web mapping in various fields, including radon risk management [14], a risk assessment system for heavy metal pollution [15], georeferenced bibliographies [16], wastewater treatment [17], systems for monitoring and managing the aquatic environment [18], planning and emergency phases in the event of floods [19] and management of transport infrastructures [20].

In the area of geoheritage studies, GIS and web mapping technologies have already been used in various fields, including digital field mapping [10,21], visualisation of geoheritage data on open-source interfaces [10,21–24] and virtual trips [23,25,26]. In particular, the purpose of such applications is to take advantage of new technologies to enhance the value of geodiversity and archaeology through a virtual itinerary [24,27–29].

In order to obtain a usable web map, the design should be adapted not only to the preferences and needs of the user but also to provide the requested information efficiently and easily. Third-generation mobile phones constitute an infrastructure with strong potential for the interpretation and promotion of geosites. They combine the necessary operating system (Android or iOS) with GPS georeferencing and an internet connection, which offers a wide range of opportunities such as location-based information [21,30], QR codes or multimedia content.

In this study, we would like to promote a new synergy among all the stakeholders involved (nonprofit associations, management authorities, museums, natural park authorities, research entities, tourism industry operators, local communities, tourists, etc.) in order to propose forms of archaeological and environmental tourism in an area characterised by high archaeological and natural heritage, not yet adequately known and visited.

All stakeholders are involved in the creation of real and virtual tours based on palaeo-geographic maps and connecting the most important archaeological sites, museums and natural assets of the study area. These real and virtual networks are enhanced by the activation of an open-source web application for the visibility and consultation of different data typologies useful to improve the knowledge of the study area.

2. Study Area

2.1. Geological Background

The study area is located in a part of the so-called 'Versilia plain', administratively belonging to the Lucca Province and to the municipalities of Pietrasanta, Forte dei Marmi, Seravezza, Stazzema, Camaiore and Viareggio.

The area is morphologically divided into three altimetric bands, which are parallel to the coast. Indeed, we can distinguish a mountain sector with maximum elevations higher than 1500 m a.s.l. (M. Altissimo, M. Corchia, M. Pania della Croce), and characterised by a diffuse outcrop of Mesozoic metamorphic and carbonate rocks positioned on a Paleozoic schist-philladic basement [9–11,31–33].

This territory is historically characterised by rare and sporadic settlements of small villages associated with pastoral or mining activities (Volegno, Pruno, Azzano, Stazzema, Casoli) [34]. The second altimetric band is characterised by small hills and large alluvial fans formed in the Pleistocene and characterised by a low slope (generally less than 5%). In this altimetric band, some important villages (e.g., Massarosa, Pian di Mommio, Querceta) are located along the historic road axis (Aurelia and railroad), and others at the confluence with transversal valleys (Seravezza, Pietrasanta, Camaiore).

The third altimetric band is characterised by the coastal plain, whose formation took place in the Holocene as a result of the progressive juxtaposition of beach ridges formed thanks to the sediments brought to the sea by major rivers (Magra to the north and Serchio and Arno to the south) and reworked by coastal currents [35]. Locally there are wetlands (e.g., Porta Lake and Massaciuccoli Lake), which can be interpreted as relics of lagoon areas progressively isolated from the sea by the formation of seaside ridges [35–38]. These wetlands are strategic ecosystems: in terms of biodiversity, 40% of all plants and animals live or breed in wetlands and are quickly disappearing due to human impact and climate change. The importance of these areas is set off by the Ramsar Convention that protects both (https://www.regione.toscana.it/-/aree-ramsar (accessed on 11 May 2022)). Thus, this action would contribute to helping vulnerable regions and communities to improve their resilience in agreement with the goals of the UN 2030 Agenda: 13—Climate Action and 15—Life on Land.

Towards the sea, the system of beach ridges extends for about 4 km from the lake to the sea and ends with a large beach characterised by an uncommon advancement trend. An extraordinary system of coastal dunes is preserved in proximity to the current beach [37,39].

Since the early 1900s, progressive urbanisation has taken place in this area, starting from the city of Viareggio (located south of the study area; Figure 1). This phase of progressive urbanisation of the coastal plain has rapidly led to huge transformations, as far as the so-called 'continuous coastal city', which is the almost seamless urbanisation of the entire coastal strip, typical of several Mediterranean coastal plains. The increase in anthropogenic pressure in this area could threaten the ecosystems of the wetlands and of the dunes, which are particularly fragile.

Figure 1. Study area. Background: OpenStreetMap (Cowley Road, Cambridge, United Kingdom) and ESRI ((Redlands, California, USA) Shaded Relief (insert).

2.2. Archaeological Background

Human settlements in the study area are testified by many archaeological findings that have been recorded since the 16th century.

Since the Neolithic period, human presence has been confirmed by the findings of several artefacts, especially those of the site of Grotta all'Onda (Figure 1) [40].

During the Etruscan Age, the relevance of maritime commercial activity is testified by the settlement of San Rocchino (Figure 1), located on the ancient coast. This site was continuously inhabited from the end of the eighth century B.C. to the third century B.C. and was located on the border of an ancient lagoon area, whose remains at present are identifiable in the Massaciuccoli Lake.

In the same period, the Migliarina site (Figure 1), a productive and specialised settlement dating back to the end of the seventh century B.C., developed on the beach ridge, separating the lagoon from the sea and marking the coastline position of that time.

Moreover, the Acquarella site (Figure 1) shows a remarkable continuity of human settlement from the Etruscan Archaic period to the early Middle Ages [41]. The burial site found at Villa Mansi, close to Acquarella (Figure 1), perhaps part of a necropolis, suggests the existence of a village and demonstrates the vitality of the area since the end of the seventh century B.C. [42].

As for the Roman Age, weak traces of centuriation, as well as occasional discoveries on the beach ridge (Figure 1), confirm the dense settlement of this area [35]. The most important evidence of Roman times is the land roads attested in this area by ancient sources, maps, aerial photograph analyses and archaeological remains [43]. An important road, which could be identified with the Aurelia/Aemilia Scauri, ran along the past coast, another one along the foothill. Both roads linked Pisa to Luni. A third road connecting the inland (i.e., Lucca) to the coast is documented in the Camaiore valley near the Acquarella site.

Montecastrese, the most populated castle of Versilia, was built between the 10th and 11th centuries and occupied an important position in the control of the territory (Figure 1).

The Fortress of Motrone (Figure 1), erected in *maris litore* [44], and the Old Castle of Viareggio (Figure 1), built during the 12th century A.D., are important reference points for dating the shift of the coastal position during the Middle Ages. The construction in the 16th century of Tower Matilde in Viareggio and of Forte Lorenese in Forte dei Marmi (Figure 1), in a more advanced position than the old fortress, demonstrates the gradual advance of the coastline.

3. Materials and Methods

Our study aims to create a network among people with different jobs and competencies to intertwine different knowledge. This approach can help to create a synergy with researchers of different disciplines (e.g., geology, archaeology) and with people working on the territory, which could enhance the dissemination of cultural and natural heritage (archaeological museums, hiking associations, etc.). This holistic method would make it possible to complete and increase the great amount of knowledge already available in the territory and to put different subjects in synergy with one another. Figure 2 shows the workflow and stakeholders involved in each section.

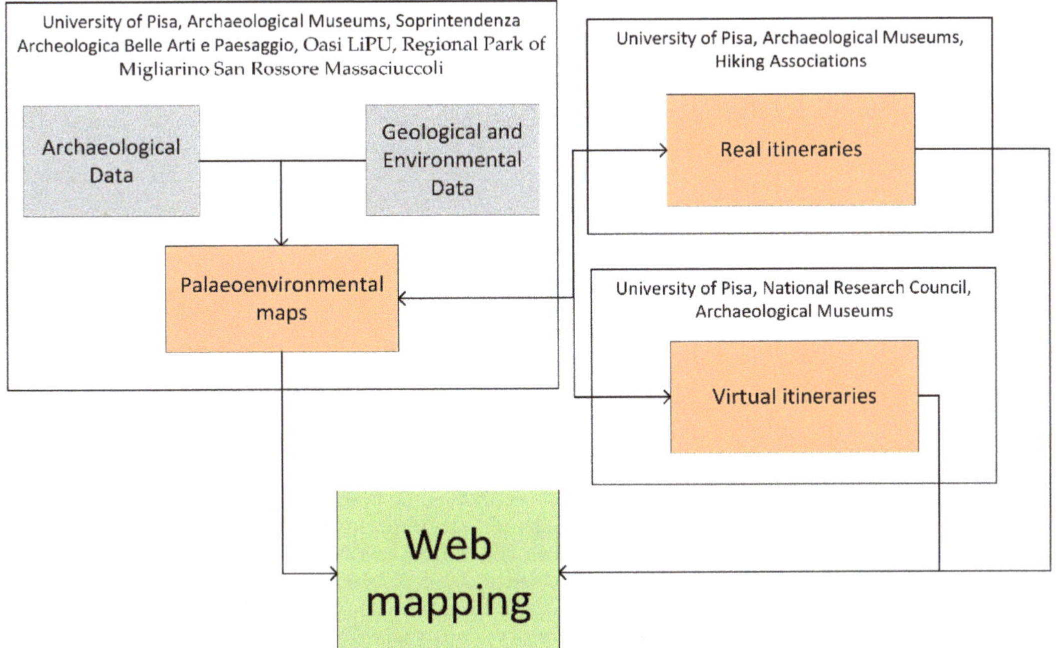

Figure 2. Research flowchart. The diagram shows the stakeholders involved in each section of the work. The grey rectangles are the bibliographic data, the red rectangles are the intermediate results and the green rectangle is the final result.

By the integration of different disciplines (geology, archaeology, biology), we reconstructed seven palaeoenvironmental maps describing seven chronological periods during which the area was settled. The periods investigated are the Last Glacial Maximum dated about 20,000 years B.P., the end of deglaciation about 7000 years B.P. and the Etruscan,

Roman, Middle, Modern and Contemporary Ages. The palaeoenvironmental maps were derived from a review of bibliographic analyses [31,35,36,45] and from newly integrated data coming from core analysis, historical cartography, remote sensing analysis, morphological analyses and geomorphological surveys. The analyses were conducted in a GIS environment using the open-source software QGIS v3.x. In particular, according to [46], we reconstructed the shoreline position at about 120 m below current sea level for the Last Glacial Maximum period and at about 8 m above current sea level for 7000 years B.C. [47,48]. In the maps, the estimation of the shoreline is based on the bathymetric measurement provided by the National Geoportal (http://www.pcn.minambiente.it/mattm/ (accessed on 11 May 2022) and on the digital elevation model (DEM) provided by the Tuscany region, with a resolution of 10 m.

The maps related to the other chronological periods were obtained from the integration of geological with archaeological data. For example, we used subsoil data to identify the extension of the Massaciuccoli Lake over time, and geomorphological combined with subsoil and archaeological data to identify the beach ridges.

This study is based on the implementation of a PostgreSQL database with a PostGIS extension to store and standardise the data produced and used in this project. The PostgreSQL database is installed on a hosting virtual machine accessible with the HTTP protocol and usable for the consultation of the data by desktop software and web applications.

A web application was developed by using HTML, JavaScript, PHP languages and exclusively open-source libraries. The front end of the web application is responsive and operates on both desktop and mobile devices. The web application is characterised by a web GIS implemented by the open-source geospatial library 'Open Layers'. The application is interfaced with the PostgreSQL database through the employment of the Geoserver application. The Open Layers library allows to load Web Feature Services (WFSs) and Web Map Services (WMSs) on a map by querying the Geoserver service. All layers are stylised according to the types of geometries and the database of the vectors. The web mapping allows to create an interaction between the user and the data of this research with two specific functionalities. The former consists of the possibility of clicking on one or more features to obtain some important information with the initialisation of a table on the interface. The visualised data can be texts, images or hyperlinks, and this allows to explain the geographic data and create an interaction between the user and the data of this research. The latter consists of the possibility of exporting the WFS in KML format, compatible with several GIS applications, such as desktop software (e.g., QGIS), web applications (e.g., Google Earth) and mobile apps for smartphones and smartwatches. This functionality makes it possible to apply the fundamentals of the open-data concept. The web application contains the palaeo maps, the virtual tours and the main natural and cultural heritage, together with several other layers that can help improve knowledge of the territory, as well as cultural and archaeological sites such as museums, environmental data, cartographies and aerial photos. In collaboration with some hiking associations, we identified nine itineraries that could link the different natural and cultural sites together, for example, an archaeological museum with its relative site and with other museums. The hiking associations will organise excursions so as to evaluate the real practicability of the itineraries in terms of distance and travel time. These itineraries are achieved in the PostgreSQL database with PostGIS extension using QGIS v3.x.

We have realised some information panels of the main archaeological and natural sites (such as Grotta all'Onda, LiPU, etc.) in both Italian and English. Some of these information panels have been installed along the real itineraries, and all the information panels are available for consultation through the web application.

The network between the sites and the museums is also incentivised with the development of virtual itineraries available for consultation on the web application.

The elaboration of the virtual tours is based on photos taken in indoor and outdoor environments by using two photographic instruments: (i) Canon 550D APS-C camera with 16 mp, 8 mm fisheye optic, panoramic head with eight sequential photoshoot positions for a resolution of 12K; and (ii) six images Kandao ObsidianR 8K 360° camera with pixel-level synchronisation technology. The images are processed by the open-source software RawTherapee for calibration of the chromatic dominant and for luminosity control. We applied the high dynamic range (HDR) technique to the 360° photos by combining a minimum of three images acquired with different exposition values: normal (0 EV), overexposed (+2 EV) and underexposed (−2 EV). The HDR technique allows for the creation of environment images with illumination levels that are more extensive than those resulting from a single photograph: the shadows and highlights are made legible by bringing the brightness and contrast values of the image closer to the perception level of the human eye. The virtual tours are based on 360° images with the Krpano software, which uses XML language. XML language allows to manage the graphic layout, modality of fruition, hyperlinks, organisation of the various graphic elements, text contents, multimedia elements, photos, video and audio of the virtual tour.

4. Results

Figure 3 shows the palaeoenvironmental maps referred to as 20,000 years B.P. (a), 7000 years B.P. (b), Etruscan Age (c), Roman Age (d), Middle Ages (e), Modern Age (f) and Contemporary Age (g). The oldest period is characterised by a sea level lower than the current one by about 120 m [46]. This corresponds to a shoreline position located about 20 km westwards of the current one. The area between the current shoreline and the Last Glacial Maximum shoreline was occupied by a wide coastal plain. With progressive deglaciation, we observe a sea level rise in the Versilia plain about 7000 years ago, when the sea reached the foothill of the mountains [47]. In the Etruscan Age, owing to the sediment carried by the main rivers and reworked by coastal drift, the first beach ridge system formed, isolating coastal lagoons behind. Successively, we observe a progressive shift of the beach ridge system towards the west, which allowed the development of the coastal plain in its current conformation. In some periods, the Massaciuccoli Lagoon gradually became a coastal lake and wetland developed in its surroundings. Specifically, between the Roman and the Middle Ages, the Massaciuccoli Lake assumed a shape that was very similar to the current one. For each historical period, Figure 3 reports the main human settlements and coastal towers located near the shore so as to be able to precisely identify the shift of the lake over time.

Figure 4 shows the nine natural and cultural itineraries identified in synergy with the local hiking associations. The first, second, third and fourth itineraries link the archaeological sites of Montecastrase and of Grotta all'Onda with the Camaiore Museum. These routes create a link between the archaeological sites and the Camaiore Archaeological Museum, where the findings and the documentation of these important sites are stored. The fifth itinerary links the archaeological area of the Roman Massaciuccoli with the Chiarone Nature Reserve situated within the Migliarino San Rossore Massaciuccoli Natural Park and then to the small village of Torre del Lago Puccini. A track around the lake starts from the small village and ends on a wide natural beach characterised by a system of well-preserved coastal dunes. The sixth itinerary is a little city route linking the Pietrasanta Archaeological Museum with the cultural site of Rocca di Sala. The seventh itinerary starts from the civic museums of Villa Paolina in Viareggio and ends at the Camaiore Archaeological Museum, going past the geoarchaeological site of Buca delle Fate. The eighth itinerary links the Museum of Camaiore with the geosite of Buca delle Fate through a mountain and hill track. Finally, the ninth itinerary links the museums of Pietrasanta and Camaiore through the Via Francigena route, the most famous of the medieval tracks.

The real itineraries are associated with the virtual tours, and some example images are reported in Figure 5. The virtual tours are available for consultation at the link (http://tlweb.it/pantarei/gis/ (accessed on 11 May 2022). The virtual itineraries are based on the creation of a link between the archaeological sites and the museums. We have also increased the multimedia and content information on the cultural and natural places that are worth visiting. With the same purposes as the natural tours, we have realised five virtual tours according to the different chronological periods analysed on the basis of the relative palaeogeographical maps. The five virtual tours are, specifically:

1. 'Grotta all'Onda', which shows the prehistoric environment in Versilia;

2. 'Museo di Viareggio', which allows to create a transition from the prehistoric to the historic period;

3. 'Museo di Pietrasanta', which regards the Etruscan Age in Versilia;

4. 'Villa Mansio, Villa Venulei Apropriani', which is referred to the Roman Age;

5. 'Il paesaggio naturalistico attuale', which describes the LIPU site and the importance of the wet areas and their preservation.

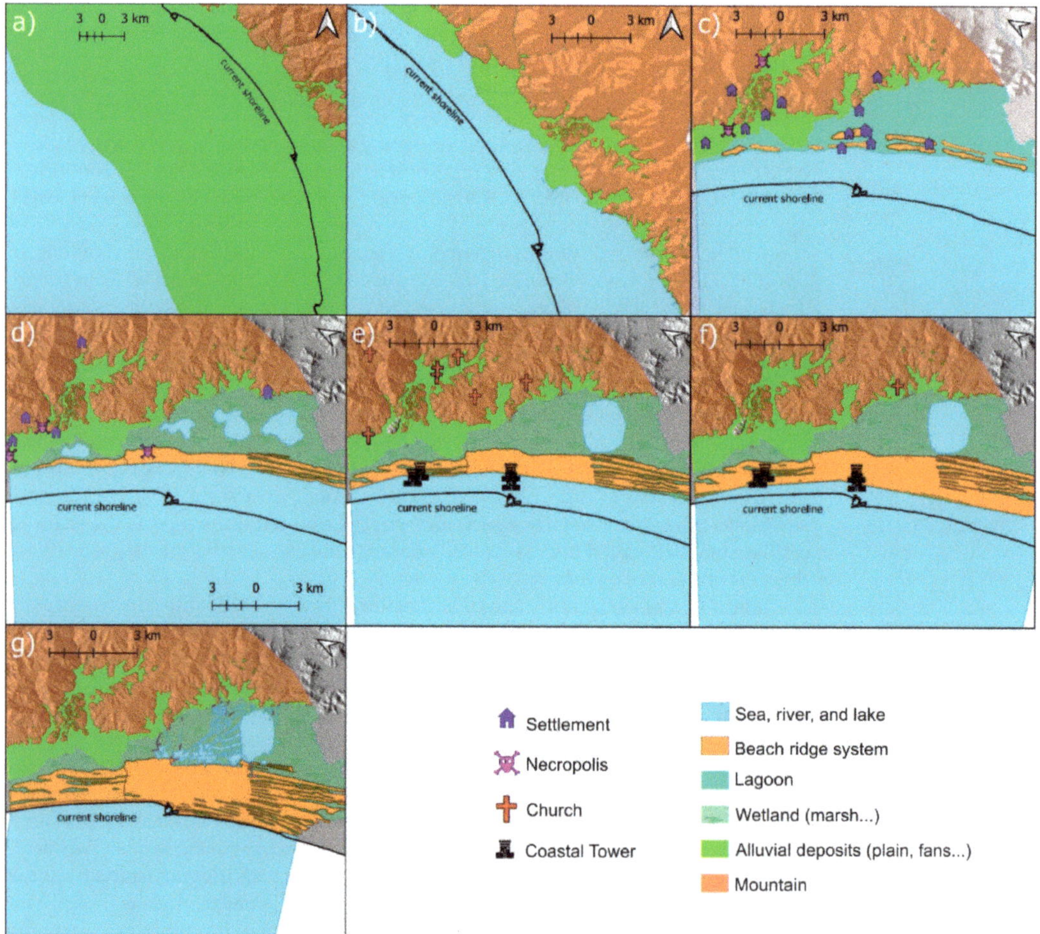

Figure 3. Reconstruction of the palaeoenvironment of (**a**) about 20,000 years B.P., (**b**) about 7000 years B.P., (**c**) Etruscan Age, (**d**) Roman Age, (**e**) Middle Ages, (**f**) Modern Age and (**g**) Contemporary Age.

Figure 4. Real itineraries linking the archaeological sites with the museums, as proposed in this study. These initiatives can help to incentivise inland tourism. Background: OpenStreetMap.

Figure 5. Some 360° images used to create the virtual tours: (**a**) Grotta all'Onda; (**b**) Museo Archeologico Camaiore; (**c**) Massaciuccoli Domus from a viewpoint; (**d**) Oasi Lipu; (**e**) Museo Archeologico Versiliese 'Bruno Antonucci' (Pietrasanta); (**f**) Massaciuccoli Terme.

The web application interface is composed of a toolbar on the left, a map in the centre and a menu on the right (Figure 6). The web application is available for consultation at the following link: http://tlweb.it/pantarei/gis (accessed on 11 May 2010). The toolbar is provided with some functions for the navigation, selection and management of the map. In the menu, the layer tree view in the menu permits to manage the layers, with the possibility of making them visible, moving them, downloading the dataset and viewing the style of the layers. By using the web application, it is possible to visit both the real and the virtual itineraries and to gain territorial as well as cultural information.

Figure 6. Web application developed in this study. The application is a web GIS reporting all the results of this study (palaeoenvironmental maps, heritage and virtual tours). The web application aims to improve the link between results and to provide an interface for querying and disclosure of the data.

5. Discussion and Conclusions

The use of web mapping for disclosure of territorial information and the enhancement of cultural and natural heritage was a valid instrument in agreement with several previous studies (i.e., [10,21–25]). The web mapping, usable by different devices (PC desktop, smartphone, etc.), can provide visibility to the territory, and this is accentuated with the creation of virtual itineraries [24–28].

This study takes a step forward compared to others by creating a triple link between real and virtual tours and making them available on a platform that allows you to consult geographic information but also other information such as information panels, tabulated data, images, etc.

The Versilia plain witnesses the history of a long human–environment interaction dating back to the Prehistoric Age. This history can still be read through the cultural and natural assets characterising the area, which tell the story of both environmental evolution and the history of settlement.

Thanks to a strong integration of data from different sources, it is possible to describe a story that can intrigue the visitor, going beyond administrative and cultural limitations.

Visitors receive information about the landscape during prehistoric times, the location of the sites that have returned prehistoric finds and the museums where these findings can be observed. Similarly, it is possible to accompany the visitors by describing the discovery of the Etruscan landscape and of the museums closely related to the history of the Etruscan period. The same can be applied to the Roman period.

The visit to the LIPU oasis shows a picture of the current environment and allows reflections on the most likely future evolutions.

The creation of both natural and virtual itineraries can help to easily connect the different historical periods in an overall evolutionary framework. Broad sharing of data allows for the dissemination of knowledge and increased awareness among both tourists and local communities.

Access to these data (through web mapping and the virtual tour) increases the visibility of a territory that, although very close to areas of great tourist flow, is actually excluded. The availability of free and open-source informatic instruments makes it possible to employ replicable instruments and methodologies in other contexts as well. The involvement of all stakeholders right from the design stages has led to a growth in synergies among the different bodies involved in the project. This collaborative effort and multidisciplinary approach have contributed to the dissemination of knowledge and to the achievement of more holistic results.

The future use of the platform and future integrations of data and functionalities will make it possible to refine the study and provide an increasingly advanced solution for the dissemination of cultural and natural heritage.

This study has proved useful for the creation of IT tools that potentially allow the dissemination of knowledge of the environmental and cultural heritage contained in the territory. However, real enhancement of the territory can only be achieved through the willingness and commitment of the managing bodies to the use and advertising of these tools.

Author Contributions: Conceptualisation, Monica Bini and Valerio Noti; methodology, Monica Bini, Marco Luppichini, Valerio Noti and Danilo Pavone.; software, Marco Luppichini and Danilo Pavone; validation, Monica Bini, Marco Luppichini, Valerio Noti, Danilo Pavone, Marzia Bonato, Francesco Ghizzani Marcìa, Stefano Genovesi, Francesca Lemmi, Lisa Rosselli, Neva Chiarenza, Marta Colombo, Giulia Picchi and Andrea Fontanelli.; resources, Monica Bini; writing—original draft preparation, Monica Bini and Marco Luppichini.; writing—review and editing, Monica Bini, Marco Luppichini, Valerio Noti, Danilo Pavone, Marzia Bonato, Francesco Ghizzani Marcìa, Stefano Genovesi, Francesca Lemmi, Lisa Rosselli, Neva Chiarenza, Marta Colombo, Giulia Picchi and Andrea Fontanelli.; project administration, Monica Bini.; funding acquisition, Monica Bini. All authors have read and agreed to the published version of the manuscript.

Funding: This research was funded by project no. 249792, "Dalla Preistoria all'Antropocene: Nuove Tecnologie per la valorizzazione dell'eredità culturale della Versilia (PANTAREI)" Tuscany region (call POR FSE 2014- 2020, Resp. M. Bini) and by project *"Cambiamenti globali e impatti locali: conoscenza e consapevolezza per uno sviluppo sostenibile della pianura Apuo-versiliese"* Fondazione Cassa di Risparmio di Lucca (call 2018 years 2019-2022- Resp. M. Bini).

Data Availability Statement: The data are available at the following link: http://tlweb.it/pantarei/gis/# (accessed on 11 May 2022).

Acknowledgments: We thank the various excursion associations that have allowed their support in the identification and dissemination of the real itineraries. We thank all the public bodies involved for their support and dissemination of the project.

Conflicts of Interest: The authors declare no conflict of interest.

References

1. World Tourism Organization UNWTO. *Tourism Highlights*; World Tourism Organization UNWTO: Madrid, Spain, 2018.
2. World Tourism Organization UNWTO. *Tourism Highlights*; World Tourism Organization UNWTO: Madrid, Spain, 2011.
3. Koens, K.; Postma, A.; Papp, B. Is Overtourism Overused? Understanding the Impact of Tourism in a City Context. *Sustainability* **2018**, *10*, 4384. [CrossRef]
4. Butler, R.W. Tourism, Environment, and Sustainable Development. *Environ. Conserv.* **1991**, *18*, 201–209. [CrossRef]
5. UNEP; UNWTO. *Making Tourism More Sustainable—A Guide for Policy Makers*; UNWTO Publications: Madrid, Spain, 2005.
6. Ingaldi, M.; Dziuba, S.T. Sustainable Tourism: Tourists' Behaviour and Their Impact on the Place Visited. *Vis. Sustain.* **2021**, *17*, 8–38. [CrossRef]

7.   Petrella, A.; Torrini, R.; Barone, G.; Beretta, E.; Breda, E.; Cappariello, R.; Ciaccio, G.; Conti, L.; David, F.; Degasperi, P.; et al. *Turismo in Italia: Numeri e Potenziale Di Sviluppo*; Banca d'Italia: Rome, Italy, 2019.

8.   Viciani, D.; Dell'Olmo, L.; Vicenti, C.; Lastrucci, L. Natura 2000 Protected Habitats, Massaciuccoli Lake (Northern Tuscany, Italy). *J. Maps* **2017**, *13*, 219–226. [CrossRef]

9.   Kraak, J.-M.; Brown, A. *Web Cartography*; CRC Press: Boca Raton, FL, USA, 2003; ISBN 0429182104.

10. Costantino, D.; Angelini, M.G.; Alfio, V.S.; Claveri, M.; Settembrini, F. Implementation of a System WebGIS Open-Source for the Protection and Sustainable Management of Rural Heritage. *Appl. Geomat.* **2020**, *12*, 41–54. [CrossRef]

11. de Castro, A.F.; Amaro, V.E.; Grigio, A.M.; Cavalcante, R.G. Modeling and Development of a WebGIS for Environmental Monitoring of Coastal Areas That Are Influenced by the Oil Industry. *J. Coast. Res.* **2011**, *SI64*, 1643–1647.

12. Tang, W.; Selwood, J. *Connecting Our World: GIS Web Services*; ESRI, Inc.: Redlands, CA, USA, 2003; ISBN 1589480759.

13. Opdam, P. Learning Science from Practice. *Landsc. Ecol.* **2010**, *25*, 821–823. [CrossRef]

14. Lopes, S.I.; Moreira, P.M.; Cruz, A.M.; Martins, P.; Pereira, F.; Curado, A. RnMonitor: A WebGIS-Based Platform for Expedite in Situ Deployment of IoT Edge Devices and Effective Radon Risk Management. In Proceedings of the 2019 IEEE International Smart Cities Conference (ISC2), Casablanca, Morocco, 14–17 October 2019; IEEE: Piscataway, NJ, USA, 2019; pp. 451–457.

15. Pan, S.; Wang, K.; Wang, L.; Wang, Z.; Han, Y. Risk Assessment System Based on WebGIS for Heavy Metal Pollution in Farmland Soils in China. *Sustainability* **2017**, *9*, 1846. [CrossRef]

16. Howell, R.G.; Petersen, S.L.; Balzotti, C.S.; Rogers, P.C.; Jackson, M.W.; Hedrich, A.E. Using Webgis to Develop a Spatial Bibliography for Organizing, Mapping, and Disseminating Research Information: A Case Study of Quaking Aspen. *Rangelands* **2019**, *41*, 244–247. [CrossRef]

17. Seng, B.; Liang, H.; Zhao, Y.; Tang, Y. Design and Implementation of Visualization System for Wastewater Treatment in Dianchi Lake Based on WebGIS. In Proceedings of the E3S Web of Conferences, Shanghai, China, 16–18 August 2019; EDP Sciences: Les Ulis, France, 2019; Volume 118, p. 04032.

18. Luo, K.; Zhang, T.; Liu, Y. Design and Implementation of the Water Environment Monitoring and Management System Based on WebGIS for the Four-Lake Basin of Hubei Province. In Proceedings of the E3S Web of Conferences, Hefei, China, 1–3 November 2019; EDP Sciences: Les Ulis, France, 2019; Volume 136, p. 06036.

19. Villani, G.; Nanni, S.; Tomei, F.; Pasetti, S.; Mangiaracina, R.; Agnetti, A.; Leoni, P.; Folegani, M.; Mazzini, G.; Botarelli, L. The RainBO Platform for Enhancing Urban Resilience to Floods: An Efficient Tool for Planning and Emergency Phases. *Climate* **2019**, *7*, 145. [CrossRef]

20. Santoso, M.I.; Gumelar, R.G.; Irawan, B. Development of the WebGIS Application for Transport Infrastructure Management in the City of Serang. In Proceedings of the IOP Conference Series: Materials Science and Engineering, Bali, Indonesia, 7–8 August 2019; IOP Publishing: Bristol, UK, 2019; Volume 673, p. 012072.

21. Tommasi, A.; Cefalo, R.; Zardini, F.; Nicolaucig, M. Using Webgis and Cloud Tools to Promote Cultural Heritage Dissemination: The Historic up Project. In Proceedings of the International Archives of the Photogrammetry, Remote Sensing and Spatial Information Sciences—ISPRS Archives, Florence, Italy, 22–24 May 2017; International Society for Photogrammetry and Remote Sensing: Christian Heipke, Germany, 2017; Volume 42, pp. 663–668.

22. Suma, A.; de Cosmo, P. Geodiv Interface: An Open Source Tool For Management And Promotion Of The Geodiversity Of Sirra De Grazalema Natural Park (Andalusia, Spain). *GeoJournal Tour. Geosites* **2011**, *8*, 309–318.

23. Lerario, A.; Maiellaro, N.; Zonno, M. Remote Fruition of Architectures: R&D and Training Experiences. In Proceedings of the 2010 Second International Conferences on Advances in Multimedia, Washington, DC, USA, 13–19 June 2010; pp. 49–54.

24. Santos, C.A.D.J.; Campos, A.C.; Rodrigues, L.P. GIS and Touristic Itineraries. The Case of São Cristóvão, Sergipe, Brazil. *Cad. De Geogr.* **2019**, *39*, 29–39. [CrossRef]

25. Cayla, N.; Hoblea, F.; Gasquet, D. Place de La Géomorphologie Dans l'offre Géotouristique de l'arc Alpin: Du Réel Au Virtuel. *Géomorphosites* **2009**, 65–71.

26. Medyńska-Gulij, B.; Zagata, K. Experts and Gamers on Immersion into Reconstructed Strongholds. *ISPRS Int. J. Geo-Inf.* **2020**, *9*, 655. [CrossRef]

27. Sâvulescu, I.; Mihai, B. Mapping Forest Landscape Change in Iezer Mountains, Romanian Carpathians. A GIS Approach Based on Cartographic Heritage, Forestry Data and Remote Sensing Imagery. *J. Maps* **2011**, *7*, 429–446. [CrossRef]

28. Weng, Y.-H.; Sun, F.-S.; Grigsby, J.D. GeoTools: An Android Phone Application in Geology. *Comput. Geosci.* **2012**, *44*, 24–30. [CrossRef]

29. Bailey, J.E.; Chen, A. *The Role of Virtual Globes in Geoscience*; Elsevier Science Publishers: Amsterdam, The Netherlands, 2011.

30. Dias, E.; Rhin, C.; Haller, R.; Scholten, H. Adding Value and Improving Processes Using Location-Based Services in Protected Areas: The WebPark Experience. *E-Environ. Prog. Chall. Spec. Ed. e-Environ. Inst. Politécnico Nac. Mex. Mex. City* **2004**, 291–302.

31. Baroni, C.; Pieruccini, P.; Bini, M.; Coltorti, M.; Fantozzi, P.L.; Guidobaldi, G.; Nannini, D.; Ribolini, A.; Salvatore, M.C. Geomorphological and Neotectonic Map of the Apuan Alps (Tuscany, Italy). *Geogr. Fis. E Din. Quat.* **2015**, *38*, 201–227.

32. Molli, G.; Giorgetti, G.; Meccheri, M. Tectono-Metamorphic Evolution of the Alpi Apuane Metamorphic Complex: New Data and Constraints for Geodynamic Models. *Boll. Della Soc. Geol. Ital.* **2002**, *1*, 789–800.

33. Carmignani, L.; Kligfield, R. Crustal Extension in the Northern Apennines: The Transition from Compression to Extension in the Alpi Apuane Core Complex. *Tectonics* **1990**, *9*, 1275–1303. [CrossRef]

34. Regione Toscana Piano Di Indirizzo Territoriale Con Valenza Di Piano Paesaggistico, Ambito Versilia e Costa Apuana. 2015. Available online: https://www.regione.toscana.it/-/piano-di-indirizzo-territoriale-con-valenza-di-piano-paesaggistico (accessed on 30 April 2022).
35. Bini, M.; Sarti, G.; da Prato, S.; Fabiani, F.; Paribeni, E.; Baroni, C. Geoarcheological Evidences of Changes in the Coastline Progradation Rate of the Versilia Coastal Plain between Camaiore AndViareggio (Tuscany, Italy): Possible Relationships with Late Holocene High-Frequency Transgressive Regressive Cycles. *Il Quat. Ital. J. Quat. Sci.* **2009**, *22*, 257–266.
36. Bini, M.; Baroni, C.; Ribolini, A. Geoarchaeology as a Tool for Reconstructing the Evolution of the Apuo-Versilian Plain (NW Italy). *Geogr. Fis. E Din. Quat.* **2013**, *36*, 215–224. [CrossRef]
37. Federici, P.R. L'ex Lago Di Porta in Versilia (Toscana): La Storia Di Una Irresistibile Pressione Ambientale. *Mem. Soc. Geogr. Ital.* **1998**, *55*, 397–414.
38. Federici, P.R. The Versilian Transgression of the Versilia Area (Tuscany, Italy) in the Light of Drillings and Radiometric Data. *Mem. Della Soc. Geol. Ital.* **1993**, *49*, 217–225.
39. D'Amato Avanzi, G.; Giannecchini, R. Eventi Alluvionali e Fenomeni Franosi Nelle Alpi Apuane (Toscana): Primi Risultati Di Un'indagine Retrospettiva Nel Bacino Del Fiume Versilia. *Riv. Geogr. Ital.* **2003**, *110*, 527–559.
40. Berton, A.; Bonato, M.; Campetti, S.; Fabbri, P.F.; Mallegni, F.; Perrini, L. *Nuove Indagini Del Deposito Preistorico Di Grotta All'Onda*; Camaiore (LU): Toscana, Italy, 2006.
41. Fabiani, F.; Rovai, E.P. *Il Frantoio Romano Dell'Acquarella*; Felici Editore: Pisa, Italy; ISBN 9788860195920.
42. Ciampoltrini Villa Mansi (Camaiore). In *Etruscorum Ante Quam Ligurum. La Versilia tra VII e III Secolo a.C.*; Paribeni (Ed.) Bandecchi e Vivaldi: Pontedera, Italy, 1990; pp. 119–121.
43. Fabiani, F. " . . . *Stratam Antiquam Que Est per Palude Set Boscos* . . . ". *Viabilità Romana Tra Pisa e Luni*; Edizioni Plus: Pisa, Italy, 2006.
44. Giannotti, S. Pietrasanta (LU). Controllo Archeologico a Motrone. *Not. Della Soprintend. Per I Beni Archeol. Della Toscana* **2006**, *2*, 23–25.
45. Pasquinucci, M.; Menchelli, S.; Mazzanti, R.; Marchisio, M.; D'Onofrio, L. Coastal Archaeology in North Etruria [North Coastal Etruria. Géomorphologie, Archaeological, Archive, Magnetometric and Geoelectrical Researches. *Rev. D'archéométrie* **2001**, *25*, 187–201. [CrossRef]
46. Lambeck, K.; Purcell, A. Sea-Level Change in the Mediterranean Sea since the LGM: Model Predictions for Tectonically Stable Areas. *Quat. Sci. Rev.* **2005**, *24*, 1969–1988. [CrossRef]
47. Sestini, A. Un'antica Ripa Marina Nella Pianura Costiera Apuana. *Mem. Della Soc. Toscana Di Sci. Nat.* **1950**, *57*, 1–6.
48. Vacchi, M.; Marriner, N.; Morhange, C.; Spada, G.; Fontana, A.; Rovere, A. Multiproxy Assessment of Holocene Relative Sea-Level Changes in the Western Mediterranean: Sea-Level Variability and Improvements in the Definition of the Isostatic Signal. *Earth-Sci. Rev.* **2016**, *155*, 172–197. [CrossRef]

Article

Thematic Content and Visualization Strategy for Map Design of City-Specific Culture Based on Local Chronicles: A Case Study of Dengfeng City, China

Xiaohui He, Chuan Liu, Lili Wu *, Yongji Wang and Zhihui Tian

School of Geoscience and Technology, Zhengzhou University, Zhengzhou 450001, China
* Correspondence: wll_dqkxxy@zzu.edu.cn

Abstract: Local chronicles are a kind of historical record in China that are written in detail and play an important role in the transmission of local history and culture. Due to the single-text-carrier form of local chronicles, people have limited access to information on urban characteristics and culture; therefore, based on the cultural gene theory and Hofstede model, also known as the cultural onion model, this paper develops a "Spirit–Sign" content framework with the themes of urban characteristics and culture. Based on this framework, we map the urban characteristics and culture (visualization strategy and map design) of local chronicles. Taking the historic city of Dengfeng in the Central Plains as an example, the spatial information of the four historical city characteristics of Dengfeng was mined for the map design using the content framework of the city characteristics proposed in this study. The results of the study found that (1) there is a certain overlap in the spatial distribution of the four characteristic cultures of Dengfeng, indicating that the spiritual (traditional customs and famous people and events) and material (famous buildings and products) are complementary and mutually reinforcing to a certain extent; and (2) with the iterative development of Chinese dynasties, the material characteristic cultures of Dengfeng show strong temporal and spatial differences, which laterally reflect the changes in human activities and urban changes of each dynasty and also reflect the very important historical position occupied by Dengfeng in China. Compared with the traditional text-carrier form of local chronicles, the content construction and map visualization of the city's historical and cultural information proposed in this study can effectively explore more potential cultural characteristics, as well as spatial and temporal connections of Dengfeng and thus help people better understand the historical characteristics of the city.

Keywords: local chronicles; map design; cultural genes; Hofstede model; Dengfeng

Citation: He, X.; Liu, C.; Wu, L.; Wang, Y.; Tian, Z. Thematic Content and Visualization Strategy for Map Design of City-Specific Culture Based on Local Chronicles: A Case Study of Dengfeng City, China. *ISPRS Int. J. Geo-Inf.* **2022**, *11*, 542. https:// doi.org/10.3390/ijgi11110542

Academic Editors: Beata Medynska-Gulij, David Forrest, Thomas P. Kersten and Wolfgang Kainz

Received: 20 August 2022
Accepted: 25 October 2022
Published: 30 October 2022

Publisher's Note: MDPI stays neutral with regard to jurisdictional claims in published maps and institutional affiliations.

1. Introduction

Local chronicles are a kind of historical record of local details, and they are a product of the city's historical and cultural heritage. Local chronicles record events from ancient times to the present day. They contain a large number of city-specific historical and cultural resources, including cultural relics and monuments, culture, customs and dialects, celebrities, art, and literature. As a kind of public cultural product, local chronicles assume the functions of storing history and educating people. They are also closely related to the historical continuity, cultural inheritance, current construction, and future development of the city; however, due to the vast amount of local chronicles resources and the single form of text carrier, it is difficult for people to process and understand them, and they have not fully entered the public's view.

Since the 1990s, when information technology first entered the field of local chronicles, local chronicle informatization has made great progress in the establishment of local chronicle full-text databases [1], the construction of local chronicle historical geographic information systems [2], and the automatic identification and mining of local chronicle texts [3,4]. Among them, in the area of automatic recognition and mining of local chronicle texts, Bai et al. [5] constructed a mining recognition system for local- chronicle-citation books based on the pattern of the citation books of the local chronicle material, combined with functional algorithms such as N-gram word separation recognition. Xu et al. [6] performed automatic recognition of four types of entities, including property aliases, characters, origins, and citation books in ancient chronicle material. This system has also been used for the construction of the knowledge base of the material. In addition, previous research on local chronicle informatization has mainly used national historical documents combined with a geographic information system (GIS). The study of the boundaries and population of a country was conducted by establishing a historical GIS, such as the Great Britain Historical GIS [7], the United States National Historical GIS [8], and the Canadian Century Research Infrastructure (CCRI) [9], which is based on a historical database of census data. The above research has solved the problem of the huge amount of local chronicle data, but there are still challenges in little-access and low-utilization. Also, the research remains at the level of textual informatization. The large amount of city-specific historical and cultural resources recorded in local chronicles lack a richer carrier to display it; therefore, there is a need to improve the single-text-carrier form of local chronicles. By doing so, local chronicles can be made available to society as a whole so that people can read them easily and quickly.

Information visualization, as an important form of presentation of information today, can present the core content of complex data [10–12]. Mapping, as the most basic tool for humans to understand the world, is the best language for geographic information expression in information visualization [13,14]. With the development of society and technology, the content and form of maps are constantly evolving. As a constant topic in map research, the current way forward for map design is to break through in terms of functions and modes [15], the need for both artistic and scientific map design [16,17], the comprehensiveness of map design themes, and the richness of information. Among them, the representative and excellent atlas products published worldwide in the past ten years, such as the Arctic Nautical Atlas [18], Atlas of Chinese Natural Disaster System [19], Atlas of Switzerland [20], and Coral Triangle Atlas [21], have explored the richness of information, the comprehensiveness of topics, and the diversity of forms in their atlases and have created valuable examples for map design. Based on these excellent atlases, Weng et al. [22] selected the city of Xi'an, China, as a research case to design tourism maps through spatial narratives to reflect its artistic characteristics and put forward some operational suggestions for improving the functionality and artistry of tourism maps. Xu et al. [23] took a series of maps of innovation development in the new Shenzhen as an example to establish a content structure and visual representation by sorting out the innovation characteristics of Shenzhen. In addition, many papers in academic journals on urban maps have also focused on the themes of natural resources [24] and economic development [25] and proposed the corresponding content framework, map design, and production based on comprehensive data such as those from web searches and fieldwork; however, the current discussion on the production of maps of urban characteristics and culture based on local chronicles is rarely mentioned. In addition, the use of thematic maps to display information, as a common visualization method, helps to mine information in spatial analysis with more emphasis on spatial distribution compared to tables and charts [26,27]. Therefore, it is necessary to convert data on natural sciences [28] or social sciences [29] from local chronicles into thematic maps and use the maps for analysis.

In response to the local and informational characteristics of local chronicles, complex, hidden, and difficult-to-express information needs to be represented in visual form, for which this study started from two aspects: content framework construction and map design. In terms of content framework construction, previous authors have adopted content framework methods based on cultural gene theory [30] and the Hofstede model, also known as the cultural onion model [31]. For example, Weng et al. [32] subdivided cultural genes into ecological-cultural genes, historical-cultural genes, and folklore-cultural genes according to cultural gene theory and excavated urban cultural resources based on these three types of cultural genes. Su et al. [33] used the Hofstede model as a theoretical basis to dissect the hierarchical structure of campus-culture systems and established the thematic content system of campus-culture atlases based on it. The above research shows that both cultural gene theory and the Hofstede model can achieve the collation and construction of the corresponding content system; however, in practice, the cultural gene theory or the Hofstede model alone is relatively imperfect in the collation of urban cultural resources, while the combination of them is more effective in the collation of urban cultural resources of local chronicles. Therefore, based on cultural gene theory and the Hofstede model, this paper developed a "Spirit–Sign" content-framework model with the theme of urban characteristic culture. Based on this model, we extracted and sorted the urban cultural resources of local chronicles, which mainly focused on the textual information elements in the chronicles and how to extract and express them according to different thematic categories of textual information. For the map design, we proposed grasping the cultural characteristics of the city as a whole from the triple perspective of "time, space and people". Then, we followed the basic map design strategy for optimizing the visual expression by figurativeness and symbolism [34,35] and explored the spatial information of the cultural characteristics of the city in terms of layout design, color emotion, and symbol creativity. Among them, expression methods such as geometric-pattern composition and artistic processing based on real objects are adopted in terms of symbol creativity. These methods enable symbols to have pictographic patterns, which are then used to indicate the real objects they represent [32,36].

In summary, this paper developed a "Spirit–Sign" content-framework model with the city's characteristic culture as the main information source and illustrated the main theories, steps, and methods of the content framework and visualization strategies with the example of the thematic map group of Dengfeng's characteristic culture. Finally, the pattern of cultural activities in the human society of Dengfeng city was analyzed and the map results were evaluated. We expect to improve the single-text-carrier form of local chronicles and to explore and display more potential spatial information and laws of urban characteristic culture to further develop the function of local chronicles in promoting urban construction, cultural industry, and tourism development and to provide a reference for similar map designs.

2. Materials and Methods

2.1. Study Area and Dataset

Dengfeng (34°35′–34°15′ N, 112°49′–113°19′ E), a historical city in central China, is part of China's Henan Province (Figure 1). It is located in the core of the Yellow River Basin, the birthplace of Chinese civilization. With a long history of more than 8000 years, Dengfeng also has a rich cultural heritage. For example, the city has 11 World Heritage Sites, among which the Shaolin Temple in Songshan is the birthplace of Shaolin Kung Fu in China.

Urban local chronicles are a series of books that comprehensively describe the natural, geographical, political, economic, social, and cultural history and current situation of cities, with their local, contemporary, extensive, informative, scientific, and authoritative characteristics [37].

The data source of this study comes from the series of the Dengfeng City Chronicle, which is updated and revised every 15 years. The latest edition of the series, which was written in 2008, is used as the data source. The Dengfeng City Chronicle exists mainly in paper form, which one can find and buy in libraries or online stores. It is also available in electronic form, which one can find and read on online databases. Similar to the local chronicles of other cities, Dengfeng City Chronicle provides a comprehensive account of all aspects of the city's history and current situation; however, because of its unique Shaolin martial arts culture, the excavation of many historical products, and the many famous historical events, Dengfeng City Chronicle covers unique and richer content. Also, due to its professional editorial board, this series of books covers a long period, and it records events from ancient times to the present day. In addition, it is systematic and comprehensive; therefore, we chose this area as the study area and wanted to preserve and inherit the cultural treasure of the Dengfeng City Chronicle, which will have profound significance for the development of the city's unique culture.

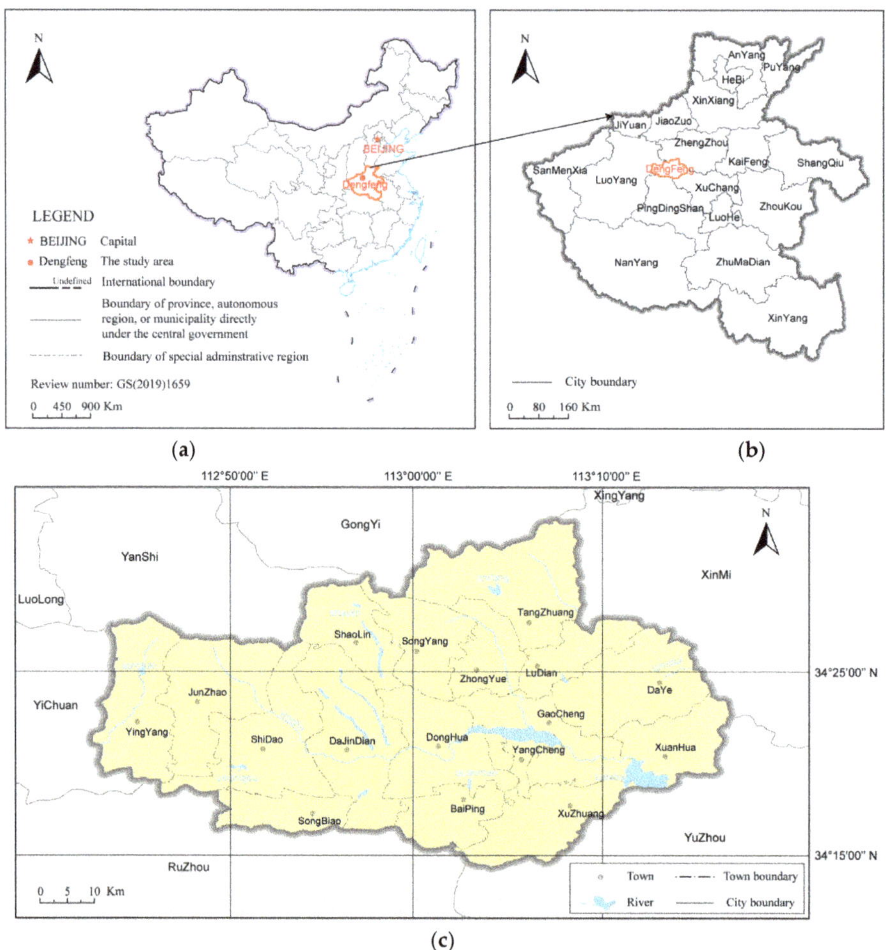

Figure 1. Location and administrative division of Dengfeng: (**a**) the location of Dengfeng in China, this map of China is mainly provided by the Ministry of Natural Resources of China; (**b**) the location of Dengfeng in Henan Province; (**c**) the administrative division map of Dengfeng city.

2.2. Main Steps and Methods

This study started from two aspects: content-framework construction and map design. Content-framework construction mainly refers to how to build a thematic content framework of urban characteristics and culture based on local chronicles and the selection of contents; the map design studied how to establish visualization strategies to convey local chronicle information scientifically and accurately. The flow chart of the method is shown in Figure 2 below.

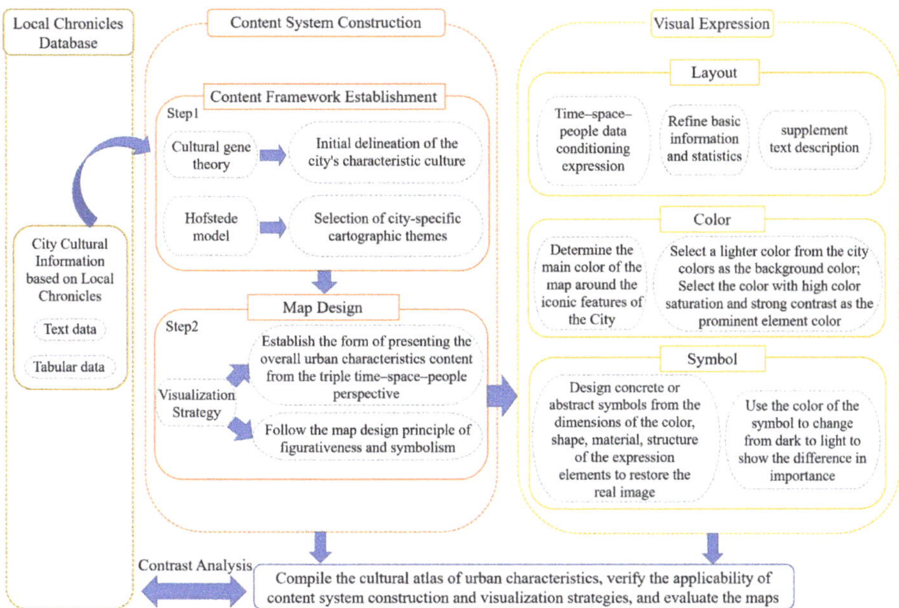

Figure 2. Flow chart.

2.2.1. Step 1: Establishment of the Urban Cultural Content Framework of Local Chronicles Based on Cultural Gene Theory and the Hofstede Model

The city has many rich and diverse representative cultures, including its unique Shaolin martial arts culture, famous individuals, local cultural and artistic works, and excavated cultural relics. To clarify the connotation of each characteristic culture in the local chronicles, we carried out the following steps: first, we used cultural gene theory to initially classify the city's distinctive cultures, and then we used the Hofstede model to select the city's distinctive mapping themes. The specific steps are as follows.

(1) Initial delineation of the city's characteristic culture by cultural gene theory

The term "cultural genes" was created by analogy with biological genes, which initiated the exploration and study of the laws of cultural evolution and transmission. The resulting cultural gene theory divides cultural genes into material and immaterial cultural genes according to the mode of transmission, in which material cultural genes are mostly presented and transmitted materially, while immaterial cultural genes exist in the form of "mental states" and are transmitted through oral narratives and behavioral expressions [38]. Based on this theory, this study classified urban characteristics into material cultural characteristics and immaterial cultural characteristics by the different ways of urban cultural transmission, then classified the characteristic cultural information in local chronicles and named them figuratively "Spirit" and "Sign" to produce the initial framework, and further excavated and sorted urban characteristic resources on this initial framework.

(2) Selection of city-specific cartographic themes based on the Hofstede model

The Hofstede model was proposed by Geert Hofstede, who argued that culture can be likened to an onion composed of different layers that interact with each other [31,39,40]: the top layer of the onion is the material "symbols", including language, artworks, architecture, and clothing; the second layer is the character of the heroic figure; the third layer is the customs and rituals; the core layer is the values. In the general local chronicles, for the material cultural characteristics of cities (Sign), information on famous buildings and famous products is recorded; for the immaterial cultural characteristics of cities (Spirit), information on traditional customs and famous individuals and events is recorded. The different layers of the Hofstede model are compared as follows. Sign: "famous buildings" and "famous products" correspond to the material symbols on the surface of the onion. Spirit: "famous people" corresponds to the heroic character in the second layer of the onion, and "traditional customs" corresponds to the customs and rituals in the third layer of the onion model. Therefore, this study used these four types of characteristics as the map theme to establish the cultural content framework of local chronicles, and because the innermost layer of the Hofstede model, "values", is more abstract, this study did not perform visual mapping of this layer.

The content framework of the city's characteristic culture based on local chronicles is shown in Figure 3.

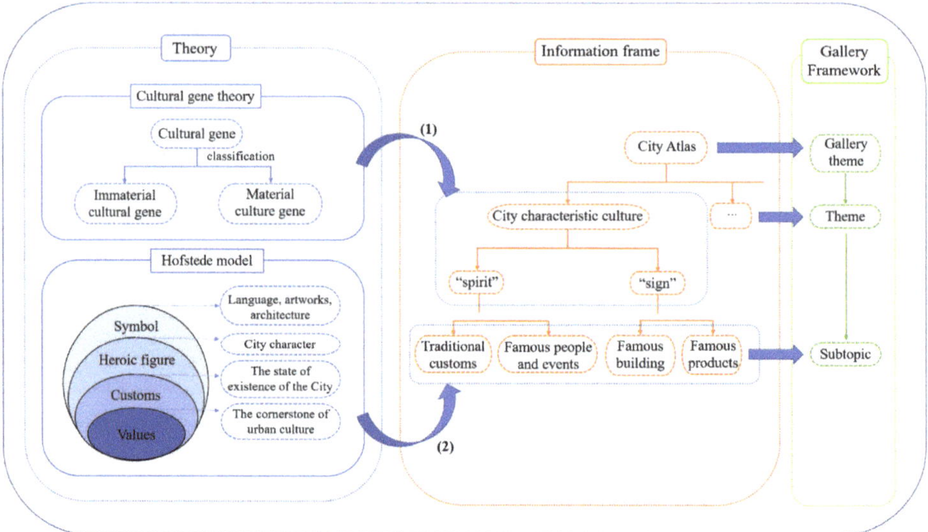

Figure 3. The content framework of city characteristic culture. (**1**) Initial delineation of the city's characteristic culture by cultural gene theory; (**2**) Selection of city-specific cartographic themes based on the Hofstede model.

2.2.2. Step 2: Propose Visualization Strategies for the Content Framework of the City's Characteristic Culture

After establishing the framework of urban characteristic cultural content based on local chronicles, it is necessary to establish the principles of a visualization strategy to fully realize the portrayal of multiple characteristics, such as regional uniqueness, recognizability, systematization, and temporality of urban local chronicle characteristic cultural information. This part consists of two main aspects: (1) establishing the form of presenting the overall urban characteristics content from the triple time–space–people perspective, and (2) following the map design principles of figurativeness and symbolism.

(1) Establish the form of presenting the overall urban characteristics content from the triple time–space–people perspective

The urban environment should be seen as the result of multiple roles in space and time, the core of which is the role of human socialization [41]; therefore, to perceive and feel the characteristics of the city as a whole from the triple time–space–people perspective, we focused on the expression of the "time, space, and people" dimensions of the city's distinctive culture in the visualization of local chronicles to help readers effectively immerse themselves in the city's distinctive culture in all aspects.

(2)   Follow the map design principle of figurativeness and symbolism

Figuration is the expression of what people see and hear in a specific context according to their cognition, while symbolization means the generalization and summary of experience through a symbolic medium [42]. When designing maps (such as color and symbol design), it should closely focus on the style of the city's characteristics and capture its features to symbolize the city culture, promote the formation of "imagery" through figurativeness, and then elicit emotional resonance through symbolism to achieve a better visualization effect.

2.3. Map Evaluation

Cartography itself shares some common goals, i.e., whether the results of map design are effective in providing information to readers; however, the evaluation of map design is subjective and different people have different opinions; therefore, we evaluated the effectiveness of the map products of this study using a questionnaire [43]. We selected 40 respondents, who were chosen mainly because some of them were teachers engaged in map research, some were graduate students majoring in maps, and some were college students who claimed to have a strong interest in maps. The details of the respondents are shown in Figure 4. There were 24 men and 16 women who completed the survey, and the ratio of male to female respondents was 3 to 2. The age of respondents was mainly between 20 and 25 years old, and the mean age of all respondents was 26.5 years (SD: 6.404). In terms of professional status, 10 teachers, 25 graduate students, and 5 university students participated in this survey, and the proportion of respondents who were familiar with and proficient in cartography was 90%. At their own will, we invited them to participate in this questionnaire. We chose the Likert scale format, which means that the respondents responded with their level of agreement with different statements [44]. In this study, the following statements and options were set to achieve the evaluation of map products.

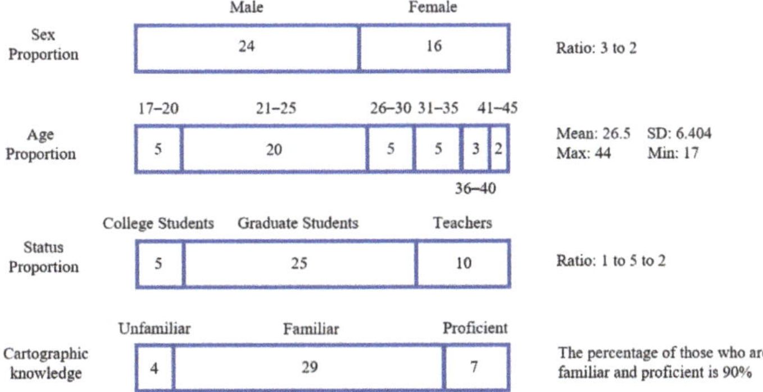

Figure 4. Details of the respondents.

Statements:

1.   The three-dimensional layout of these map products in terms of time, space, and people is a way to understand the city in all aspects;
2.   The symbolic design and color design of these map products can convey the characteristic style of the city;
3.   The pictorial point symbols are a very expressive representation in these visualizations;
4.   Text descriptions in rectangular frames are a good complement to cartographic visualization;
5.   These map products help access local-chronicle information, and they provide a richer carrier.

Options:

A. I strongly agree; B. I agree; C. neutral; D. I disagree; E. I strongly disagree.

3. Results

3.1. The Cultural Content Framework of Dengfeng City's Chronicles

Taking the 2008 edition of the Dengfeng City Chronicle as the data source, based on cultural gene theory and the Hofstede model, we established a framework of the city's characteristic cultural contents under the four themes of "traditional customs", "famous people and events", "famous buildings", and "famous products" in Dengfeng city, as shown in Figure 5. Among them, the expression of "traditional customs" mainly includes folk-culture resources such as the Dengfeng opera and folk activities; the expression of "famous people and events" includes important people and related events from ancient times to the present; the expression of "famous buildings" mainly includes temples, watchtowers, palaces, and altars in Dengfeng; and the expression of "famous products" mainly includes national grade I cultural relics, national grade II cultural relics, and other grade relics excavated in Dengfeng. These three classes of relics are classified according to their preciousness, rarity, and archaeological value. National grade I cultural relics are the most valuable, rare, and archaeologically valuable; national grade II cultural relics are second in value, rarity, and archaeological value to national grade I cultural relics; while other grade relics are the least valuable, rare, and archaeological value of the three.

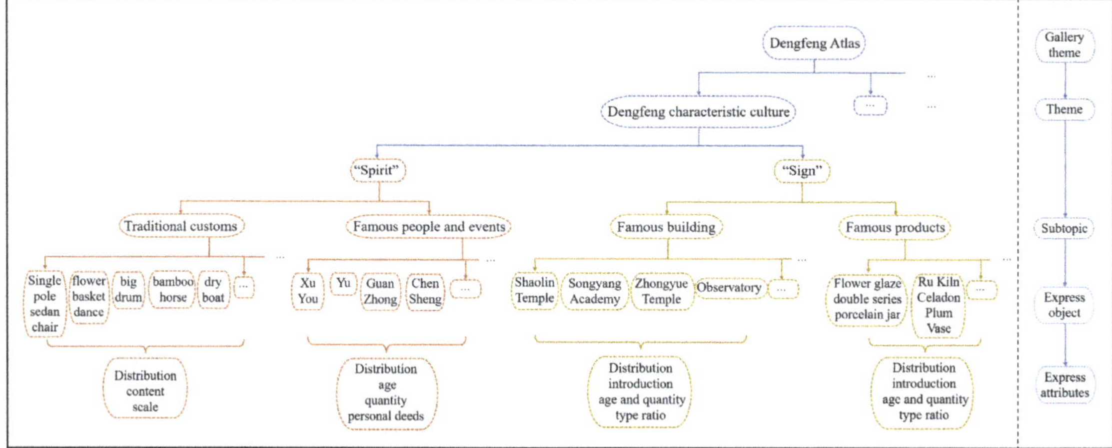

Figure 5. "Spirit–Sign" content framework of the Dengfeng City Chronicle.

3.2. Visual Expression

3.2.1. Construction of Presentation Form of Overall Urban Characteristic Content from the Triple Time–Space–People Perspective

The four thematic maps in this study are of size A2 (420 × 594 mm), with famous buildings and products on the front and back of the first "Sign" map, and traditional customs and famous people and events on the front and back of the "Spirit" map. The layout of the maps is divided into three subspaces based on the principle of grasping the overall city characteristics from the triple time–space–people perspective. Among them, the spatial scale (Panel A) shows the relationships among various features of Dengfeng (traditional customs, famous people and events, famous buildings, and products) and their surroundings; the temporal scale (Panel B) shows the historical axis of cultural changes in various features of Dengfeng, and the human scale (Panel C) emphasizes the spatial and temporal scale of human awareness of the city's distinctive contents. In addition, the primary function of the map is to show spatial location relationships. Therefore, the spatial scale display occupies the main position in the layout. Figure 6 shows the layout diagram.

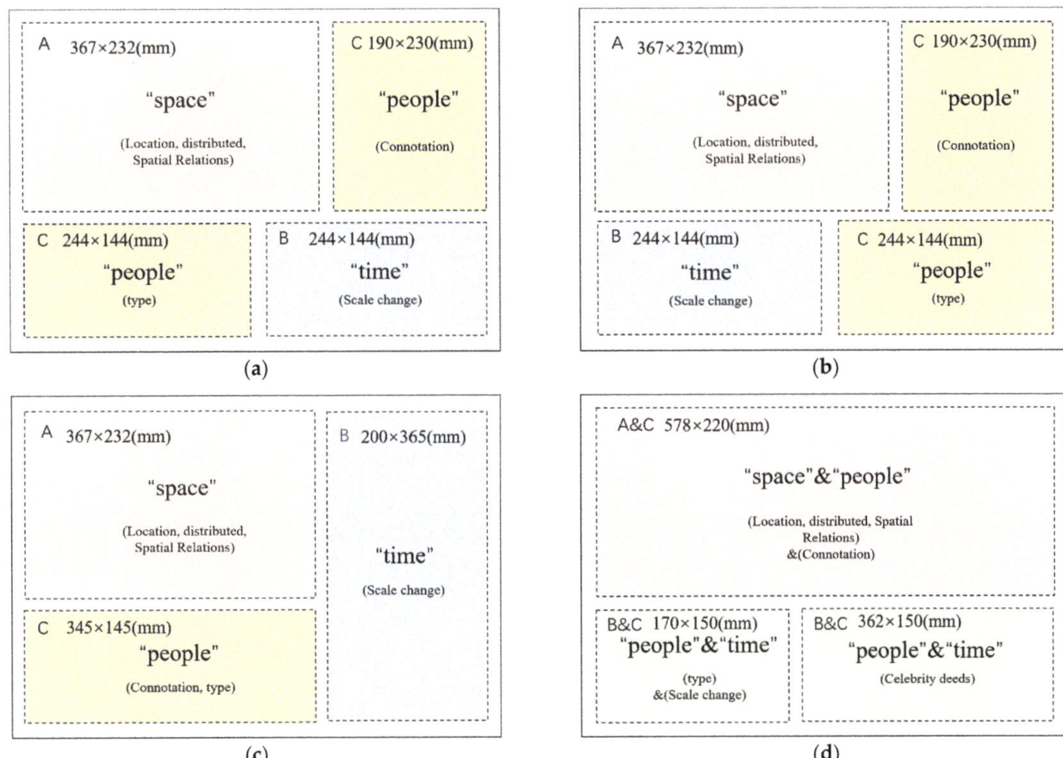

Figure 6. Layout settings for the layout system: (**a**) famous buildings layout; (**b**) famous products layout; (**c**) traditional customs layout; (**d**) famous people and events layout.

3.2.2. Follow the Map Design Principle of Figurativeness and Symbolism

(1)   Color System

Visualization-narrative research considers color to have the same narrative function as language and text [10]. The establishment of the color system of map design in this study was based on the principle of figurativeness and symbolism, and the main colors used in the map were determined around the iconic features of Dengfeng [45,46]. Among them, the buildings and landscapes (e.g., Shaolin Temple, Stargazer Terrace, etc.), which are symbols of the city, have strong local characteristics, and adopting their main colors, such as red and brown, can better reflect the culture and temperament of the city [22]. At the same time, following the principle that color intensity presents different visual levels to the expression elements [47,48], lighter and more contrasting colors from the city colors were selected as the background colors; the colors with high color saturation and strong contrast were selected as the prominent element colors, which in turn formed the whole color system. According to the principle of wholeness, all four thematic maps used a uniformly established color system, as shown in Table 1 below.

Table 1. The color system of the thematic map.

Map Elements	Color	R	G	B	Map Elements	Color	R	G	B
Background		254	244	244	Word		204	210	188
		206	206	155			201	188	156
Outstanding elements		160	183	84	Chart		116	156	169
		239	130	78			192	156	37

(2) Symbol System

The core themes of the four thematic maps are the customs, people, buildings, and products of Dengfeng. Based on the principles of figurativeness and symbolism [34], this study designed concrete or abstract symbols to restore the real imagery from the dimensions of color, shape, material, and structure of the expression elements and used expression methods such as geometric pattern composition and physical objects for artistic processing. These methods enable symbols to have pictographic patterns, which are then used to indicate the real objects they represent [32,49–52]. Among them, the famous products were represented by abstract symbols, but according to their different degrees of importance, the color of the symbols was changed from dark to light to represent national grade I cultural relics, national grade II cultural relics, and other grade relics, respectively. The specific illustration of the symbol design is shown in Table 2.

Table 2. The symbol design of the thematic map.

Express Content	Symbol Type	Expression Method	Symbol Example
Traditional customs	Concrete symbol	Based on the physical art processing method	
Famous people and events	Concrete symbol	Based on the physical art processing method	
Famous buildings	Abstract symbol	Geometric pattern composition	
Famous products	Abstract symbol	Geometric pattern composition	

3.2.3. Full Map

In this study, after taking the Dengfeng City Chronicle as the information source and establishing the content framework of the city's characteristic culture based on cultural gene theory and the Hofstede model, we followed the visualization strategy to generate maps using the software of ArcGIS 10.6 and Adobe Illustrator 2020.

ArcGIS, as a GIS platform software, facilitated the processing of the administrative division map of Dengfeng. In addition, Adobe Illustrator, as a vector drawing software, was convenient for the design and creation of the maps. ArcGIS was used to realize the import and presentation of the administrative division map data of Dengfeng city, the addition of scale and compass, cartographic synthesis, and the export of the administrative division map. Adobe Illustrator was used to lay out the map content, and this layout considered that the map must show the information of the time–space–people scale. Moreover, the design of the color and symbols of the map was achieved using Adobe Illustrator software. Specifically, we added the spatial locations of four special cultures to the administrative map and used symbols to represent them. Creative diagrams were designed for related thematic information, and necessary text annotations were added to the map to make it easier to fully understand the information presented on the map. The final map results are shown in Figures 7 and 8 below.

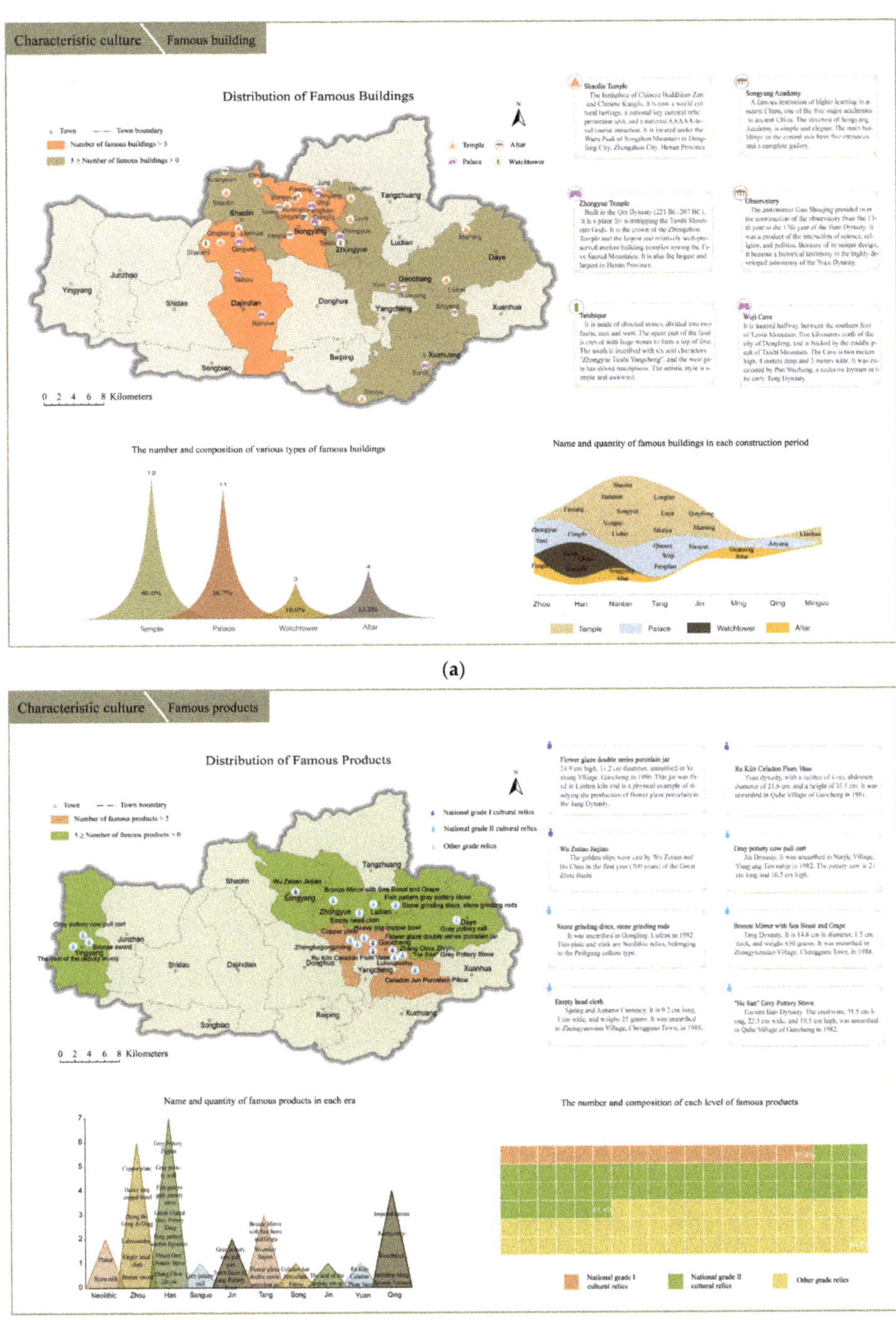

Figure 7. The front and back of the "Sign" map: (**a**) front side: map of famous buildings; (**b**) backside: map of famous products.

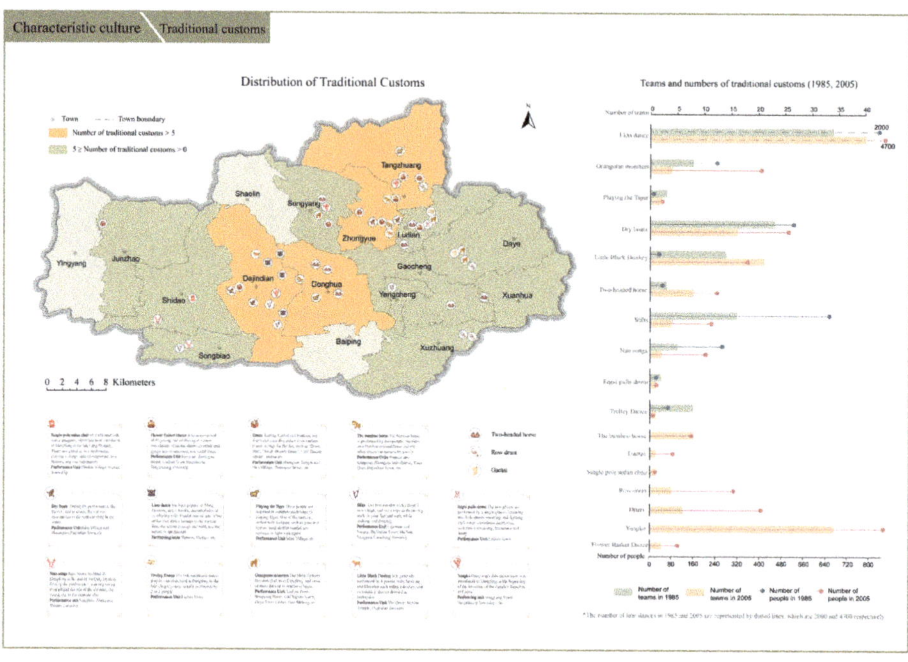

(**a**)

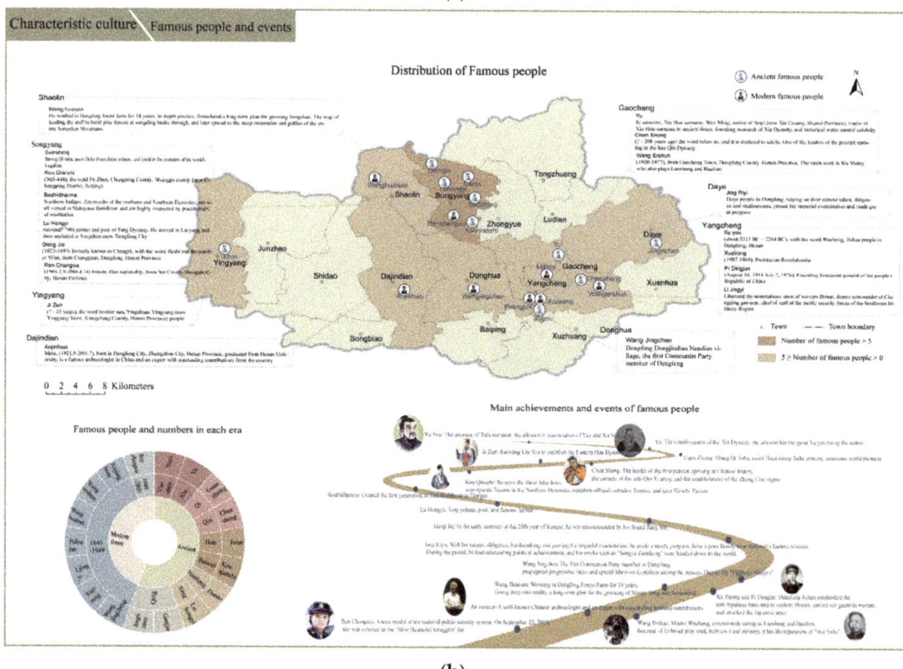

(**b**)

Figure 8. The front and back of the "Spirit" map: (**a**) front side: map of traditional customs; (**b**) backside: map of famous people and events.

3.3. Evaluation Results

During the survey, we showed the map results of this study to 40 respondents and then invited them to fill out the questionnaire. At the end of the survey, we collected all respondents' answers to the questionnaire, tabulated them (Table 3), and created a statistical chart (Figure 9). The statistical chart clearly shows the percentage of respondents' choices for each question, with blue indicating positive options (I strongly agree, or I agree) and red indicating negative options (I disagree, or I strongly disagree). The chart shows that the percentage of positive options exceeds 85% for all questions, with the maximum percentage being 97.5%, the minimum percentage being 85%, and the median being 95%. Ultimately, we believe that a median percentage of positive options greater than 90% confirms that the map products in this study present local chronicle information well and that the typography, colors, symbols, and supplemental text of these maps are reasonable and well-liked by users.

Table 3. Questionnaire answers from all respondents.

Id	Q1	Q2	Q3	Q4	Q5	Id	Q1	Q2	Q3	Q4	Q5	Id	Q1	Q2	Q3	Q4	Q5	Id	Q1	Q2	Q3	Q4	Q5
1	A	A	B	A	B	2	A	B	B	A	A	3	B	B	C	A	A	4	A	B	A	A	A
5	A	A	A	A	A	6	A	C	C	B	A	7	B	B	C	B	B	8	B	B	A	C	A
9	A	A	A	B	C	10	A	B	B	A	B	11	A	A	A	A	A	12	A	A	A	A	A
13	A	A	A	A	B	14	A	A	A	A	B	15	A	A	B	A	B	16	A	B	A	A	B
17	B	C	C	C	A	18	A	C	B	A	B	19	B	A	C	A	B	20	A	A	A	A	A
21	A	A	A	B	A	22	B	B	A	A	A	23	B	B	A	B	A	24	A	A	A	A	A
25	A	A	B	A	A	26	A	A	A	A	A	27	A	A	A	B	A	28	B	B	A	A	A
29	B	B	A	B	A	30	A	A	A	A	A	31	A	A	B	A	A	32	A	A	A	A	A
33	A	B	B	B	A	34	C	D	B	A	B	35	B	B	D	B	B	36	B	A	B	A	B
37	A	B	A	A	A	38	A	B	B	A	A	39	B	B	B	A	A	40	B	B	B	A	A

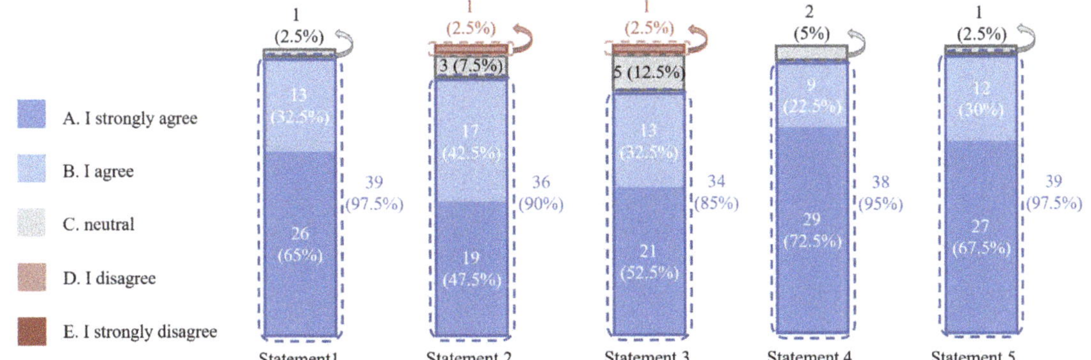

Figure 9. The statistical chart of questionnaire results for map evaluation.

4. Discussion

Based on cultural gene theory and the Hofstede model, this paper developed a "Spirit–Sign" content-framework model with the theme of urban characteristic culture. Based on this model, the urban cultural resources of local chronicles were excavated and sorted out. In addition, this study proposed a series of visualization strategies for the information of urban culture characteristics in local chronicles, i.e., grasping urban culture characteristics from the triple time–space–people perspective and following the map design principle of figurativeness and symbolism to optimize the visual expression. Finally, the feasibility and effectiveness of the above content-framework model and visualization strategies were verified by Dengfeng's distinctive cultural group.

The designed maps were compiled as a special-cultural-thematic map group together with the rest of the thematic maps in a paper atlas. The whole atlas will be published as a map visualization result of the Dengfeng City Chronicle and in the form of electronic maps for people to browse on the internet. The readers of this atlas will range from politicians to ordinary citizens. Perhaps through it, politicians will have easier access to information on various aspects of the city for planning and decision making. Ordinary citizens can learn more about the city through it and increase their love for the city.

The results of cartographic design are potentially important for understanding and effectively providing information to decision makers. Our specific goal in this section is to demonstrate the necessity and value of cartography for local chronicles. The case analysis shows that the rapidity and effectiveness of this study in comparing the original books of local chronicles to obtain the characteristic cultural information of the city can be reflected in the following aspects. First, this paper combined the cultural gene theory with the Hofstede model, which can effectively meet the demand for map visualization of historical and cultural data of local chronicles, help to fully complete the sorting and structure of relevant thematic information, and help local chronicles to change a single-text carrier and visualize maps. Second, the proposed visualization strategies can effectively separate the urban cultural information of the Dengfeng City Chronicle from the textual method, facilitate the use of a map to show the distribution and spatial relationships of the elements, help us to design visual elements such as symbols and colors and fully adopt the novel graphical form to intuitively express the thematic information, facilitate access to the overall distribution of urban characteristic culture, and effectively explore more potential Dengfeng characteristic cultural features, as well as spatial and temporal connections. In addition, we analyzed the information of each panel of "Sign" and "Spirit" from the perspectives of "space" and "time, space, and humanity" and drew the following conclusions so that we can better understand and grasp the historical characteristics of Dengfeng.

4.1. Spatial Scales

The spatial scales of "Sign" and "Spirit" (Panel A) both show the distribution of various features of Dengfeng (traditional customs, famous people and events, and famous buildings and products) and their relationships with their surroundings and as the spatial scale panel occupies the main position of the layout, we can obtain much information and rules from it through analysis. Taking the "Sign" panel as an example, the main distribution of famous buildings and products on its front and back is located in Gaocheng town, Songyang Street, Zhongyue Street, Daye town, etc. Combined with the distribution of traditional customs and famous people's addresses in the "Spirit" panel, it can be seen that there is an overlap among the four main areas of Dengfeng's characteristic culture; for example, there are four characteristic cultural sites in Songyang Street, Gao Cheng, and Daye. This result can be explained by the fact that Songshan is located on Songyang Street, which has been an important place to visit and worship at since ancient times; it has been visited by kings, generals, scholars, and monks, leaving more temples, palaces, and other buildings. The result further implies that the activity areas of celebrities in Dengfeng in the past and present have brought activity to the corresponding areas, thus promoting the completion of ancient buildings and the output of famous products and eventually forming a unique traditional culture that has been inherited to this day.

4.2. Time, Space, and Humanities Scales

The temporal scales of "Sign" and "Spirit" (Panel B) show the historical axis of the cultural changes in each characteristic of Dengfeng, while the human scale (Panel C) emphasizes the spatial and temporal scales of human awareness of the city's distinctive contents. Combining the spatial scale (Panel A) with the temporal scale (Panel B) and referring to the human scale (Panel C) helps us better grasp the patterns of cultural change in the relevant urban characteristics. To this end, we took the front and back panels of famous buildings and products of "Sign" as an example, and we further analyzed them in the A, B, and C (space and time and humanities) panels, as follows.

Figure 10 shows the time scale (Panel B) chart of the famous buildings and famous products. Based on this chart, it can be observed that the different dynasties to which the famous buildings and famous products belong paint the famous buildings and famous products of the same dynasty with the same color. Then, the spatial information of Panel A was used to spatially connect the famous buildings and products of the same dynasty. If there is only one famous building or product in the same dynasty, we will cover it with an independent circle. The result is several colorful polygons with different dynasties, and then we can obtain the analysis chart shown in Figure 11. Based on this chart, the spatial range of famous buildings and excavated products with different dynasties was visualized and analyzed.

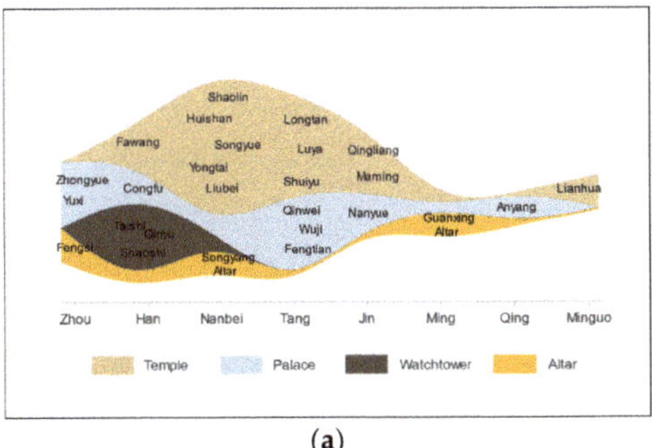

(**a**)

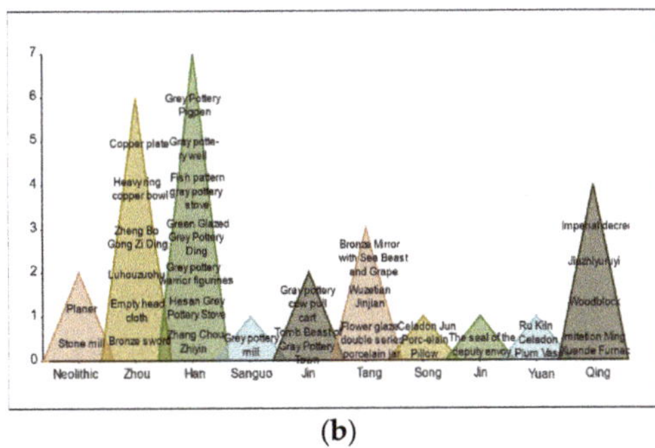

(**b**)

Figure 10. Time scale (Panel B) results: (**a**) changes in famous buildings over time; (**b**) changes in famous products over time.

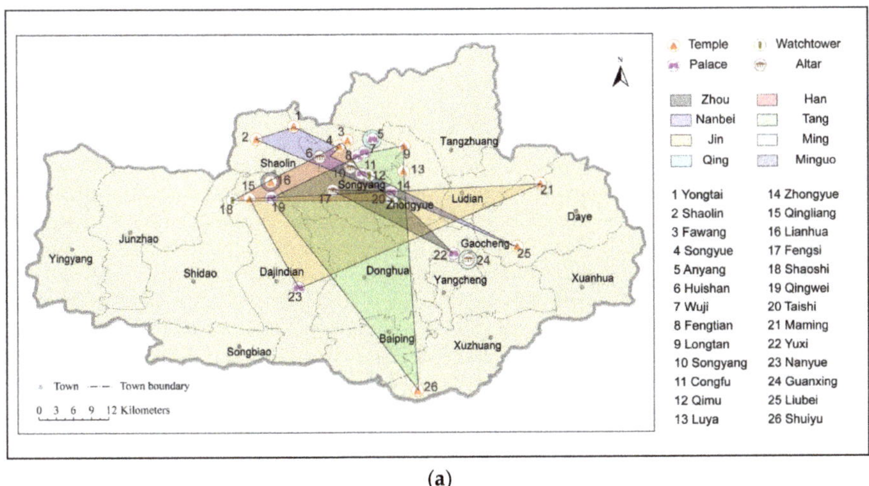

(**a**)

Figure 11. *Cont.*

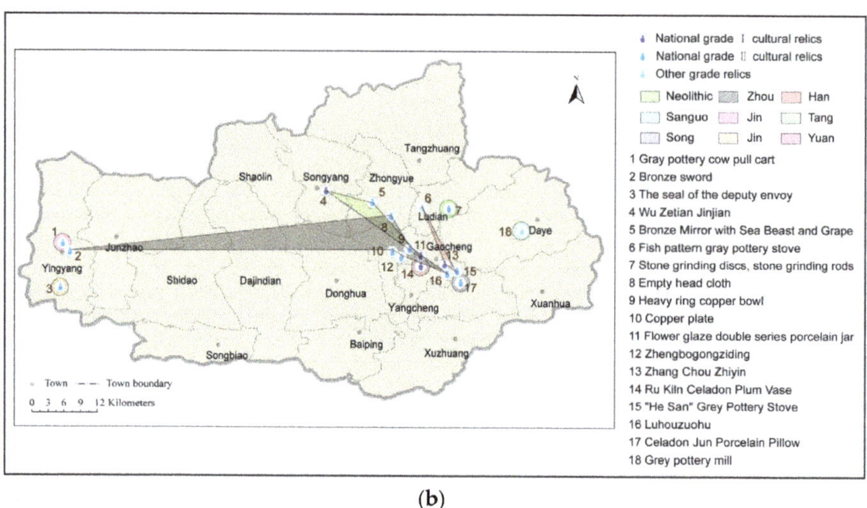

(b)

Figure 11. Graphs of the results of the spatial and temporal analysis: (**a**) spatial share of famous buildings over time; (**b**) spatial share of famous products over time.

For example, the Zhou Dynasty began to build famous buildings that were mainly located on Songyang Street and Zhongyue Street in a few places, and the Han Dynasty began to build famous buildings in Dajindian, Songyang Street, and Gaocheng town in a few places. Also, the Zhou Dynasty, Han Dynasty, and Tang Dynasty excavated famous products that were mainly located in Songyang and Gaocheng town, showing a relatively concentrated situation. Combining the graph with the above analysis, we can conclude that as the dynasties proceeded, the spatial scope of the construction of the famous buildings expanded, which implies that the changes in human activities were also expanding and the different directions of urban changes in each dynasty, and the more concentrated scope of the excavation of the famous products, may be related to archaeological human factors.

5. Conclusions

In the context of urban history and culture gradually attracting public attention, this study used maps to visualize the information of local chronicles given the limitations of accessing the information on urban characteristics and culture caused by the single-text carrier form of local chronicles. This study took content-framework construction and map design as breakthrough points and provided new ideas and successful practical cases on how to effectively dig out and sort local chronicle information and how to optimize the visual-expression effect from layout design, color emotion, symbol creativity, and other elements; then, after analyzing the four completed maps of this study in combination with the advantage of being able to express geographical relationships that the maps themselves have, the following conclusions were found. On the one hand, there is an overlap among the four main concentrated areas of Dengfeng's distinctive culture, which indicates that spirit and matter exist in a complementary and mutually reinforcing relationship; however, there are also certain differences in their spatial distribution, which indicates that there is also a contingency between material and immaterial culture, not absolute interdependence. On the other hand, as the dynasties proceeded, the spatial scope of the famous buildings built in different dynasties expanded, which implies that the changes in human activities were also expanding and that there were different directions of urban changes in each dynasty, reflecting that Dengfeng had a certain historical status in each historical dynasty due to its splendid culture and rich resources. The concentration of the excavated range of famous products may be related to archaeological anthropogenic factors, etc. Finally, the feasibility and validity of the above content-framework model and visualization strategies for map design were verified by evaluating the map results of this study with a questionnaire.

The local-chronicles-content framework and visualization strategies for urban cultural characteristics proposed in this study are operable and universal and can be used in other cities to carry out the work related to local-chronicle-information visualization and urban-cultural-map design; however, the local chronicles cover a wide range of contents, and the excavation and collation of urban characteristics in this study are biased toward historical and cultural elements, while the collation of natural ecological and industrial economic elements of urban characteristics needs further research. Given the above limitations, we will carry out follow-up studies and make more efforts to better integrate local chronicles with maps.

Author Contributions: Conceptualization, Xiaohui He; methodology, Yongji Wang and Lili Wu; validation, Chuan Liu; formal analysis, Chuan Liu and Yongji Wang; investigation, Chuan Liu; data curation, Chuan Liu; writing—original draft preparation, Chuan Liu, Xiaohui He, and Lili Wu; writing—review and editing, Chuan Liu, Xiaohui He, and Yongji Wang; visualization, Chuan Liu; supervision, Xiaohui He; project administration, Zhihui Tian; funding acquisition, Zhihui Tian and Xiaohui He. All authors have read and agreed to the published version of the manuscript.

Funding: This work was supported by the Science and Technology Major Project of Henan Province under Grant 201400210900.

Data Availability Statement: The data presented in this study are available from the author upon reasonable request.

Acknowledgments: The authors would like to acknowledge all the reviewers and editors.

Conflicts of Interest: The authors declare no conflict of interest.

References

1. Wang, F.; Zhu, H.; Zhang, L.; Long, M.; Liao, X. The Full-Text Search on Local Chronicles' Database. Available online: http://en.cnki.com.cn/Article_en/CJFDTOTAL-CHTB200802020.htm (accessed on 1 August 2022).
2. Zhu, S.L.; Bao, P. The use of Geographic Information Systems in the development and utilization of ancient local chronicles. *Libr. Hi Tech.* **2015**, *33*, 356–368. [CrossRef]
3. Li, N. Automatic Extraction of Alias in Ancient Local Chronicles Based on Conditional Random Fields. *J. Chin. Inf. Process.* **2018**, *32*, 41.
4. Aguilar, J.; Vizcarrondo, J. Distributed chronicle for the fault diagnosis in distributed systems. *Int. J. Commun. Netw. Distrib. Syst.* **2020**, *24*, 284–315. [CrossRef]
5. Bai, Z.; Hou, H. Design and implementation of a text categorization system based on word-extracting dictionary and multi-categorizing algorithms. *J. China Soc. Sci. Tech. Inf.* **2008**, *27*, 337–343.
6. Xu, C.; Ye, H.; Bao, P. Automatic Recognition of Produce Entities from Local Chronicles with Deep Learning. *Data Anal. Knowl. Discov.* **2020**, *4*, 86–97.
7. Southall, H. Rebuilding the Great Britain Historical GIS, Part 3. *Hist. Methods* **2014**, *47*, 31–44. [CrossRef]
8. Fitch, C.A.; Ruggles, S. Building the National Historical Geographic Information System. *Hist. Methods* **2003**, *36*, 41–51. [CrossRef]
9. Gaffield, C. Conceptualizing and constructing the Canadian Century Research Infrastructure. *Hist. Methods* **2007**, *40*, 54–64. [CrossRef]
10. McKenna, S.; Riche, N.H.; Lee, B.; Boy, J.; Meyer, M. Visual Narrative Flow: Exploring Factors Shaping Data Visualization Story Reading Experiences. *Comput. Graph. Forum* **2017**, *36*, 377–387. [CrossRef]
11. Tang, J.; Liu, Z.; Sun, M. A Survey of Text Visualization. *J. Comput. Aided Des. Comput. Graph.* **2013**, *25*, 273–285.
12. Oliveira, E.C.; Oliveira, L.C.; Cardoso, A.; Mattioli, L.; Lamounier, E.A. Meta-model of information visualization based on Treemap. *Univers. Access Inf. Soc.* **2017**, *16*, 903–912. [CrossRef]
13. Pang, P.C.I.; Biuk-Aghai, R.P.; Yang, M.Y.; Pang, B. Creating realistic map-like visualizations: Results from user studies. *J. Vis. Lang. Comput.* **2017**, *43*, 60–70. [CrossRef]
14. Biuk-Aghai, R.P.; Yang, M.Y.; Pang, P.C.I.; Ao, W.H.; Fong, S.; Si, Y.W. A map-like visualization method based on liquid modeling. *J. Vis. Lang. Comput.* **2015**, *31*, 87–103. [CrossRef]
15. Bartling, M.; Havas, C.R.; Wegenkittl, S.; Reichenbacher, T.; Resch, B. Modeling Patterns in Map Use Contexts and Mobile Map Design Usability. *Isprs Int. J. Geo-Inf.* **2021**, *10*, 527. [CrossRef]
16. Qin, Y.; Pang, X.; Zhao, X.; Ren, F. Design Process of Map Cognitive Function of Atlas: A Case Study of Atlas of Pesticide Residue Levels in Fruits and Vegetables in China. *J. Geomat.* **2021**, *46*, 110–113.
17. Liu, Y.; Ma, C.; Su, Z. Visual Art Design of Atlas of Shenzhen Based on Users' Cognition. *J. Geomat.* **2021**, *46*, 143–147.

18. Wu, L.; Liu, Z. Design and Compilation of Atlas of Arctic Navigation. *Hydrogr. Surv. Charting* **2016**, *36*, 63–66.
19. Qi, C. Design and Compilation of Atlas of Major Natural Disasters in China. 2010. Available online: http://en.cnki.com.cn/CJFD_en/Detail.ashx?t=e&url=/Article_en/CJFDTOTAL-CHTB201301031.htm (accessed on 1 August 2022).
20. Schulz, T. Graphical-statistical Atlas of Switzerland, 1914. *Int. J. Cartogr.* **2021**, *7*, 226–232. [CrossRef]
21. Shi, P.J.; Yang, X.; Xu, W.; Wang, J.A. Mapping Global Mortality and Affected Population Risks for Multiple Natural Hazards. *Int. J. Disaster Risk Sci.* **2016**, *7*, 54–62. [CrossRef]
22. Weng, M.; Song, X.Y.; Wang, L.Q.; Xie, H.; Zhang, P.; Su, S.L.; Kang, M.J. A tourist map of Xi'an: Combining historical city characteristics with art. *J. Maps* **2020**, *16*, 195–202. [CrossRef]
23. Xu, Y.; Ma, C.; Zhang, D.; Du, Q. Content Construction and Cartographic Expression of Urban Innovation: A Case Study of Atlas of Shenzhen. *J. Geomat.* **2021**, *46*, 96–99.
24. Jia, M.C.; Huang, Q.H.; Li, M.C.; Hu, W. Illegal land use risk assessment of Shenzhen City, China. *J. Maps* **2015**, *11*, 798–805. [CrossRef]
25. Romanillos, G. Collaborative mapping of emerging cities in developing countries: The Leon Emergente project. *J. Maps* **2016**, *12*, 584–590. [CrossRef]
26. Guo, R.; Chen, Y.; Ying, S.; Lu, G.; Li, Z. Geographic visualization of pan-map with the context of ternary spaces. *Geomat. Inf. Sci. Wuhan Univ.* **2018**, *43*, 1603–1610. [CrossRef]
27. Qin, Y.; Pang, X.P.; Zhao, X.; Liu, H.Y.; Ren, F.; Bi, C.; Zhang, D.J.; Wu, J. Visualization of pesticide residue data. *J. Maps* **2017**, *13*, 892–899. [CrossRef]
28. Cevasco, A.; Brandolini, P.; Scopes, C.; Rellini, I. Relationships between geo-hydrological processes induced by heavy rainfall and land-use: The case of 25 October 2011 in the Vernazza catchment (Cinque Terre, NW Italy). *J. Maps* **2013**, *9*, 289–298. [CrossRef]
29. Beconyte, G.; Alekna, V.; Rociute, I. A Map of 21st Century Conflicts in Europe. *J. Maps* **2011**, *7*, 1–8. [CrossRef]
30. Aoki, K. A stochastic model of gene-culture coevolution suggested by the culture-historical hypothesis for the evolution of adult lactose absorption in humans. *Proc. Natl. Acad. Sci. USA* **1986**, *83*, 2929–2933. [CrossRef]
31. Hofstede, G.J.; Jonker, C.M.; Verwaart, T. Cultural Differentiation of Negotiating Agents. *Group Decis. Negot.* **2012**, *21*, 79–98. [CrossRef]
32. Weng, M.; Song, X.; Du, Q.; Kang, M.; Su, S. Tourism map design for historical and cultural cities: Thematic content and visualization strategy. *Sci. Surv. Mapp.* **2021**, *46*, 178–185.
33. Su, S.; Wang, L.; Du, Q.; Li, L.; Kang, M.; Weng, M. The design of campus cultural atlas—A case of Luojia Walking Atlas. *Sci. Surv. Mapp.* **2020**, *45*, 153–160.
34. Su, S.; Wu, L.; Du, Q.; Kang, M.; Weng, M. Theoretical framework and case study of urban cultural atlas design. *Sci. Surv. Mapp.* **2021**, *46*, 145–152.
35. Wang, J.; Jiang, N.; Zhang, W.; Zhang, X.N. Study and Practice of Mobile Map Symbols Design for Urban Traffic Map Users. In Proceedings of the International Conference on Information Electronic and Computer Science, Zibo, China, 26–28 November 2010; pp. 180–183.
36. Li, J.X.; Cao, Y.N.; Li, A.G. Design of Emergency Map Symbol. In Proceedings of the 18th International Conference on Geoinformatics, Beijing, China, 18–20 June 2010.
37. Christian-smith, L. *Chinese Local History*; Taylor and Francis: Abingdon, UK, 2019.
38. Huo, Y. Theory Analysis of the Culture Gene Extraction and Transmission Path of Beijing-Hangzhou Grand Canal. *Archit. Cult.* **2017**, 59–62.
39. Degens, N.; Hofstede, G.J.; McBreen, J.; Beulens, A.; Mascarenhas, S.; Ferreira, N.; Paiva, A.; Dignum, F. *Creating a World for Socio-Cultural Agents*; Springer International Publishing: Cham, Switzerland, 2014; pp. 27–43.
40. De Mooij, M.; Hofstede, G. The Hofstede model Applications to global branding and advertising strategy and research. *Int. J. Advert.* **2010**, *29*, 85–110. [CrossRef]
41. Smith, D.M. Human-geography-a welfare approach-response. *Prog. Hum. Geogr.* **1995**, *19*, 393–394.
42. Taverne, E.R.M. Genius-loci-towards a phenomenology of architecture-Norberg Schulz, c. *J. Gard. Hist.* **1981**, *1*, 414–416.
43. Zagata, K.; Gulij, J.; Halik, L.; Medynska-Gulij, B. Mini-Map for Gamers Who Walk and Teleport in a Virtual Stronghold. *Isprs Int. J. Geo-Inf.* **2021**, *10*, 96. [CrossRef]
44. Medynska-Gulij, B.; Zagata, K. Experts and Gamers on Immersion into Reconstructed Strongholds. *ISPRS Int. J. Geo-Inf.* **2020**, *9*, 655. [CrossRef]
45. Skopeliti, A.; Stamou, L. Online Map Services: Contemporary Cartography or a New Cartographic Culture? *ISPRS Int. J. Geo-Inf.* **2019**, *8*, 215. [CrossRef]
46. Guo, L.; Li, L.; Zhang, Y. Color Design of the Hillshading Map Based on the Visual Image. *Geomat. Inf. Sci. Wuhan Univ.* **2004**, *29*, 492–495.
47. Niu, Y.; Ma, G.; Xue, W.; Xue, C.; Zhou, T.; Gao, Y.; Zuo, H.; Jin, T. Research on the Colors of Military Symbols in Digital Situation Maps Based on Event-Related Potential Technology. *Isprs Int. J. Geo-Inf.* **2020**, *9*, 420. [CrossRef]
48. Cui, H.; Jiang, N.; Bai, X.; Hu, Y. The color design scheme of a map for the color vision impaired. In Proceedings of the 2008 International Conference on Computer Science and Software Engineering (CSSE 2008), Wuhan, China, 12–14 December 2008; pp. 272–274. [CrossRef]

49. Horbinski, T.; Zagata, K. Interpretation of Map Symbols in the Context of Gamers' Age and Experience. *ISPRS Int. J. Geo-Inf.* **2022**, *11*, 150. [CrossRef]
50. Bartoněk, D.; Andělová, P. Method for Cartographic Symbols Creation in Connection with Map Series Digitization. *ISPRS Int. J. Geo-Inf.* **2022**, *11*, 105. [CrossRef]
51. Gong, X.Y.; Li, Z.L.; Liu, X.T. Structures of Compound Map Symbols Represented with Chinese Characters. *J. Geovis. Spat. Anal.* **2022**, *6*, 14. [CrossRef]
52. Wielebski, Ł.; Medyńska-Gulij, B. Graphically supported evaluation of mapping techniques used in presenting spatial accessibility. *Cartogr. Geo-Inf. Sci.* **2019**, *4*, 311–333. [CrossRef]

International Journal of
Geo-Information

Article

3D Visualisation of the Historic Pre-Dam Vltava River Valley—Procedural and CAD Modelling, Online Publishing and Virtual Reality

Michal Janovský *, Pavel Tobiáš and Vojtěch Cehák

Department of Geomatics, Faculty of Civil Engineering, Czech Technical University in Prague, Thákurova 7, 166 29 Praha, Czech Republic; pavel.tobias@fsv.cvut.cz (P.T.); vojtech.cehak@fsv.cvut.cz (V.C.)
* Correspondence: michal.janovsky@fsv.cvut.cz

Abstract: As a part of "The Vltava River" project, it was necessary to create a visualisation of the historic Vltava River valley before the construction of the so-called Vltava Cascade (nine dams built in the Vltava River basin between 1930 and 1992). Vectorisations of the *Imperial Obligatory Imprints of the Stable Cadastre*, and a terrain model created from contour lines from the *State Map 1:5000-Derived (SMO-5)*, prepared in an earlier phase of the project, were used as a basis for this visualisation. Due to the extent of the modelled area, which is approximately 1670 km^2, and the available underlying data realistically usable for the visualisation, mainly procedural modelling with the use of the CGA shape grammar was chosen for the creation of 3D objects. These procedurally created 3D models were completed with more detailed models of landmark buildings created in CAD. The outcomes were used to establish a virtual reality (VR) application in the Unreal Engine software. The results are a 3D scene created in a form corresponding approximately to the state of the Vltava River valley in the 19th century, which is available for viewing via a web application, and a VR scene used for demonstration at exhibitions.

Keywords: procedural 3D modelling; large-scale 3D modelling; virtual reality (VR); Unreal Engine; CityEngine; historical landscape; 3D visualisation of landscapes and urban scenes

Citation: Janovský, M.; Tobiáš, P.; Cehák, V. 3D Visualisation of the Historic Pre-Dam Vltava River Valley—Procedural and CAD Modelling, Online Publishing and Virtual Reality. *ISPRS Int. J. Geo-Inf.* **2022**, *11*, 376. https://doi.org/10.3390/ijgi11070376

Academic Editors: Wolfgang Kainz, Beata Medynska-Gulij, David Forrest and Thomas P. Kersten

Received: 16 May 2022
Accepted: 4 July 2022
Published: 6 July 2022

Publisher's Note: MDPI stays neutral with regard to jurisdictional claims in published maps and institutional affiliations.

1. Introduction

The three-dimensional visualisation of the Vltava River valley is a part of the project "The Vltava River—Changes in Historical Landscape due to Floods, Construction of Dams and Changes in Land Use with Links to Cultural and Social Activities in the Surroundings" (DG18P02OVV037), hereinafter referred to as the "Vltava Project". The Vltava Project is a part of the Ministry of Culture of the Czech Republic programme "NAKI II—Support of Applied Research and Experimental Developments of National and Cultural Identity 2016–2022". The creation of the 3D visualisation within the Vltava Project is focused primarily on extinct villages in the Vltava River valley. Moreover, several important landmark buildings were modelled in more detail using a classical modelling approach and CAD software. Due to the time-consuming nature of detailed 3D modelling, and the fact that some important buildings changed significantly during the period under review, the 3D scene was created in a compromise form corresponding approximately to the state in the 19th century. This compromise form depicts the significant buildings (modelled in CAD software) in their form as depicted on the *Imperial Obligatory Imprints of the Stable Cadastre* from 1826–1843. Since there are no 19th century maps with elevation with sufficient accuracy to cover the entire Vltava Valley, SMO-5 maps created after 1950 were used to create the DTM. Because of this difference in the years of origin of the materials used, we refer to the resulting vectorisation as a compromise form, corresponding approximately to the state in the 19th century. Later visualisations, which will show other time periods (during and after the construction of the dams), where major changes in important buildings will be visible, for example, due to the influence of the war, will already be in the "proper form of the state in the 20th century".

The 3D visualisation of Vltava River valley (not yet in the final version) is accessible online [1] and will be available in the final version directly from the project website [2] by the end of the year 2022. Furthermore, on the project website you can find information about the exhibition *Vltava—Transformations of the Historical Landscape* [3], which took place in the atrium of the Faculty of Civil Engineering of the CTU in Prague from 8 February to 7 April 2022.

The studies focused directly on the Vltava River basin were most often devoted to (historical) floods, chemical and biological research (content and changes in the content of elements in sediments and soil, etc.), or focused only on smaller sections of the Vltava River and its tributaries. Studies of the Vltava River include the "3D model of the historical Vltava Valley in the area of the Slapy reservoir" [4], which can be considered as a predecessor of the Vltava Project, as it deals with the same issues, but only on a small part of the Vltava; in addition, the study by Yiou et al. on floods in Bohemia since 1825 [5], which deals with floods on the Vltava and Elbe rivers.

Other projects involving research on water bodies include the works of Zlinszky and Timár [6], which focused on Lake Balaton (597 km^2) and Lake Balaton watershed (5700 km^2) using similar mapping data, from which they drew information on the state of Lake Balaton over a wider time horizon. Another study addressing a similar issue was the Italian case study of Surian [7]. In this study, the authors looked at changes in the Piave River (222 km) channel in the eastern Alps of Italy due to the construction of hydroelectric power plants. However, while the river channel changes and other hydrological changes caused by dam construction and other human activities were investigated here, the impacts of these changes on the area along the river and its surroundings were not examined. Other similar studies are described and compared in the discussion section.

The two main parts of the preparation of our 3D scene include the creation of a digital terrain model (DTM), and the modelling of 3D objects, both of which are based on the vectorisations of old maps. The vectorisation of the *State Map 1:5000-Derived* is used, together with other sources, such as the longitudinal profile of the river, to create a DTM [8], mainly using automatic vectorisation of elevation contours from the map [9,10]. The vectorisation of the *Imperial Obligatory Imprints of the Stable Cadastre 1:2880* [11] is used as a basis for procedural modelling.

There were mainly two technologies employed to create the visualisation. The first of the technologies used is 3D modelling. Our project utilises a well-established approach, where a common conurbation is modelled procedurally based on old maps and historical iconographic material. Landmark buildings are then reconstructed in more detail by employing simple CAD software. A similar approach is applied for the creation of well-known, 3D reconstructions, such as the model of ancient Rome (original link to Rome Reborn project website: https://www.romereborn.org; Rome Reborn is now included at virtual tourism website: https://www.flyoverzone.com) or Pompei [12,13]. The procedural modelling approach deals with the creation of 3D models based on sets of rules (rule files), according to which these models are algorithmically modelled. In our project, we employ the CGA shape grammar [14] because shape grammars are described as "presently the most developed, used and compact method for building representation" [15]. This technology is widely used in historical visualisations based on map data [11,16,17]. Archival drawings, photographs and other materials can also be used as data sources, from which individual attributes can be better determined, and set in a rule file for the creation of more specific or more realistic objects [17]. It is also used as a method for generating models in the vicinity of areas of interest, which no longer need to be processed in such high detail, and serve, for example, as a supplement to the visualisation of the surroundings. Most procedural modelling techniques often use the results from other technologies, such as GIS [11,16]; conversely, the results of procedural models are used by other technologies, such as game engines [18,19]. Even at official ESRI conferences (Esri Developer Summit), there are regular contributions dealing with the connection of CityEngine (procedurally created 3D models and scenes) with game engines (Unity and Unreal) to improve the visualisation of scenes and the use of virtual reality [20,21].

The second technology used is the game engine, which is a software framework primarily designed for video game development, and generally includes relevant libraries and supporting programmes, which may include a rendering engine for 2D or 3D graphics, a physics engine, collision detection, sound, scripting, animation, artificial intelligence, networking, streaming, memory management, threads, localisation support, scene graphs and video support for cinematography [20,21]. Due to the large number of processing options, game engines are increasingly used for data management [22]. In cartography, one of the most frequently used functions of game engines is the creation of VR applications [19]; they are also used in the processing of 3D models and GIS data, with a focus on (realistic) visualisation of the results [23]. This is further demonstrated in the virtual reconstruction of the ancient city of Karakorum [24]. The functionality of game engines can be fully exploited in the reconstruction of ancient buildings, often for which, little documentation remains [25]. One of the drawbacks of game engines may then be their hardware requirements, but these can be reduced using appropriate input data [26].

2. Methodology

This section will describe the main parts of the processing dealing mainly with 3D modelling and 3D web scene creation. This work is part of a larger project called Vltava, during which many other works were carried out prior to the 3D modelling and creation of the resulting visualisations described in this article. This work is illustrated in the following flowchart (Figure 1).

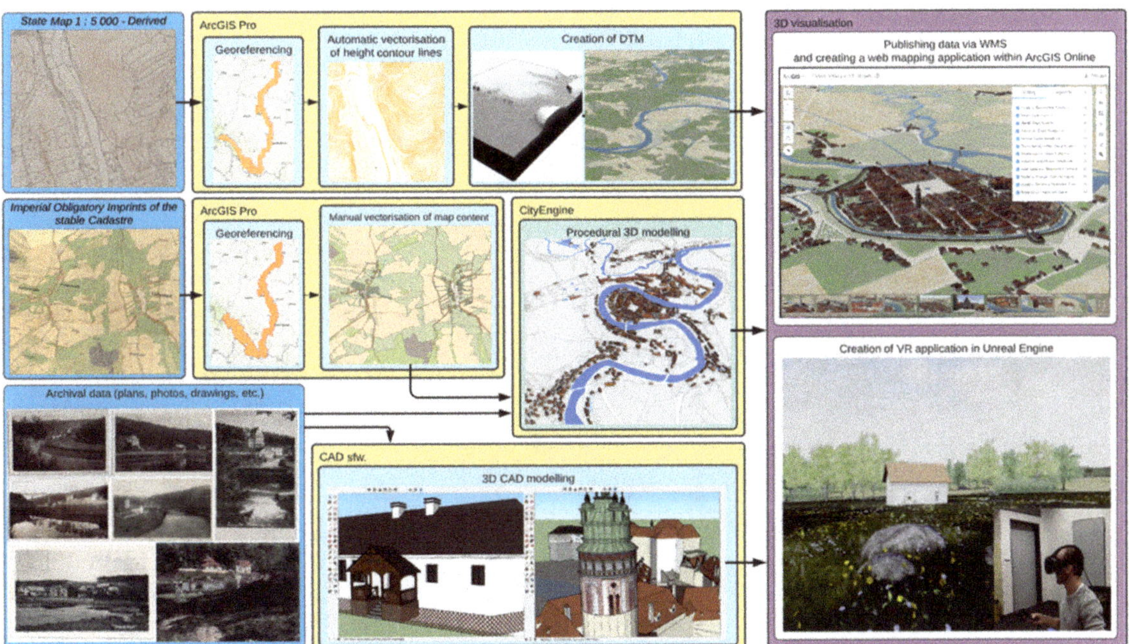

Figure 1. Flowchart showing the processing procedure with image examples.

2.1. Procedural Modelling

The manual vectorisation of the *Imperial Obligatory Imprints of the Stable Cadastre 1:2880* (Figure 2a) provided input data for the procedural modelling of 3D objects. For building modelling, not only vectorised footprints of buildings were used, but also other data readable from underlying maps, such as building types (on the Imprints, buildings are categorised by the material used as either combustible-wooden or non-combustible constructions). The procedural modelling of buildings and vegetation was subsequently performed based on parts of this vectorised map, specifically on the vectorised footprints of buildings and polygons of forests, fruit orchards and other areas with dense vegetation (Figure 2b). The automatic vectorisation of elevation contours (Figure 3b) from the State Map 1:5000-Derived (Figure 3a) was used, together with other sources, such as the longitudinal profile of the river, to create a historical DTM. The vectorisation of the *Imperial Obligatory Imprints* (Figure 4a) and *SMO-5* (Figure 4b) map sheets was conducted for the entire area of interest of the Vltava River valley.

(**a**)   (**b**)

Figure 2. *Imperial Obligatory Imprints of the Stable Cadastre 1:2880*: (**a**) example of a georeferenced scan of a map sheet used for vectorisation; (**b**) vectorisation of the map sheet shown in (**a**).

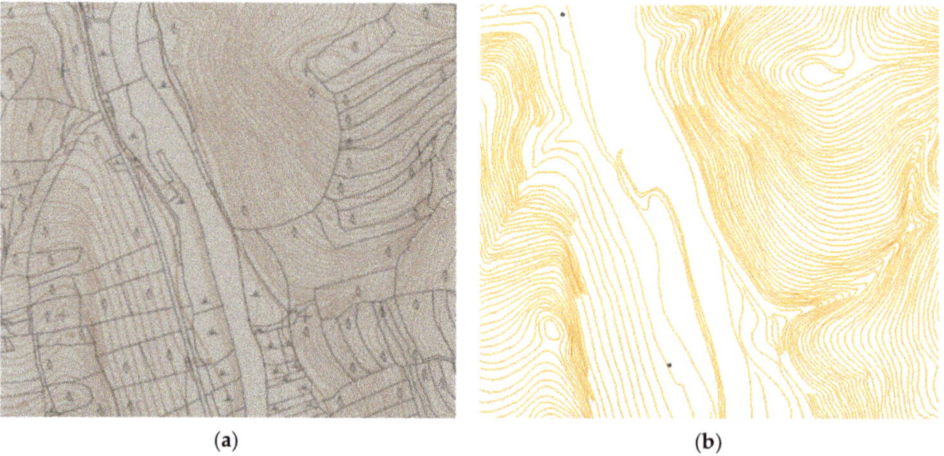

(**a**)   (**b**)

Figure 3. *State Map 1:5000-Derived*: (**a**) example of a part of a georeferenced scan of a map sheet used for the automatic vectorisation of elevation contours; (**b**) vectorised elevation contours from the map piece shown in (**a**) [10].

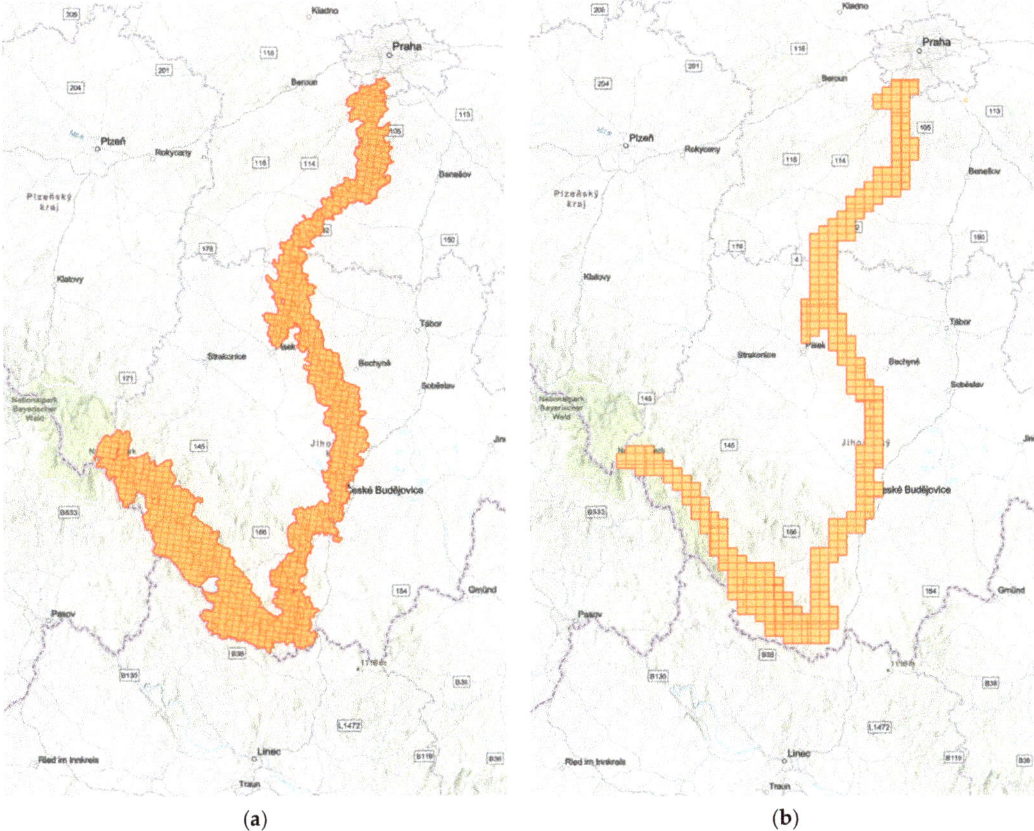

Figure 4. Display of the map sheets of old maps in the Vltava River valley from the capital of the Czech Republic, Prague, to the area bordering on Germany and Austria, used for vectorisation. (**a**) *Imperial Obligatory Imprints of the Stable Cadastre 1:2880* used as input data for procedural modelling; (**b**) *State Map 1:5000-Derived* used as input data for the creation of a DTM.

Other input data comprised archival data, such as written sources, drawings, photographs and postcards, according to which, statistical data of modelled objects were determined (average number of floors, average number of windows, type of roofing, roofing angles, etc.). These parameters were further used in the rule file as attribute values for procedural modelling (Figure 5).

Modelling, testing, editing rule sets and other procedural modelling work was performed primarily in the ESRI CityEngine software. This software application implements the CGA shape grammar, and is, therefore, well suited for the procedural modelling of large urban areas. In addition, ArcGIS Pro was used, mainly for editing procedurally modelled objects and for publishing in the online environment. However, as ArcGIS Pro still does not allow the authorising of procedural modelling rules, the main part of the preparation of rule files was conducted in CityEngine.

```
attr typ_CE = "nespalna" //spalna, nespalna, vyznamna - differentiates  the type of a building, source: SHP
attr Area = geometry.area() // the area of shapes is calculated automatically
const min_shape_area = 25//minimal shape area to generate two storeys

attr floor_count = case Area < min_shape_area : 1

                   else: case typ_CE == "spalna" : (99%: 1 else: 2) //floor count depending on the type of each
                         case typ_CE == "nespalna" || typ_CE == "": (50%: 1 else: 2)
                         case typ_CE == "vyznamna" : 3
                         else : 0


attr floor_height = (25%: 2.4 25%: 2.5 25%: 2.6 else: 2.7)  //height of floors

attr socle_height = case typ_CE == "spalna" && Area < min_shape_area: 0
                    else: 0.5//height of a socle

attr socle_type = (33%: "kamen1" 33%: "kamen2" else: "kamen3" ) //for multiple socle textures

attr tile_width = 34%: 4 33%: 5 else: 6 //width of facade tiles

////////////////////////////////////////////////////////////////////////////////////////////////////
// ROOFS ///////////////////////////////////////////////////////////////////////////////////////////

//Roof and roofing attributes
attr roof_type =    case typ_CE == "spalna" : "gable"
                    case typ_CE == "nespalna" || typ_CE == "": (90%: "gable" else: "hip")
                    case typ_CE == "vyznamna" : "hip"
                    else : ""
                    //"gable"= roof angled from 2 sides,
                    //"hip"= roof angled from 4 sides

attr angle = case (typ_CE == "spalna" && Area < min_shape_area) || Area > 1000 : 20
             else: (10%: 42 10%: 41 30%: 40 10%: 39 10%: 38 10%: 37 10%: 36 else: 35) //roof angle

attr overhangX = case typ_CE == "spalna" && Area < min_shape_area: 0
                 else: (34%: 0.6 33%: 0.5 else: 0.4)
                 //overhang in the slope direction of the roof, works for both hip and gable

attr overhangY = case typ_CE == "spalna" && Area < min_shape_area: 0
                 else: (34%: 0.35 33%: 0.25 else: 0.15)
                 //only for Gable: overhang on the sides of the roof
```

Figure 5. Example of used attributes for the procedural modelling of buildings.

The main goal of our visualisations is not to create super-realistic models of every single house or other objects, but to create such objects that will allow one to better orient themself in the whole Vltava River valley model, especially focusing on the older generation, who experienced the Vltava River before the construction of dams, and on young people in the field of geography teaching. Furthermore, it was necessary to look at the volume of data, due to the large area of interest, with an area of approximately 1670 km^2, which contains over 28,000 procedurally generated buildings. For this reason, the procedurally modelled objects are simple in nature (Figure 6a), without a large amount of spatial fragmentation; however, these apparent shortcomings are mainly replaced using high-quality textures with high resolution.

The use of simple models is also related to the limitations of procedural modelling. Since we are modelling more than 28,000 buildings, one of the limitations is the hardware requirements for the processing, procedural generation, display and editing of 3D models on such a large area containing a large number of objects, villages and cities (Figure 6b). Procedural modelling on this scale cannot be performed on regular computers; it is necessary to use a workstation computer. A workstation with 32 GB of RAM was used to conduct procedural modelling; any smaller value of RAM resulted in the "freezing" and "crashing" of the software, and even when using this workstation, actions involving the layer of models (modelling, seed changes, application of rule file changes, export of models, etc.) took up to tens of minutes. Due to the iterative process of editing rule files, finding errors, testing, publishing and editing for further use (e.g., for VR), the use of more spatially fragmented and complex models would mean considerable increases in the difficulty and prolongation of the work.

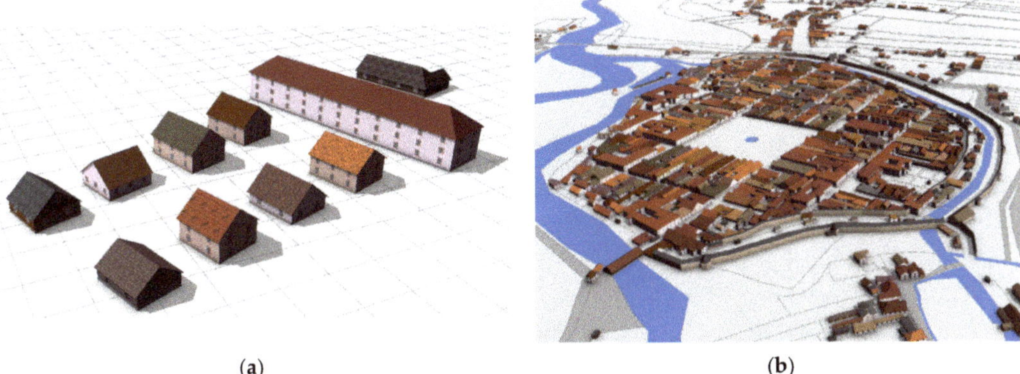

(a)   (b)

Figure 6. Example of procedurally modelled buildings: (**a**) testing scene for debugging the rule file; (**b**) buildings and city walls of České Budějovice, which were procedurally modelled based on vectorisations of the *Imperial Obligatory Imprints of the Stable Cadastre 1:2880*.

Another limitation of procedural modelling is working with the terrain, respectively placing the created buildings on the historical terrain model. Since the data sources used for procedural modelling, and for creating the digital terrain model, are heterogeneous, with different levels of precision, the two models (procedurally modelled buildings and the terrain model) may not always fit together properly. This corresponds with the fact that building models are regular 3D objects that have to be aligned with the continuous 2.5D digital terrain model.

Since the options for local DTM editing are limited in GIS software (at most, various filtering methods are suitable for smoothing the terrain), we have to use workarounds. Although the CityEngine software offers several functions for working with terrain (placing objects at the lowest, highest and average footprint touch points, fitting terrain to models), these functions cannot fully eliminate problems with terrain–model intersections, and on the contrary, may exaggerate these problems (height shifting of objects does not eliminate embedding/floating problems, fitting terrain to objects appears very artificial and unrealistic). In order to prevent the building models from floating above the terrain or, on the contrary, being excessively embedded into it, which is most often the case on slopes, hills, rocks or banks of the Vltava River, a second set of models was procedurally modelled. This new set of models was oriented downwards, i.e., below the terrain, and serves as the foundations for the buildings, on which the models of the buildings themselves are located (Figure 7). These foundations have a uniform texture, and a roof angle of 0°. In addition to these minor changes, they use the same set of rules as the buildings above ground.

This solution is not perfect, but it can be applied once to the entire Vltava Valley area, and most building models will gain a more natural appearance due to the removal of floating, which is replaced by foundations. Another advantage of this solution is that it only needs to be applied once, even if the building models change, which happens frequently due to the continuous tuning and optimisation of the models.

Another limitation of the procedural modelling method, or rather, the CityEngine software in which the procedural modelling was performed, was the inclusion of external 3D models. CityEngine has problems with importing external 3D models (texture inversion, importing models in incorrect location or scale), and problems with exporting scenes to ArcGIS Online, even though it is an ESRI product (general error messages). For these reasons, piecing together these parts of the visualisation was conducted in ArcGIS Pro.

Figure 7. Procedurally modelled foundations for buildings.

In order to combine the procedurally modelled buildings and other objects with other 3D models (photogrammetric models, models of significant buildings, churches, etc.), and import them into ArcGIS Pro, they had to be exported in geodatabase format (*.gdb). However, the procedurally created building models exported in this way were too large, over 90 GB, and, therefore, the textures were compressed. Due to the selection of high-quality textures, this reduction in quality is not noticeable on the models.

The last limitation of procedural modelling we encountered was the creation of spatially complex objects, such as churches and castles. Since their creation by procedural modelling would probably require the creation of separate rule sets for each such object, and since we have the necessary documentation for these objects, it was decided that these objects would be modelled using CAD software. This procedure is described in the next section.

2.2. Detailed Manual Modelling in CAD

In addition to the procedural model of a common conurbation in the whole area of interest, we also created more detailed models of landmark buildings in CAD software. As manual 3D modelling is rather time consuming, we carefully chose specific buildings to be modelled. First, thirteen church areas were selected, primarily in disappeared settlements. Churches were preferred as they are the natural distinguishing features of individual municipalities.

Second, historic buildings in the town of Český Krumlov and the city of České Budějovice were also modelled in more detail. The reason for this is that the first-mentioned municipality is famous worldwide (on the UNESCO World Heritage List since 1992), and it was necessary to depict it at least in a reasonable amount of detail. At the same time, it was possible to use the already existing models of heritage buildings in the town (the castle and monument reserve), which are used by the municipal authority. Therefore, it was sufficient to alter the building models to be in a state corresponding to the 19th century (Figure 8a).

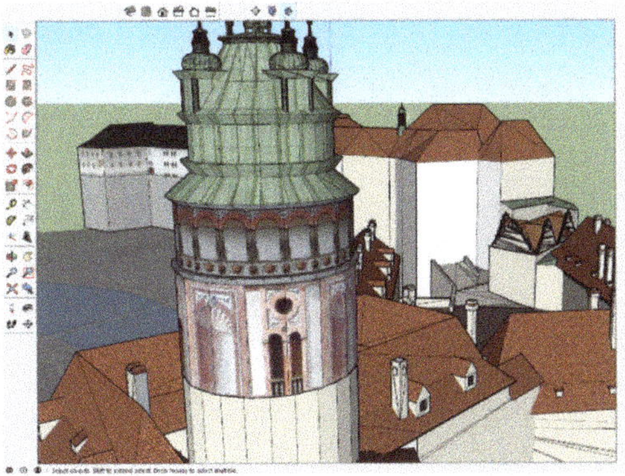

(a)   (b)

Figure 8. Manual 3D modelling in CAD software (Trimble SketchUp): (**a**) Český Krumlov Castle; (**b**) disappeared school building in the parish complex in Červená nad Vltavou.

Finally, several other heritage buildings were selected, such as castles, chateaus and monasteries in the Vltava River valley, which are continuously modelled. This was facilitated by enthusiastic 3D modellers who had prepared their models originally for Google Earth and agreed to the use and modification of their models for the needs of our 3D scene. Thus, our results overlap with those of crowdsourcing-based techniques.

It is apparent that detailed 3D modelling requires more accurate input data sources. Therefore, we performed additional archival research to find existing archival drawings and textual documents from structural historical surveys. The amount of data sources found varied for each modelled object. However, the most important buildings were reasonably well documented. For example, this was the case for probably the most valuable parish complex in Červená nad Vltavou (with an originally Romanesque single-nave church from the end of the 12th century), which was relocated before the dam was filled in 1960, and its condition before the relocation is very well documented (Figure 8b).

All input drawings had to be carefully processed by the means of digital cartography. Materials in paper form had to be scanned at a sufficient resolution, and then georeferenced in, at least, a local coordinate system. The latter was crucial to allow for the measuring of dimensions directly on the drawings. The actual modelling was performed in simple 3D CAD software. Our modelling team mainly utilised the Trimble SketchUp application, which is highly suitable for this task. An alternative could be, for example, Bentley MicroStation, which, however, is a bit more cumbersome for 3D modelling. For the clarity of our results, it was essential to divide individual building parts into components and appropriate layers.

The textures of the resulting building models were chosen to match archival iconographic materials (historical postcards and photographs). Unfortunately, most of the period photographs could not be directly used for texturing as they are of too low resolution, and are often taken from a distance, most often across a river (Figure 9a) or from vantage points, such as cliffs, valley tops, etc. (Figure 9b). For these reasons, parts of the buildings were textured with modern textures that are consistent with the archival source.

(a)

(b)

Figure 9. (**a**) Těchce as seen from the pine grove on the right bank, 1930s, photo Plichta Pečice, period postcard, archive of Vojtěch Pavelčík, (**b**) Overall view of Orlice Zlákovice from the hillside on the left bank, 1930s, photo by Plichta Pečice, period postcard, archive of Vojtěch Pavelčík.

2.3. Conversion to the GIS Environment and Publication of the 3D Web Scene

The outcomes of the procedures described in the previous section are textured 3D models in the native format of the modelling software (e.g., SKP files in the case of Trimble SketchUp). To transform them into a form suitable for publication in the resulting web scene, we first had to export them to the GIS spatial database. This step was facilitated by the KMZ file format. The KMZ files were first exported from SketchUp, and then imported into the ESRI file geodatabase using the KML To Layer geoprocessing tool.

The resulting multipatch features were then spatially located, as the approximate position from SketchUp (Geo-location tool) is not particularly accurate. Georeferencing was performed manually based on the building footprints obtained from old maps. This was sufficient for our purpose, as the detailed models had the same level of accuracy for georeferencing as the procedurally modelled surrounding situation, whose automatic modelling was based on the same input data.

After georeferencing, we proceeded to attribute management. To enhance the information value of the resulting 3D scene, we added basic attributes to all the detailed models of landmark buildings (building type, name, dates of construction, reconstructions, or deconstruction, brief textual description of the history of the object, URL to supplementary information sources). The georeferenced models enriched with attribute data were embedded into the context of procedurally generated surroundings. All procedurally modelled buildings are also identifiable and contained the basic attributes (based on the specification of the used material derived from the *Imperial Obligatory Imprints*). The entire resulting 3D scene was therefore consistent, and all models were queryable (Figure 10).

While publishing the 3D models, we had to ensure that the results would be available to a wide range of users directly via a web browser. Encouraging users to install any third-party browser plugins is no longer acceptable. Currently, there are just two options for publishing models from a geodatabase to be viewable online.

The first alternative is exporting the whole 3D scene as one physical file (3WS) using the CityEngine procedural modelling tool. The advantage of this choice is the ease of handling and transferring the single file; however, the disadvantages prevail. The scene must be loaded into the memory as a whole before the users can view it, and the loading is quite slow; connected with this is the limited extent of the displayed area. Moreover, the export of textured surfaces from CityEngine is highly unreliable (randomly missing textures and/or inverted normals), and subsequent corrections make this method of publication a time-consuming affair.

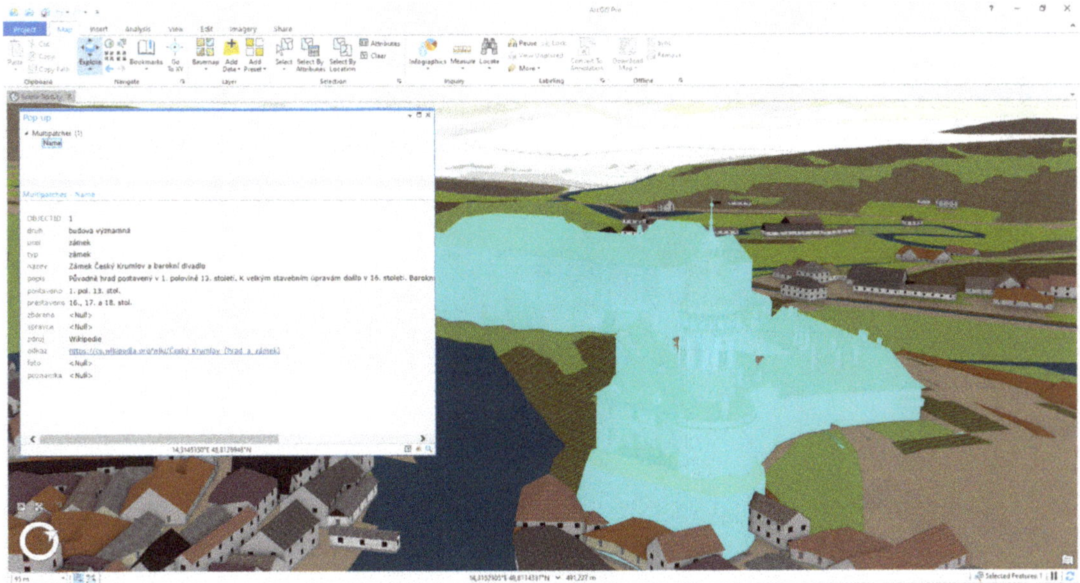

Figure 10. Both CAD and procedural models are queryable for attribute data (values of textual attributes are in Czech).

Therefore, we decided to use the second alternative and publish the results using a more modern approach, i.e., via the 3D web service of ArcGIS Pro. This option enables dynamic content loading depending on the displayed territory in the same manner as 2D web mapping applications. For this reason, this method is also suitable for 3D scenes of large areas, as in our case. The building models were exported as 3D Object Scene Layer Packages—SLPKs. The landmarks were exported individually, while the procedurally modelled common conurbation was exported as a whole. Although this was very demanding in terms of computing power, it allows one to display the complete conurbation model as a whole more clearly within the resulting web scene. Subsequently, all SLPKs were published as hosted services on ArcGIS Online. The digital terrain model was published ibid as an image service (elevation). Textures for the terrain were created from the vectorised old maps (*Imperial Obligatory Imprints*). The vector model was symbolised with semi-photorealistic textures, and then published in the form of raster tiles. Individual published layers were combined online in the form of a 3D web scene via Scene Viewer.

3. Virtual Reality (VR)

The above-mentioned 3D web scene is suitable for presentation to the general public, as it can be accessed online via a web browser. On the other hand, it achieves only a limited level of detail. Therefore, another 3D output is a realistic 3D model created by Unreal Engine (game engine) and prepared for presentation at close range in a virtual reality (VR) headset.

A VR headset is a device consisting of a head-mounted display and two controllers sensing head and hand movements, and changing the image to give the user a sense of being present in a different environment. This provides an immersive experience of viewing non-existent places in life-size. The main VR headset used in this project was the Oculus Rift S (Figure 11a), using cameras to capture head motion (without the need to place so-called VR lighthouses); other headsets were also used during testing (Figure 11b).

(a)

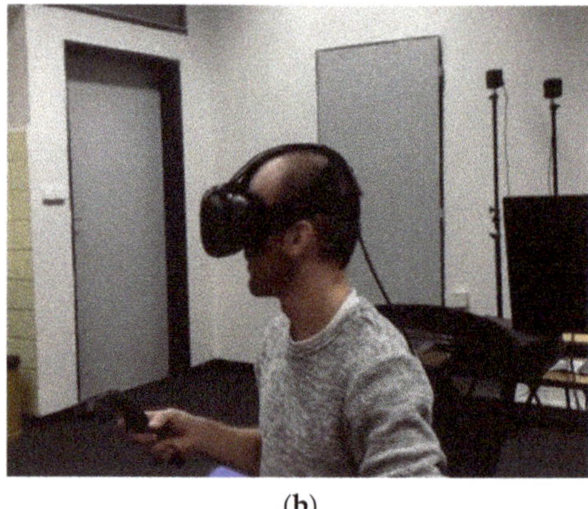

(b)

Figure 11. (**a**) Photo of most-used VR headset—Oculus Rift S; (**b**) Vojtěch Cehák wearing a VR headset.

A realistic 3D model of the historical landscape with terrain relief, vegetation and the basic shapes of buildings (for easy visibility, procedurally modelled buildings only have a white texture) was created in the area around three water reservoirs (Vrané, Štěchovice, Slapy) (Figure 12). Unfortunately, Unreal Engine, in the version we used (4.26), does not directly support most GIS formats, and it is necessary to use specialised plugins. In this project, the TerraForm Pro plugin (version 2.1.5, Horizon Simulation Ltd., Maidenhead, UK) was used to import GIS data in TIFF or SHP file formats (including the coordinate system). The following GIS data were used for Unreal Engine: a digital terrain model (raster), a polygon layer with features distinguished according to land cover categories and a line layer of the Vltava River axis.

Figure 12. Close-up on the vegetation in a virtual reality scene.

It is also necessary to cover the landscape with material. For this purpose, the "Procedural Landscape Ecosystem" set from the UE Marketplace online store was used, containing complex realistic materials (textures) and models of European vegetation in a high level of detail, and in a multi-seasonal form. Another dataset used contained realistic representations of standing and flowing water. The vegetation was placed using procedural generation—i.e., at random locations according to set rules (CGA rule files are not supported by Unreal Engine). The water level was inserted twice—in the form of before

(Figure 13a) and after (Figure 13b) the reservoir was filled. The model was also supplemented with artificial elements to facilitate orientation; for example, village name labels or a small map showing the user's current location.

Figure 13. Virtual landscape model (**a**) before and (**b**) after reservoir filling.

The last step was to connect the model with a VR headset and set up functions for the controller buttons in order to achieve the best user experience. The basic kind of movement in VR is called "teleportation" (a beam emanated from the hand controller sets new location). However, after initial user testing, which was conducted on colleagues from the Department of Geomatics at CTU (approximately ten people, comprising graduate students and professors with no experience using virtual reality), this "teleportation" proved to be unsuitable, as many users found the controls to be unintuitive and the landscape viewing experience was greatly compromised. In the end, a fixed-trajectory fly-through of the landscape was chosen (the user can only pause and resume the movement) at an altitude of about 500 metres (with the ability to look around in all directions). Due to hardware requirements, several levels of detail (LOD) had to be created. At higher altitudes, smaller objects (especially vegetation) are not loaded into the memory, reducing the required computer power to function in virtual reality.

Using all the above-mentioned geographic data and 3D models, a landscape model was created with land cover according to the categories distinguished in the *Imperial Obligatory Imprints* (including buildings). The viewing of the landscape takes the form of a fly-through, during which it is possible to pause the flight, view the current river level or view the map using buttons on the controllers. As multiple river sections were modelled, one initial fictional room was created to serve as a place for the user to learn the controls and select one of the river sections for the subsequent fly-through.

4. Results

During the work on the visualisation of the Vltava River valley, the possibilities of processing a vast amount of data were explored, namely, the large area modelling of 3D objects (over 28,000 3D models), creation of extensive DTMs from old maps (around 1670 km^2), CAD modelling of landmark buildings, and merging these models into one scene.

Procedural modelling based on the CGA shape grammar in the CityEngine software was used for common conurbation modelling as it allowed us to model relatively quickly and easily, and also, because CityEngine is compatible with other software applications used in the Vltava Project. We have identified and solved problems with this method of processing, from software problems, such as hardware requirements, and problems with importing, exporting and publishing data online (these problems were mostly solved using the ArcGIS Pro software), to technical problems, most often related to terrain and 3D object interactions.

Furthermore, the possibility of using the "Unreal Engine" game engine to work with the GIS data and 3D models created within the project was also explored, for the purpose of achieving more realistic visualisations and creating virtual reality on such a large scale. Due to time and technical problems, only a few selected localities (Slapy, Štěchovice, Vrané) were processed to the form of VR. Further work on virtual reality is considered within the planned *Vltava 2* Project.

The results of this work are 3D visualisations of the historic Vltava River valley, which were also presented at the exhibition *Vltava—Transformations of the Historical Landscape*, which took place in the atrium of the Faculty of Civil Engineering CTU in Prague on 8 February to 7 April 2022 [3]. The exhibition presented the visualisations of selected localities of the Vltava River valley in virtual reality (Figures 14 and 15), and a 3D model of the complete Vltava River valley in the form of a web application (Figures 16–18).

Figure 14. Co-author Vojtěch Cehák demonstrating virtual reality during a tour of a primary school at the exhibition *Vltava—Transformations of the Historical Landscape*.

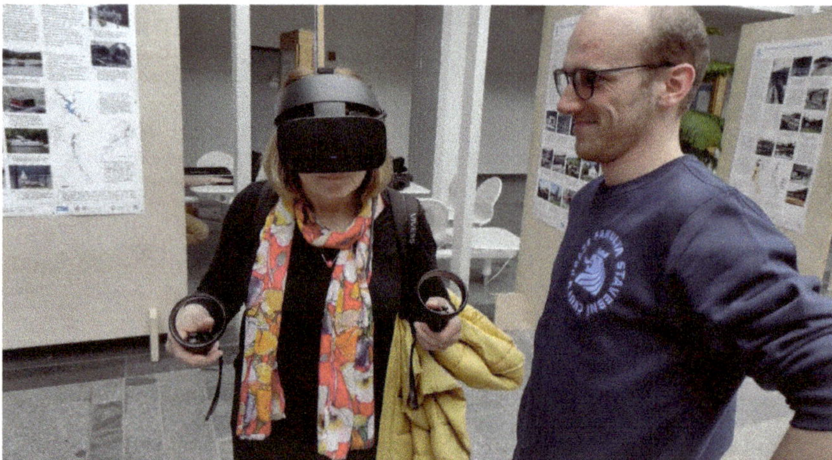

Figure 15. Co-author Vojtěch Cehák assisting a visitor at the exhibition *Vltava—Transformations of the Historical Landscape*.

Figure 16. Demonstration of a 3D web application created by a member of the Vltava Project.

Figure 17. Using a touch screen to control a 3D web map during a tour of an elementary school at the exhibition *Vltava—Transformations of the Historical Landscape*.

All results (findings from socio-hydrological research, 3D models and scenes, historical photos, period orthophotos, processed maps and other documents) will be available free of charge on the project website [2] (the site is already up and running, but not finished) by the end of 2022. This site will also contain other supplementary material about the Vltava River and its history, such as drawings of buildings and dams, documents on history, floods, boating, transport and other interesting information.

The 3D web scene (not yet in the final version, completion set to be by December 2022) is available online [1] and already contains several layers with individual landmark models, procedurally modelled buildings, and a digital terrain model with semi-photorealistic land cover textures. This 3D web scene will be further supplemented by additional layers, completed in the form of a 3D web application, and it will continue to be worked on in the follow-up Vltava 2 Project.

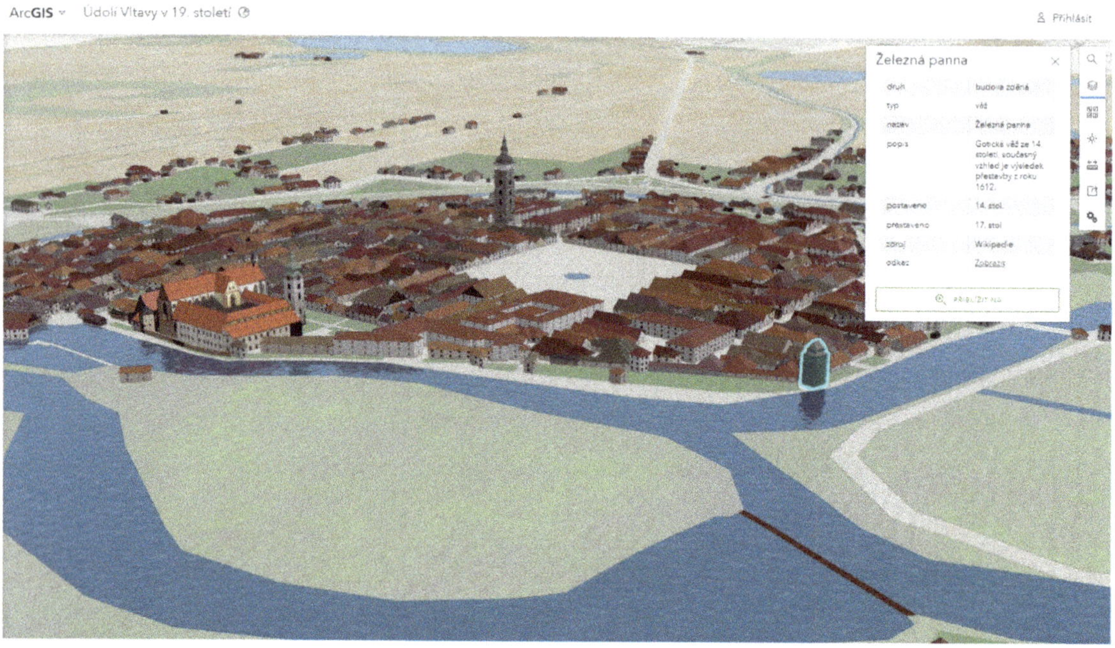

Figure 18. Preview of a part of the resulting 3D web scene—the city of České Budějovice on the banks of the Vltava River in the 19th century. The fortification tower was identified to access its attribute information. The modelling of the city was based on *Imperial Obligatory Imprints of the Stable Cadastre 1:2880* and period plans and photographs.

The results of the whole project, including the created 3D visualisations, will not only be available to the general public, but will also be used by historians, museums, the Vltava River Basin (state enterprise), the Ministry of Culture, local towns and villages, and will be used in history education and other applications. Furthermore, this project and its results will also serve as a basis/example for the creation of other similar projects in the Czech Republic.

5. Discussion

This project focuses on the longest river in the Czech Republic, the Vltava River, and its entire valley (1670 km^2), which makes it one of the largest projects of its kind in the Czech Republic (if we consider only the Czech part of the second-longest river Elbe River (294 km), which is shorter than the Vltava River (434 km)).

Although no new methods or innovative solutions for visualising large areas have been developed, as mentioned in the results section, this work is still significant in the circles of Czech history, science and education. The initial results of the project were presented at a month-long exhibition and were very well received by representatives of the collaborating (state) companies and institutions, as well as by the general public. Due to this, other ambitious projects have already started to emerge based on the Vltava Project, such as the follow-up project Vltava 2, the project Extinct Šumava—virtual reconstruction of landscapes and settlements, the Sázava River project (river history, boaters, water management, tramping), the project Pre-industrial Landscape of Bohemia, and many others.

The 3D visualisations of the Vltava River valley will continue within the follow-up "Vltava 2" Project, in which new 3D scenes will be created that will extend the existing results by new, more specific time periods corresponding to the Vltava River valley in the first half of the 19th century (circa 1840), and just before filling the dams and flooding parts of the Vltava River valley in the first half of the 20th century.

These stages will correspond to both the state of common development and the appearance of important landmark buildings around the river, which will also be shown in the scene in more detail. In most cases, changes to the buildings will be rather minor (typically, a different type of roof for sacral buildings), on the other hand, there are also buildings near the river that have undergone major changes between the two stages. This applies, for example, to Hluboká nad Vltavou and Orlík nad Vltavou Castles, whose appearances before the neo-Gothic reconstructions during the 19th century are poorly known to the public. For the purposes of adjustments, it will be necessary to carry out a careful archival survey to determine the specified appearance of the objects in this period. A survey of archives at the mentioned Hluboká and Orlík Castles will be crucial. The image in the first half of the 20th century will be generated procedurally, probably based on the first edition of the *State Map 1:5000-Derived*, and detailed models created into a specific form in the given period will be transferred onto the scene. The project will use an advanced digital model of the pre-dam topography as a basis for the placement of 3D models.

Compared to other corresponding projects, the Vltava Project focuses not only on the analysis and research of the impact of dam construction on the river basin, as in the Italian case study of Surian [7], but also on extensive modelling of the now-flooded areas and the whole Vltava Valley. Compared to Surian's case study, the Vltava Project does not incorporate hydrological changes per se (flow regime and sediment supply), but focuses mainly on changes in the river surroundings (land use changes, formation and disappearance/flooding of municipalities) and the use of the river itself (timber rafting).

Another theme that appears in the project is flooding (not meant as flooding due to dam construction, but flooding as a natural disaster). Although flooding is documented in the project, it is not the subject of in-depth research, as it is, for example, in the work of Yiou [5].

Similarly, in the work of Zlinszky and Timár [6], historical maps were also georeferenced and further used as a source of information for research and analysis, and as a basis for 3D modelling of the Vltava Valley. The difference here is mainly in the maps used, where Zlinszky and Timár use Krieger Map and First, Second and Third Habsburg military surveys. Since, in the Vltava Project, we are exploring the Vltava Valley before the construction of the Vltava Cascade (built 1930–1992), we use more recent historical maps, which are the successors of the Habsburg military surveys maps, the *Imperial Obligatory Imprints of the Stable Cadastre 1:2880*. The georeferencing procedures of these maps can then be considered similar.

In the next part of the work, 3D object modelling was carried out. This was similar to the well-known Rome Reborn project, which focused on the recreation of ancient Rome into digital form. As in Rome Reborn, we adopted a proven modelling approach, where a common conurbation is modelled procedurally based on old maps and historical iconographic material. Landmark buildings were then reconstructed in more detail employing simple CAD software. It should be noted that the quality of the models within Rome Reborn is considerably higher than ours. Therefore, it is better to compare the procedurally created models with the work of D. Kitsakis, E. Tsiliakou, T. Labropoulos and E. Dimopoulou [17]. In this work, the authors describe the creation of a 3D model of a traditional settlement in the region of central Zagori in Greece. The models created by these authors are also quite simple; however, they have more spatial detail, such as niches, balconies, etc. Unfortunately, we cannot afford these details due to the volume of data and the time required to generate more than 28,000 buildings. However, what we have in addition to the mentioned work is our solution for the intersection of buildings with the terrain. You can notice that the models created by them are quite embedded in the ground. We solved this (along with the floating buildings) by generating the foundations on which the buildings now stand. The last part of our work involved creating a virtual reality application using the Unreal Engine game engine. In contrast to the work of Kersten T, Drenkhan D and Deggim S [26], our work was mainly concerned with creating visualisations, while performance optimisation was only performed during procedural modelling. Although the ability to import GIS data into game engines has been around for a long time [23], we faced a rather unique problem. Although the Unreal Engine does not natively support importing GIS data, this can be solved by installing plugins. However, since the Czech Republic uses its own S-JTSK coordinate system (EPSG 5514), which has coordinate values of $X = -430,000$ to $-905,000$ and $Y = -935,000$ to $-1,230,000$, most of these plugins did not work, or the imported data was not in the right places. Fortunately, this was solved by installing a specialised plugin TerraForm Pro (version 2.1.5, Horizon Simulation Ltd., Maidenhead, UK). Similar to Khorloo O., Ulambayar E., and Altantsetseg E. [24], we imported GIS data, 3D models and terrain into Unreal Engine, where these data were processed into an interactive virtual reality environment; however, unlike these authors, we visualised three rural areas with several small villages as opposed to one large and detailed city. Moreover, for a large part of these villages, little has been preserved in the form of underlying data, and so their reconstruction can be compared to the work of Günay S. [25], where many of their objects are also created from limited data sources.

Author Contributions: The work can be split into three parts, where each part was mainly worked on by one of the authors. Procedural modelling—Michal Janovský, creation of VR application— Vojtěch Cehák and assembling 3D models and publishing them onto the online 3D web scene—Pavel Tobiáš. For the article, the following author contributions apply: conceptualisation, Pavel Tobiáš; methodology, Pavel Tobiáš; software, Michal Janovský and Vojtěch Cehák; validation, Pavel Tobiáš; formal analysis, Pavel Tobiáš; investigation, Michal Janovský and Vojtěch Cehák; resources, Pavel Tobiáš; data curation, Michal Janovský, Pavel Tobiáš and Vojtěch Cehák; writing—original draft preparation, Michal Janovský; writing—review and editing, Michal Janovský; visualisation, Pavel Tobiáš; supervision, Pavel Tobiáš; project administration, Pavel Tobiáš; funding acquisition, Michal Janovský. All authors have read and agreed to the published version of the manuscript.

Funding: This project was funded by the MINISTRY OF CULTURE CR, grant number DG18P02OVV037, and the APC was funded by the STUDENT GRANT COMPETITION 2022, grant number SGS22/048/ OHK1/1T/1.

Data Availability Statement: Due to the constant development and improvement of the presented visualisations, all the most up-to-date outputs are available only on request until the official end of the project. Some underlying data used to create visualisations cannot be (individually) provided for licensing reasons (for example, due to textures used in procedural modelling, maps used for vectorisation), but may be provided in the form of final visualisations, rule packages, etc. More information is available on request. All visualisations, results and available materials of the Vltava Project will be freely available on the website of the project http://vltava.fsv.cvut.cz by the end of the project, i.e., in December 2022.

Conflicts of Interest: The authors declare no conflict of interest. The funders had no role in the design of the study; in the collection, analyses, or interpretation of data; in the writing of the manuscript; or in the decision to publish the results.

References

1. The 3D Visualisation of Vltava River Valley. Available online: https://arcg.is/1yDr9i0 (accessed on 3 July 2022).
2. Official Website of the Vltava Project. Available online: http://vltava.fsv.cvut.cz (accessed on 3 July 2022).
3. Website of the Exhibition Vltava—Transformations of the Historical Landscape. Available online: vltava.fsv.cvut.cz/VYSTAVA/ (accessed on 3 July 2022).
4. Canthal, J.; Tobiáš, P.; Kratochvílová, D. 3D model of the historic Vltava River valley in the area of Slapy Reservoir. In *Advances and Trends in Engineering Sciences and Technologies III*; CRC Press: Boca Raton, FL, USA, 2019; pp. 679–684. [CrossRef]
5. Yiou, P.; Ribereau, P.; Naveau, P.; Nogaj, M.; Brázdil, R. Statistical analysis of floods in Bohemia (Czech Republic) since 1825. *Hydrol. Sci. J.* **2006**, *51*, 930–945. [CrossRef]
6. Zlinszky, A.; Timár, G. Historic maps as a data source for socio-hydrology: A case study of the Lake Balaton wetland system. *Hydrol. Earth Syst. Sci.* **2013**, *17*, 4589–4606. [CrossRef]
7. Surian, N. Channel changes due to river regulation: The case of the Piave River, Italy. *Earth Surf. Process. Landf.* **1999**, *24*, 1135–1151. [CrossRef]

8. Pacina, J.; Cajthaml, J.; Kratochvílová, D.; Popelka, J.; Dvořák, V.; Janata, T. Pre-dam valley reconstruction based on archival spatial data sources—Methods, accuracy and 3D printing possibilities. *Trans. GIS* **2022**, *26*, 385–420. [CrossRef]
9. Cajthaml, J.; Tobiáš, P.; Kratochvílová, D. Creating of 3D model of the historical landscape from contour lines displayed on old maps: Vltava River valley. In *ESaT*; Technická univerzita v. Košiciach, Stavebná Fakulta: Košice, Slovakia, 2018; pp. 1–4. ISBN 978-80-553-2982-6.
10. Kratochvílová, D.; Cajthaml, J. Using the automatic vectorisation method in generating the vector altimetry of the historical Vltava River valley. *Acta Polytech.* **2020**, *60*, 303–312. [CrossRef]
11. Janovský, M.; Janata, T.; Cajthaml, J. Visualization of the Vltava River valley: Illustration of work procedures on data from the Kamyk reservoir surroundings. In Proceedings of the 20th International Multidisciplinary Scientific GeoConference SGEM STEF92, Albena, Bulgaria, 18–24 August 2020; Technology Ltd.: Sofia, Bulgaria, 2020; pp. 469–476, ISBN 978-619-7603-07-1. [CrossRef]
12. Watson, B.; Muller, P.; Wonka, P.; Sexton, C.; Veryovka, O.; Fuller, A. Procedural Urban Modeling in Practice. *IEEE Comput. Graph. Appl.* **2008**, *28*, 18–26. [CrossRef]
13. Haegler, S.; Müller, P.; Gool, L.V. Procedural Modeling for Digital Cultural Heritage. *EURASIP J. Image Video Process.* **2009**, *2009*, 852392. [CrossRef]
14. Schwarz, M.; Müller, P. Advanced Procedural Modeling of Architecture. *ACM Trans. Graph.* **2015**, *34*, 107. [CrossRef]
15. Smelik, R.M.; Tutenel, T.; Bidarra, R.; Benes, B. A Survey on Procedural Modelling for Virtual Worlds. *Comput. Graph. Forum* **2014**, *33*, 31–50. [CrossRef]
16. John, D.M.A.; Don, L. Creating a longitudinal, data-driven 3D model of change over time in a postindustrial landscape using GIS and CityEngine. *J. Cult. Herit. Manag. Sustain. Dev.* **2018**, *8*, 434–447. Available online: https://www.historicalgis.com/uploads/6/8/8/2/68821567/arnold_and_lafreniere_2018.pdf (accessed on 10 May 2022). [CrossRef]
17. Kitsakis, D.; Tsiliakou, E.; Labropoulos, T.; Dimopoulou, E. Procedural 3D modelling for traditional settlements. The case study of central Zagori. *Int. Arch. Photogramm. Remote Sens. Spat. Inf. Sci.* **2017**, *42*, 369–376. [CrossRef]
18. Pascal, M. High-End 3D Visualization with CityEngine, Unity and Unreal. In Proceedings of the 2017 Esri Developer Summit, Palm Spring, CA, USA, 7–10 March 2018. Available online: https://proceedings.esri.com/library/userconf/devsummit18/index.html; https://www.youtube.com/watch?v=qdo2WSm_5UU (accessed on 8 March 2022).
19. Pascal, M.; Stefan, A. Building VR/AR Experiences with CityEngine, Unity, and Unreal. In Proceedings of the 2020 Esri Developer Summit, Palm Spring, CA, USA, 10–13 March 2020. Available online: https://proceedings.esri.com/library/userconf/devsummit20/index.html; https://www.youtube.com/watch?v=c1wq1wy01b4 (accessed on 8 March 2022).
20. Unreal Engine 4 Documentation. Available online: https://docs.unrealengine.com/4.27/en-US/ (accessed on 28 November 2021).
21. Unity Documentation. Available online: https://docs.unity.com (accessed on 28 November 2021).
22. Ma, Y.-P. Extending 3D-GIS District Models and BIM-Based Building Models into Computer Gaming Environment for Better Workflow of Cultural Heritage Conservation. *Appl. Sci.* **2021**, *11*, 2101. [CrossRef]
23. Mat, R.C.; Shariff, A.R.M.; Zulkifli, A.N.; Rahim, M.S.M.; Mahayudin, M.H. Using game engine for 3D terrain visualisation of GIS data: A review. *IOP Conf. Ser. Earth Environ. Sci.* **2014**, *20*, 012037. Available online: https://iopscience.iop.org/article/10.1088/1755-1315/20/1/012037 (accessed on 16 March 2022). [CrossRef]
24. Khorloo, O.; Ulambayar, E.; Altantsetseg, E. Virtual reconstruction of the ancient city of Karakorum. *Comput. Anim Virtual Worlds* **2022**, e2087. [CrossRef]
25. Günay, S. Virtual reality for lost architectural heritage visualization utilizing limited data. *Int. Arch. Photogramm. Remote Sens. Spat. Inf. Sci.* **2022**, 253–257. [CrossRef]
26. Kersten, T.; Drenkhan, D.; Deggim, S. Virtual Reality Application of the Fortress Al Zubarah in Qatar Including Performance Analysis of Real-Time Visualisation. *KN J. Cartogr. Geogr. Inf.* **2021**, *71*, 241–251. [CrossRef]

International Journal of
Geo-Information

Article

A Technical and Operational Perspective on Quality Analysis of Stitching Images with Multi-Row Panorama and Multimedia Sources for Visualizing the Tourism Site of Onshore Wind Farm

Jhe-Syuan Lai, Yi-Hung Tsai, Min-Jhen Chang, Jun-Yi Huang and Chao-Ming Chi *

Department of Civil Engineering, Feng Chia University, Taichung 40724, Taiwan; jslai@fcu.edu.tw (J.-S.L.); d0610158@mail.fcu.edu.tw (Y.-H.T.); d0676303@mail.fcu.edu.tw (M.-J.C.); m0806485@mail.fcu.edu.tw (J.-Y.H.)
* Correspondence: cmchi@mail.fcu.edu.tw; Tel.: +886-4-2451725 (ext. 3120)

Abstract: A virtual tour of the onshore wind farm near Gaomei Wetland, Taichung City, Taiwan, was produced by producing panoramic images of the site by stitching images captured with a full-frame digital single-lens reflex camera and a multi-row panorama instrument, which automatically and precisely divided each scene into several images. Subsequently, the image stitching quality was improved by calculating the root mean square error (RMSE) of tie point matching and adjusting the tie points. Errors due to eccentricity attributed to the camera's relative position to the rotational axis of the multi-row panorama instrument were examined and solved; the effect of the overlap rate on image stitching quality was also investigated. According to the study results, the overlap rate between the original images was inversely proportional to the RMSE and directly proportional to the time required for photography and image processing. The stitching quality was improved by resolving eccentricity and by increasing the number of tie points. The RMSEs of the panoramas of all stations were all less than 5 pixels. Subsequently, multimedia materials providing information on wind turbine attributes were combined with the panorama platform to establish a virtual reality tour platform. The content of the platform could be accessed with a smartphone and viewed with a virtual reality device and could promote both tourist attractions and wind energy.

Keywords: green energy promotion; image stitching; multimedia; panorama; virtual tour

Citation: Lai, J.-S.; Tsai, Y.-H.; Chang, M.-J.; Huang, J.-Y.; Chi, C.-M. A Technical and Operational Perspective on Quality Analysis of Stitching Images with Multi-Row Panorama and Multimedia Sources for Visualizing the Tourism Site of Onshore Wind Farm. *ISPRS Int. J. Geo-Inf.* **2022**, *11*, 362. https://doi.org/10.3390/ijgi11070362

Academic Editor: Wolfgang Kainz

Received: 6 May 2022
Accepted: 23 June 2022
Published: 24 June 2022

Publisher's Note: MDPI stays neutral with regard to jurisdictional claims in published maps and institutional affiliations.

1. Introduction

1.1. Challenge and Motivation

The global COVID-19 outbreak since November 2019 has severely disrupted everyday life. In response to the pandemic, businesses have established new models of interaction by incorporating virtual reality (VR) technology. Oscar et al. [1] proposed a virtual objective structured clinical examination (OSCE) method as an alternative learning option for students who could not participate in in-person examinations due to the pandemic. Gao et al. [2] developed a VR-based exercise platform to mitigate physical deterioration in people, particularly older adults, during COVID-19 quarantines, which prohibit outdoor activities. Van et al. [3] used VR to explore the role of human–machine interactive devices to revitalize the tourism industry in post-pandemic Vietnam.

1.2. Green Energies and the Tourism Site

With the increasing global focus on climate changes and sustainable development, renewable energy and green energy—such as solar, tidal, and particularly wind power—have garnered growing attention. Graabak and Korpås [4] suggested that wind power would be a primary source of power generation in Europe's future electricity system. According to Thayumanavan et al. [5], rapid development in power electronic converters over the past two decades has given wind power technology advantages over other renewable and green energy technologies, such as high power generation capacity, high conversion efficiency, a pollution-free power generation process, and high availability. Studies on wind power development have been conducted in places such as Greece, China, and Taiwan [6–9]. This extensive research into wind power reveals its importance as a source of green energy. According to the Global Offshore Wind Speed Ranking report released by 4C Offshore, a market intelligence organization [10], 12 of the top 20 wind farms with the best wind conditions were located in the Taiwan Strait. The Taiwanese government plans to complete the development of 5.5 GW of offshore wind farms by 2025; moreover, they have implemented a series of policies promoting the development of industries related to onshore wind farms since 2000. Gaomei Wetland, located in Qingshui District, Taichung City, is a critical ecosystem conservation area and tourist attraction. Due to the proximity of the site to the large-scale onshore wind farms in the Port of Taichung, visualizing the site has excellent promotional value for both green energy and tourism.

1.3. Information-Technology-Based Visualizations

Due to technological advancements, information technologies and multimedia tools now facilitate incorporating sound, images, and text into VR. Burdea [11] proposed the three I's of VR—interaction, immersion, and imagination—that create a sense of presence for the user. Weng [12] proposed that people are the key factor for creating a sense of space and presence in a virtual space; changes in people's senses of space and presence prompt them to develop different perceptions in the space. VR has a wide range of applications. In education research, Bricken [13] reported that people exhibited a stronger motivation to learn in VR-based learning compared with conventional learning. Michitaka [14] discussed how virtual panoramas could be used to increase public interest in exhibitions and theaters. Lai et al. [15] employed VR and panoramic photography to establish a website for students to take a surveying practice course and reported that 92% of the study participants considered the website helpful to the course learning process. Studies on VR applications in engineering education, mechanical design, and physical education training [16–18] have verified the utility of VR in education. Regarding VR applications in tourism, the Louvre Museum provides VR panoramas of its interior to enable people to appreciate its paintings without visiting the museum in person [19]. A VR view of the Great Wall of China is publicly accessible on its official website [20]. Lee [21] established a three-dimensional virtual sightseeing platform of Taiwan's Yehliu Geopark to familiarize visitors with the park environment; 75% of the park visitors reported that the platform was helpful on a questionnaire survey. Harrington et al. [22] integrated data on the terrain, water, and plants in a botanical garden to design a realistic and immersive VR travel experience. Antonio et al. [23] employed the light detection and ranging (LIDAR) technology to digitally preserve historic sites in Cordova, Spain, and adopted VR technology to provide virtual tours of the sites. Other applications of VR in tourism include VR-based geoheritage tours in Iceland and icebreaking cruises in Hokkaido [24,25]; these VR projects enable those who cannot visit in person to immerse themselves in the experiences. Through a questionnaire survey and in-depth interviews, Chao [26] concluded that VR-based scientific experiments can expose people to events and objects that they would not otherwise experience in their everyday lives, thereby enhancing autonomous learning motivation. Li et al. [27] reported that VR increased tourist wellbeing and positively affected perceived value. Chiao et al. [28] examined the impact of virtual-tour-based platforms with different statistical indices (e.g., p-value and R-squared) for cultural tourism education. Given the aforementioned positive effects of VR in education and tourism, VR is an indispensable part of the evolution of the social model in response to the COVID-19 pandemic.

This study established a multimedia platform using panoramic technology due to its advanced development and low cost. In a panorama, the surrounding environment is presented in a cylindrical or spherical projection in VR space, providing the user with a 360° view of their position. In accordance with the approach proposed by [29], the 360° panoramas with a horizontal angle of 360° and a vertical angle of less than 180° were captured in this study. Panoramas are typically taken with one of two methods [15], namely by using a 360° spherical camera to directly produce panoramas or by placing one or more cameras on a platform to capture images and stitching the images together. Compared with the second strategy, the first is simpler and more convenient but usually has a lower resolution and limited capability of customized functionality. The quality of image stitching in the second strategy is affected by the photography environment; for example, the production of high-quality images requires that the subject remains in the same position, that the subject's features are not overly monotonous, and that adequate lighting is present. These requirements are the main challenges in this strategy. The primary objectives of this study were to produce high-resolution panoramas and to offer customizable VR functionality. Therefore, the stitching strategy was employed. Ernie and Mark [30] suggested that capturing numerous images with high rates of overlap could facilitate producing high-resolution panoramas and that a 25% overlap is sufficient for generating 1-billion-pixel panoramas. Zheng et al. [31] detailed photography strategies and image stitching techniques. Liu et al. [32] improved the quality of image stitching through the use of extracting feature points and matching tie points. Lyu et al. [33] not only reviewed the pixel- and feature-based stitching methods for images, but also included video materials obtained from static and moving cameras. Lee et al. [34] proposed the improved optical flow algorithm to refine the task of image stitching.

1.4. Research Scopes

Accordingly, this study aimed to enable users to visit the Gaomei Wetland and the wind farm nearby the Port of Taichung virtually online and to understand the compositional structure and related information of wind turbines. First, multimedia visualization was performed on the onshore wind turbines; a low-cost, convenient process was developed that enables the user to view content by using a simple VR device (e.g., Google CARDBOARD) and clicking on a link or scanning a QR code, lowering the barrier to use for the system. From a technical perspective, two factors affecting the quality of high-resolution panoramic photography and image stitching are discussed: (1) the eccentricity effect attributable to the camera's relative position to the rotational axis of the panoramic instrument and (2) rates of overlap between images. The discussion could provide insight into the future development of VR technology, the promotion of tourist attractions and wind power, and the establishment of a panoramic visualization.

2. Materials and Methods

2.1. Study Area

Gaomei Wetlands, located to the south of the estuary of Dajia River, is a crucial ecosystem conservation area due to its expansive land area of 701.3 ha, rich natural resources, and status as one of the few places in Taiwan that is home to breeding colonies of *Anatidae* duck species. The wind farm near the Port of Taichung was dubbed "the wind turbine boulevard" due to the neat arrangement of its wind turbines along the flood-control road (Figure 1). An internationally renowned tourist destination, Gaomei Wetland attracts approximately 3.5 million tourists every year. The wind farm comprises 18 wind turbines manufactured by the Dutch company Zephyros. The total installed capacity is 2000 kW, composing 30% of the total capacity of the Taiwan Power Company Offshore Wind Farm Phase 1 Project. In sum, the site is critical for wind power generation in Taiwan.

2.2. Procedure

Figure 2 presents the research process. First, parameters on the panoramic instrument and camera were set and images were captured onsite. Subsequently, the overlapping images were stitched to produce panoramas, which were then imported to the proposed virtual tour platform. Next, a VR mode and other interactive functions were added to the platform to complete the panoramic visualization. The process is detailed in Sections 2.2.1 and 2.2.4. All image processing steps were performed on the Windows 10 (64-bit) computer with Core i9-9900KF @ 3.60 GHz CPU, 32 GB RAM, and NVIDIA GeForce RTX 2080 GPU (graphics size: 8010 MB).

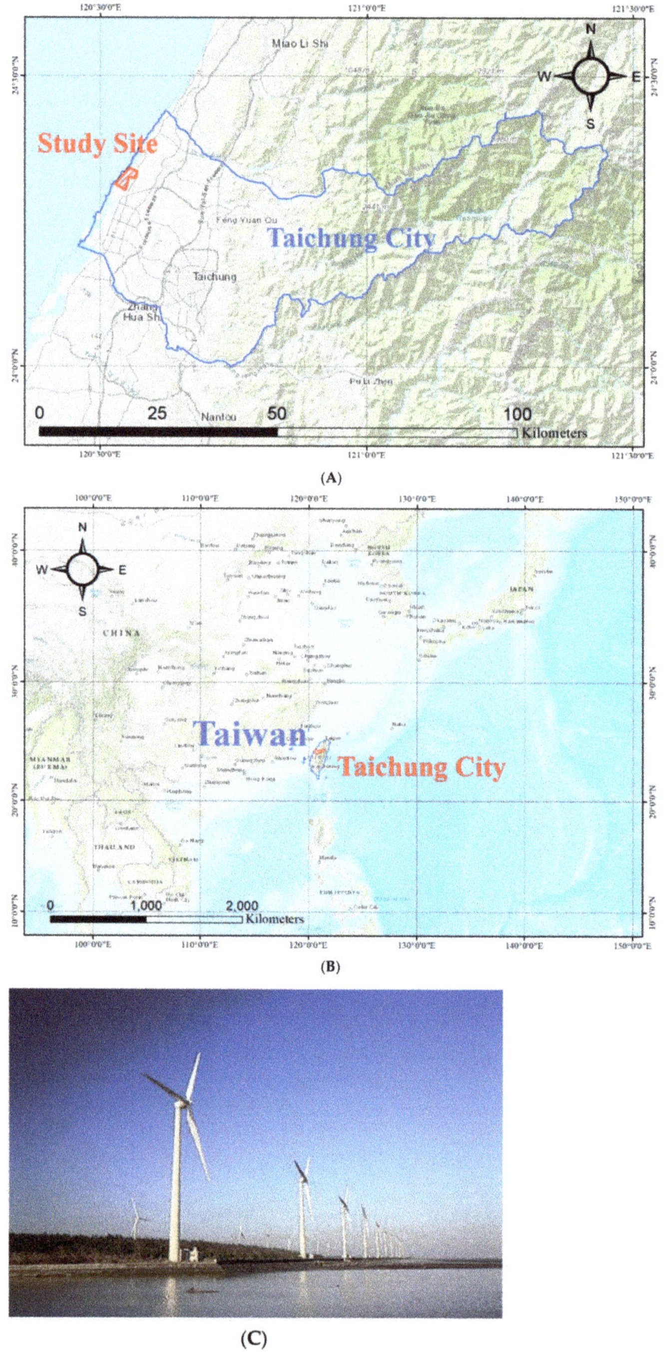

Figure 1. Study site (**A**) in Taichung, Taiwan (**B**) and onshore wind farms (**C**).

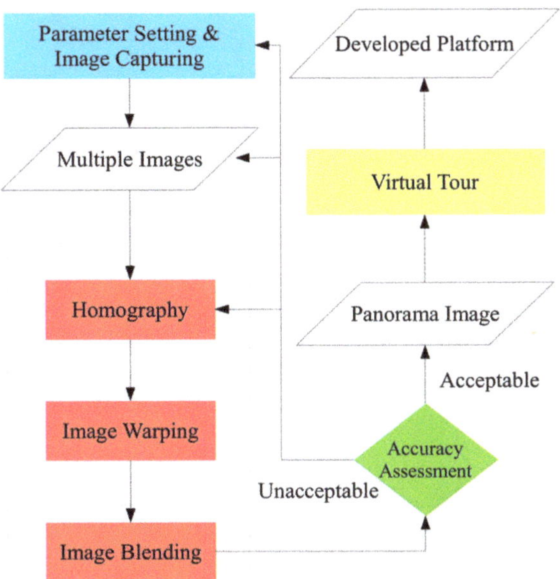

Figure 2. Study procedure (each color will be presented in the same section).

2.2.1. Parameter Setting and Image Capturing

The GigaPan EPIC Pro V panorama device (hereafter "GigaPan"), which produces multi-row panoramas (Figure 3), and Canon EOS 6D Mark II camera equipped with the Canon EF 16–35 mm F/4L IS USM, an ultra-wide-angle zoom lens, were used for panoramic photography. A test shoot was performed in Feng Chia University, Taichung City, before the panoramic photography at the wind turbine boulevard. The test shoot was performed to test the parameter settings and to use the panoramic instrument to control the overlap rate to calculate the root mean square error (RMSE) after tie point stitching. The test shoot results served as a basis for the parameter settings used at the actual shoot at the wind turbine boulevard. GigaPan offers a range of shooting modes; the 360 Panorama mode selected in this study requires the user to set the upper and lower image limits and the number of images captured from each angle before connecting to the camera for automatic 360° panoramic photography. In this mode, the instrument rotates automatically and precisely divides a scene into several grid images. The camera was set to manual mode, and its parameters were adjusted to obtain the optimal parameter combination for the environment. The exposure value was set to bracketing mode and three shots—bright, normal, and dark—were taken with the camera from each angle. An optimal image from the three shots was then selected for stitching.

For cameras to adapt quickly to various environmental and brightness conditions, several exposure control modes have been developed since the development of microelectronic technology in the 1980s [35]. In this study, the manual mode was adopted in which all shooting parameters are set by the user, including the aperture size and shutter speed. For this study, the parameters were set in accordance with the study area's bright, outdoor environment and the constantly changing external factors (e.g., clouds or other moving objects). The camera aperture size, in addition to its basic function of regulating the amount of light passing through the lens, can also be adjusted to improve image quality and control the depth of field. A large aperture size enables the camera to capture details in a short time (high shutter speed) assuming that adequate lighting is present. With a small aperture size, the camera takes longer (low shutter speed) to capture the scene. Shutter speed is a critical component that controls the length of time that light can pass through the lens and determines the camera's ability to freeze a moving object. Moreover, shutter speed controls the amount of light to which the image sensor is exposed and can be used to minimize the effect of camera vibrations on images, preventing unwanted movements or blurs from appearing in the images [36]. Shutter speed directly affects the brightness of an image; a lower shutter speed causes the camera to capture more reflected light or light sources. If lighting is sufficient, a higher shutter speed enables the camera to capture more subject details. ISO represents the sensitivity of the camera to light, and the ISO setting must be adjusted particularly carefully because an overly high ISO results in grainy images and blurs the image content. This is because a high ISO enlarges the electronic signals and creates noise. To minimize noise, the photographer must determine the optimal ISO value based on their experience and repeated testing.

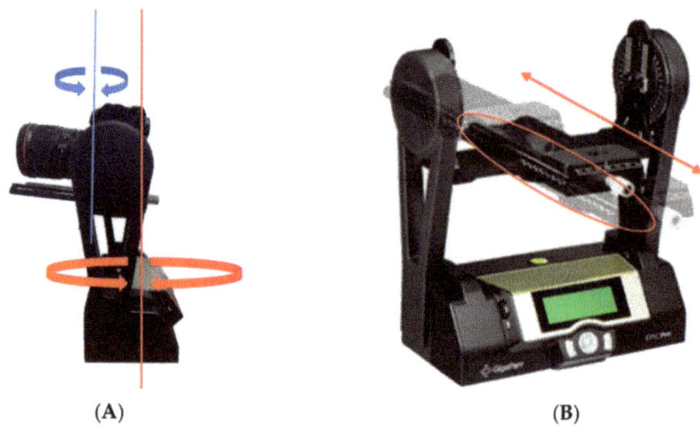

(A)   (B)

Figure 3. Panoramic instrument of GigaPan EPIC Pro V. (**A**) Eccentric issue caused by different axes of rotation between panoramic instrument (red part) and camera (blue part); (**B**) Adjustable position (scale marks) to put camera (red part).

In addition, the rate of image overlap controlled by GigaPan affects the quality of image stitching. A high overlap rate requires more images to be captured in both the vertical and horizontal directions, which facilitates the extraction of feature points in the overlapping areas and the matching of these points as tie points. Therefore, a high overlap rate improves the quality of image stitching, but requires increased time for shooting and image processing. Conversely, a low overlap rate requires a short shooting time but has low stitching quality; to improve the quality, the tie point positions must be adjusted manually. Therefore, the overlap rate is correlated positively to the shooting time, image processing time, and number of tie points and negatively to labor cost. Assuming that image stitching quality must be high, a trade-off exists between the overlap rate and labor cost. The geometrical relationship between the rotational axes of the image center and GigaPan is also a variable. If the perpendicular axes of these two rotational axes are not in the same axial direction (Figure 3A), eccentricity may occur, which causes the actual overlap rate to deviate from the GigaPan's overlap rate setting, thus reducing the stitching quality. In this study, an eccentric calibrated field was established; the camera's position was adjusted in accordance with the scale on GigaPan (Figure 3B) to compile a look-up table for focal length. Figure 4A presents the eccentric calibrated field; the camera, benchmark rod, and the center of the benchmark paper form a straight line. Lengths I and II are of the same length. Images were captured with the camera with the GigaPan rotated to determine whether the benchmark rod completely overlaps with the axis perpendicular to the benchmark paper center in the images to determine whether eccentricity was present. Eccentricity did not exist if the benchmark rod and the axis perpendicular to the benchmark paper center remained completely overlapped as the GigaPan rotated right or left (Figure 4B). Figure 4C,D depicts images demonstrating the presence of eccentricity. If the benchmark rod shifted to the left or right when GigaPan rotated rightward or leftward, respectively, the camera should be moved to a greater mark value on the scale. Conversely, if the benchmark rod shifted to the right or left when GigaPan rotated rightward or leftward, respectively, the camera should be moved to a smaller mark value on the scale.

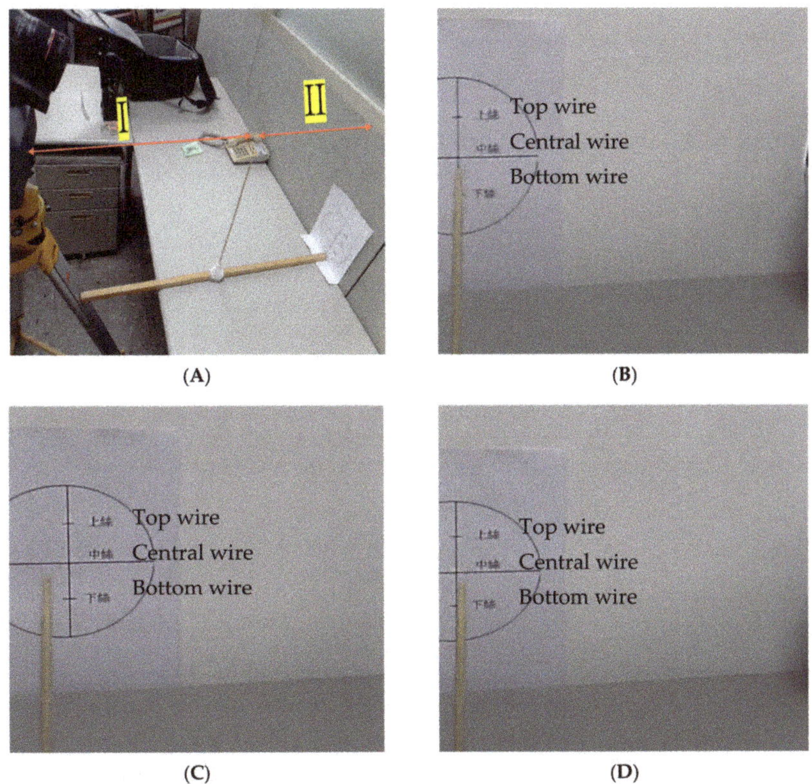

(A)

(B)

(C)

(D)

Figure 4. Developed field for calibration of eccentric issue between camera and panoramic instrument. (**A**) Calibrated field; (**B**) Non-eccentric issue; (**C**) Eccentric issue with left shift; (**D**) Eccentric issue with right shift.

2.2.2. Image Stitching

After the completion of a fixed-point shoot, the individual captured images were stitched into a panorama by using Kolor Autopano Giga version 4.4 (hereafter "Autopano"). The image stitching procedure is marked with a dotted box in Figure 2. Stereoscopy was employed for photography in this study; feature points were extracted from images captured from different angles. Image matching was performed to identify pairs of feature points and subsequently create tie points. Capturing numerous images with a high overlap rate, despite the required long image processing time, provides many tie points and thus improves the image stitching quality. Subsequently, the stitched images were rendered to remove motion trails (e.g., of wind turbine blades or people) to simplify the subsequent image-editing procedure. Image stitching comprises three stages: homography, warping, and blending; these are detailed in the following text.

- Homography

Homography involves the transformation of two images captured from different angles by identifying corresponding feature points in the same plane and the overlapping areas between the images. Autopano was used for image stitching; the algorithm employs a scale-invariant feature transform (SIFT) to reduce the effects of image rotation, scale, and grayscale value differences on feature points in image pairs [37], thus facilitating the extraction and matching of feature points for tie point generation. Subsequently, homogeneous coordinates were used to obtain a 3 × 3 matrix h to document the image translation, scaling, and rotation in different spaces as a preparation for subsequent steps. For example, if the feature point set of one image is $(x, y, 1)$, and that of another is $(x', y', 1)$, the mapping relationship between the tie points of the two images can be presented as a 3 × 3 matrix h as in Equation (1).

$$\begin{bmatrix} x \\ y \\ 1 \end{bmatrix} = \begin{bmatrix} h_{00} & h_{01} & h_{02} \\ h_{10} & h_{11} & h_{12} \\ h_{20} & h_{21} & h_{22} \end{bmatrix} \times \begin{bmatrix} x' \\ y' \\ 1 \end{bmatrix} \tag{1}$$

The establishment of the transformation matrix H between the two images requires determining the relationships between four pairs of feature points in the two images [38]. However, because more than four feature point pairs between two images are typically identified, the optimal pairs—those with the least error—should be used. Autopano employs the random sample consensus (RANSAC) algorithm to remove outliers [31]. In the algorithm, a random subset of the population of numerical values is selected to estimate the optimal model parameters. The numerical values comprise inliers that conform to the numerical model and outliers that do not fit the model. The outliers are detected by the suggested threshold of Autopano in this study. Different inlier points are substituted into Equation (1) to determine the set of inliers with the least error (the optimal solutions).

- Image Warping

According to homography calculations, the identification of corresponding tie points requires the use of homogenous coordinates for determining the level of image distortion in the space in the subsequent image warping step. For example, to stitch two images, one image is used as the reference image, and the other is projected through algorithmic computations and warping, onto the coordinate space of the reference image to ultimately superimpose the two images. The projection procedure, comprising forward and backward warping, involves determining the positions of target pixels in the space before placing the original image pixels over the target image. Subsequently, repeated calculations are conducted by using an inverse function to restore damages in the forward warped image. Distortion that often occurs in the non-overlapping areas of the image during projection can be effectively reduced using cylindrical and spherical projections.

In this study, a spherical projection was used to produce 360° panoramas without excessive image distortion. Equation (2) is the transformation between spherical and Cartesian coordinate systems. Given an image coordinate (x, y) and its projection on a spherical surface (x, y, f), the spherical coordinate is presented as (r, θ, φ), where r represents the distance from the center of the sphere and the target, θ is the angle between r and the zenith with a range of $[0, \pi]$, and φ is the angle between the planar projection line of r and the x-axis with a range of $[0, 2\pi]$.

$$(r \sin\theta \cos\varphi, r \sin\theta \sin\varphi, r \cos\theta) \propto (x, y, f) \tag{2}$$

- Blending

Image blending is the last step of image stitching. Various image blending methods have been proposed, including feather blending and multi-band blending. The purpose of blending is to synthesize the warped images to create a gradient effect in the overlapping area of the two images. By using color balancing algorithms, multiple images can be modified to look like a single image. There is no general consensus at present as to which is the best blending method to apply. It is subjective, as well as depending on purpose. This process is further detailed in [39–41].

2.2.3. Accuracy Assessment

To produce panoramas, Autopano presents the RMSE of tie points in each panorama to demonstrate its stitching quality. RMSE is a common measure of numerical differences and is often used to evaluate model predictions or observed estimates. A root mean square deviation indicates an error between the predicted and observed values (i.e., the presence of sample standard deviation). An error derived from a sample-based estimation is often known as residual error. In the context of panoramas, RMSE is defined as in Equation (3), where N denotes the number of tie points, and *Diff* is the pixel displacement between matched tie points.

$$\text{RMSE} = \sqrt{\frac{\sum_{i=1}^{N} Diff_i^2}{N}} \tag{3}$$

RMSE, calculated using the pixel distance between tie points after image warping, is highly sensitive to excessively large or small errors in a group of measurements and thus effectively reflects the preciseness of measurements [42]. In this study, the quality of image stitching was evaluated with the RMSE color index, which was calculated using Equation (3). The index value offers a precise indication of the image stitching quality for users to plan a subsequent image processing procedure. In this study, an index value of less than 5 pixels indicates good image stitching quality; larger and smaller values indicate lower and higher quality, respectively. Image stitching quality was determined entirely on the basis of numerical values without naked-eye evaluations.

2.2.4. Virtual Tour

Overlapping original images were stitched into panoramas in Autopano; these were then imported to the Kolor Panotour version 2.5.8 (hereafter "Panotour") platform to establish a virtual tour environment. The platform was customized by adding relevant attributes and multimedia materials, including videos, menus, and audio navigation. Next, the platform was exported as a webpage, and its address was converted into a QR code. Users could visit the webpage simply by scanning the QR code on their phone and access the established VR environment with a VR headset for tourism purposes and improving learning outcomes. Koehl and Brigand detailed the functions and applications of Autopano in [43].

3. Results

3.1. Camera Settings

Daylight during our shoot was sufficient; thus, the aperture size was set to F/8–F/11. Because the strong wind at the wind farm caused camera shake, which may result in blurry images, the shutter was set to high speeds throughout the shoot, apart from some slight adjustments to compensate for the lighting conditions; for example, the shutter speed was set at 1/300 when a large aperture size was used. Given the sufficient daylight, ISO was set between 100 and 200.

3.2. Test for Eccentric Issue and Rate of Overlap

The Canon EF24–70 mm F/2.8L II USM was used for the shoot. Table 1 presents the relationship between the camera's focal length and its position (presented using the scale mark on GigaPan), which was determined according to the calibrated field displayed in Figure 4A. Because the calibrated field was located in an indoor space, the ISO, aperture size, and shutter speed were set at 800, F10, and 1/30, respectively. Table 1 reveals that the optimal scale mark value varied across focal lengths. Specifically, the optimal scale mark values for the 24, 35, and 70 mm focal lengths were 105, 100, and 80, respectively. If the camera was placed at a smaller scale mark value, the benchmark rod in the captured image would shift to the left in the clockwise rotation of GigaPan; if the camera was placed at a larger scale mark value, the benchmark rod in the image would shift to the right.

Table 1. Look-up table for eccentric issue with different focal lengths (bold indicates the suggested scale marks).

24 mm		35 mm		70 mm	
Scale Mark	**Object Shift in Clockwise**	**Scale Mark**	**Object Shift in Clockwise**	**Scale Mark**	**Object Shift in Clockwise**
30	left	30	left	30	left
55	left	55	left	55	left
100	left	**100**	-	**80**	-
105	-	115	right	100	right
110	right				
115	right				

The effects of focal length, scale mark, and overlap rate on stitching quality were further investigated. A test shoot was conducted at the square outside the Science and Aeronautical Engineering Building of Feng Chia University by using 27 combinations of focal length, scale mark, and overlap rate. Two levels of overlap rate for the half-sampling were considered, i.e., Level 1 (50%) and Level 2 (25% and 75%). Thus, three overlap rates (25% (low), 50% (moderate), and 75% (high)) and two focal lengths (24 and 35 mm) were used. Table 2 presents the stitching error obtained using these parameter combinations; RMSE represents the average RMSE of all tie point stitching in a single panorama. With the 24 mm focal length, the stitching quality was the highest at scale mark 105 for all overlap rates, and the RMSE was 2.56 for the 25% overlap rate, 2.45 for the 50% overlap rate, and 2.15 for the 75% overlap rate. If the focal length was 35 mm and the scale mark value was 100, the RMSE was 2.56 for the 25% overlap rate, 2.69 for the 50% overlap, and 2.15 for the 75% overlap. In addition, the camera's optical center was the farthest from the rotational axis of GigaPan when the scale mark value was 55 or 33 with a focal length of 24 or 35 mm, respectively, which resulted in the highest RMSE and thus the poorest stitching quality. The test shoot results (Table 2) were in agreement with the recommended scale mark positions in Table 1. Accordingly, the presence of eccentricity was strongly negatively associated with the quality of panorama stitching. Regarding processing loading, the duration of the (1) capturing time, (2) degree of stitching period, and (3) number of images with different overlap rates is 3 min, short, and 60 for 25%, 6 min, middle, and 105 for 50%, and 16 min, long, and 273 for 75%, respectively.

Table 2. Quality analysis with different rate of overlap and scale marks (bold indicates the suggested scale marks).

24 mm			35 mm		
Rate of Overlap	**Scale Mark**	**Avg. RMSE (Unit: Px)**	**Rate of Overlap**	**Scale Mark**	**Avg. RMSE (Unit: Px)**
	55	2.82		30	2.96
	100	2.74		55	2.89
25%	**105**	**2.65**	25%	**100**	**2.56**
	110	2.87		115	2.74
	115	2.75			
	55	2.72		30	2.87
	100	2.56		55	2.82
50%	**105**	**2.45**	50%	**100**	**2.69**
	110	2.55		115	2.69
	115	2.70			
	55	2.51		30	2.47
	100	2.26		55	2.45
75%	**105**	**2.15**	75%	**100**	**2.21**
	110	2.19		115	2.27
	115	2.22			

Figure 5 visualizes the stitching results with a focal length of 24 mm, an overlap rate of 25%, and scale mark values of 10, 55, and 105. Table 3 presents the numbers of tie points and average RMSEs. Figure 5A presents the panorama; Figure 5B–D shows zoomed views of the area marked by the red box in panoramas produced using different scale mark values for a comparison of the stitching quality and to verify the effect of the scale mark position on the stitching result. Figure 5B,C has displacement and blurs, whereas the result in Figure 5D is clear and superior. The test shoot results (Table 2 and Figure 5) obtained using recommended scale mark positions in Table 1 reveal that the number of tie points was positively correlated with stitching quality and that greater eccentricity was associated with a larger RMSE.

Figure 5. Visualizing the stitched errors for 25% of overlap with different scale marks. (**A**) Stitched image and area of interest (red part) for comparison; (**B**) Zoom-in (**A**)'s red part with the case of scale mark 10; (**C**) Zoom-in (**A**)'s red part with the case of scale mark 55; (**D**) Zoom-in (**A**)'s red part with the case of scale mark 105.

Table 3. Comparison between number of tie points and avg. RMSEs with different scale marks.

Rate of Overlap	Scale Mark	Number of Tie Points	Avg. RMSE (Unit: Px)
	55	161	2.82
25%	**105**	**194**	**2.65**
	115	166	2.75

3.3. Real Case of the Onshore Wind Farms

Images were captured from top to bottom and left to right; each image had a 50% overlap with its adjoining images. The scale mark at which the camera was placed was determined in accordance with Table 1. Feature points were extracted from the overlapping area and subsequently matched into tie points for image stitching. When stitching was complete for panoramas, Autopano presents the stitching quality of each tie point in colors (Figure 6). Green indicates an RMSE of less than 5 pixels (high-quality stitching); orange represents an RMSE of between 5 and 10 pixels (moderate quality); and red denotes an RMSE greater than 10 pixels, indicating that the tie point be adjusted or removed. If the photography subject is glass, a surface with a solid color, or an object lacking distinct features, the photograph lacks feature points for extraction, resulting in few or even incorrect tie points. Consequently, human intervention may be necessary to adjust and add new tie points. We drag-selected incorrectly matched tie points to correct them, extracted new feature points, or stitched unmatched parts together until all stitches were of high quality (green). If the modification was complete, the images were rendered to reduce or remove motion trails and blurs and output the final complete panoramas. The panoramas were then imported into Panotour for functional customization.

Figure 6. Quality illustration of tie points (left) and the edited interface (right) in Autopano after stitching a panoramic image.

Figure 7 presents the locations at which the panoramas were captured. The shoot began at Station A and continued along the wind turbine boulevard to Station F for a total of six stations. The stitching quality obtained at each stop and the corresponding panoramas are presented in Table 4 and Figure 8, respectively. According to Table 4, Station A (Figure 8A) had the smallest RMSE standard deviation; that is, the stitching quality remained similar across the entire panorama compared with those of other stations; however, the average RMSE was higher than at other stations. Stations C and D (Figure 8C,D) had the lowest average RMSEs, indicating that their panoramas had the highest overall stitching quality among all stations. Moreover, the number of tie points was proportional to the stitching quality and inversely proportional to RMSE; therefore, increasing the number of tie points could effectively improve stitching quality. The number of tie points was also associated with the weather and the number of features on the subject at the shooting site. According to the RMSE classification of Autopano, all panoramas produced in this study were of high quality, and all of their RMSEs satisfied the basic requirement (less than five pixels) after manual correction.

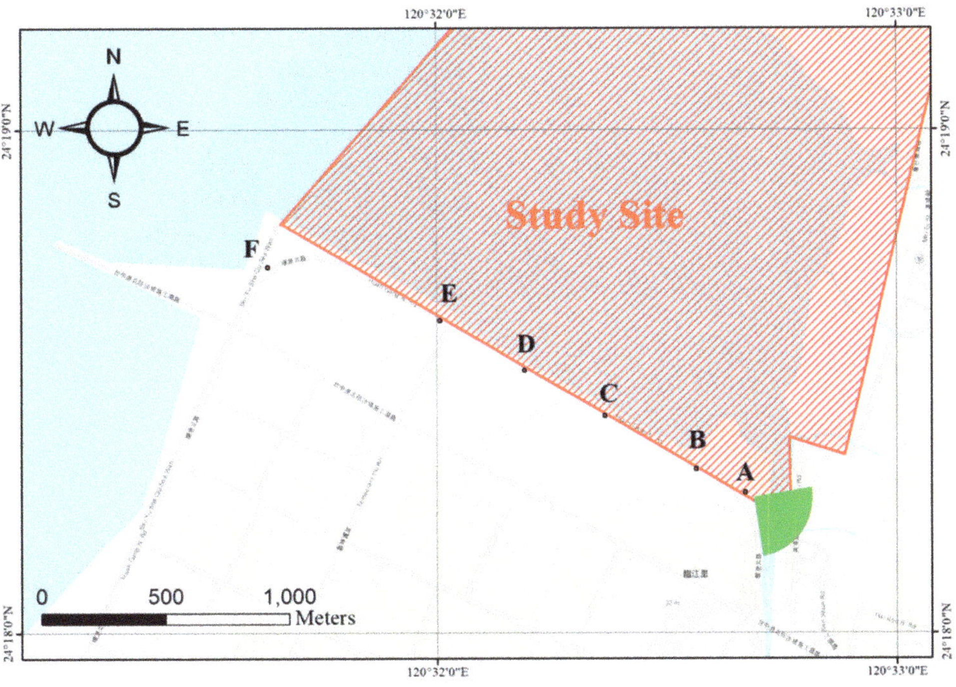

Figure 7. Locations of panoramic stations (points indicate panoramic stations, and the green symbol represents user's current field of view).

Table 4. Statistics of RMSEs with different stations (unit: pixel).

Station	Sd.	Avg.	Max	Min	Number of Tie Point
A	0.50	3.36	4.59	2.10	171
B	0.95	3.04	5.62	1.33	238
C	0.72	2.65	5.74	1.39	366
D	0.56	2.65	5.11	1.65	355
E	0.64	2.78	5.28	1.50	393
F	0.52	3.24	4.58	1.94	155

(A)

Figure 8. *Cont.*

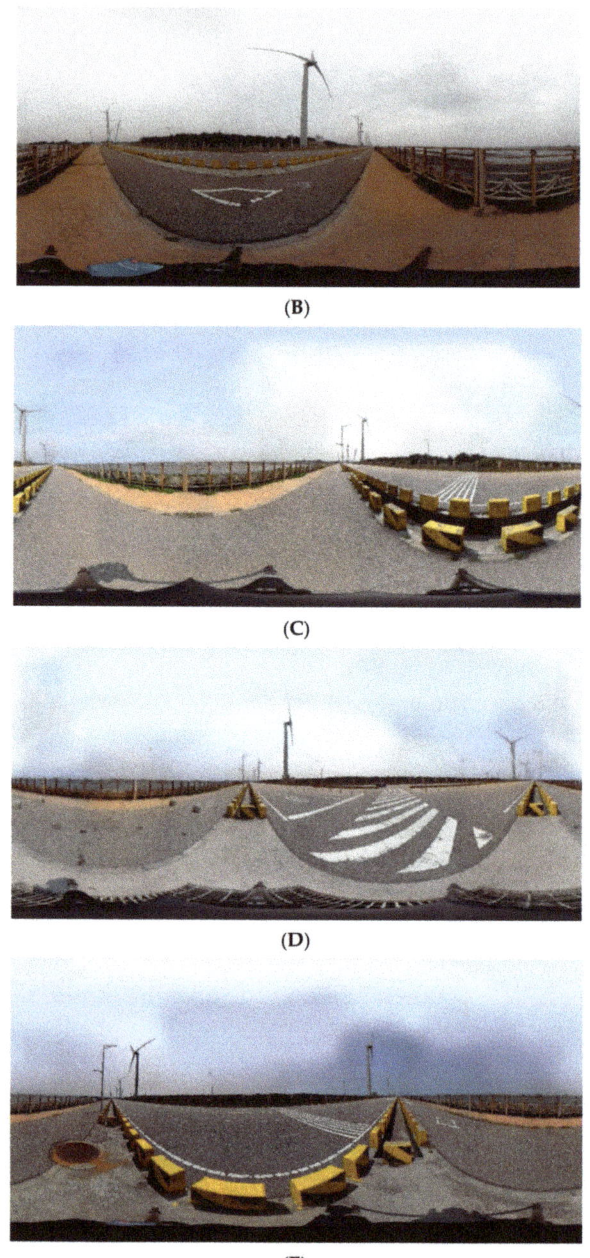

Figure 8. *Cont.*

(F)

Figure 8. Stitched images station by station. (**A–F**) indicate Stations A to F, as shown in Figure 7.

The panoramas for all stations were stitched in Autopano (Figure 8) before being imported to Panotour to establish the virtual tour platform. To ensure that users could engage in a travel experience at any time and could learn from their experience with multimedia technologies without being interrupted by external factors, the platform design was begun by considering the compositional structure of and relevant information regarding wind turbines. The platform (Figure 9) was designed with various functions using Panotour. The relationship among the stations was set in accordance with the tour route for learning purposes during the tour. When users enter the developed environment of a station, they first listen to a spoken commentary, such as an introduction to the history of the Gaomei wind turbine boulevard or navigation instructions for the platform. The audio ends with the sound of wind turbine operation to enhance the authenticity of the virtual tour. The green arrow in Figure 9 is a connector that leads to the panorama of the next station; these arrows enable users to virtually visit the stations one by one as if they were actually walking along the wind turbine boulevard. With a simple VR headset, users can view the VR environment simply by clicking on VR mode in the control bar and visiting the website on their phones (Figure 9F). However, the field of vision is fixed, capturing from the six panoramic stations (Stations A to F). Participants can only browse at a horizontal angle of 360° from these panoramic stations. If participants want to move to other stations, there are two approaches. First is to click the connectors (e.g., green arrows in Figure 9), or to click the point in Figure 7. After that, the field of view (i.e., the green area in Figure 7) will be changed by the user instantaneously.

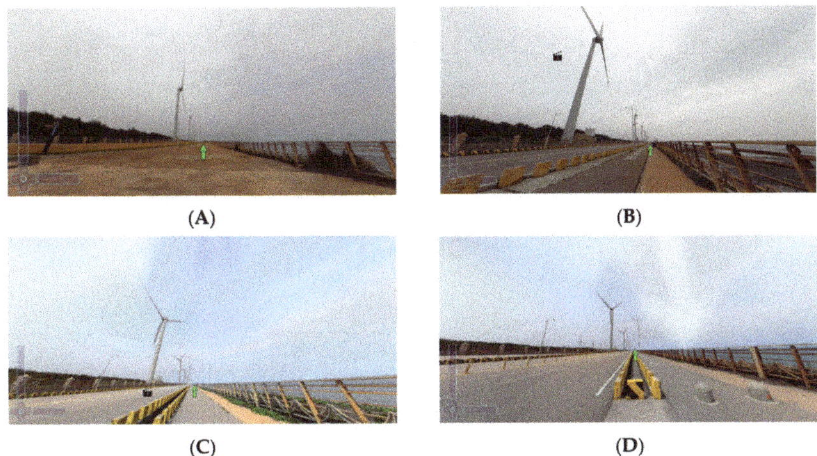

(A)   **(B)**

(C)   **(D)**

Figure 9. *Cont.*

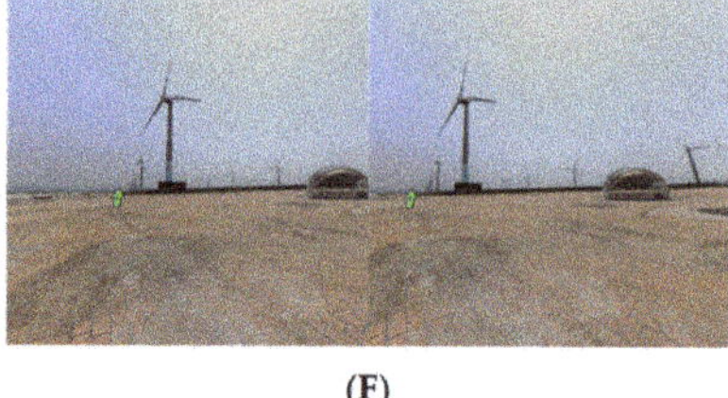

(E)   (F)

Figure 9. Scenes in the developed virtual tour environment with the connectors for browsing multiple panoramic stations. (**A–F**) indicates Stations A to F, and (**F**) shows the VR mode.

Figure 10 presents the scene for the main station (Station A). Interacting with Symbol 1 brings users to an introduction of the wind turbine attributes, including the manufacturer, type of power generator, model, installed capacity, diameter of blade, and total weight; this introduction serves to provide an overview of the wind turbines at the park (Figure 11). Symbol 2 is a button that activates an education video illustrating a wind turbine concept; one illustration video was prepared for each of the five stations (Table 5). The video for Station A presents the operation of a wind turbine; that for Station B details the transportation of electricity; that for Station C describes the construction of the gravity-based foundation; that for Station D provides an overview of wind turbine construction and electricity transportation; and that for Station E introduces the monopile foundation for offshore wind turbines. These videos served to familiarize users with the principles involved in the construction and operation of wind turbines (Figure 12). Symbol 3 in Figure 10 presents the control bar with which users can adjust the zoom level, travel to the previous or next station, or turn on the VR mode. Symbol 4 is a mini map that tells users their current position and the direction they are heading along the tour route; users can also click on a station icon to travel to a particular station. The cross mark in the VR headset, pointing (few seconds) the connectors for browsing multiple panoramic stations (e.g., green arrows in Figure 9) and videos (e.g., Symbol 2 in Figure 10), can be able to interact with these customized functions.

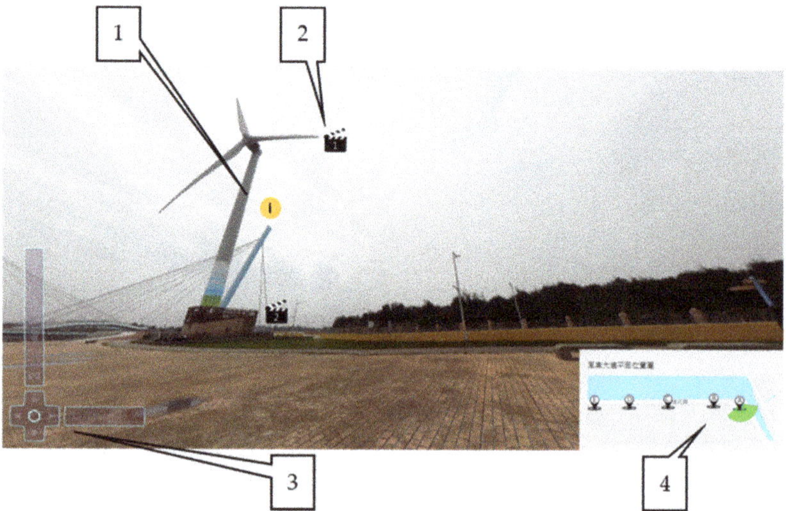

Figure 10. Main station (Station A) of the developed virtual tour environment. Symbol 1: more information, Symbol 2: video for illustration, Symbol 3: control bar, Symbol 4: guide map.

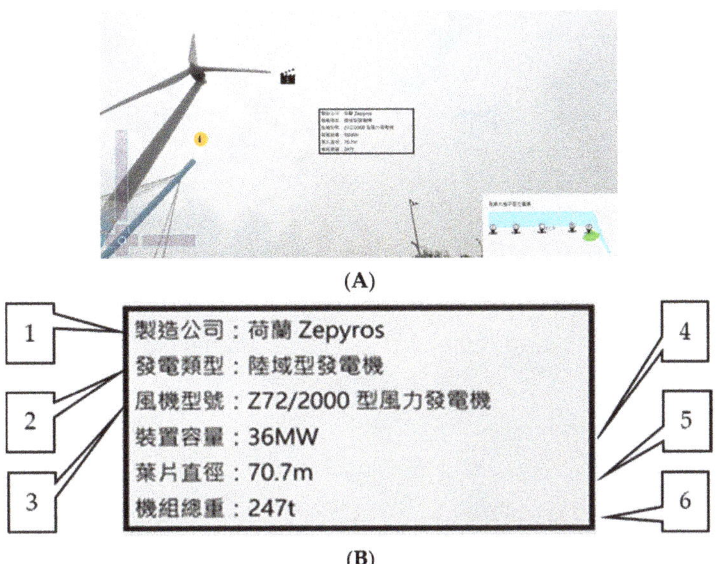

Figure 11. An example of the attributes in the main station. (**A**) Attributes of the object; (**B**) Zoom-in the attributes in (**A**), where Symbol 1: manufacturer, symbol 2: type of power generation, Symbol 3: specification, Symbol 4: capacity, Symbol 5: diameter of fan, Symbol 6: weight.

Table 5. Illustrated issues from the videos.

Station	Illustrated Issues from the Videos
A	Operational procedure, substation, and power storage
B	Wind power and electricity transportation process
C	Construction of gravity-based foundations
D	Transportation of wind turbine tower and construction methods
E	Monopile foundation for offshore wind turbines (extension topic)

Figure 12. *Cont.*

deos in the same or

Figure 12. Embedded vider as Table 5 for illustrating the issues related to wind farm knowledge. (**A**) Station A (video source from https://youtu.be/4Ha5vSlpW2w, accessed on 29 March 2022); (**B**) Station B (video source from https://youtu.be/DILJJwsFl3w, accessed on 29 March 2022); (**C**) Station C (video source from https://youtu.be/B4jEFPdUiMs, accessed on 29 March 2022); (**D**) Station D (video source from https://youtu.be/QK1hr_xCa_4, accessed on 29 March 2022); (**E**) Station E (video source from https://youtu.be/xKYClcExkMY, accessed on 29 March 2022).

4. Discussion

Developing the science and technology of geographic information with Cartography and Geomedia (including Multimedia) is one of the critical topics in the GIS domain. This direction demonstrates that applying different forms such as text, photography, images, animation, video, sound, music, virtual environment (e.g., 3D models), computer games, and others is useful for visualization or geovisualization [44]. These achievements presented in this study are highly correlated to the topics of interest (e.g., information potential of geomedia, geomedia for cartography, cartographic materials for construction and reconstruction in virtual space) in the special issue of Cartography and Geomedia. While comparing with the existing results for VR (e.g., Refs. [1–3]) and panorama (e.g., Refs. [22–24]) developments, our contributions not only constructed a virtual tour platform for the spatial and interactive visualization, but also explored the data quality with two major factors in multi-row panorama capturing (i.e., overlap rate and eccentric issue). Furthermore, visualizing the outside environment for onshore wind farms and the tourism site is rarely observed in the literature because of its complexity and specialization. Thus, readers can treat this article as one of the references for the development of cartography, geomedia, and geovisualization.

From a technical viewpoint in this study, an outdoor shoot can be influenced by various conditions. These factors make the displacement of the same target causing the stitching errors. More precisely, the SIFT algorithm cannot find the invariant location for the same target between different overlapped images. For example, image stitching requires identifying tie points between images; however, a moving photography subject may cause gaps or motion trails in the stitched image. In this study, the moving subjects include the clouds and wind turbine blades. Problems caused by cloud movement were resolved by setting the photography mode to the row-down setting and then slightly adjusting the images in Autopano. For problems caused by the rotation of turbine blades, images containing the blades were manually captured before using GigaPan for an automatic panoramic shoot (the same angle of view and instrument were used in the manual and automatic photography). In image stitching, images captured using GigaPan that contained blade rotation were replaced with the corresponding manually captured images to reduce incorrect tie point matching and thus increase the success rate of image stitching. Moreover, backlighting was a problem because the day of the shoot was sunny; thus, we took multiple shots at each station and rendered the images in Autopano to minimize chromatic aberration.

From the visualization and application-based perspective, the Street View function of Google Maps enables the public to explore popular tourist attractions; however, not all locations are covered by this function. The constructed platform for onshore wind farms established in this study offers high-quality panoramas, detailed introductions to wind turbine functions, and videos illustrating wind turbine technology. This integration between panoramas and wind turbine introductions could expedite learning about wind power. From a tourist viewpoint, the information of provided attributes and videos is suitable for public users. From a civil engineering perspective, the installation and construction of wind turbines comprise various work stages, and a comprehensive plan is required to determine which methods, equipment, and machinery are used in accordance with the surrounding environment. For example, the turbine presented in Figure 11 has a diameter of 70.7 m, which requires a careful choice of crane and transportation tools for installation in accordance with the environment surrounding the construction site. The proposed platform provides opportunities for students and learners to practice making comprehensive construction plans. Therefore, this platform can be directly used to promote the tourism site and onshore wind farms, or be further extended for engineering education.

A wind turbine has an average useful life of 25 years and is operated and maintained by professionals. The Global Wind Organization has established a series of training standards for wind turbine maintenance technicians to foster the professional competence of the trainees. However, conducting intensive and large-scale physical training in an offshore wind farm is difficult. Therefore, panoramas of the internal structure of wind turbines could be incorporated into the proposed platform to familiarize maintenance technicians with the internal information, equipment, and related maintenance procedures of wind turbines. Additionally, wind turbines are constantly subject to lateral wind loads, which, when transferred to the pile foundation, cause it to undergo repeated axial tension or pressure, potentially resulting in the foundation structure tilting. A tilt angle of over $0.5°$ undermines the operational stability of a wind turbine [45]. Accordingly, fixed-point photography may be performed on a regular or irregular basis to document the tilt angle of wind turbines for subsequent engineering analyses.

5. Conclusions

Advancements in information technology (e.g., multimedia and cloud computing) and communication engineering (e.g., the establishment of 5G networks) have prompted new developments in tourism. People can now engage in realistic virtual travel experiences without being restricted by external factors (e.g., the COVID-19 pandemic, weather, or finances) and gain knowledge from these experiences. In this study, a multi-row panorama instrument and a single-lens reflex camera were used to capture images which were then stitched into panoramas. A multimedia-based travel platform was established with these panoramic images and equipped with interactive functions; the platform enables users to view panoramas in $360°$. The stitching quality of panoramas was analyzed and improved to optimize the viewing experience.

The results reveal that the panorama image quality was affected by overlap rate, the relative position between the camera and GigaPan, brightness, and the number of identified tie points. Because the images were captured in an outdoor environment, weather was the primary factor directly affecting the photography result. Taking the shoot on a sunny, cloudless day could simplify the subsequent image processing tasks. Additionally, a higher overlap rate was associated with a lower RMSE and fewer stitching errors assuming that the camera was positioned appropriately. On average, the time required for the shoot doubled per 25% increase in overlap rate. Therefore, not only stitching quality but also the time required for shooting and subsequent image processing are key considerations. The optimal overlap rate should be selected according to the equipment performance and the purpose of these images to optimize outcomes. Panoramas were produced for six stations in this study; all of the stitching results had an average RMSE of less than five pixels. Wind power has been a focus of development in recent years in Taiwan; increasing attention has been paid to technological development and fostering talent in related fields. The proposed multimedia platform enables people confined to their homes to engage in travel experiences and to learn about wind turbines.

In summary, the contributions in this study not only take operations of multimedia production for spatial visualizations into consideration, but also test technical options with the overlap rate and the eccentric issue for improvement of data quality. The proposed framework can be also utilized to map the multi-row panoramic images for different users and cases. Future research could extend the applications of the platform to the internal structure of wind turbines or to offshore wind turbines to increase the diverse VR travel experiences on the platform and enhance public interest in wind turbines. On the other hand, the virtual-tour-based platform constructed by panoramic images in this study still provides the two-dimensional geometry causing that the accurate measurement is limited. Multiple panoramic stations for the photogrammetric task [46–49] to produce a truly 3D virtual tour can be explored in the future.

Author Contributions: Jhe-Syuan Lai and Chao-Ming Chi conceived and designed this study; Jhe-Syuan Lai, Yi-Hung Tsai and Min-Jhen Chang performed the experiments and analyzed the results; Jhe-Syuan Lai, Yi-Hung Tsai, Min-Jhen Chang, Jun-Yi Huang and Chao-Ming Chi wrote the paper. All authors have read and agreed to the published version of the manuscript.

Funding: This study was supported, in part, by (1) the Ministry of Education in Taiwan under the Project MOE-NEEMEC-B-108-02 for New Engineering Education Method Experiment and Construction, and (2) the Ministry of Science and Technology in Taiwan under the Projects 110-2221-E-035-029- and 109-2813-C-035-075-M.

Acknowledgments: The authors would like to thank (1) Fuan Tsai and Yu-Ching Liu in the Center for Space and Remote Sensing Research, National Central University, and (2) Ting-Yu Wu, Kuan-Lin Ho, Pei-Jun Su, and Ping-Tsung Chang in the Department of Civil Engineering, Feng Chia University, Taiwan, for supporting the software and panoramic instrument operation.

Conflicts of Interest: The authors declare that they have no conflicts of interest.

References

1. Oscar, A.; Eva, M.; Laura, C.; Alberto, P.; Diana, J. High-Fidelity Virtual objective structured clinical examinations with standardized patients in nursing students: An innovative proposal during the COVID-19 pandemic. *Healthcare* **2021**, *9*, 355.
2. Gao, Z.; Lee, J.; McDonough, D.; Albers, C. Virtual reality exercise as a coping strategy for health and wellness promotion in older adults during the COVID-19 pandemic. *J. Clin. Med.* **2020**, *9*, 1986. [CrossRef] [PubMed]
3. Van, N.; Vrana, V.; Duy, N.; Minh, D.; Dzung, P.; Mondal, S.; Das, S. The role of human–machine interactive devices for post-COVID-19 innovative sustainable tourism in Ho Chi Minh City, Vietnam. *Sustainability* **2020**, *12*, 9523. [CrossRef]
4. Graabak, I.; Korpås, M. Variability characteristics of European wind and solar power resources—A review. *Energies* **2016**, *9*, 449. [CrossRef]
5. Thayumanavan, P.; Kaliyaperumal, D.; Subramaniam, U.; Bhaskar, M.; Padmanaban, S.; Leonowicz, Z.; Mitolo, M. Combined harmonic reduction and DC voltage regulation of a single DC source five-level multilevel inverter for wind electric system. *Electronics* **2020**, *9*, 979. [CrossRef]
6. Ganea, D.; Amortila, V.; Mereuta, E.; Rusu, E. A joint evaluation of the wind and wave energy resources close to the Greek Islands. *Sustainability* **2017**, *9*, 1025. [CrossRef]
7. Feng, T.; Yang, Y.; Yang, Y.; Wang, D. Application status and problem investigation of distributed generation in China: The case of natural gas, solar and wind resources. *Sustainability* **2017**, *9*, 1022. [CrossRef]
8. Chang, C. Wind Power Generation Prediction Using Data Mining Technique: The Case of Chang Kung Wind Farm. Master's Thesis, National Chung Hsing University, Taichung, Taiwan, 2018.
9. Cheng, K.-S.; Ho, C.-Y.; Teng, J.-H. Wind and see breeze characteristics for the offshore wind farms in the central coastal area of Taiwan. *Energies* **2022**, *15*, 992. [CrossRef]
10. 4C Offshore. Available online: https://www.4coffshore.com/windfarms/windspeeds.aspx (accessed on 8 April 2022).
11. Burdea, G. Virtual reality systems and applications. In Proceedings of the Electro/93 International Conference, Short Course, Edison, NJ, USA, 28 April 1993.
12. Weng, C. A Study on Sense of Space and Presence in Virtual Space. Ph.D. Thesis, National Chiao Tung University, Hsinchu, Taiwan, 2007.
13. Bricken, M. Virtual reality learning environments: Potentials and challenges. *Comput. Graph.* **1991**, *25*, 178–184. [CrossRef]
14. Michitaka, H. Virtual reality technology and museum exhibit. *Int. J. Virtual Real.* **2006**, *5*, 31–36.
15. Lai, J.; Peng, Y.; Chang, M.; Huang, J. Panoramic mapping with information technologies for supporting engineering education: A preliminary exploration. *ISPRS Int. J. Geo-Inf.* **2020**, *9*, 689. [CrossRef]
16. Soliman, M.; Pesyridis, A.; Dalaymani-Zad, D.; Gronfula, M.; Kourmpetis, M. The application of virtual reality in engineering education. *Appl. Sci.* **2021**, *11*, 2879. [CrossRef]

17. FernandoJ, F.; Rebeca, M.; Manuel, C. Constructionist learning tool for acquiring skills in understanding standardised engineering drawings of mechanical assemblies in mobile devices. *Sustainability* **2021**, *13*, 3305.
18. Lee, H.; Lee, J. The effect of elementary school soccer instruction using virtual reality technologies on students' attitudes toward physical education and flow in class. *Sustainability* **2021**, *13*, 3240. [CrossRef]
19. Louvre. Available online: https://www.louvre.fr/en/visites-en-ligne (accessed on 29 March 2022).
20. Great Wall of China. Available online: https://www.thechinaguide.com/destination/great-wall-of-china (accessed on 29 March 2022).
21. Lee, C. Applying Orient-Object Technique to Construct 3D Virtual Tourism—A Case Study in the Yeliou Geo-Park. Master's Thesis, National Taiwan University, Taipei, Taiwan, 2007.
22. Harrington, M.; Bledsoe, Z.; Jones, C.; Miller, J.; Pring, T. Designing a virtual arboretum as an immersive, multimodal, interactive, data visualization virtual field trip. *Multimodal Technol. Interact.* **2021**, *5*, 18. [CrossRef]
23. Antonio, M.; Alberto, R.; Gasparini, M.; Hornero, A.; Iraci, B.; Martín-Talaverano, R.; Moreno-Escribano, J.; Muñoz-Cádiz, J.; Murillo-Fragero, J.; Obregón-Romero, R.; et al. A heritage science workflow to preserve and narrate a rural archeological landscape using virtual reality: The Cerro del Castillo of Belmez and its surrounding environment (Cordoba, Spain). *Appl. Sci.* **2020**, *10*, 8659.
24. Mariotto, F.; Bonali, F. Virtual geosites as innovative tools for geoheritage popularization: A case study from eastern Iceland. *Geosciences* **2021**, *11*, 149. [CrossRef]
25. Garinko Ice-Breaker Cruise 360 Experiences. Available online: https://www.youtube.com/watch?v=W3OWKEtVtUY (accessed on 29 March 2022).
26. Chao, T.-F. The Study of Virtual Reality Application in Education Learning—An Example on VR Scientific Experiment. Master's Thesis, National Kaohsiung First University of Science and Technology, Kaohsiung, Taiwan, 2009.
27. Li, Y.; Song, H.; Guo, R. A study on the causal process of virtual reality tourism and its attributes in terms of their effects on subjective well-being during COVID-19. *Int. J. Environ. Res. Public Health* **2021**, *18*, 1019. [CrossRef]
28. Chiao, H.-M.; Chen, Y.-L.; Huang, W.-H. Examining the usability of an online virtual tour-guiding platform for cultural tourism education. *J. Hosp. Leis. Sport Tour. Educ.* **2018**, *23*, 29–38. [CrossRef]
29. Panos. Available online: https://www.panosensing.com.tw/faq6/ (accessed on 29 March 2022).
30. Ernie, H.; Mark, K. Implementing GigaPan technology into an airport's foreign object debris management program. *Transp. Res. Rec. J. Transp. Res. Board* **2013**, *2336*, 55–62.
31. Zheng, J.; Zhang, Z.; Tao, Q.; Shen, K.; Wang, Y. An accurate multi-row panorama generation using multi-point joint stitching. *IEEE Access* **2018**, *6*, 27827–27839. [CrossRef]
32. Liu, X.C.; Guo, X.T.; Zhao, D.H.; Cao, H.L.; Tang, J.; Wang, C.G.; Shen, C.; Liu, J. Integrated velocity measurement algorithm based on optical flow and scale-invariant feature transform. *IEEE Access* **2019**, *7*, 153338–153348. [CrossRef]
33. Lyu, W.; Zhou, Z.; Chen, L.; Zhou, Y. A survey on image and video stitching. *Virtual Real. Intell. Hardwar* **2019**, *1*, 55–83. [CrossRef]
34. Lee, H.; Lee, S.; Choi, O. Improved method on image stitching based on optical flow algorithm. *Int. J. Eng. Bus. Manag.* **2020**, *12*, 1–17. [CrossRef]
35. Lee, Y.A. Comparative Analysis on the Traditional and Digital Interface of Single-Lens Reflex Cameras—A Case Study of Exposure Control. Master's Thesis, National Taiwan University of Science and Technology, Taipei, Taiwan, 2009.
36. Cooper, J.D.; Abbott, J.C. *Exposure Control and Lighting (Nikon Handbook)*; Amphoto: Kansas, MI, USA, 1979.
37. Lowe, D.G. Object recognition from local scale-invariant features. In Proceedings of the 7th IEEE International Conference on Computer Vision, Kerkyra, Greece, 20–29 September 1999.
38. Wang, G.; Wu, Q.M.J.; Zhang, W. Kruppa equation based camera calibration from homography induced by remote plane. *Pattern Recogn. Lett.* **2008**, *29*, 2137–2144. [CrossRef]
39. Sakharkar, V.S.; Gupta, S.R. Image stitching techniques—An overview. *Int. J. Comput. Sci. Appl.* **2013**, *6*, 324–330.
40. Blending in AutoPano Video Pro. Available online: https://www.youtube.com/watch?v=knqT-2UJOao (accessed on 29 March 2022).
41. What Is Blending? Autopano Video Pro. Available online: https://www.youtube.com/watch?v=H6Tjq0rB9UA (accessed on 29 March 2022).
42. Zeng, B.; Huang, Q.; Saddik, A.E.; Li, H.; Jiang, S.; Fan, X. Advances in multimedia information processing. In Proceedings of the 19th Pacific-Rim Conference on Multimedia, Hefei, China, 21–22 September 2018.
43. Koehl, M.; Brigand, N. Combination of virtual tours, 3D model and digital data in a 3D archaeological knowledge and information system. In Proceedings of the XXII ISPRS Congress, Melbourne, Australia, 25 August–1 September 2012.
44. Medynska-Gulij, G.; Forrest, D.; Cybulski, P. Modern cartographic forms of expression: The renaissance of multimedia cartography. *ISPRS Int. J. Geo-Inf.* **2021**, *10*, 484. [CrossRef]
45. Huang, T.-Y. Deformation Responses of Piles in Cohesionless Soil under Cyclic Axial Tension Loads. Master's Thesis, National Cheng Kung University, Tainan, Taiwan, 2014.
46. Lee, I.-C.; Tsai, F. Applications of panoramic images: From 720° panorama to inerior 3D models of augmented reality. In Proceedings of the Indoor-Outdoor Seamless Modelling, Mapping and Navigation, Tokyo, Japan, 21–22 May 2015.
47. Tsai, V.J.D.; Chang, C.-T. Three-dimensional positioning from Google street view panoramas. *IET Image Process.* **2013**, *7*, 229–239. [CrossRef]
48. Huang, T.-C.; Tseng, Y.-H. Indoor positioning and navigation based on control spherical panoramic images. *J. Photogram. Remote Sens.* **2017**, *22*, 105–115.
49. Teo, T.-A.; Chang, C.-Y. The generation of 3D point clouds from spherical and cylindrical panorama images. *J. Photogram. Remote Sens.* **2018**, *23*, 273–284.

 International Journal of *Geo-Information*

Article

BiodivAR: A Cartographic Authoring Tool for the Visualization of Geolocated Media in Augmented Reality

Julien Mercier [1,2]*, Nicolas Chabloz [1], Gregory Dozot [1], Olivier Ertz [1], Erwan Bocher [2] and Daniel Rappo [1]

[1] Media Engineering Institute (MEI), School of Engineering and Management Vaud, HES-SO University of Applied Sciences and Arts Western Switzerland, 1400 Yverdon-les-Bains, Switzerland
[2] Lab-STICC, UMR 6285, CNRS, Université Bretagne Sud, F-56000 Vannes, France
* Correspondence: julien.mercier@heig-vd.ch

Abstract: Location-based augmented reality technology for real-world, outdoor experiences is rapidly gaining in popularity in a variety of fields such as engineering, education, and gaming. By anchoring medias to geographic coordinates, it is possible to design immersive experiences remotely, without necessitating an in-depth knowledge of the context. However, the creation of such experiences typically requires complex programming tools that are beyond the reach of mainstream users. We introduce *BiodivAR*, a web cartographic tool for the authoring of location-based AR experiences. Developed using a user-centered design methodology and open-source interoperable web technologies, it is the second iteration of an effort that started in 2016. It is designed to meet needs defined through use cases co-designed with end users and enables the creation of custom geolocated points of interest. This approach enabled substantial progress over the previous iteration. Its reliance on geolocation data to anchor augmented objects relative to the user's position poses a set of challenges: On mobile devices, GNSS accuracy typically lies between 1 m and 30 m. Due to its impact on the anchoring, this lack of accuracy can have deleterious effects on usability. We conducted a comparative user test using the application in combination with two different geolocation data types (GNSS versus RTK). While the test's results are undergoing analysis, we hereby present a methodology for the assessment of our system's usability based on the use of eye-tracking devices, geolocated traces and events, and usability questionnaires.

Keywords: location-based augmented reality; augmented reality authoring tool; cartographic authoring tool; user-centered design; user experience; usability; educational technology; nature exploration; geolocated media spatial visualization; cartographic symbols visualization; immersive cartography; open source web tools; nature exploration; biodiversity education; educational technology

Citation: Mercier, J.; Chabloz, N.; Dozot, G.; Ertz, O.; Bocher, E.; Rappo, D. BiodivAR: A Cartographic Authoring Tool for the Visualization of Geolocated Media in Augmented Reality. *ISPRS Int. J. Geo-Inf.* **2023**, *12*, 61. https://doi.org/10.3390/ijgi12020061

Academic Editors: Beata Medynska-Gulij, David Forrest, Thomas P. Kersten and Wolfgang Kainz

Received: 29 December 2022
Revised: 24 January 2023
Accepted: 6 February 2023
Published: 9 February 2023

1. Introduction

1.1. Goals and Structure of the Paper

The goal of this paper is to describe the development of the location-based augmented reality (AR) web application *BiodivAR* and associated user-centered design (UCD) methods [1] for geolocation data enhancement. It also proposes a methodology to further evaluate the usability of the system. As such, it does not present new research results, but merely a new tool and the methods used to develop it. In 2017, the first iteration in the form of a proof of concept in educational technologies (edTech) saw light under the name of *BioSentiers*. This paper aims to present the output of the second iteration of this project and the methods behind it. The first iteration both highlighted some benefits of using location-based AR for nature exploration and identified key challenges that needed to be addressed to make this pairing more meaningful. In the then prototype, the augmented objects were hard coded in the application: editing or adding points was therefore impossible. The AR interface also showed unsteadiness, which caused usability problems. Our new prototype consists of an authoring tool for custom location-based AR experiences while attempting to overcome these issues.

The structure of the paper is divided into the following sections: The current Section 1 introduces the goals of the paper and the project's research goals. The Section 2 contains a review of existing projects leveraging AR in the field of education, as well as key findings of the *BioSentiers* project that provided the ground for this iteration. It also lists the specific challenges of using location-based AR for outdoor biodiversity education; The Section 3 presents descriptions of the conceptual and formal stages of designing the proposed system and their associated challenges; The Section 4 details the setup considers the possible benefits of using Real-Time Kinematic positioning technology (RTK); The Section 5 outlines the methodology that we intend to implement to assess our system's usability; The Section 6 discusses the progress of the established goals and the steps ahead.

1.2. Research Goals of the Project

The *BiodivAR* project aims to explore the role location-based AR can play to support nature exploration and biodiversity education. Its research goals are therefore manifold and can be synthesized as follows:

Create an authoring tool that is easy to use The application we developed features a cartographic authoring tool for the creation of custom AR learning experiences (AR-LEs). It was designed based on information retrieved from interviews and co-design sessions with target users and it is primarily intended to facilitate the organization of field trips in nature. It should be actionable by anyone with little prior technological knowledge, [2] which comes with great utilisabilty requirements. It is also targeted to be implemented in the context of citizen science (CS) projects, through visualizing existing geolocated data, or contributing new data.

Design a location-based AR interface that supports nature exploration The application features an AR immersive mode for visualizing the text, photography, graphics, 3D animation, and sound in virtual environments. When these spatial media contain information about the geographical space they're anchored to, they have the potential to engage users in a singular way [3–8]. We aim to leverage this particular type of immersive interface's attractiveness and efficiency to engage users to actively explore outdoors.

Try to better the usability of location-based AR interfaces Because location-based AR depends on geolocation data, the accuracy of that data is of direct consequence on the interface's stability [4,5,8–11]. We propose the use of an external module for satellite navigation with real-time kinematics (RTK) that achieves centimeter-accuracy. We also lay down the concept for a data fusion method using simultaneous localization and mapping (SLAM) data and geolocation data to further improve heading data, which has an equally important impact on location-based AR interfaces.

Evaluate the tool's usability As a UCD project, its successes and failures are to be assessed primarily on the basis of its usability, acceptance and user satisfaction as measured by self-reported user data, and on the improvements observed with respect to the prior iteration [1]. We have conducted a comparative user test (n = 54) to assess the tool's usability in combination with two geolocation data sources. The results have not been fully processed and analyzed yet and will be presented in a subsequent paper.

2. Background

2.1. Augmented Reality for Education

Augmented reality (AR) describes a type of interface that overlay real and virtual objects in the field of view in real-time, whether it is through the screen of a mobile device or a head-mounted display. It is interactive and recorded in 3D [12]. Users' perception stems primarily from real-world objects, while virtual objects appear to be spatially or semantically related to the real world. In AR, virtual and real objects are combined in order to create the impression of an enriched environment. In an educational setting, AR can make traditional educational material more attractive by overlaying virtual information onto them, hence motivating students to learn [13]. Studies have shown that using AR in education fosters students' motivation [14]. ARLEs hold much potential for improving learning processes and they are undergoing intensive development and research [15]. Their usage is not very widespread yet. EdTech researchers suggested that this was due to the difficulty of developing augmented reality learning experiences, as it usually requires programming skills. They call for the creation of ARLE authoring tools that are easy to use for teachers and students [2]. Another possible reason is that there aren't many tools designed specifically for education. They rather tend to be developed for entertainment, civil engineering or general purpose [16]. However, it has been observed that the number of AR applications for education is increasing since 2010 [17,18]. A systematic literature covering 32 papers written between 2003 and 2013 revealed that the most reported advantages of AR in education are learning gains and motivation. It also showed that the typical target group is bachelor students, and the most common topics are Mathematics and Natural Sciences [19]. A meta-analysis conducted on 64 "AR in education" experiments has reveled that AR had had a medium effect on students' learning gains [20].

Similar to their marker-based or SLAM-based counterparts, location-based ARLEs are assumed to foster immersion and support learning[5,21–24]. Positive learning experiences and high motivation and engagement levels have been repeatedly reported [3–8]. Researchers have investigated ways to design effective and attractive immersive experiences for visualizing geographical virtual spaces based on traditional cartographic data and materials [25]. It was also reported that location-based AR supported contextualization, [26,27] ecological engagement, [28] and caused users to experience a positive interdependence with nature, which fosters improved immersion and learning [29]. Because it uses geolocation data–gathered through a Global Navigation Satellite System (GNSS) sensor–location-based AR only functions outdoors. Compared to marker-based or SLAM–which are go-to technologies for indoor environments–location-based AR does not operate at the same scale and does not cater to the same needs. location-based AR is suitable for outdoor use over distances of several dozen meters. The growing popularity of outdoor education [30] makes it a go-to technology, although leveraging AR technologies for outdoor learning can be challenging because of the informal setting it often takes place in [13]. On the other hand, location-based AR in particular can draw on learning theories that promote learning in context, or inquiry-based learning, [27] which is the valued teaching philosophy in the natural sciences field [31]. When aligned with sound pedagogical theory, AR will be more effective [32].

The use of location-based AR in an educational setting is also justified for the physical activity it is able to stimulate. In a study, researchers have found that using the location-based AR application *Pokémon Go* [33] has led to substantial physical activity increases across genders, ages, weight status, and prior activity levels, unlike many health apps that were primarily designed to that end [34].

Despite the aforementioned reported benefits, AR in education is still at an early stage compared to other digital technologies. The majority of studies are short-term one-time experiments, with little to no longitudinal reports on its impact on learning outcomes [13]. In the specific case of location-based AR, jittery interfaces caused by the use of inaccurate GNSS data make it the weak leg of AR, [4,5,8,10] and has thus not been widely investigated.

2.2. Proof of Concept: The BioSentiers Mobile Application

The application presented in this manuscript is the second iteration of a research project started in 2016. The Media Engineering Institute (MEI) and the Territorial Engineering Institute (INSIT) developed a location-based AR application named *BioSentiers* [9]. The goal of the project was to improve pupils' understanding of biodiversity and establish a connection with nature. The application featured geolocated points of interest (POIs) positioned adjacent to plant specimens. When users approached the specimen in real-life, the POI's color changed on the graphical interface and they were able to prompt an informational card on the plant species. The data collection and selection of the species presented were carried out by a biologist, and the application was tested with fifteen ten-year-old pupils during field trips in a natural reserve, as visible in Figure 1.

Figure 1. The *BioSentiers* location-based AR application. The species presented were prepared by a biologist so as to create an educational trail for pupils in a natural reserve. A video showing the application in use during a field trip is available on a Zenodo publicly accessible repository https://zenodo.org/record/6501843 (accessed on 28 December 2022).

The test emphasized the benefits of using AR for wayfinding in nature: users seemed more efficient when using the AR view rather than the 2D interactive map [9]. They were also more motivated and engaged with the activity, which concurs with conclusions drawn by comparable projects [3,5–8,26–28]. The test also revealed significant usability problems, caused by the instability of the anchoring of the POIs in the AR user interface. Because the display of POIs is directly controlled by the device's measured position, the GNSS lack of accuracy was held to be the cause [9]. At times, the virtual objects either disappeared completely or were so out of position that users got lost. Several other researchers have also underlined the usability problems caused by the imprecision of the geolocation data, which is often regarded as location-based AR's bottleneck [4,5,8,10,11]. In consequence, some have chosen to use marker-baser AR over location-based AR because their experience led them to consider location-based AR to be frustrating and distracting to users [35,36]. It was observed in the *BioSentiers* test that pupils mostly interacted with the tablet us screen (88.5% of the duration of the experience on average) rather than with the surrounding nature [9]. This imbalance is, at least in part, believed to be a collateral effect of inaccurate geolocation data: in the video captures, users are seen spending considerable time reorienting and repositioning themselves. These usability issues caused the users to deviate from the application's primary goal of exploring nature, and they even caused some to trip on branches and fall on the ground. In other instances of AR being used in an outdoor educational setting to promote the exploration of nature, researchers have made similar observations on how technology tends to monopolize users' attention. [37,38]. Mobile devices often require constant attention and interaction from the students, which leads them to focus on the device more than expected. In a wide review of mobile learning projects, the authors have found that the technology dominated the experience in a problematic way in 70% (28/38) of the cases [39].

Following the test, an evaluation [40] of the application offered further improvement suggestions. It recommended to add a function that would allow the users to publish their observations in the form of texts or photographs, rather than being restricted to a passive viewing role. It also suggested the application be partitioned between a "student" and "teacher" version, and the possibility for teachers to create their own learning experiences.

As a primarily technical proof of concept, *BioSentiers* did not leverage a UCD methodology during its conceptual phase, and as such failed to consider several usability aspects. Moreover, as this was an initially modest project, the user tests carried out with users were not governed by a strict and complete methodology. Because of this, its results and conclusions are limited in scope and generalization. However, they sufficed to provide the basis for a new iteration, in which we attempted to address the reported challenges–which appear to be consistent with those of other edTech projects on location-based AR.

3. Materials and Methods

3.1. User-Centered Design

Since weaknesses and challenges had been identified in the previous iteration, we considered that a User-Centered Design would be most likely to help us overcome them. An application developed through UCD should focus and be optimized for the needs of end users so that they adapt to these needs, rather than have the users need to change their habits to use it [1]. During the early phases of conception, we attempted to formulate responses to the challenges raised by the *BioSentiers* project while also updating our goals: to offer a cartographic authoring tool for the creation of location-based AR experiences. We wanted to create a lightweight and flexible tool that would be easy to maintain and evolve, and suitable for different uses. The idea was to integrate the possibility of interacting with the content, and to make the creation of AR experiences accessible to non-specialists, so that the content could be adjusted to meet the specific needs of any field trip. The research goals (1.2) were set based on findings made since the beginning of the project (i.e., through interviews and co-design sessions). In order to come up with relevant solutions, we leveraged several UCD methods in order to gain a more thorough understanding of our target users' requirements and the context in which our system may be used. This entailed participant observation, interviews, co-design sessions, defining main use cases, rapid prototyping, ad hoc testing, and ideation, as described in Sections 3.1.1–3.1.6.

3.1.1. Participant Observation

Following the UCD framework, the conception began with a research phase that helped better understand the end users' environment, their requirements and their expectations. Our main use case is educational, for example in schools. Consequently, the system should be easy to handle by non-specialists such as teachers and students [2]. Other target groups have been included in the design process, including higher-education-level students in nature engineering, and citizen scientists. During this early stage, three participant observation sessions took place with biodiversity professionals. During these sessions, the first author of this paper followed educational field trips organized by Switzerland's main nature preservation organization for elementary and secondary school pupils (aged 8–12 y.o.). It was the opportunity to witness how biodiversity education is carried out outdoors, in informal learning contexts. One key observation was that tablets were already used to watch photographs, videos and listen to animal sounds that could not be witnessed on command. It was also observed that the younger audience was highly motivated by outdoor activities, even without the use of mobile technology. It may indicate that they are not the target audience that could benefit the most from the use of mobile technologies. Slightly older pupils may be more subject to becoming unengaged in educational and outdoor activities, yet very eager to use mobile devices. This led us to focus on a slightly older age group (12–15 y.o.) than the one that was previously targeted in the *BioSentiers* project.

3.1.2. Interviews

Interviews were conducted with four actors likely to be concerned by the use of the application: two education specialists from different disciplines (Outdoor education/French, Natural Sciences); a biodiversity educator; and an elementary school teacher. The interviews were conducted in an informal setting (three over videoconference and one face-to-face) while the interviewer took notes on paper or on the computer. These interviews helped refine the research goals (1.2) and to define specific features that would be relevant to each end-user profile. It also gave some participants the opportunity to voice their concerns about the use of screens for outdoor education. This step reaffirmed our goal of better usability leading to a better balance between screen interaction versus nature interaction. The biodiversity educator from a nature preservation organization in particular expressed that the hypothetical motivation provided by the use of AR was irrelevant in some cases. In their experience, the children were already sufficiently motivated during the activities. However, they observed both the occasional loss of motivation and interest in mobile technologies in teenagers and therefore acknowledged the use of AR could help. This concurred with findings made during the participatory observation stage and encouraged us to narrow our target audience to the 12–15 years old pupils age group, which tends to lose interest in school and science [41] while being avid mobile technology users [42], according to some studies.

3.1.3. Co-Design

In addition to the interviews, three co-design workshops were organized with education and nature engineering specialists. These took place during the design development stage with the goal to ensure the resulting system would meet their needs and expectations, as potential end users. The first participant was a middle school teacher who had previously used the *BioSentiers* application with their pupils; the second one was an outdoor learning didactician; and the third was a higher education teaching assistant. The goal was to elaborate outdoor biodiversity learning sequences for different age groups that comprised the use of a mobile AR application, based on their own assumptions of what features such an app may provide. We came up with pedagogical scenarios for the yet hypothetical application in order to imagine how biodiversity learning could be integrated into the curriculum of different school subjects (French, Geography, Science, History). From the imagined scenarios, several new features were added to the application's architecture.

1. The workshop with the middle-school teacher resulted in the formulation of a scenario suitable for pupils in this age group (8–11 y.o.). It entails an excursion during which the pupils visualize information about various plant species in AR, some of which are useful to humans (edible, in medicine, for manufacturing, etc.), and others are not directly useful or even harmful. Both categories would be presented successively, in two distinct geographical zones. Upon returning to the classroom, a period of reflection would enable the pupils to notice the differences by themselves. The goal is for them to realize that the utilitarian relationship humans hold with nature as a commodity sometimes collides with the wish to protect and place all plant species on an equal footing. The workshop led to the implementation of:

    (a) A style customization feature for the POIs (colors, shapes, volumes, fills, etc.)
    (b) A "geofencing" feature that would allow users to draw walls visible in AR
    (c) Data visualization features on a 2D map that would allow users to visualize the media anchored within POIs remotely, after the field trip

2. The workshop with the outdoor learning didactics led to the creation of a learning scenario where the excursion was framed as a mission where pupils would be presented with an environmental problem. Specifically, the scenario would be that of the construction of a real estate project on a natural reserve. After exploring the milieu, pupils would have to write an essay discussing the pros and cons of the project and how they think the situation should be handled, based on their observations during the field trip. The didactician suggested various perspectives be offered within the ARLE so that pupils may learn to express themselves from different points of view, which is thematized in the French curriculum for this age group. This specific workshop did not result in the inclusion of new features within the app, but it provided insight on how to broaden the possible pedagogical framing that happens outside of the application. It also supported the relevance of data visualization features on a 2D map, by planning to use these tools after the field trip, upon returning to the classroom.

3. The workshop with the higher-education teaching assistant was oriented towards the creation of a use case for students (18–25 y.o.) in nature engineering. The system

would be used to collect data on the field as they explore nature. They would be tasked with collecting data, creating POIs, uploading photographs, identifying and monitoring specimens, writing comments, and making observations. In this scenario, the goals would both be for the students to learn about biodiversity by applying theoretical knowledge they have previously studied, and also to actively contribute to biodiversity data that can be harnessed for research. Thanks to the use of interoperable GeoJSON files, participants can further valorize the data they collected by uploading it to existing CS databases such as the Global Biodiversity Information Facility https://www.gbif.org/ (accessed on 28 December 2022) or Infoflora (in Switzerland) https://www.infoflora.ch/fr/ (accessed on 28 December 2022) projects. Because the objectives of this age group are aligned with–and better known as–those of CS, they have been modeled according to this nomination in the application's use cases. The workshop led to the inclusion of the following features that allow users to achieve CS goals:

(a)   A feature for creating custom data attributes;
(b)   The possibility to create/edit POIs from the mobile interface;
(c)   The ability to import/export data (interoperability with other geoportals);
(d)   The ability to make ARLEs available/editable by everyone (wiki);
(e)   The possibility to switch rapidly between ARLEs, to use them as quiz/answer key.

3.1.4. Use Cases

The previous steps (participant observation, interviews, co-design) provided grounds to model the use cases for each of our target users, as visible in Figure 2. Our primary users are pupils who will use the system to learn about biodiversity by exploring a milieu while visualizing media in AR (sounds, images, 3D models) during field trips. For pupils to effectively use AR for educational purposes, the augmented content should be tailored to their needs, which implies that their teachers have to be able to use the system in the first place. To help pupils learn a new task or concept, teachers routinely use instructional scaffolding or target their zone of proximal development. AR contents should be adaptable and their management should be easy, as identified by Cubillo et al. [2]. Therefore, our system's secondary users are teachers whose goal is to teach biodiversity and use the desktop client of the system to develop or adapt an ARLE prior to a field trip. CS pursues the dual objectives of learning about biodiversity and collaboratively collecting data, which overlap with the goals of biodiversity education. We therefore included citizen scientists, who may use the system occasionally, as another secondary user group. The system should be a useful resource for CS initiatives, notably through the ability to visualize any collaboratively collected (interoperable) geodata in an AR interface.

3.1.5. Rapid Prototyping and Ad Hoc Testing

We used a rapid prototyping method [43] to create an early version of the application with ARIS, [44] a user-friendly, open-source authoring tool for AR experiences, as a method to test our initial ideas and gather user feedback. It included testing storytelling scenarios using analogies to convey abstract concepts related to biodiversity. We also wanted to test the use of gamification features such as characters, points, and missions, as visible in Figure 3. Due to the limitations of the underlying AR engine (Vuforia), the AR was marker-based, which means that visual targets had to be recorded in the application and placed in the field.

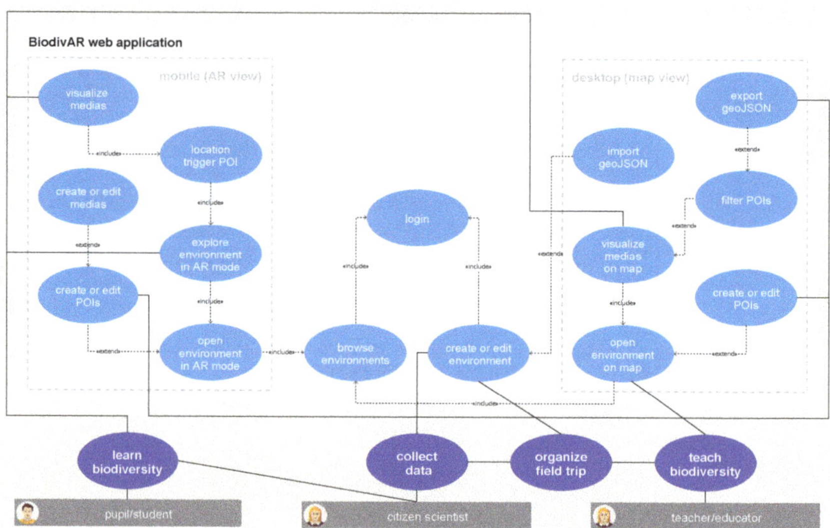

Figure 2. Use Cases: system's scope, end users and their interactions with the system, main expected scenarios and the goals that the system may help end users achieve.

Figure 3. Screenshots of the prototype made with ARIS. Location-trigger POIs prompt dialogs with fictional characters who send users on missions to go explore newly appeared POIs on the map. When reaching the POIs' location, users either had to listen to sounds, identify or photograph a specimen, or watch a video before returning to the character to earn a reward.

In July 2021, we tested this prototype during a one-day continuing education outdoor learning training session. As visible in Figure 4, twenty teachers from elementary and middle school individually explored a patch of forest with our AR game for about twenty minutes on a smartphone [45]. After the tests, we conducted group discussions to learn about what the end users thought of the system and its features.

Figure 4. Teachers watching a video in AR about a plant species, overlaying a visual marker placed next to a specimen.

Overall, a majority of the teachers thought that the system could help their pupils engage with nature and teach them contextual information about the plants they encountered. They discovered plant specimens and their taxonomy. They appreciated that the system prompted them to walk about and explore areas of interest. However, some of the participants failed to see the link between the augmented objects and plant specimens in the real world, which made them confused about the system's main goal. Based on our observations, this was due to the inaccuracy of the geolocation data, which concurred with what had been observed during the *BioSentiers* test. Text-based media in the system were deemed to be generally too long. The audio contents (animal sounds) were appreciated. This feedback helped us refine our specs and set our priorities to make our concept more user-friendly and engaging.

3.1.6. Ideation

As visible in Figure 5, an ideation session was organized as a method to leverage collective intelligence, ideate thoughts and come up with solutions to some of the problems we have faced while conceiving the application. The participants received a short introduction about the application, its target users and intended use cases. A wireframe was also presented. During the first activity, each participant had to identify any difficult, complex, or unclear points on post-its. They were then clustered by topic and displayed, and the participants discussed possible workarounds. In a second activity, participants were asked to draw on paper six different versions of a proposed interface that had previously been identified as complex for users. Both activities were helpful in simplifying the issues and coming up with the most logical options: The external observations made during the activity led to the removal of several redundant layers in the final application. They also allowed us to which of the system features were instrumental in achieving its set goals.

Figure 5. UX/UI designers and developers during an ideation session.

3.2. Architecture and Implementation

3.2.1. Concept

The *BiodivAR* mobile application aims to support nature exploration for pupils, students, and concerned citizens and allow them to visualize and interact with geolocated POIs containing media in AR. Additionally, it aims to enable teachers to author location-based AR experiences. Our concept thus assumes the two following main types of utilization: media visualization in AR on location/outdoors ; and the authoring of AR environments that can be carried out remotely/indoors. As such, the application comprises two main modules: a mobile component, which mainly allows AR visualization of geographic data, and a desktop cartographic authoring geoportal that allows users to create and manage environments and visualize data on a 2D interactive map. This two-part design is shown in Figure 6. The latter may also be used in a learning context. The translation and correspondence between the cartographic authoring tool and its 3D, augmented counterpart is at the core of the concept.

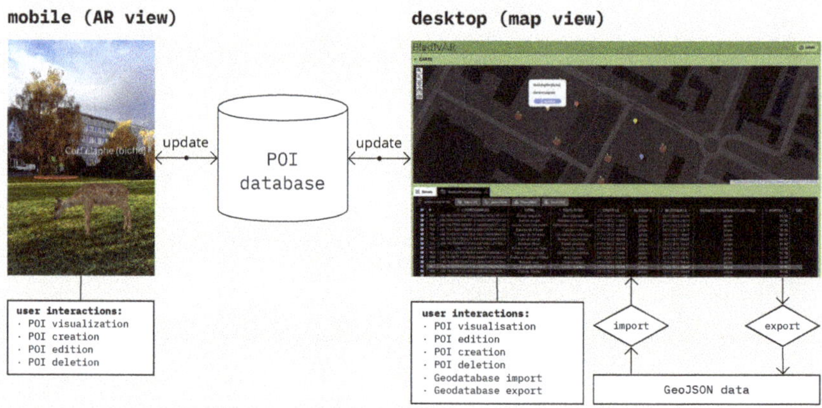

Figure 6. The *BiodivAR* concept: the mobile AR interface allows on-site, spatial interaction with media anchored to geographic coordinates. The desktop interface displays the same geodata on a 2D interactive map (leaflet). GeoJSON data can be imported or exported.

The data are stored within *bioverses*, which are thematic augmented reality learning environments (ARLEs). They can be set to public/private, editable or non-editable, duplicated, merged, and visualized on the 2D map. They contain geolocated points of interest (POIs), which contain various media and/or data:

1. A *bioverse* is an ARLE containing geolocated POIs. It can be created by any registered user and set to public or private. If public, it can be visited by any user. A bioverse may also be set to editable in order to allow users to contribute or edit its content, or non-editable to grant them visualization rights only. Bioverses can be exported as geoJSON files for archiving/re-importing, or to visualize and analyze their data in other geoportals. They can also be duplicated or deleted within the application. They have a title, a short description, and may be associated with a location to help grasp their range (i.e., "The World", "Switzerland", "Yverdon-les-Bains", "my garden", etc.). The managing of bioverses is meant to support various use cases, which came up during co-design sessions. A teacher could conceive several bioverses and assign them to different groups of students, or create one as a quiz and a second one as an answer key. Basically, it enables the creation of thematic virtual environments, and facilitates navigation across them, such as TV channels.

2. A *POI* is a 3D scene anchored to a set of geographic coordinates. Users may add as many media (3D, audio, image files, texts) as they such as in the scene and position each one of them relative to the origin (= the geographic coordinates). The creator thus decides to present media separately or jointly to transmit information about a location, depending on what makes sense in a given situation. In AR mode, any media attached to the POI appears somewhat realistically, as if it belonged to the real world: it scales up or down as users approach or move away. A POI may also contain user-defined custom and stylistic attributes. Because the design of 3D/spatial cartographic symbols directly impacts usability, [46] our system allows users to customize 3D and 2D symbols to better suit their specific needs. Each POI needs to have a visibility/audibility *scope* (in meters): it is the distance from which the POI becomes visible/audible in AR mode. This allows authors to choose the best-fitting settings for any given scenario: where the visibility is short (urban zone or forest), POIs should not be visible from too far to avoid them overlapping with landscape components. Each POI also contains a *radius* (in meters), which is the distance within which users will trigger the visibility/audibility of additional media, depending on their attributes. The radius may be stylized (shape, size, stroke width, stroke color, fill color, elevation, animation) to be visible as a colored circle/cylinder/sphere/hemisphere in AR mode. This helps users realize when they "enter" a POI. While users are located within a POI's radius, the tracking mode will stop relying on geolocation data and switch to WebXR immersive-AR mode's tracking capabilities only, which rely on computer vision and inertial sensor data. This prevents jumpy geolocation data from kicking users out of a POI while they are visualizing media within it. A POI may also feature a 50×50 px image that will be displayed as a cartographic symbol on the 2D map.

3. A *media* can be a 3D file (gltf, glb), a picture (.jpg, .png), a sound file (.m4a, .mp3), or plain text that will be converted into a 3D mesh. Audio media can be set to play spatially or not and can be looped or only play once. Each media is imported individually in the POI editor and placed in space relatively to the origin, which is also the geographic coordinates the POI is anchored to. This flexibility allows users to create a wide range of POI types, from photo galleries to audio-only invisible ones. Each media can be set to be triggered by a user's location at different thresholds:

   - It can be visible/audible from the moment a user is within the "scope," and disappear when they enter the "radius." Use case: a 3D cartographic symbol; a sound-effect that plays when users enter/exit the "radius."
   - It can be invisible/silent until the user reaches the "radius". Use case: a gallery of media about a specimen that should only be visible once users reach a specific area; a podcast that plays when users are facing a landmark).

- A media may be set to be visible/audible both within "scope" and "radius" (use case: a 3D symbol that should remain visible while another sound-based media is triggered at close range; an ambient spatial sound).
- In a more peripheral way, a media may be set to be invisible and silent at all times, when AR visualization features are not required. Use case: a CS data-collecting field trip.

3.2.2. Tools, Frameworks and Standards

After the conceptual grounds were set, the actual development of the application began and the adoption of suitable frameworks came into question. After reviewing various options (Unity, Ionic, Wikitude, WebXR) we opted for a web-based application for the numerous advantages it offered. It can be accessed through an URL in a WebXR-enabled Web Browser without downloading any file. A single deployment allows the targeting of handheld and head-mounted AR devices. As an applied research project, continuity is important, and this choice meant our code would run on new devices without the need for important updates. However, it also came with limitations. Because the code and assets are stored on a distant server, web-based applications tend to be slower and less responsive than native apps, which can be critical for AR usability. Access to the device's hardware features may also be limited. Last but not least, users are required to have access to an active network connection.

BiodivAR was thus conceived as an application using a client–server structure with web technologies exclusively. We had to adapt to the current WebXR Device API (Application Programming Interface) limitations, which is the standard for VR/AR/XR experiences on the web. WebXR uses Google's ARCore SDK and enables access to a device's inertial sensors and camera data in order to handle the AR cameras in compatible (chromium-based) browsers. Although experimental support for the WebXR API is offered by Mozilla to iOS users through their "WebXR Viewer" application, Apple has yet to enable global support for the API. WebXR is thus officially supported on Android devices (as well as non-android AR glasses, i.e., Hololens 2). However, WebXR features have recently appeared in iOS 15.4 (June 2022) Safari browser, which tends to indicate that Apple is ready to adopt the standard in the close future, despite their own parallel development of the ARkit SDK and USDZ format in collaboration with Pixar.

We used the A-Frame https://aframe.io/ (accessed on 28 December 2022) open-source web framework to support the building of WebXR-compatible 3D scenes. It uses an entity component system architecture which is familiar to game development. It is convenient to use by both developers and designers to create 3D scenes without having to struggle with webGL. It also benefits from an enthusiastic and dedicated community which allows it to be enhanced with new features at a rapid pace. To our knowledge, the only existing open-source web framework that featured location-based AR is AR.js, which also features an authoring tool [47]. However, AR.js covers a wide range of AR types that we would not use in our system, and its geolocation components had not been updated or maintained in a while. Consequently, we created LBAR.js https://github.com/MediaComem/LBAR.js/ (accessed on 28 December 2022), a minimalist A-Frame library for creating WebXR location-based anchors. LBAR.js places entities in a virtual 3D space based on geolocation data and inertial sensor data. It includes one system (*gps-position*) and three components (*faces-north*; *gps-position*; *pitch-roll-look-controls*) [48]. The *gps-position* system defines the accuracy threshold over which (after a movement of n meters) the geolocated entities' anchoring is refreshed. The *faces-north* component forces an entity to always be facing north (as measured by the compass bearing, or absolute device orientation). It lets users create a parent entity aligned with reality under which all geolocated entities are anchored as children. The *gps-position* component allows the position of an entity (a geolocated POI) to be expressed in WGS84 coordinates, by converting those into local coordinates so that it may be anchored in the A-Frame scene. It needs to be a child entity of an entity with the *faces-north* component in order for it to work. The *pitch-roll-look-controls* component disables the default yaw control of the AR camera, since the *faces-north* entity (and its children entities) rotates on that axis instead. Other open-source dependencies that were used for the client–server structure include SQLite (database), PrismaJS (ORM), hapi.js (API), vue.js (frontend), and leaflet.js (2D interactive map). The full dependencies' diagram is visible in Figure 7.

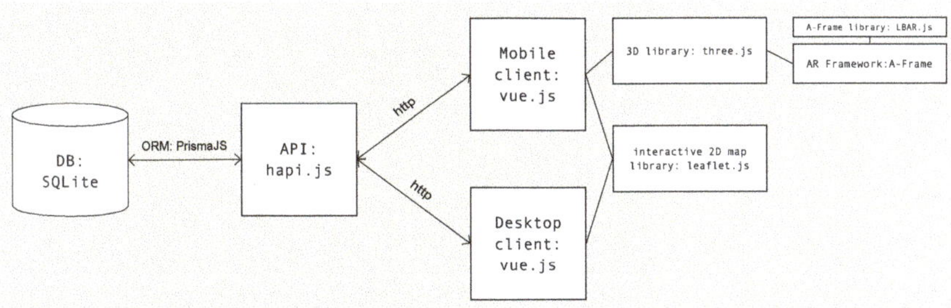

Figure 7. *BiodivAR* structure and dependencies. The user interfaces (desktop and mobile) are managed by the Vue.js framework, which sends a request to a Rest API built with hapi.js. The Rest API gets the requested data from an SQLite database through a Prisma.js object–relational mapping (ORM). The mobile and desktop 2D interactive maps are powered by the Leaflet.js library and the AR cameras and objects are managed by the WebXR API, the three.js 3D library and the A-Frame framework, with the LBAR.js [48] additional custom library we designed to power geolocated POIs. All of theses dependencies are open-source.

3.2.3. Desktop User Interface

In *BiodivAR*, users may browse bioverses they authored as well as publicly accessible ones. They may be opened in AR mode on mobile devices only because it requires geolocation and angular data that only they can provide. However, bioverses can also be visualized on desktop devices on an interactive map, as visible in Figure 8. On this interface, a user may create a new POI (1), import external geodata (GeoJSON) or export what is currently displayed on the map (2). They may browse available bioverses (3) and open several ones for parallel visualization, analysis, or merging data. In any of the available tabs (POIs, paths, tracelog and events), geodata may be toggled (4) to be made visible or invisible on the map. An object's title and description will appear in an infobox (5) when hovered or selected on the map. If the current bioverse can be edited by the current user, an "edit" button will appear. POIs' radius will display on the map (6) with the same custom style (color, stroke, fill, etc.) they were assigned for their AR, 3D counterpart. POIs' scope will show as a white dashed circle (7). If a POI contains audio media, the audio scope, or audibility threshold will appear on the map as a dark blue dashed circle (8). Events (9) and traces (10) is user-generated geodata that can be displayed on the map. The traces tab (12) holds a table with the trajectory a user has followed from the moment they opened a bioverse in AR mode to the moment they exit it. If selected, the paths will appear as randomly colored polylines on the map. The events tab (13) stores user-actions that occur during regular use of the application: entering/exiting a bioverse, entering/exiting a POI, creating/editing a POI, and opening/closing the 2D map. The events appear as red circles on the map, with infoboxes specifying their type. Traces and events are only accessible by users who authored them, in a logic of data transparency and awareness. This data may be collected during user testing for evaluation and analysis purposes. Finally, an additional menu (14) allows users to export, copy, paste, delete and reorganize data. POIs can be copied and pasted (individually or in bulk) from one bioverse to another. Bioverses can also be exported as GeoJSON files nd opened with other geoportals, or for archiving.

Figure 8. A view of *BiodivAR*'s desktop user interface with its main features highlighted. Various types of geodata (POIs, traces, events) may be displayed on the map by toggling them on their respective tabs.

When a POI is created or edited, the POI editor modal window is opened, as seen in Figure 9. It lets users manage and previsualize the POI as an embedded A-Frame 3D scene that can be navigated with the "WASD" keyboard keys. Some parameters are POI-wide (i.e., geographic coordinates, title, description). Style attributes can be assigned to a POI's radius and will reflect both on the 2D map and in AR mode. The scope (visibility/audibility threshold) can be adapted. Users may add new media (image/sound files, texts) and define their position relatively to the origin (= arrow helper). Various behaviors can be assigned to each object in the scene, such as whether the media should always face users or hold an absolute position; whether they should appear when users enter the scope *or* the radius; objects' position can be animated; and so on. The image on the left (Figure 9) previews the state of a POI before the user has entered the perimeter of the radius. The image on the right previews its state after the user has entered the radius.

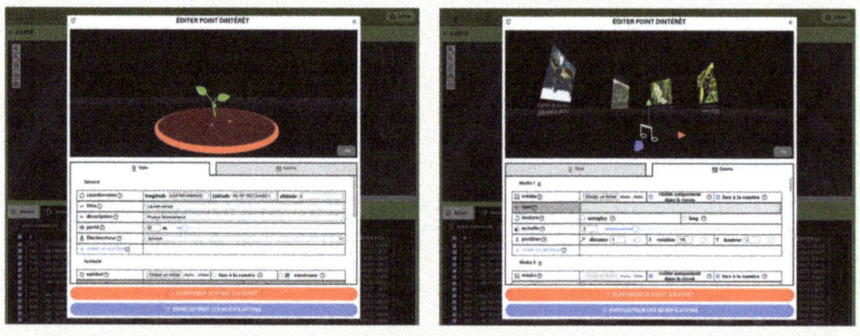

Figure 9. The POI editor appears when a POI is created or edited. It lets users manage and customize POIs. Users can upload media (3D models, pictures, sound, or plain text) and position them in the embedded 3D scene preview. Each media may be assigned individual behaviors so that they appear at specific user-triggered events.

3.2.4. Mobile User Interface

BiodivAR's mobile user interface is accessed with the same URL. After login in, (Figure 10a) users may browse bioverses similarly to the desktop counterpart (b). The "enter AR" blue button initiates webXR's AR mode and loads the geodata contained in a bioverse: the POIs are displayed in superposition with the camera view background, simulating their presence in the real world (c). Users may swipe the interactive 2D map from the bottom of the screen to help them navigate toward POIs (d). When they enter a POI's radius, author-scheduled events are triggered and new media may appear (e). In the example shown in Figure 10, a distribution map of the tree specimen sitting next to the POI appears, while a short audio podcast delivers information about the species.

Figure 10. *BiodivAR*'s mobile user interface: (**a**) login; (**b**) all available bioverses sorted by categories (authored by user, public, bookmarked); (**c**) AR view after opening a bioverse; (**d**) a collapsible 2D map for navigation; (**e**) view of a media showing the distribution map of an adjacent specimen's species, after entering a POI's radius.

3.2.5. Database Structure

The data are organized around the concept of *bioverse*, consistently with the application principle of managing this learning environment on an interactive 2D map and visualizing them in AR. A bioverse is stored in a GeoJSON object with the type "FeatureCollection" [49]. Each POI is a feature that contains a geometry type (point), geographic coordinates, and a bunch of user-defined properties related to the POI's style, contents, media, positions, and events. As visible in Figure 11, eight tables structure all the data in the database:

- *UserTrace*: an object that stores the *Coordinates* of a given *User* at a given time rate. All logged *Coordinates* include a timestamp, the accuracy of the geolocation data, and the currently open *Bioverse*.
- *Event*: an object that stores user-triggered events while they are using the system in AR mode. All logged events include a timestamp, *Coordinates* and the accuracy of the measured location. There are eight types of logged events:
  - *biovers-open* | *biovers-close*: a user opens or closes a bioverse.
  - *biovers-enter-poi-708* | *biovers-exit-poi-708*: a user enters or exits the radius of a POI (+ POI id).
  - *open-map* | *close-map*: a user opens or closes the map while in AR mode.
  - *create* | *update-poi-933*: a user creates or edits a POI (+ POI id).
- *User*: an object that stores a *User*'s profile information (username, email, password…) and the bioverses, POIs and paths they have authored.

- *Coordinate*: an object that contains geographic coordinates (longitude, latitude, altitude). *POI*, *Path*, *UserTrace*, and *Event* all include a *Coordinate* object.
- *Path*: an object that contains a series of *Coordinates* that constitute a polyline entity in the A-Frame scene.
- *POI*: an object that contains *Coordinates* and *Media* that constitute a POI entity in the A-Frame scene.
- *Media*: an object that contains a media's url (3D, sound, image, text) as well as its local coordinates relative to a *POI*'s geographic *Coordinates*.
- *Bioverse*: an object that contains all other objects as well as user-defined parameters defining the virtual environment's visibility and editability to other users.

As visible in Figure 11, the *Bioverse* table is the central element. It includes all of the other tables and is therefore aware of the whole data. It contains *POIs*, *Paths*, *Events* and *UserTraces*. Each one of these elements may include its own subdata: a *POI* may include and be enriched by as many *Media* elements as desired. Most objects contain one or more *Coordinate* to either position it in the AR interface (*POI*, *Path*) or to log the location an object was generated by the user (*UserTrace* and *Event*). Eventually, these elements are linked to the *User* table to manage creation, deletion and editing rights.

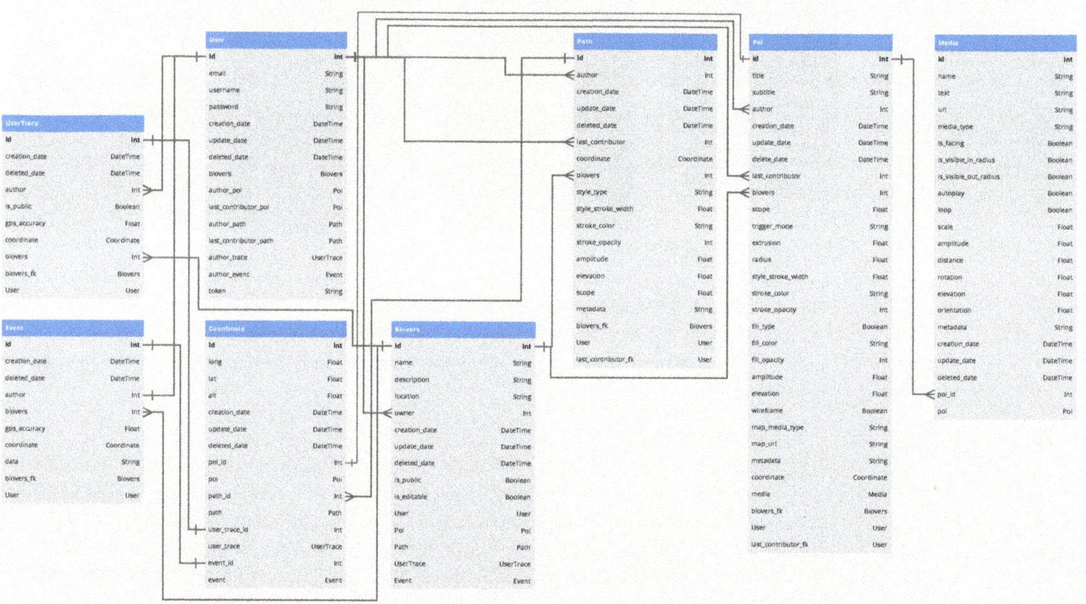

Figure 11. *BiodivAR*'s data structure: The *Bioverse* table in the center contains all the other elements and notably *POIs*, which contain one *Coordinate* and *n Media*.

The data structure aims at supporting and leveraging interoperable standards so as to enable the use and creation of open data sets. The following data can be exported and downloaded as GeoJSON files by users (see Figure 8, 14):

- *POIs*: a GeoJSON file that contains the data to build geolocated A-Frame point entities in the AR scene, including media URL and their positions relative to the geographic coordinates.
- *Paths*: a GeoJSON file that contains the data to build geolocated A-Frame polyline entities in the AR scene.
- *Tracelog*: a GeoJSON file that contains user-generated geographic coordinates of a user (at 1 Hz).

- *Events*: a GeoJSON file that contains user-generated events that are most of the possible interactions between a user and the system while in AR mode.

4. Geolocation Data Enhancement

4.1. Location-Based AR Bottleneck

One of location-based AR's main advantages over other types of AR is the possibility to create and anchor augmented objects in a global spatial referential. They can thus be anchored remotely, as opposed to having to be physically on site to anchor content. It can leverage existing geodatabases, and it is possible to enrich them for further visualization or analysis of other geoportals. However, as observed during the *BioSentiers* field test [9] as well as by other researchers who have tried to use this technology for education purposes, [4,5,8,10] location-based AR's main drawback is the instability of the anchoring of augmented objects which results in a degraded user experience. Indeed, a typical mobile device-embedded GNSS sensor has limited accuracy [50]. Under favorable conditions, it can be as low as one meter. However, as conditions worsen, this range can routinely reach 15 to 30 meters [51]. Unlike 2D interactive maps, which tolerate somewhat imprecise geolocation data and remain usable, running location-based AR interfaces on such data impairs the user experience to the point of being inoperable. In comparison, marker-based or SLAM-based AR usually anchor virtual objects with millimeter-accuracy [52].

We believe that the combined use of location-based AR with consumer-grade mobile devices' geolocation data may be responsible for some part of the usability issues that are routinely associated with this technology. Furthermore, we believe that this causes an additional adverse side effect: an increase in the time spent interacting with the screen rather than with the surrounding nature (88.5% versus 11.5% during the *BioSentiers* tests). While it is of course necessary that users interact with the screen to benefit from the potential positive aspects of the ARLE, any added time they spend repositioning and reorienting themselves does not serve the purpose of the learning process and is to be prevented, if possible. We are under the impression that geolocation data accuracy may be the primary limiting factor for the normal use of location-based AR. In the course of our project, we will thus explore the use of external GNSS modules in combination with our application. We expect that more accurate data will result in an improved positioning of geolocated objects in the AR interface.

4.2. Geolocation Data Enhancement with RTK

In an effort to acquire more accurate geolocation data than that provided by mobile devices' embedded GNSS modules, we propose to use external hardware modules for satellite navigation with Real-Time Kinematic Positioning (RTK). RTK is a type of differential positioning system that works with a base station whose actual and accurate position is known. The difference between the known position and the one continuously measured by the base station is known as the range error. The station broadcasts the range error over a standard internet protocol (NTRIP caster), which makes it accessible by any mobile device equipped with a multi-GNSS receiver and an internet connection (also known as "rover station"). The receiving mobile device needs to run a third-party open-source NTRIP client application (i.e., SW maps, Lefebure) to retrieve the range error or correction data. By removing the range error from each satellite distance measurement made by the multi-GNSS receiver, the application computes adjusted geolocation data and exposes it device-wide as a "mock-location." In developer mode, Android devices feature an option that replaces the default geolocation data with mock-location data. For optimal results, the base station and the rover need to be subjected to similar phenomena (atmospheric, meteorological, environmental etc.), which assumes that they must be in a radius of up to twenty kilometers. Thanks to this process, the majority of GNSS ambiguities are fixed and the rover station retrieves centimeter-accurate positioning.

4.2.1. Ardusimple RTK Surveyor Kit

After considering various options, we decided to use a low-cost (399€) ready-made GNSS/RTK module manufactured by Ardusimple for Android mobile devices [53]. It includes a Funduino processor with a u-blox ZED-F9P multi-GNSS receiver, a helical GNSS antenna (SY-301) and a Bluetooth module. Once the kit is powered and connected to a mobile device by Bluetooth or USB, the NTRIP client application retrieves the correction data from the base station and streams it to the kit's multi-GNSS receiver, which additionally measures satellite distances on its own. The correction data and measured data are integrated into the navigation engine, and the adjusted geolocation data are sent by the processor and its Bluetooth module to the NTRIP client application. The application exposes that data to the entire mobile device through the "mock location" feature. This setup is visible in Figure 12.

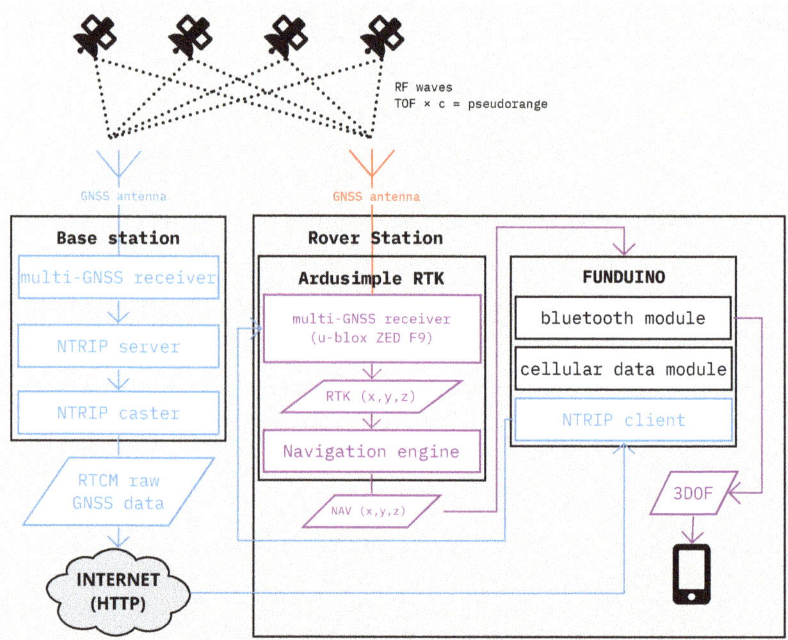

Figure 12. RTK geolocation data integration from the Ardusimple RTK kit to a mobile device. This diagram details the setup for real-time kinematic positioning: A reference/base station broadcasts correction data (RTCM raw format) through an NTRIP caster to an NTRIP client (mobile application). The data are integrated by the multi-GNSS receiver, which corrects most of the GNSS biases, resulting in centimeter-accurate geolocation. The Funduino processor broadcasts the geolocation data (3DOF) to the mobile device through a Bluetooth module.

4.2.2. Data Fusion: SLAM and GNSS

Although geolocation data is of importance in the context of location-based AR, orientation tracking and the pitch, yaw (also known as azimuth or heading) and roll data may play an equally important role in the accurate positioning and anchoring of augmented objects. Azimuth in particular is known to be relatively inaccurate on mobile devices' built-in compasses [54]. This is due to various causes ranging from magnetic anomalies caused by electronic devices, ferrous materials, and electrical or mechanical infrastructures. As a result, an augmented object will start diverging from its intended position in the AR interface as the user moves away from its exact location, even when used in combination with accurate geolocation data. Inversely, the placement of an augmented object in the interface will become increasingly accurate as the user approaches its geographic coordinate. Because any small-scale microelectromechanical systems (MEMS) magnetic field sensor (magnetometer) will suffer the same perturbations, the use of a more accurate, external magnometer module does not seem to be an option. Instead, we are laying down a concept for data fusion using SLAM data and geolocation data to improve all at once the positioning and anchoring of the virtual layer. SLAM is the process of simultaneously mapping moving objects with optical sensors, achieving self-localization and modeling the surrounding environment while in motion. When a user moves while holding a device running a WebXR-powered system in AR mode, they keep track of their displacement, relative to a *local* origin (the place where AR mode was initiated) with great accuracy. Simultaneously, they are able to estimate their *absolute* geolocation with the Geolocation API, based on the device's GNSS sensors or an external RTK module. As soon as at least seven (3D) coordinates have been measured in both datums (*local* and *global*), it is possible to use a geometric similarity transformation method (also known as seven-parameter transformation or Helmert transformation) to derive a rotation matrix and three translation vectors. These allow us to seamlessly convert coordinates from one datum to another. This would enable continuous updates on the position of the augmented space as new measurements are made. Thanks to the rotation matrix, the heading data would be generated without relying on the magnetometers, but only from positions measured in both datums. This could potentially solve both the geolocation and heading accuracy issues at once. The implementation of this concept is currently in a pilot phase, in collaboration with Geographic Information System (GIS) specialists from the Territorial Engineering Institute (INSIT).

5. Proposed Methodology for Evaluation

One of our project's research goals (1.2) is to try and improve the overall user experience of location-based AR interfaces so that they may achieve their full potential when it comes to nature exploration and biodiversity education. In the *BioSentiers* experiment, we observed that the geolocation data caused imprecise placement of the augmented objects in the interface, and that users seemed to spend a considerable amount of time interacting with the screen only to redirect themselves. We want to determine if the use of more accurate geolocation data provided by RTK positioning might help stabilize the AR interface and if it has an impact on usability, exploration, and interaction time with the screen.

Comparative User Test: Location-Based AR Combined with GNSS Versus RTK

Between November and December 2022, we conducted a comparative user test (n = 54) with two groups. Half of the participants used the *BiodivAR* application in combination with the Ardusimple RTK kit, while the other half used it with data provided by the devices' embedded GNSS sensor as a control group. The participants used the app to explore a biodiversity-themed ARLE for 15 min. During the test, in-app events and geolocated traces were recorded by the application. 47 of the participants also agreed to wear an eye-tracking device (Tobii Pro Glasses 3) that captured their gaze direction in order to measure how much time they spent interacting with the screen versus nature. Directly after the test, participants answered an online survey containing a demographic questionnaire, an open question, and three different usability questionnaires:

- *System Usability Scale* (SUS), [55] for a generic evaluation of the system.
- *User Experience Questionnaire* (UEQ), [56] for a comprehensive measure of user experience in terms of attractiveness, efficiency, reliability, stimulation, and novelty.
- *Handheld Augmented Reality Usability Scale* (HARUS), [57] a mobile AR-specific questionnaire.

With the data collected through the questionnaires, we intend to obtain an overall evaluation of the system we developed as well as more specific observations on the impact of different geolocation data. The eye-tracking data will allow us able to compare how much time users interacted with the screen versus nature within each group. The in-app events and geolocated traces will allow us to compute variables such as the total distance traveled, the time spent in or in-between POIs, and how long users have been using the interactive 2D map. We will investigate the role played by these independent variables (interaction time, total distance, amount of POIs visited, etc.) on user-reported usability by means of multiple linear regression. Thanks to this process, we expect to further observe the impact of geolocation data on usability. Finally, thanks to the unstructured feedback gathered through the open question, we are hoping to improve the *BiodivAR* application as much as possible before it is made available to a learning audience. The data collected during this test are currently being processed and the findings are not yet known. The design of the test is as visible in Figure 13.

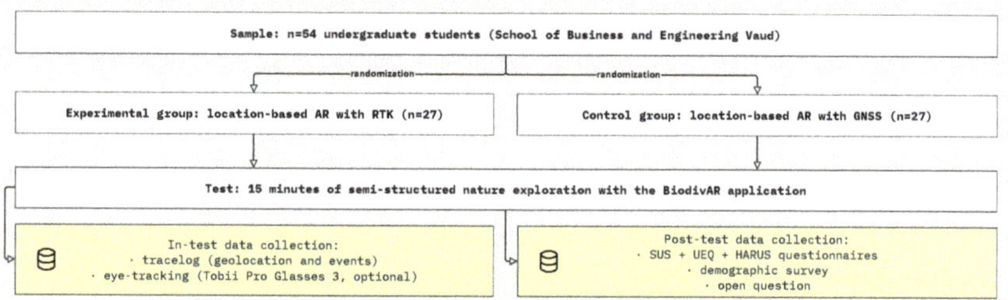

Figure 13. Comparative user test design: an experimental group used the *BiodivAR* application for 15 min in combination with RTK data while the control group used it with GNSS data. Various data were collected during and after the test to help observe the impact of geolocation data on usability.

6. Conclusions

In this paper, we present the UCD approach that has shaped the development of an original authoring application for location-based AR experiences. Its architecture handles various types of media for the customization of interactive, augmented points of interest triggered by user location. The system allows an immersive visualization of media in the geographical space, through a weighted use of geolocation, computer vision and inertial sensor data. Its end users may use it in various contexts such as tourism, education, infrastructure engineering, immersive cartography for navigation, geofencing, etc. The UCD approach presented allowed us to tailor the system to meet the end-users' needs while achieving the first two of our project's goals, namely the creation of a versatile authoring tool usable by non-specialists, and to use it to design engaging AR experiences featuring with geolocated media that populate the geographical space. While only a complete analysis of the comparative user test results will unlock the fulfillment of the remaining goals, the fact that we conducted the test keeps us on a steady track and allows us to expect quick results. Furthermore, running the test has put the system into practice, which was an enriching experience in itself. Considering previous efforts for location-based AR in education, several meaningful outcomes were achieved:

1. Firstly, the exclusive use of open-source frameworks makes it suitable for practices such as remote collaboration, co-creation and community development. These open-source tools and frameworks additionally comply with geographical and web standards, which enables data interoperability. In particular, the use of an actively emerging open web standard (WebXR) allows for easy maintenance and contributions from the web development community, which will allow the application to be progressively improved and enriched with new AR features. Beyond the open software elements that compose it, the system focuses on participation by the public by allowing anyone to contribute to the bioverse directory, making them discoverable and/or editable by other users, in the spirit of CS initiatives.

2. Secondly, focusing on problems and challenges identified in a previous—proof of concept—iteration [9] (AR usability; geolocation data accuracy; the lack of interaction with augmented objects; and the lack of a flexible authoring tool) allowed us to build on top of an existing knowledge base. By using a UCD methodology, we attempted

to tailor the application to fit the needs and expectations of our target users. We introduce the use of an external RTK module to try and address the inaccuracy of geolocation data and the usability issues it induces. While the previous prototype had hard-coded geolocated POIs, the current application allows the creation or editing of augmented objects by unskilled users. The cartographic interface with online editing features makes it possible for anyone without advanced programming skills to create AR experiences. The data structure we conceptualized enables the creation and editing of advanced, customizable and interactive geolocated POIs.

3. Finally, we conducted a comparative test to assess our system's usability and understand the impact of different geolocation data on usability. Eye-tracking data and geolocated traces will provide us with a multitude of unique points of view on how the application is actually used by participants. In a typical iterative design methodology, we will use user feedback to help us further refine and improve the application in a subsequent iteration.

In the coming months, we will process and analyze the results of the comparative user test we conducted. This will be informative in many ways, beginning with the relevance of using external RTK modules in our future experiments. As of now, we do not know whether the use of RTK data in combination with location-based AR has brought any enhancement in usability. If there is any difference between our two groups, they may still be statistically non-significant, because the anchoring of the POIs in the AR interface may also be impacted by the orientation data. Because we think the use of location-based AR in combination with standard GNSS data can cause usability problems and is the main limiting factor for broad adoption, our future efforts will focus on developing solutions to these. There are encouraging signs, such as the recent publication of Google's Geospatial API and Visual Positioning System (VPS), which is based on trillions of point clouds harvested over the last 15 years for the Google Street View service [58]. It retrieves accurate positioning by comparing the device's camera view with this huge database, after initial filtering based on the user's approximate geolocation. While this approach is very promising for solving the anchoring accuracy problem, it will only be available in locations where Google Street View cars were able to collect data. There will thus always remain large non-urban areas where this method will be unavailable. We therefore remain hopeful that the development of a fusion algorithm for SLAM and GNSS data will meet a certain type of need.

Following the analysis of our initial comparative user test results, we have planned to conduct a new test with pupils to evaluate the effectiveness of location-based AR in supporting nature exploration and biodiversity education. We are currently recruiting middle-school teachers to that end and we plan to begin the tests from June 2023 onward.

Several questions remain open that may provide guidance for future research. For example, we have not investigated the importance of visual variables for the differentiation of various POIs, although their impact on usability is documented [46]. When visualized on an AR interface, 3D cartographic symbols simulate their belonging to the user's spatial referential. Prior knowledge on the semiology of graphics may not apply in the same way it does to two-dimensional maps. While it is currently beyond our study's scope, we believe our tool would lend itself well to investigating the impact of cartographic design on usability. Researchers have made observations on the usefulness of various animated cartographic symbols by comparing objective data (effectiveness in completing a task) with expert opinion [59]. With a comparable approach in mind, it would be interesting for us to compare the usability scores obtained from self-reported user data with objective effectiveness, as measured by quantitative data generated by those same users (geolocated trace logs and eye-tracking data).

Author Contributions: Conceptualization, Julien Mercier; methodology, Julien Mercier; software, Nicolas Chabloz, Gregory Dozot and Julien Mercier; validation, Erwan Bocher, Olivier Ertz and Daniel Rappo; formal analysis, Julien Mercier; investigation, Julien Mercier; resources, Olivier Ertz and Daniel Rappo; data curation, Julien Mercier; writing—original draft preparation, Julien Mercier; writing—review and editing, Julien Mercier, Nicolas Chabloz, Olivier Ertz, Erwan Bocher and Daniel Rappo; visualization, Julien Mercier; supervision, Erwan Bocher, Olivier Ertz and Daniel Rappo; project administration, Olivier Ertz and Daniel Rappo; funding acquisition, Olivier Ertz and Daniel Rappo. All authors have read and agreed to the published version of the manuscript.

Funding: This research was funded by the Swiss National Science Foundation (SNSF) as part of the NRP 77 "Digital Transformation" (project number 407740_187313) and by the University of Applied Sciences and Arts Western Switzerland (HES-SO): Programme stratégique "Transition numérique et enjeux sociétaux".

Informed Consent Statement: Study participation was voluntary, and written informed consent to publish this paper was obtained from all participants involved in the study. Participants were informed that they could withdraw from the study at any point.

Data Availability Statement: The data presented in this study are openly available on Zenodo at https://doi.org/10.5281/zenodo.6542781 (accessed on 28 December 2022).

Acknowledgments: Thanks to Sébastien Guillaume and Kilian Morel of the Institute of Territorial Engineering (INSIT) for their help with implementing the RTK positioning systems and for their continuous collaboration on geographic information science matters.

Conflicts of Interest: The authors declare no conflict of interest. The funders had no role in the design of the study; in the collection, analyses, or interpretation of data; in the writing of the manuscript, or in the decision to publish the results.

Abbreviations

The following abbreviations are used in this manuscript:

API	Application Programming Interface
AR	Augmented Reality
ARLE	Augmented Reality Learning Environment
CS	Citizen science
GNSS	Global Navigation Satellite System
JSON	JavaScript object notation
ORM	Object–relational mapping
POI	Point of interest
RTCM	Radio Technical Commission for Maritime Services
RTK	Real-time kinematic positioning
SLAM	Simultaneous localization and mapping
UCD	User-Centered Design
UI	User Interface
UX	User experience
VPS	Visual Positioning System
WGS	World Geodetic System
3DOF	Three degrees of freedom

References

1. Rubin, J.; Chisnell, D. *Handbook of Usability Testing: How to Plan, Design, and Conduct Effective Tests*, 2nd ed.; Wiley Pub: Indianapolis, IN, USA, 2008.
2. Cubillo, J.; Martin, S.; Castro, M.; Boticki, I. Preparing augmented reality learning content should be easy: UNED ARLE-an authoring tool for augmented reality learning environments: *Comput. Appl. Eng. Educ.* **2015**, *23*, 778–789. [CrossRef]
3. Rosenbaum, E.; Klopfer, E.; Perry, J. On Location Learning: Authentic Applied Science with Networked Augmented Realities. *J. Sci. Educ. Technol.* **2007**, *16*, 31–45. [CrossRef]
4. Chiang, T.H.C.; Yang, S.J.H.; Hwang, G.J. An Augmented Reality-based Mobile Learning System to Improve Students' Learning Achievements and Motivations in Natural Science Inquiry Activities. *J. Educ. Technol. Soc.* **2014**, *17*, 352–365.

5. Dunleavy, M.; Dede, C.; Mitchell, R. Affordances and Limitations of Immersive Participatory Augmented Reality Simulations for Teaching and Learning. *J. Sci. Educ. Technol.* **2009**, *18*, 7–22. [CrossRef]
6. Facer, K.; Joiner, R.; Stanton, D.; Reid, J.; Hull, R.; Kirk, D. Savannah: Mobile gaming and learning?: Mobile gaming and learning. *J. Comput. Assist. Learn.* **2004**, *20*, 399–409. [CrossRef]
7. Perry, J.; Klopfer, E.; Norton, M.; Sutch, D.; Sandford, R.; Facer, K. AR gone wild: Two approaches to using augmented reality learning games in Zoos. In Proceedings of the 8th on International conference for the Learning Sciences, Utrecht, The Netherlands, 24–28 June 2008; Volume 3, pp. 322–329.
8. Ryokai, K.; Agogino, A. Off the paved paths: Exploring nature with a mobile augmented reality learning tool. *J. Mob. Hum. Comput. Interact.* **2013**, *5*, 21–49. [CrossRef]
9. Ingensand, J.; Lotfian, M.; Ertz, O.; Piot, D.; Composto, S.; Oberson, M.; Oulevay, S.; Da Cunha, M. Augmented reality technologies for biodiversity education—A case study. In Proceedings of the 21st Conference on Geo-Information Science, AGILE, Lund, Sweden, 12–15 June 2018.
10. Admiraal, W.; Huizenga, J.; Akkerman, S.; Dam, G.T. The concept of flow in collaborative game-based learning. *Comput. Hum. Behav.* **2009**, *27*, 1185–1194. [CrossRef]
11. Lee, G.; Duenser, A.; Kim, S.; Billinghurst, M. CityViewAR: A Mobile Outdoor AR Application for City visualization. In Proceedings of the 2012 IEEE International Symposium on Mixed and Augmented Reality—Arts, Media, and Humanities (ISMAR-AMH), Atlanta, GA, USA, 5–8 November 2012; 64p. [CrossRef]
12. Azuma, R.T. A Survey of Augmented Reality. Presence: Teleoperators and Virtual Environ. *Presence: Teleoperators and Virtual Environments* **1997**, *6*, 355–385. [CrossRef]
13. Geroimenko, V. *Augmented Reality in Education: A New Technology for Teaching and Learning*; Springer International Publishing: Berlin/Heidelberg, Germany, 2020.
14. Lee, S.; Choi, J.; Park, J.I. Interactive E-Learning System Using Pattern Recognition and Augmented Reality. *IEEE Trans. Consum. Electron.* **2009**, *55*, 883–890. [CrossRef]
15. Alnagrat, A.; Ismail, R.; Syed Idrus, S.Z. A Review of Extended Reality (XR) Technologies in the Future of Human Education: Current Trend and Future Opportunity. *J. Hum. Reprod. Sci.* **2022**, *1*, 81–96. [CrossRef]
16. Martin, S.; Diaz, G.; Sancristobal, E.; Gil, R.; Castro, M.; Peire, J. New technology trends in education: Seven years of forecasts and convergence. *Comput. Educ.* **2011**, *57*, 1893–1906. [CrossRef]
17. Antonioli, M.; Blake, C.; Sparks, K. Augmented Reality Applications in Education. *J. Technol. Stud.* **2014**, *40*, 96–107. [CrossRef]
18. Diegmann, P.; Schmidt-Kraepelin, M.; Eynden, S.; Basten, D. Benefits of Augmented Reality in Educational Environments— A Systematic Literature Review. *Wirtsch. Proc.* **2015**, *2015*, 103.
19. Bacca, J.; Baldiris, S.; Fabregat, R.; Graf, S.; Kinshuk. Augmented Reality Trends in Education: A Systematic Review of Research and Applications. *J. Educ. Technol. Soc.* **2014**, *17*, 133–149.
20. Garzón, J.; Acevedo, J. Meta-analysis of the impact of Augmented Reality on students' learning gains. *Educ. Res. Rev.* **2019**, *27*, 244–260. [CrossRef]
21. Georgiou, Y.; Kyza, E.A. Relations between student motivation, immersion and learning outcomes in location-based augmented reality settings. *Comput. Hum. Behav.* **2018**, *89*, 173–181. [CrossRef]
22. Cabiria, J. Augmenting Engagement: Augmented Reality in Education. *Cut. Edge Technol. High. Educ.* **2012**, *6*, 225–251. [CrossRef]
23. Dede, C. Immersive Interfaces for Engagement and Learning. *Science* **2009**, *323*, 66–69. [CrossRef]
24. Raber, J.; Ferdig, R.; Gandolfi, E.; Clements, R. An analysis of motivation and situational interest in a location-based augmented reality application. *Interact. Des. Archit.* **2022**, *52*, 198–220. [CrossRef]
25. Medyńska-Gulij, B.; Zagata, K. Experts and Gamers on Immersion into Reconstructed Strongholds. *ISPRS Int. J. Geo-Inf.* **2020**, *9*, 655. [CrossRef]
26. Arvola, M.; Fuchs, I.E.; Nyman, I.; Szczepanski, A. Mobile Augmented Reality and Outdoor Education. *Built Environ.* **2021**, *47*, 223–242. [CrossRef]
27. Chiang, T.H.C.; Yang, S.J.H.; Hwang, G.J. Students' online interactive patterns in augmented reality-based inquiry activities. *Comput. Educ.* **2014**, *78*, 97–108. [CrossRef]
28. Barclay, L. Acoustic Ecology and Ecological Sound Art: Listening to Changing Ecosystems. In *Sound, Media, Ecology*; Droumeva, M., Jordan, R., Eds.; Palgrave Studies in Audio-Visual Culture; Springer International Publishing: Berlin/Heidelberg, Germany, 2010; pp. 153–177. [CrossRef]
29. O'Shea, P.M.; Dede, C.; Cherian, M. Research Note: The Results of Formatively Evaluating an Augmented Reality Curriculum Based on Modified Design Principles. *Int. J. Gaming Comput. Mediat. Simulat.* **2011**, *3*, 57–66. [CrossRef]
30. Bloom, M.A.; Holden, M.; Sawey, A.T.; Weinburgh, M.H. Promoting the Use of Outdoor Learning Spaces by K-12 Inservice Science Teachers Through an Outdoor Professional Development Experience. In *The Inclusion of Environmental Education in Science Teacher Education*; Bodzin, A.M., Shiner Klein, B., Weaver, S., Eds.; Springer: Dodreht, The Netherlands, 2010; pp. 97–110. [CrossRef]
31. Capps, D.K.; Crawford, B.A. Inquiry-Based Instruction and Teaching About Nature of Science: Are They Happening? *J. Sci. Teach. Educ.* **2013**, *24*, 497–526. [CrossRef]
32. Bitter, G. The Pedagogical Potential of Augmented Reality Apps. *Int. J. Eng. Sci. Invent.* **2014**, *3*, 13–17.

33. Rauschnabel, P.A.; Rossmann, A.; tom Dieck, M.C. An adoption framework for mobile augmented reality games: The case of Pokémon Go. *Comput. Hum. Behav.* **2017**, *76*, 276–286. [CrossRef]

34. Althoff, T.; White, R.W.; Horvitz, E. Influence of Pok\'emon Go on Physical Activity: Study and Implications. *arXiv* **2016**, arXiv:1610.02085. https://doi.org/10.48550/arXiv.1610.02085.

35. Bressler, D.M.; Bodzin, A.M. A mixed methods assessment of students' flow experiences during a mobile augmented reality science game. *J. Comput. Assist. Learn.* **2013**, *29*, 505–517. [CrossRef]

36. Debandi, F.; Iacoviello, R.; Messina, A.; Montagnuolo, M.; Manuri, F.; Sanna, A.; Zappia, D. Enhancing cultural tourism by a mixed reality application for outdoor navigation and information browsing using immersive devices. *IOP Conf. Ser. Mater. Sci. Eng.* **2018**, *364*, 012048. [CrossRef]

37. Alakärppä, I.; Jaakkola, E.; Väyrynen, J.; Häkkilä, J. Using nature elements in mobile AR for education with children. In Proceedings of the 19th International Conference on Human–Computer Interaction with Mobile Devices and Services, Association for Computing Machinery, MobileHCI '17, Vienna, Austria, 4–7 September 2017; pp. 1–13. [CrossRef]

38. Eliasson, J.; Knutsson, O.; Ramberg, R.; Cerratto-Pargman, T. Using Smartphones and QR Codes for Supporting Students in Exploring Tree Species. In *Proceedings of the Scaling up Learning for Sustained Impact*; Lecture Notes in Computer Science; Hernández-Leo, D., Ley, T., Klamma, R., Harrer, A., Eds.; Springer: Berlin/Heidelberg, 2013; pp. 436–441. [CrossRef]

39. Goth, C.; Frohberg, D.; Schwabe, G. The Focus Problem in Mobile Learning. In Proceedings of the 2006 Fourth IEEE International Workshop on Wireless, Mobile and Ubiquitous Technology in Education (WMTE'06), Athens, Greece, 16–17 November 2006; pp. 153–160. [CrossRef]

40. Cimbali, F. Adaptation de l'application BioSentiers pour une utilisation pédagogique [Bachelor thesis]. School of Engineering and Management Vaud, HES-SO University of Applied Sciences and Arts Western Switzerland. 2018. Available online: http://tb.heig-vd.ch/6531 (accessed on 28 December 2022).

41. Potvin, P.; Hasni, A. Analysis of the Decline in Interest Towards School Science and Technology from Grades 5 Through 11. *J. Sci. Educ. Technol.* **2014**, *23*, 784–802. [CrossRef]

42. Lenhart, A.; Ling, R.; Campbell, S.; Purcell, K. *Teens and Mobile Phones*; Technical Report; Pew Research Center's Internet & American Life Project: Washington, DC, USA, 2010.

43. Tripp, S.; Bichelmeyer, B. Rapid Prototyping: An Alternative Instructional Design Strategy. *Educ. Technol. Res. Dev.* **1990**, *38*, 31–44. [CrossRef]

44. Gagnon, D. ARIS: An Open Source Platform for Developing Mobile Learning Experiences [Computer Software]. Madison, WI: Field Day Lab. 2020 Available online: https://fielddaylab.org/make/aris/ (accessed on 28 December 2022).

45. Mercier, J.; León, L. edTech, réalité augmentée et exploration. *Educateur* **2021**, *10*. Available online: https://www.le-ser.ch/lecole-dehors-vers-une-pedagogie-du-rapport-au-monde (accessed on 28 December 2022).

46. Halik, Ł.; Medyńska-Gulij, B. The Differentiation of Point Symbols using Selected Visual Variables in the Mobile Augmented Reality System. *Cartogr. J.* **2017**, *54*, 147–156. [CrossRef]

47. Carpignoli, N. AR.js Studio: An Authoring Platform to Build Web Augmented Reality Experiences [Computer Software]. 2020. Available online: https://ar-js-org.github.io/studio/ (accessed on 28 December 2022).

48. Chabloz, N. LBAR.js: A Minimalist Library for Creating WebXR Location-Based Markers with A-Frame [Computer software]. 2020. Available online: https://github.com/MediaComem/LBAR.js/ (accessed on 28 December 2022).

49. Butler, H.; Daly, M.; Doyle, A.; Gillies, S.; Schaub, T.; Hagen, S. The GeoJSON Format. Request for Comments RFC 7946, Internet Engineering Task Force, 2016. p. 28. [CrossRef]

50. GPS.gov: Official U.S. Government Information about the Global Positioning System (GPS) and Related Topics [Internet Website]. Available online: https://www.gps.gov (accessed on 28 December 2022).

51. Abdi, E.; Mariv, H.S.; Deljouei, A.; Sohrabi, H. Accuracy and precision of consumer-grade GPS positioning in an urban green space environment. *For. Sci. Technol.* **2014**, *10*, 141–147. [CrossRef]

52. Morar, A.; Băluțoiu, M.A.; Moldoveanu, A.; Moldoveanu, F.; Butean, A.; Asavei, V. Evaluation of the ARCore Indoor Localization Technology. In Proceedings of the 2020 19th RoEduNet Conference: Networking in Education and Research (RoEduNet), Bucharest, Romania, 11–12 December 2020; pp. 1–5. ISSN: 2247-5443. [CrossRef]

53. ArduSimple: RTK Handheld Kit [Internet Website]. Available online: https://www.ardusimple.com/product/rtk-handheld-surveyor-kit/ (accessed on 28 December 2022).

54. Renaudin, V.; Afzal, M.; Lachapelle, G. Magnetic perturbations detection and heading estimation using magnetometers. *J. Locat. Based Serv.* **2012**, *6*, 161–185. [CrossRef]

55. Brooke, J. Usability Evaluation In Industry. *SUS: A 'Quick and Dirty' Usability Scale*; CRC Press: Boca Raton, FL, USA, 1996; pp. 207–212. [CrossRef]

56. Construction and Evaluation of a User Experience Questionnaire. Available online: Available online: https://www.researchgate.net/publication/221217803_Construction_and_Evaluation_of_a_User_Experience_Questionnaire (accessed on 28 December 2022).

57. Santos, M.E.; Polvi, J.; Taketomi, T.; Yamamoto, G.; Sandor, C.; Kato, H. Toward Standard Usability Questionnaires for Handheld Augmented Reality. *IEEE Comput. Graph. Appl.* **2015**, *35*, 66–75. [CrossRef]

58. Google. Build Global-Scale, Immersive, Location-Based AR Experiences with the ARCore Geospatial API [Computer Software]. 2022. Available online: https://developers.google.com/ar/develop/geospatial (accessed on 28 December 2022).
59. Medyńska-Gulij, B.; Wielebski, Ł.; Halik, Ł.; Smaczyński, M. Complexity Level of People Gathering Presentation on an Animated Map—Objective Effectiveness Versus Expert Opinion. *ISPRS Int. J. Geo-Inf.* **2020**, *9*, 117. [CrossRef]

International Journal of
Geo-Information

Article

Usefulness of Plane-Based Augmented Geovisualization—Case of "The Crown of Polish Mountains 3D"

Łukasz Halik * and Łukasz Wielebski

Adam Mickiewicz University, Poznań, Poland, Department of Cartography and Geomatics,
61-712 Poznań, Poland
* Correspondence: lukasz.halik@amu.edu.pl; Tel.: +48-61-829-6246

Abstract: In this article, we suggest the introduction of a new method of generating AR content, which we propose to call plane-based augmented geovisualizations (PAGs). This method concerns cases in which AR geovisualizations are embedded directly on any plane detected by the AR device, as in the case of the investigated "Crown of Polish Mountains 3D" application. The study on the usefulness of the AR solution against a classic solution was conducted as part of an online survey of people from various age and social groups. The application in the monitor version showing 3D models of mountain peaks (without AR mode) was tested by the respondents themselves. The use of the application in the AR mode, which requires a smartphone with the appropriate module, was tested by the respondents based on a prepared video demonstrating its operation. The results of the research on three age groups show that the AR mode was preferred among users against all compared criteria, but some differences between age groups were clearly visible. In the case of the criterion of ease of use of the AR mode, the result was not so unambiguous, which is why further research is necessary. The research results show the potential of the AR mode in presenting 3D terrain models.

Keywords: mountain geovisualizations; augmented reality; mobile devices; 3D visualization of landscapes; geomedium efficiency; relief representation; cartographic products

Citation: Halik, Ł.; Wielebski, Ł. Usefulness of Plane-Based Augmented Geovisualization—Case of "The Crown of Polish Mountains 3D". *ISPRS Int. J. Geo-Inf.* **2023**, *12*, 38. https://doi.org/10.3390/ijgi12020038

Academic Editors: Wolfgang Kainz and Florian Hruby

Received: 2 December 2022
Revised: 11 January 2023
Accepted: 17 January 2023
Published: 22 January 2023

1. Introduction

Along with the development of technology and the introduction of mobile devices, such as tablets or smartphones, to the market, new opportunities to use maps and the derivatives of cartographic visualizations that apply the third dimension to present spatial data have arisen [1–3]. New map products have been emerging as a result of technological advances. Digital geospatial representation has been evolving in three areas: three-dimensional (3D) spaces, real-time dynamics, and fusion of virtual and real objects [4,5]. Augmented reality (AR) has similar characteristics. The concept of AR was first introduced by scientists working for Boeing Company in the early 1990s to assist in mechanical assembly [6]. The important step in laying the theoretical foundation for the augmented reality system was to create a reality–virtuality continuum [7] in which AR could find its place. This paved the way for a further process of defining three fundamental features of AR by Azuma [8]: combining real and virtual, interaction in real time, and registered in 3D.

From a geographical point of view, visualizations that use AR can be divided into: augmented virtual environments (AVEs) or as augmented geographic reality (AGR) [9]. According to Cheng et al. [9] (p. 427), AGR should be divided, based on field experiences and maps, into augmented reality environments (AREs) and augmented maps (AMs), respectively. In AREs, visual information (e.g., digital tags) is added to objects in the real world to perform functions, such as navigation and illustration. AMs are maps on which multiple types of geographic information (e.g., models and multimedia files) are superimposed to enhance cartographic information transfer and users' spatial cognitive ability. AMs are a subgroup of cartographic products that use the marker-based tracking method [10].

The campus tour system prototype by Feiner et al. [11] is referred to as the earliest prototype of the ARE. The opportunities offered by ARE were also examined in the context of using classic visual variables [12] and those that were gradually appearing as a result of technological advance related to digital cartography [13] for designing signatures that demonstrate the distance of the topographic object from the observer [14]. Another example of an ARE is provided by the research of Fukuda et al. [15], who demonstrated a solution that uses handheld augmented reality systems for urban landscape simulation. Kourouthanassis et al. [16] presented the use of mobile AR on the smartphone, testing a travel guide named CorfuAR.

In 2002, Bobrich and Otto [17], for the first time, developed AMs that integrated paper maps and virtual geographic information. In this system, the map becomes the reference frame of multiuser interaction, and cards with a quick response (QR) code become the identification anchor points of virtual interaction tools. When listing examples of the opportunities that augmented maps (AMs) bring, one also needs to mention the research by Morrison et al. [18], in which the employment of AR technology in combination with traditional paper maps was compared. De Almeida Pereira et al. [19] noted that adding 3D AR representation at the top of a paper map can enhance users' abilities to perform spatial positioning and to read map data.

Two aforementioned methods of generating AR content (ARE and AM) fail to consider cases in which AR geovisualizations are located on a freely chosen plane detected by the AR device. It was possible to single out the new method of generating AR content thanks to the progress in the evolution of algorithms of detecting planes in the real world. For this reason, we suggested introducing the third group of geovisualizations, known as plane-based augmented geovisualization (PAG), to the classification system. In such solutions, geovisualization is mounted on a freely selected plane identified by the mobile device. It releases the user from the necessity of handling a physical object (e.g., a map) that displays virtual content on its screen. PAGs are a subgroup of cartographic products that use the feature-based tracking method [10].

Until recently, AR solutions have been created and shared mainly as applications uploaded from dedicated stores and installed directly in the memory of the smartphone. Advances in technology have allowed one to establish a set of *webXR standards* that use the opportunities of web browsers of mobile devices. *WebXR* is a group of standards that are used together to support rendering 3D scenes to hardware designed for presenting virtual worlds (virtual reality, VR) or for adding graphical imagery to the real world (augmented reality, AR) on the Web (https://developer.mozilla.org/en-US/docs/Web/API/WebXR_Device_API accessed on 18 November 2022) [20].

Numerous currently available models of smartphones are equipped to handle webXR, which allows one to assume that in the near future the number of users of such solutions is going to grow. It is also the reason why such applications should be developed in cartography as well. Nevertheless, similar assumptions should be supported by the research.

In the literature, one may encounter different solutions used for cartographic visualizations, starting with static 2D ones, surface three-dimensional, and ending with interactive ones [21]. The examples of publications that present land relief with the use of 3D and AR are as follows: studies by Siqueira [22] that apply this technology to teaching topographic surfaces; papers by Templin et al. [23] in which it is used as a supporting tool in "inland and coastal water zones" navigation; works by Brejcha et al. [24] that create augmented photographs resulting from the combination of DEM and a photo taken in the field. The research on moving around the virtual world by walking and teleportation is also known [25].

The usefulness of 3D geovisualization in the AR technology displayed with the use of the PAG method was the topic discussed in this research. In the research, the usefulness was defined by examining the preferred by users' way in which the landscape and topographic traits were explored for the collection of mountain peaks, visualized in AR and using the PAG method, and without the AR mode, within one "Korona Gór Polski 3D" (The Crown of Polish Mountains 3D) application. The application tested included a 3D representation of the collection of mountain peaks, the so-called crown of mountains, i.e., the highest mountain tops in a given country, whose review and exploration is possible through the Web browser, on the smartphone (in AR), or, more conventionally, on the PC (without AR technology). The subjective feelings and impressions that application users had during the process of using both modes (AR/no AR) may decide upon what solution will be chosen more willingly; thus, we decided to test how respondents evaluated those two opportunities.

2. Aim and Questions

The main objective of the research was to determine the preferences of users from different age groups related to the opportunity to watch a 3D model of the *Tarnica* mountain peak onscreen in the AR mode (using the PAG method) and without AR. To meet this objective, researchers prepared an online survey, which allowed them to test the application, based on a set of objective and subjective questions, and ask users' opinion concerning it.

An online survey is a method of testing used on application users that offers the opportunity to reach the largest group of respondents of different ages and from different backgrounds. The survey proves to be a useful tool for obtaining opinions concerning the tested cartographic products regardless of whether researchers need the opinions of experts or amateurs [26,27]. Based on the appropriate structure of surveys, one may obtain answers from users related to their subjective evaluation of cartographic visualizations and objective effectiveness determined on the basis of the tasks completed [28]. In this research, we focused on subjective feelings that matter significantly because, for the large part, they determine whether or not a specific cartographic product meets with interest and is going to be willingly used. To make it more precise, the objective of this research raised the following questions:

— How do users evaluate applications with a 3D geovisualization displayed in AR and without AR? Which mode functions better in the specific comparative criteria (convenience and simplicity of operating the 3D model; effectiveness related to the opportunity to scrutinize the 3D model more thoroughly; attractiveness of the 3D model's exploration; opportunity to assess differences in relative terrain elevation differences, based on the 3D model)?
— Is there any correlation between the age of respondents and the way they perceive both available, comparable solutions (i.e., modes with AR and without AR)? can any principle be observed in this context?

3. Materials and Methods

To meet the objective and answer the above questions, we adopted six main research stages:

— To create and make publicly available the tested application (https://kgp3d.amu.edu.pl accessed on 29 November 2022) (Section 3.1);
— To schedule simple tasks for users to familiarize themselves with the application in the screen mode (i.e., without AR) (Section 3.2);
— To produce a demonstration video (movie) showing the use of the tested application on a smartphone in the AR mode (Section 3.2);
— To prepare an internet questionnaire and carry out surveys among the users (Section 3.2);
— To choose age groups for the result analysis (Section 3.3);
— To present a statistical analysis of the results (Section 4).

3.1. The "Korona Gór Polski 3D" Application

The "Korona Gór Polski 3D" (The Crown of the Polish Mountains 3D) application is an original application developed by the authors. The application uses webXR standards that make it work directly in the web browser, without the necessity of installing files. It is available to anybody, free of charge, at the following link: https://kgp3d.amu.edu.pl (accessed on 29 November 2022) The purpose of the application is to demonstrate 28 3D models of mountain peaks on the territory of Poland, located in different mountain ranges referred to as "Korona Gór Polski" [29,30]. The application may be used on a PC or smartphone screen. Figure 1 presents an example of the view in the application on the screen of the mobile device. Figure 1a shows the application's welcome screen, with the option to select a 3D model of the mountain peak, 1b demonstrates a given peak with the option to display it in the AR mode and 1c presents the same peak in the AR mode.

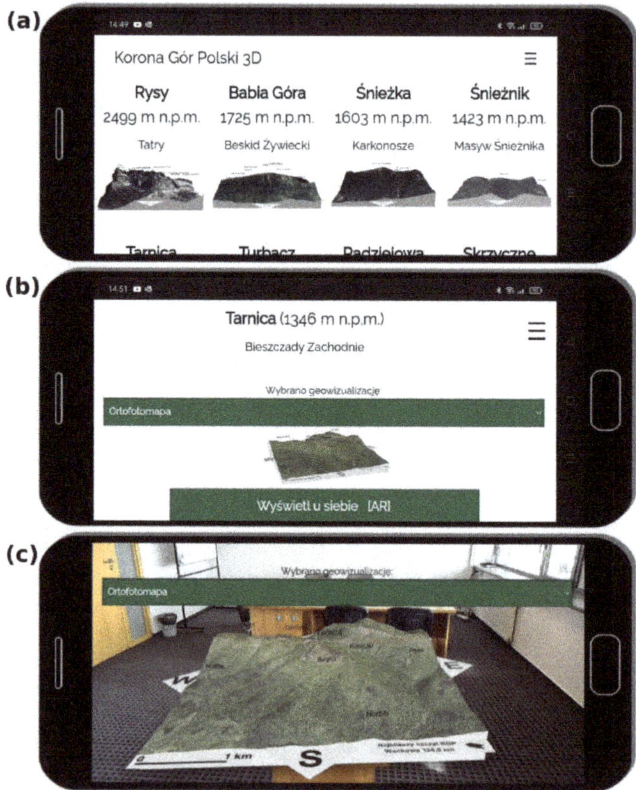

Figure 1. View of the tested application displayed on the screen of a smartphone (**a**) with the initial page enabling the selection of a mountain peak; (**b**) the 3D model of the *Tarnica* peak displayed without the AR mode; (**c**) the 3D model of the *Tarnica* peak located on the top of a coffee table in the AR mode using the PAG method.

Technologically, the application is based on the *model-viewer* library (https://modelviewer.dev accessed on 27 November 2022) [31]. It offers the opportunity to view 3D models of specific mountain peaks, both on a monitor and smartphone screen, in the classic mode or in AR, provided that the device is equipped with the ARCore module. The application has the same functions both on PCs and mobile devices, but the there are differences in how the 3D model is navigated on the mobile device with a touch screen vs. on the PC. Those differences are presented in Table 1.

Table 1. Comparison of the operations and capabilities of the tested application according to three modes of use: on a computer screen (PC), on a smartphone without AR, on a smartphone with AR.

Type and Characteristics of the Device	PC (Personal Computer) *Desktop Mode*	Smartphone *without AR Mode*	Smartphone *with AR Mode*
Screen (display) type	Traditional screen (non-touch)	Touch screen (or a tablet/laptop with a larger touch screen)	Touch screen with AR image superimposed
Screen size	Big	Small	Small
Scaling the 3D model	Using the mouse scroll	Using two fingers to slide across the screen (increase/decrease the distance between the fingers)	
Rotation of the 3D model by 360 degrees in the horizontal plane	Hold down the left mouse button and move the mouse (360 degrees for the vertical axis)/use the left or right arrow keys on the keyboard	Drag your finger across the screen (swipe)	Using two fingers on the touchscreen in opposite directions (circling)
Rotation of the 3D model by 360 degrees in other planes	Hold down the left mouse button and move the mouse (2×180 degrees for the horizontal axis) / use the up or down arrow keys on the keyboard	Drag your finger across the screen (swipe)	Impossible (the 3D model keeps its position in the horizontal plane all the time)
Changing the position of the 3D model relative to the observer	By default, the 3D model is always in the center of the screen (hold down the right mouse button to drag to another place)	By default, the 3D model is always in the center of the screen (double quick tap on the screen moves the model)	Possible (you can display the 3D model in any part of the screen by swiping it with your finger on the touch screen)
The relationship of the observer and the 3D model	Observer position fixed—change of scale and rotation of the 3D model relative to the observer		A. The position of the observer is variable (moving) - the position of the 3D model remains constant in the horizontal plane. To see the model from a different angle or to move closer/ farther away from it, the observer must change his position B. Fixed position of the observer - the observer can change the position of the 3D model on the screen, but only in the horizontal plane, rotate the model around its vertical axis or scale the model
3D model exploration	The observer can view the model only at a distance (specified maximum magnification level)		The observer has the ability to enter the area occupied by the model

The first difference is related to the size of the screen on which the 3D model is presented. For PC monitors, the difference is much greater than for smartphones. Only some monitors are equipped with a touch screen (i.e., interactive monitor) function, which is actually the fundamental feature of smartphones [32]. To change the scale of the 3D model on the classic monitor, one needs to use the scroll mouse and on the smartphone a two-finger move (changing the distance between them increases or decreases the scale). In the application displayed on the PC monitor, the rotation of the 3D model in the application used on the smartphone, in the no AR mode, is related to the scrolling move of the finger on the screen, whereas in the AR mode, the rotation is possible with the use of two fingers and making a circular movement with them.

When it comes to the opportunity to change the location of the 3D model in relation to the observer, the model remains in the center of the monitor for the whole time by default. As far as the smartphone (without AR) is concerned, a quick double tap on the screen moves the model. In the AR mode, the user may move the 3D model freely, moving the finger on the touchscreen to a freely selected part of the screen. When the application is used on a PC monitor, the location of the model remains fixed in relation to the observer like on the smartphone without the AR mode. The AR mode is related to two types of the observer–3D model relation. The first one assumes a fixed location of the operator in relation to the model. It is actually a situation, in which, initially, users find a plane on which they want to display the model, rotating and scaling it to adjust it to the plane appropriately. The second type of the observer–3D model in the AR mode relation is directly related to viewing the 3D model in that mode after it is mounted on the selected plane.

The application's modes of use differ in the ways the model may be explored. In the monitor mode, the observer may watch the model from a distance like on the smartphone without AR. However, when the AR mode is on, the observer has an opportunity to enter the model, i.e., by making a large close-up or by stepping onto the plane the model is mounted on if the model is on the floor.

3.2. The Subject of the Study and the Online Survey

The 3D application, presented in Section 3.1, works on mobile devices in the AR mode or without the AR mode (compare with Table 1). However, to be able to collect unified data, independent of the parameters of the mobile device, we decided that respondents would study the following material: the application that was displayed on the PC screen without AR and the demonstrational video on using the application on the smartphone in the AR mode. This approach allowed us to carry out the study on a bigger group of respondents. Regardless of the technical abilities of the mobile device that the respondents had, it was how they could see how to use the application with the AR technology with the PAG method applied. The preferences of the users who selected either the AR mode or no AR mode while answering questions, according to the traits that were included in the comparative criteria, constituted the object of research. In order to collect all of the data, an internet survey was designed.

The research survey was constructed in Limesurvey (internet survey tool) and consisted of four metric questions and two task sections with 6 subjective questions, based on the Likert scale [33]. Firstly, the data regarding respondents were collected (sex and age). In addition, respondents were asked about the AR term and previous experience related to the application that used that technology. The interaction between the user and the 3D model of the mountain peak *Tarnica* without AR (Figure 2b) was the next step. This was the rotation of the peak, closing it up and switching on and off the layer that showed the area visible from the peak. This was supposed to familiarize users with the way the application (and some selected elements of its functioning) is handled.

Figure 2a depicts the welcome screen of the application displayed on the monitor in the web browser in which respondents had to select a mountain peak from the list. Figure 2b presents a view of the selected 3D model.

Then, the respondents were asked to watch a video that showed how to use the application in the AR mode (Figure 3). Due to the registration of the real image and of the image displayed on the smartphone screen, functioning in the AR technology, each respondent had an opportunity to see what operating the 3D model looked like and what opportunities to view/operate it in that mode were available (https://youtu.be/E_t8ZU2OtMY accessed on 29 November 2022). To demonstrate how the AR mode worked, a Samsung Galaxy S20 smartphone was used.

Figure 2. (**a**) The welcome view of the tested application window with a list of peaks to choose from; (**b**) the view of the selected three-dimensional model of the *Tarnica* peak in the web browser displayed on the PC monitor.

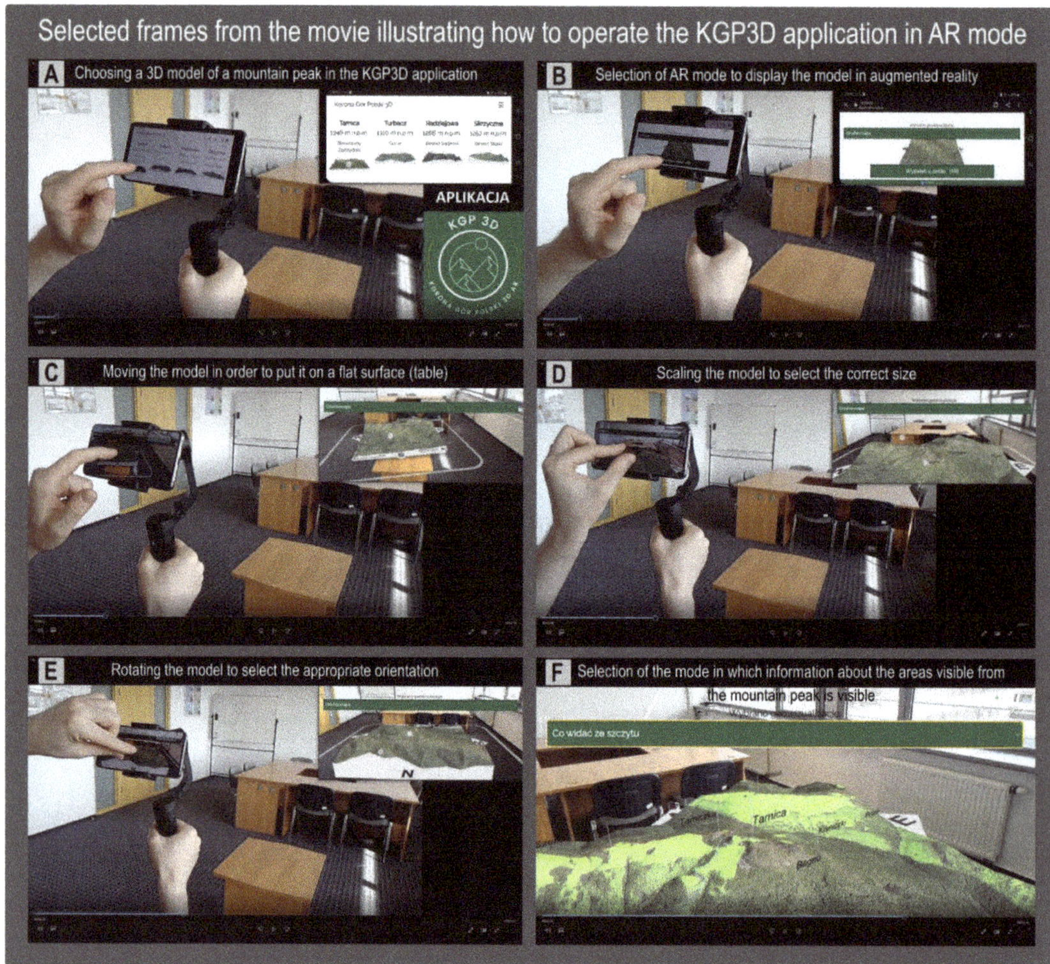

Figure 3. Selected frames from the video attached to the survey questionnaire showing the use of the tested application and activities (**A–F**) performed by the operator on a smartphone equipped with the AR mode (https://youtu.be/E_t8ZU2OtMY accessed on 28 November 2022)—compare with Figure 1a–c.

In the video, which was the main part of the test survey (Figure 3), the view of the real image, along with the view of the 3D model laid on the real-world view and visible on the smartphone screen, was registered. It was also possible to see the activities performed by the operator. Selected frames of the video are presented in the Figure 3.

In the first shot from the video (A), one can see the moment of the selection of the *Tarnica* mountain peak (1346 m above sea level). Shot B presents the moment of starting the AR mode in which the mountain peak is going to be displayed. Shot C shows the operator moving the 3D model to place it in the appropriate position on a flat surface. In shot D, one can see how the operator is scaling the model to adjust its size to the one required. Shot E demonstrates that the operator is rotating the model around its own axis. In shot F, one can clearly see that the operator has moved; the model is located in a fixed place that was presented previously, but the operator is looking at the model from a different side of the table, from a different angle and distance. The shot also demonstrates the area visible from the mountain peak (in green) that can be entered by the option "Co widać ze szczytu?" (What can you see from the mountain top?). In the top right corner of the frames (A–E), from the video one can see the same image that is visible on the smartphone screen. The short video was necessary to make it easier for the respondents to compare both modes available in the application and to help them decide which one (i.e., AR/no and AR) fits a given criterion better (in their opinion).

After watching the video, each respondent would receive six questions concerning their preferences related to the modes of the application compared, i.e., no AR mode (Figure 2b) and AR mode (Figure 3), and suggestions on which one was:

- More convenient and with a simpler manner of operating the 3D model (F1);
- More effective (the opportunity to scrutinize a fragment of the 3D model more thoroughly) (F2);
- More realistic in terms of the visual reception of the 3D model (F3);
- More natural when it comes to the manner of viewing the 3D model (F4);
- More interesting and attractive way of exploring the 3D model (F5);
- Better as far as observing and assessing relative differences in the altitude of the land that was concerned (F6).

In the brackets above, the researchers provided the number of the subsection according to consecutive factors (compare with Figure 4). Respondents had the following qualitative answers: Definitely the AR mode; Rather the AR mode; Comparably AR and no AR mode; Rather no AR mode; Definitely no AR mode; and No opinion.

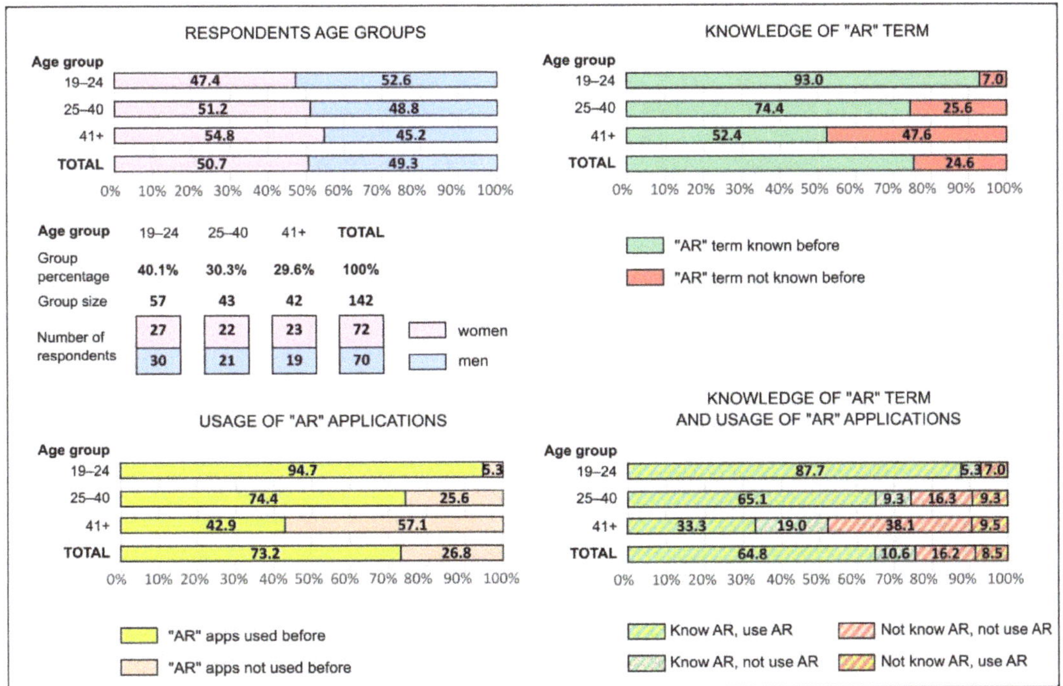

Figure 4. Structure of the respondents by age, sex and knowledge of the term "AR" and previous use of other AR applications by the respondents.

3.3. Participants

The study was carried out in the form of an online survey in the group of 142 respondents (72 women, 70 men) of different ages. The youngest participant was 19, the oldest was 83 and the largest number of respondents consisted of people aged 21 (the average age was 32, the median was 28). The respondents were also diverse in terms of social groups; there were students doing majors in Geography-related fields, researchers and people from Klub Zdobywców Korony Gór Polski (Club of the Conquerors of the Crown of Polish Mountains) active in social media. The respondents filled in the survey voluntarily, without any time constraints and gratification, and they could end it at any moment. They were divided into three age groups (19–24; 25–40; 41+), with a similar number of people in each, and with similar percentages in terms of the division into sexes. Figure 4 demonstrates the entire structure of the respondents according to the knowledge of the term "augmented reality" and according to sex, in specific age groups, and overall.

The youngest age group consisted of people aged 19–24. Many of them use a computer and mobile devices not only for studying and working but also for the entertainment, such as computer games, social media and multimedia (e.g., films and music). Another group included people aged 25–40, who remember the "era" before mobile devices with colorful touch screens, and the existence of solutions is not as obvious to them as it is for the younger age group that failed to understand or remember a different state of affairs. The third age group (over 40) consisted of both people who started their adventure with computers when computers were just beginning to come into common use and those who remembered the times when computers and mobile devices had not existed yet or maybe had been in a very early stage of development. Among people aged 19–24, 93% had heard about the term "augmented reality" before they took part in the research; in the group aged 25–40, it was 74.4%; and in the group aged over 40, it was a little bit more than half (52.4%). As much as 94.7% of the youngest respondents used an application that used AR in the past; in the group of people aged 25–40, that percentage was only 74.4%; and in the group aged 41+, the percentage was lower than half (42.9%). There were 50 people who knew the term AR and used such applications in the 19–24 group (87.7% of respondents in this age group), 28 in the 25–40 group (which is 65.1%), and 14 in the 41+ group (respectively 33.3%). When it comes to people that knew the term AR but failed to use such applications, the numbers were the following: 3 in the 19–24 group (5.3% of respondents in this age group), 4 in the 25–40 group (which is 9.3%), and 8 in the 41+ group (respectively 19.0%). There were no people in the 19–24 group that did not know the AR term and failed to use AR applications, in the 25–40 group there were 7 such people (16.3% of respondents in this age group), and in the 41+ group there were 16 such people (which is 38.1%). There were 4 people in each age group that failed to know the AR term but did use such applications (which is 7.0% for the 19–24 age group, 9.3% for the 25–40 age group and respectively 9.5% for the 41+ age group). A total of 64.8% of all respondents in the study knew the term AR before and had used AR applications; 10.6% knew the term AR before, but had not used AR apps; 16.2% declared that they did not know the term AR before and had not used AR apps as well; and 8.5% of respondents did not know the term AR, but they had used apps using this technology.

4. Results

Figure 5 demonstrates the preferences of the respondents that evaluated the two modes of using the application (i.e., AR and without AR), i.e., viewing the 3D model of the exemplary mountain top, *Tarnica*, in terms of which one works best in the context of the analyzed usage traits. The respondents had the following qualitative answers to choose from (Figure 5): Definitely the AR mode (dark blue), Rather the AR mode (light blue), Comparable between the AR mode and the no AR mode (yellow), Rather no AR mode (light purple), Definitely no AR mode (dark purple), and No opinion (grey). The black triangles show the dominating answer (if specific modes are indicated, the triangles refer to "Definitely" and "Rather" answers considered jointly). As far as age groups are concerned, there was a visible tendency to give the "No opinion" answer more often by respondents in the 41+ age group. In each case, it was the number reaching one-third of the respondents in a given age group. People from the 19–24 group were more decisive as the percentage of the "No opinion" answer was the lowest in this group. Generally, the total result for all age groups and for all criteria showed that the AR mode had more supporters.

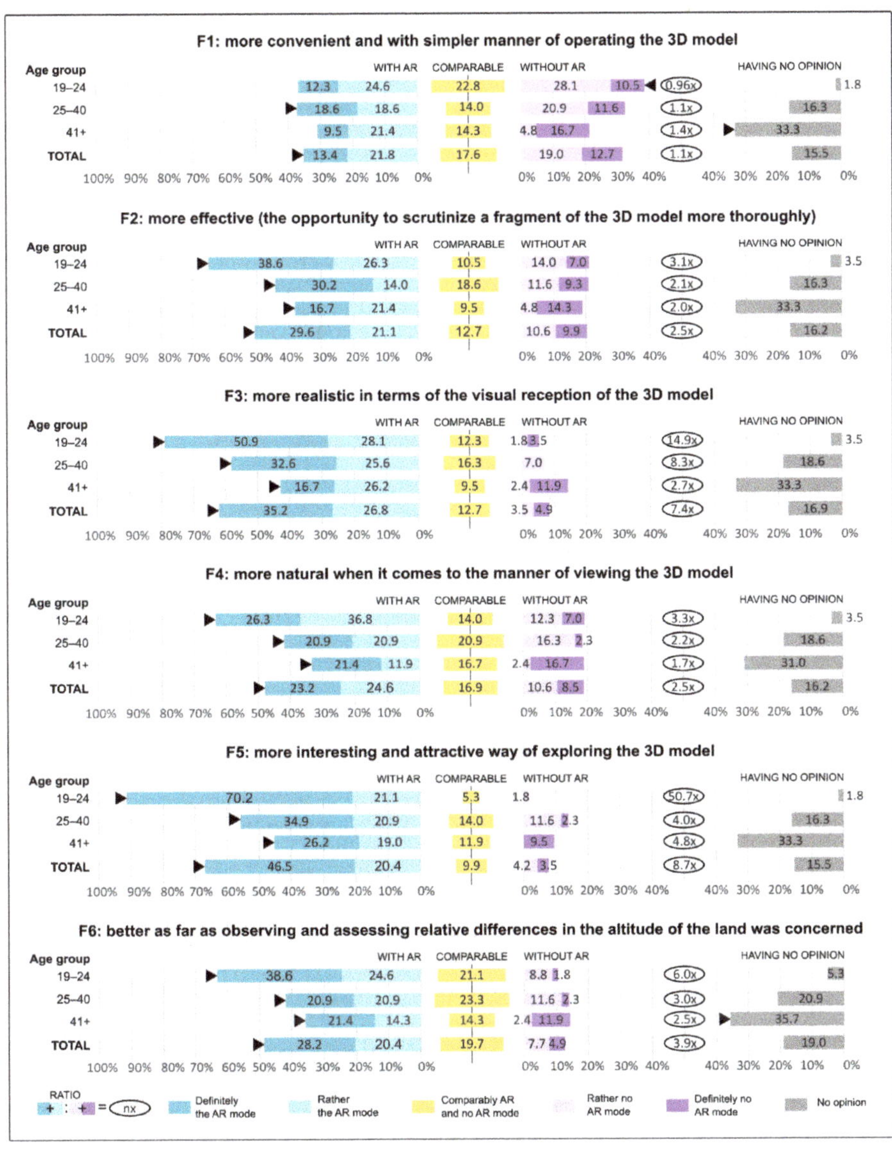

Figure 5. Respondents' preferences regarding the use of the application in the AR mode or non-AR mode, broken down by age group and in total.

In F1, in which respondents were supposed to choose the mode of using the application that was more convenient and with the simpler manner of controlling the 3D model, the polarization was the greatest; the number of answers suggesting that the AR/no AR mode was the better choice was similar in all age groups. In the youngest age group, the no AR mode scored slightly higher (the ratio of with AR/Without AR answers was 0.96), and in the middle (25–40) group the AR mode scored higher (1.1). The AR mode was preferred also in the oldest age group, but it was the "No opinion" answer that actually scored the highest there. The "Comparable between the AR mode and no AR mode" scored the highest in the 19–24 age group. Considering all of the respondents from all age groups, the AR mode seemed to score the highest but only slightly (1.1). The percentage of the respondents who thought that both modes were comparable in that respect was 17.6% and of those who had no opinion 15.5%.

The next subsection (F2) was related to the choice of the more effective mode, i.e., the one that allowed one to take a closer look at the fragment of the model. In that case, in each age group, the mode with AR won significantly. The with AR/without AR ratio was 3.1 for the 19–24 group, 2.1 for the 25–40 group and 2.0 for the 41+ group, respectively. Taking into consideration all answers, the AR mode scored slightly higher than 50% of all answers (29.6% for Definitely the AR mode, 21.2% for Rather the AR mode); for 12.7% of the respondents, both modes were comparable in this respect; slightly more than 20% of the respondents chose Without AR (10.6%—Rather no AR mode, 9.9%—Definitely no AR mode); and 16.2% of the respondents had no opinion. The people that definitely preferred the AR mode came from the youngest age group (38.6% of answers in that group).

As far as F3 is concerned, the respondents were supposed to evaluate which mode was more realistic in terms of the visual perception of the 3D model. In all age groups, there was a strong preference for the AR mode, with a relatively low percentage of "Without AR" answers. In the 19–24 age group, 80% of respondents chose the AR mode; in the 25–40 age group, nearly 60%; and in the 41+ group, it was less than 45%. The last group scored higher in terms of the "Without AR" answer, still reaching less than 15%. Considering all answers from all age groups, the AR mode scored over 60% (35.2% Definitely the AR mode, 26.8% Rather the AR mode), 12.7% of the respondents claimed that both modes were comparable, less than 10% of respondents selected the "Without AR" answer (3.5% Rather no AR mode, 4.9% Definitely no AR mode) and 16.9% of the respondents had no opinion on the matter.

When it comes to F4, according to which respondents decided which mode was the more natural when it comes to viewing the 3D model, it was characteristic that in all age groups answers favoring the absence of AR scored 20%, while in the 41+ group the answer "Definitely no AR mode" was selected more often (16.7%). In the 19–24 group, 63.1% selected the AR mode; in the 25–40, it was 41.8%; and in the 41+, it was 33.3%. In total, 46.8% of the respondents chose the AR mode (23.2%—Definitely the AR mode, 24.6%—Rather the AR mode), 16.9% of respondents believe that both modes were comparable and 19.4% of the respondents chose no AR (10.6%—Rather no AR mode, 8.5%—Definitely no AR mode), with 16.2% of the respondents choosing the "No opinion" answer.

The fifth criterion, according to which respondents selected the more interesting and attractive way of exploring the 3D model, (F5), was dominated by the supporters of AR. It was particularly visible for the 19–24 age group, in which 91.3% of the respondents believed so in total (70.2% chose the "Definitely the AR mode" answer). It was the highest score that was not achieved for any other criterion. At the same time, only 1.8% of the youngest respondents selected no AR for that criterion, and only 5.3% decided that both modes were comparable. In the 25–40 group and 41+ group, a general preference for AR was observed, while there were more "Comparable" and "No opinion" answers. In total, 66.9% of all of the respondents selected AR (46.5%—Definitely the AR mode, 20.4% Rather the AR mode). Both modes were evaluated as comparable by 9.9% of the respondents, and only 7.7% of the respondents chose the mode without AR. "No opinion" was selected by 15.5% of the respondents.

The last criterion, (F6), was related to choosing which mode was better for perceiving and estimating the relative land elevation differences. The distribution of the results looks similar to the one for the fourth criterion. There were many undecisive people in the 41+ age group. In total, 48.6% of the respondents selected the AR mode (28.2%—Definitely the AR mode, 20.4%—Rather the AR mode), 19.7% claimed that both modes seemed comparable to them, 12.6% chose no AR (7.7%—Rather no AR mode, 4.9%—Definitely the AR mode) and 19% of the respondents had no opinion on the matter.

The most unambiguous results were achieved for F5, as the with AR/Without AR ratio was 8.7 in total. This means that the respondents selected AR nearly nine times more often. Particularly, a great part was played by young people, as the number of young people who selected AR was 50 times higher than the number of those who did not select AR. F3 was the second most important criterion. The number of AR supporters was 7.4 times higher than the number of AR opponents. Again, the largest ratio, not as high as for F2, was noted for the 19–24 age group (14.9 times). F6 ranked 3rd. The general ratio was 3.9. F2 and F4 ranked 4th ex aequo. F1 ranked last, with the number of votes distributed nearly evenly, with a slight advantage of AR.

5. Discussion

It seems that, along with the development and popularization of new technologies, devices that handle webXR standards and detection of planes are going to be increasingly available. Currently, the list of such devices includes over 600 models of smartphones of different manufacturers (https://developers.google.com/ar/devices accessed on 29 November 2022) [34]. Among the multiple types of the AR user interface available, such as a backpack computer with HDM, PDA, tablet or AR glasses [35], smartphones are those that seem to be the most available and universal, as we use them for numerous other purposes and have them always at our fingertips.

The AR technology opens up a whole range of new opportunities to interact with digital, 3D models of land relief that may constitute visual aids in learning the topography of specific places. Using webXR standards and algorithms of plane detection allows one to separate a new method of displaying content in AR, i.e., plane-based augmented geovisualizations (PAGs). A PAG frees users from the necessity to be in a specific place to see virtual content, as it frequently happens in augmented reality environments (AREs). Moreover, it is not necessary to use physical objects (paper maps) to start digital content as in augmented map (AM) solutions The tested application, developed for the purposes of this research, constitutes an example of using the PAG method in creating AR solutions. The application makes it possible to display 3D models whenever the mobile device that handles webXR standard detects a plane.

The display of the 3D model on the smartphone screen, combined with the real view of the surroundings (i.e., the room) in which the observation takes place and the plane, on which the model is mounted, makes the entire image more realistic, allowing the user to become immersed in the emerging, partially fictional world. The research shows that the feelings and impressions that were accompanying users during the process of exploring the land relief model seemed to be much closer to reality than when a classic interaction with a 3D model, operated with peripheral devices (a mouse, keyboard or a joystick), takes place. The manner of viewing the model in the AR mode resembles the natural one, because it requires observers to move, change the perspective of looking, change the distance if they want to see the model from a different side or scrutinize a detail that is invisible at a larger distance. At the same time, due to the mobility and flexibility of the solution, users have the freedom to choose the location in which they want to view the model, the scale at which the model is being presented and the initial rotation angle in relation to the observation location.

This research constituted a part of the trend in publications that encourages researchers to examine and establish whether various solutions, such as AR, may be used in practice; whether they may constitute an interesting method of promoting tourism; whether they may be helpful in education (e.g., in teaching geography), specifically, topography [22]; and whether they may be useful in such specific fields as mountain rescue services. As far as education is concerned, Scholz et al. [36] and De Almeida Pereira et al. [19] noted that the 3D representation of AR can enhance users' abilities to perform spatial positioning and read map data, respectively.

The research on the usability of geomedia is frequently related to homogenous groups, e.g., students [37,38]. What is important in our study, however, was the testing of the usefulness of the AR application among different age groups. The division of respondents into three age groups (19–24; 25–40; 41+) and from different backgrounds became crucial. It seemed legitimate to divide respondents into three age groups because similarities and differences in the preferences related to the factors analyzed (F1–F6) could immediately be observed. Furthermore, significant differences in the subjective preferences of amateurs and experts were noted by Medyńska and Zagata [26] for the reconstruction of the stronghold in 3D, as well as by Medyńska et al. [27] for the complexity level of a people gathering presentation on an animated map.

Those age thresholds were arbitrary, and the respondents may significantly differ in terms of habits, skills, and knowledge of the technology of geovisualizing the geographic space of the mountains, regardless of age. However, the differences obtained seem to prove that the conditions described may be the foundation for adopting such categories for dividing respondents into subgroups.

6. Conclusions

Generally speaking, one may conclude that the AR mode in the application received positive evaluation according to different criteria (F2–F6), which means that respondents, regardless of age, can see the potential of such a solution. In some of the criteria, such as the attractiveness of that way of exploring the 3D model, a significant advantage may be observed; in others, such as easy operation (navigation), the opinions seemed a lot more divided. Quantitative differences were clearly visible in the age groups. One may actually observe certain regularities. It needs to be said that respondents in the 19–24 age group had the most positive opinion concerning presenting the 3D model in the AR mode. In all cases apart from F1, their percentage was over 65%. F1 was the only criterion in which, in that age group, there were slightly more people skeptical concerning AR but that was a statistically irrelevant difference (only 1.7%). It was also the only case in the entire research when the number of supporters of the monitor mode outnumbered the supporters of AR. In the 25–40 age group, the percentage of AR supporters was slightly lower, still reaching over 40% for all criteria, except F1, for which the opinions were almost evenly distributed (roughly 50/50) in all age groups. In the 41 and more age group, AR supporters comprised at least 35% of the respondents for all criteria (except F1). Older respondents (41+ years old) were more reserved towards that way of using the application. Those respondents, much more often than the younger ones, did not have an opinion concerning whether AR worked better for a given criterion. This was most observable for F6 (35.7%) but also for other criteria, where the percentage of indecisive answers was 31–33.3%. There were fewer respondents without an opinion in the 25–40 age group and that percentage was the lowest in the youngest group (the percentage oscillated between 1.8 and 5.3%).

The most unambiguous preferences for using AR for the presentation of the 3D model were observed for criterion F5 (the more interesting and attractive way of exploring the 3D model). F3 (the more realistic when it comes to the visual perception of the 3D model) ranked 2nd. F6 (better at perceiving relative terrain elevation differences) ranked 3rd. F2 (more effective in terms of taking a closer look at the 3D model), and F4 (more natural when it comes to viewing the 3D model) ranked 4th ex aequo. F1 (more convenient and characterized by the simpler way of controlling the 3D model) had an almost equal distribution of votes, with a slight advantage of those for AR.

It is also worth noting that there were some differences in terms of the previous experience with AR within age groups. When it came to the knowledge of the term "augmented reality" and previous experience in using similar applications, those differences were arranged according to age groups: the lower the age of respondents, the higher percentage of AR users or those who knew the term AR prior to participating in this study The higher the age of respondents, the lower the percentage of such people. Nine respondents from the 41+ age group knew the term AR but had not used similar applications before. In each group there were three respondents that shared the same experience. There were four people in each group that failed to know the term AR but did use AR applications. There was nobody aged 19–24 that used AR applications but failed to know the term. In the 25–40 age group, there were six such respondents, and there were 16 in the 41+ group.

The results show that AR technology that uses the PAG method for creating applications that allow one to view 3D models of land relief fragments is highly useful. They corroborate the fact that this solution may constitute an interesting, willingly chosen alternative to the classic approach to interactive exploration of 3D models on the computer. High scores obtained for specific criteria prove that users have a great interest in this form of geovisualization. The results of the research corroborate the potential of the AR technology and encourage researchers to conduct more studies, extending the range of research by adding the test with the application displayed in AR, with respondents using it physically on their smartphones.

Further research and the development of 3D models of land relief forms in AR may focus on special effects that could be added to the model to make the image even more realistic [39]. On historical topographic maps, the third dimension of the land relief was highlighted by means of suitably applied painterly means of expression [40]. Here, the orthophoto map was the cartographic base and the main objective was to simulate lighting and weather conditions. The opportunities to change the lighting conditions are well known from GIS applications, used for rendering 3D land relief models. Those conditions have a significant impact on the perception of the model [41]. It seems that, in this situation, smart solutions should adjust the height of the source of lighting, as well as the direction of lighting and length of the shadow cast to the position of the observer in relation to cardinal directions, the season, the time of the day and the real lighting conditions that may occur in a given location as a result of all those factors [42–44]. Furthermore, adding mist, as it takes place in the online 3D atlas of Switzerland (www.atlasderschweiz.ch accessed on 29 November 2022) [45,46], as well as rain, snow and clouds (as in 3D cities or in the digital land relief model in the TerrainView application), could create a highly interesting effect.

Author Contributions: Conceptualization, Ł.H. and Ł.W.; methodology, Ł.H. and Ł.W.; software, Ł.H.; validation, Ł.H.; formal analysis, Ł.W.; investigation, Ł.H. and Ł.W.; resources, Ł.H.; data curation, Ł.W.; writing—original draft preparation, Ł.H.; writing—review and editing, Ł.W.; visualization, Ł.H.; supervision, not applicable; project administration, Ł.H.; funding acquisition, not applicable. All authors have read and agreed to the published version of the manuscript.

Funding: This research received no external funding.

Data Availability Statement: Not applicable.

Conflicts of Interest: The authors declare no conflict of interest.

References

1. Biljecki, F.; Ledoux, H.; Stoter, J.; Zhao, J. Formalisation of the level of detail in 3D city modelling. *Comput. Environ. Urban Syst.* **2014**, *48*, 1–15. [CrossRef]
2. Gill, L.; Lange, E. Getting virtual 3D landscapes out of the lab. *Comput. Environ. Urban Syst.* **2015**, *54*, 356–362. [CrossRef]
3. Wang, W.; Wu, X.; He, A.; Chen, Z. Modelling and Visualizing Holographic 3D Geographical Scenes with Timely Data Based on the HoloLens. *IJGI* **2019**, *8*, 539. [CrossRef]
4. Lv, G.N.; Yu, Z.Y.; Yuan, L.W.; Luo, W.; Zhou, L.C.; Wu, M.G.; Sheng, Y.H. Is the Future of Cartography the Scenario Science? *J. Geo-Inf. Sci.* **2018**, *20*, 1–6. [CrossRef]
5. Kersten, T.P.; Edler, D. Special Issue "Methods and Applications of Virtual and Augmented Reality in Geo-Information Sciences". *PFG* **2020**, *88*, 119–120. [CrossRef]
6. Caudell, T.; Mizell, D. Augmented reality: An application of heads-up display technology to manual manufacturing processes. In Proceedings of the Twenty-Fifth Hawaii International Conference on System Sciences, Kauai, HI, USA, 7–10 January 1992; p. 2. [CrossRef]
7. Milgram, P.; Kishino, F. A Taxonomy of Mixed Reality Visual Displays. *IEICE Trans. Inf. Syst.* **1994**, *E77-D*, 1321–1329.
8. Azuma, R.T. A Survey of Augmented Reality. *Presence: Teleoperators Virtual Environ.* **1997**, *6*, 355–385. [CrossRef]

9.   Cheng, Y.; Zhu, G.; Yang, C.; Miao, G.; Ge, W. Characteristics of augmented map research from a cartographic perspective. *Cartogr. Geogr. Inf. Sci.* **2022**, *49*, 426–442. [CrossRef]

10. Dickmann, F.; Keil, J.; Dickmann, P.L.; Edler, D. The Impact of Augmented Reality Techniques on Cartographic Visualization. *KN J. Cartogr. Geogr. Inf.* **2021**, *71*, 285–295. [CrossRef]

11. Feiner, S.; MacIntyre, B.; Höllerer, T.; Webster, A. A touring machine: Prototyping 3D mobile augmented reality systems for exploring the urban environment. *Pers. Technol.* **1997**, *1*, 208–217. [CrossRef]

12. Bertin, J.; Barbut, M. *Sémiologie Graphique: Les Diagrammes, Les Réseaux, Les Cartes*; Éditions de l'École des hautes études en sciences sociales: Paris, France, 2013; ISBN 2713224179.

13. Halik, Ł. The analysis of visual variables for use in the cartographic design of point symbols for mobile Augmented Reality applications. *Geod. Cartogr.* **2012**, *61*, 19–30. [CrossRef]

14. Halik, Ł.; Medyńska-Gulij, B. The Differentiation of Point Symbols using Selected Visual Variables in the Mobile Augmented Reality System. *Cartogr. J.* **2016**, *32*, 147–156. [CrossRef]

15. Fukuda, T.; Zhang, T.; Yabuki, N. Improvement of registration accuracy of a handheld augmented reality system for urban landscape simulation. *Front. Archit. Res.* **2014**, *3*, 386–397. [CrossRef]

16. Kourouthanassis, P.; Boletsis, C.; Bardaki, C.; Chasanidou, D. Tourists responses to mobile augmented reality travel guides: The role of emotions on adoption behavior. *Pervasive Mob. Comput.* **2015**, *18*, 71–87. [CrossRef]

17. Bobrich, J.; Otto, S. Augmented Maps. *Int. Arch. Photogramm. Remote Sens. Spat. Inf. Sci.* **2002**, *34*, 502–505.

18. Morrison, A.; Mulloni, A.; Lemmelä, S.; Oulasvirta, A.; Jacucci, G.; Peltonen, P.; Schmalstieg, D.; Regenbrecht, H. Collaborative use of mobile augmented reality with paper maps. *Comput. Graph.* **2011**, *35*, 789–799. [CrossRef]

19. De Almeida Pereira, G.H.; Stock, K.; Stamato Delazari, L.; Centeno, J.A.S. Augmented Reality and Maps: New Possibilities for Engaging with Geographic Data. *Cartogr. J.* **2018**, *54*, 313–321. [CrossRef]

20. Mozzila.org. WebXR Device API. Available online: https://developer.mozilla.org/en-US/docs/Web/API/WebXR_Device_API (accessed on 2 December 2022).

21. Horbiński, T.; Medyńska-Gulij, B. Geovisualisation as a process of creating complementary visualisations: Static two-dimensional, surface three-dimensional, and interactive. *Geod. Cartogr.* **2017**, *66*, 45–58. [CrossRef]

22. Siqueira, P. Teaching Topographic Surface Concepts in Augmented Reality and Virtual Reality Web Environments. *International J. Innov. Educ. Res.* **2019**, *7*, 307–320. [CrossRef]

23. Templin, T.; Popielarczyk, D.; Gryszko, M. Using Augmented and Virtual Reality (AR/VR) to Support Safe Navigation on Inland and Coastal Water Zones. *Remote Sens.* **2022**, *14*, 1520. [CrossRef]

24. Brejcha, J.; Lukáč, M.; Hold-Geoffroy, Y.; Wang, O.; Čadík, M. *LandscapeAR: Large Scale Outdoor Augmented Reality by Matching Photographs with Terrain Models Using Learned Descriptors*; Springer International Publishing: Cham, Switzerland, 2020.

25. Zagata, K.; Gulij, J.; Halik, Ł.; Medyńska-Gulij, B. Mini-Map for Gamers Who Walk and Teleport in a Virtual Stronghold. *IJGI* **2021**, *10*, 96. [CrossRef]

26. Medyńska-Gulij, B.; Zagata, K. Experts and Gamers on Immersion into Reconstructed Strongholds. *IJGI* **2020**, *9*, 655. [CrossRef]

27. Medyńska-Gulij, B.; Wielebski, Ł.; Halik, Ł.; Smaczyński, M. Complexity Level of People Gathering Presentation on an Animated Map—Objective Effectiveness Versus Expert Opinion. *IJGI* **2020**, *9*, 117. [CrossRef]

28. Wielebski, Ł.; Medyńska-Gulij, B. Graphically supported evaluation of mapping techniques used in presenting spatial accessibility. *Cartogr. Geogr. Inf. Sci.* **2019**, *46*, 311–333. [CrossRef]

29. Andrzejewski, B.; Witkowska, M. *Górska Korona Polski: Wędrówka Przez Cztery Pory Roku, Wydanie II.*; Wydawnictwo STAPIS: Katowice, Poland, 2019; ISBN 8379670930.

30. Bzowski, K. *Korona Polskich Gór, Wydanie II.*; Wydawnictwo Helion: Gliwice, Poland, 2018; ISBN 8328348209.

31. Easily display interactive 3D models on the web & in AR. Available online: https://modelviewer.dev/ (accessed on 2 December 2022).

32. Medyńska-Gulij, B.; Gulij, J.; Cybulski, P.; Zagata, K.; Zawadzki, J.; Horbiński, T. Map Design and Usability of a Simplified Topographic 2D Map on the Smartphone in Landscape and Portrait Orientations. *IJGI* **2022**, *11*, 577. [CrossRef]

33. Likert, R. A technique for the measurement of attitudes. *Arch. Psychol.* **1932**, *22*, 55.

34. Google. ARCore supported devices. Available online: https://developers.google.com/ar/devices (accessed on 2 December 2022).

35. Halik, Ł. Wykorzystanie mobilnego systemu rozszerzonej rzeczywistości w robotach geodezyjnych–wywiad terenowy. *Geod. -Mag. Geoinformacyjny* **2012**, *209*, 20–25.

36. Scholz, M.A.; Huynh, N.T.; Brysch, C.P.; Scholz, R.W. An Evaluation of University World Geography Textbook Questions for Components of Spatial Thinking. *J. Geogr.* **2014**, *113*, 208–219. [CrossRef]

37. Cybulski, P. Spatial distance and cartographic background complexity in graduated point symbol map-reading task. *Cartogr. Geogr. Inf. Sci.* **2020**, *47*, 244–260. [CrossRef]

38. Cybulski, P.; Wielebski, Ł. Effectiveness of Dynamic Point Symbols in Quantitative Mapping. *Cartogr. J.* **2019**, *56*, 146–160. [CrossRef]

39. Svobodová, J.; Voženílek, V. *Relief for Models of Natural Phenomena*; Springer: Dordrecht, The Netherlands, 2009.

40. Medyńska-Gulij, B. Geomedia Attributes for Perspective Visualization of Relief for Historical Non-Cartometric Water-Colored Topographic Maps. *IJGI* **2022**, *11*, 554. [CrossRef]

41. Medyńska-Gulij, B.; Lis, M.; Wielebski, Ł. Wizualizacja wymiernych i plastycznych cech rzeźby na podstawie numerycznego modelu terenu dla Wielkopolskiego Parku Narodowego. In *BADANIA FIZJOGRAFICZNE 2012, R. III–Seria A–Geografia Fizyczna (A61)*; Wydawnictwo Poznańskiego Towarzystwa Przyjaciół Nauk: Poznan, Poland, 2012; pp. 187–207.
42. Farmakis-Serebryakova, M.; Hurni, L. Comparison of Relief Shading Techniques Applied to Landforms. *IJGI* **2020**, *9*, 253. [CrossRef]
43. Veronesi, F.; Hurni, L. A GIS tool to increase the visual quality of relief shading by automatically changing the light direction. *Comput. Geosci.* **2015**, *74*, 121–127. [CrossRef]
44. Jenny, B.; Heitzler, M.; Singh, D.; Farmakis-Serebryakova, M.; Liu, J.C.; Hurni, L. Cartographic Relief Shading with Neural Networks. *CoRR* **2020**, *27*, 1225–1235. [CrossRef]
45. Institute of Cartography and Geoinformation, ETH Zurich. Atlas of Switzerland. Available online: www.atlasderschweiz.ch (accessed on 2 December 2022).
46. Sieber, R.; Serebryakova, M.; Schnürer, R.; Hurni, L. *Atlas of Switzerland Goes Online and 3D—Concept, Architecture and Visualization Methods*; Springer International Publishing: Cham, Switzerland, 2016.

International Journal of *Geo-Information*

Article

Map Design and Usability of a Simplified Topographic 2D Map on the Smartphone in Landscape and Portrait Orientations

Beata Medyńska-Gulij [1,*], Jacek Gulij [2], Paweł Cybulski [1], Krzysztof Zagata [1], Jakub Zawadzki [1] and Tymoteusz Horbiński [1]

[1]   Adam Mickiewicz University, Poznań, Poland; Department of Cartography and Geomatics, 61-712 Poznań, Poland
[2]   Collegium Da Vinci Poznań, 61-719 Poznań, Poland
*   Correspondence: bmg@amu.edu.pl

Abstract: Map design and usability issues are crucial when considering different device orientations. It is visible, especially in exploring the topographical space in landscape or portrait orientation on the mobile phone. In this study, we aim to reveal the main differences and similarities among participants' performance in a map-based task. The study presents an original research scheme, including establishing conceptual assumptions, developing map applications with gaming elements, user testing, and visualizing results. It appears that the different phone orientation triggers different visual strategy. This transfers into decision-making about the path selection. It turned out that in landscape orientation, participants preferred paths oriented east–west. On the other hand, portrait orientation supported north–south path selection. However, considering the given task accomplishment, both mobile phones' orientations are adequate for the exploration of topographical space.

Keywords: map design; geomedia usability; smartphone; landscape orientation; portrait orientation; mapping technique; eye-tracking; cartographic visualization; Unity; gameplays

Citation: Medyńska-Gulij, B.; Gulij, J.; Cybulski, P.; Zagata, K.; Zawadzki, J.; Horbiński, T. Map Design and Usability of a Simplified Topographic 2D Map on the Smartphone in Landscape and Portrait Orientations. *ISPRS Int. J. Geo-Inf.* **2022**, *11*, 577. https://doi.org/10.3390/ijgi11110577

Academic Editors: Wolfgang Kainz and Florian Hruby

Received: 13 October 2022
Accepted: 18 November 2022
Published: 20 November 2022

Publisher's Note: MDPI stays neutral with regard to jurisdictional claims in published maps and institutional affiliations.

1. Introduction

The most popular maps on websites tend to have highly simplified graphics and symbols that create cartographic content that does not require map legends as they are interpreted intuitively [1]. Using basic map design principles enables easy map use, particularly on small display screens. These principles are: limit color choice and have perceptually clear differences [2], use conventional colors [3], ensure high contrast of point symbols [4], and low density of graphic elements [5]. Traditional topographic map design focuses on the graphical representation that can be read as a whole or separately and presented in a large format [6], whereas map design for smartphones' size of the view is limited, and the two orientations, landscape and portrait, become significant. The user's classic way of moving around the large printed cartographic image is to move one's finger on it. However, the user's way of moving around the map on a small smartphone display will necessarily be quite different. Indeed, traditional map design principles may still be employed, e.g., the focus of attention [7,8]. For centuries, map design has adapted to the methods of publication and the medium [9]; therefore, how maps are implemented on mobile devices and how users interact with them requires investigation.

Video games are one of the more interesting and littler explored mediums for presenting and researching cartographic products for smartphones [10]. The main concept that appears in such research is 'Gamification', which is the use of technology to design characteristic elements for video games in contexts not related to the game [11]. Gamification is a popular technique in many mobile applications that are content driven that require rich interaction between the user and the application [12].

To develop apps with interactive map viewing for mobile devices, one option is to use game engines such as Unity, Unreal, and Godot. These allow one to implement moving in

2D space and displaying appropriate maps in this space [13,14]. Unity, combined with the C# programming language, is an appropriate environment for creating cartographic apps. During the process of mobile map app design, it is important to minimize the interface but exploit the intuitiveness of touching the screen [1,15].

In this study, we analyze responsive map design in two contexts [16]. The first is related to the adaptation of cartographic content displayed on-screen, the so-called adaptation to the size and resolution of the device, while the other one is linked to the change in the device's orientation [17]. The small size of the display also limits the space that can be allocated to buttons and other functions [18]. In the literature, there is a gap in research on using the same content of spatial information in landscape and portrait orientations of smartphones. There is very little evidence of the differences resulting from the interaction with the geographical space in the landscape or portrait orientation on small displays.

The methods of studying map usability are often based on the experiment in which map users do the same task but using different map versions [19,20]. A homogenous group usually consists of 20–30 respondents for public users and 10–20 respondents for experts [21]. Tasks to be performed by each individual respondent become the basis for obtaining objective data on the time of completing the task and the way of moving around the geographical space [22]. Tasks should be comprehensible for the user and possible to complete in a few minutes.

Nowadays, what is gaining popularity is eye-tracking applied in empirical studies on the efficiency of maps, e.g., animated maps [23], recognizability of map symbols in video games [24], user preferences and behavior in a topographic immersive virtual environment [19,25], walking and teleporting in a virtual stronghold [26]. However, the eye-tracking technique has only been employed in a few studies using mobile cartography [27] and focused more on many kinds of human–phone interfaces and the efficiency of apps on smartphones [28,29]. Therefore, eye tracking could show the differences in the visual strategy among users who perform a map-based task in different device orientations. If landscape orientation could lead to more frequent users' activity on the east–west parts of the map, then eye tracking would help understand the visual attention distribution related to the device orientation.

Data obtained from any experimental research are saved in tables, print screens, text files, etc., which are then subject to statistical analyses and cartographic visualization. The use of graphical methods and mapping techniques for presenting results allows one to interpret the results more fully [22,30]. Appropriate statistical analyses [31,32] and data mapping techniques [33] are employed to form partial and full conclusions (cartographic research method).

2. Aim and Questions

The main objective of the research is to demonstrate the similarities and differences in moving around on the 2D map on the smartphone in the landscape and portrait orientations with attitude to map design and cartographical methods for representation of usability.

The objective of the research raises the following questions:

- How to design a simplified 2D topographic map for gameplay with the opportunity to move on the roads/paths for smartphones?
- How can one obtain data on the player's movement on the smartphone display with landscape and portrait orientations?
- What statistical analyses and mapping techniques can be applied to present the properties of moving around on the 2D map in landscape and portrait orientation?

3. Materials and Methods

To meet the objective and answer the above questions, we have adopted six main research stages (Figure 1):

- to establish the basic conceptual assumptions (Section 3.1);

- to create a 2D map for smartphones with landscape and portrait orientations (Section 3.2, Figure 2);
- to create a map application with gameplay elements (Section 3.3, Figure 3);
- to prepare and carry out surveys among map users on smartphones in landscape and portrait orientations (Section 3.4; Figures 4–6);
- statistical analysis of the results (Section 4.1; Figures 7 and 8);
- cartographic visualization of the results (Section 4.2; Figures 9–11).

3.1. Concept

Considering the cartographico-geomedial attitude (design of topographic map, map orientation, mapping techniques, geomedia efficiency, eye-tracking for cartographic content) and IT aspects (Information Technology: game engines software, programming scripts, data collection scripts, numerical data of path occupation), we adopted a concept embracing the following assumptions (Figure 1):

- Map design: the whole map area is closed in the square, centralization of the two 16:9 ratio viewing with landscape and portrait orientation, simplified content of the topographic map, eight stone numbered monuments (see Figure 2);
- App programming: moving around the roads, a user-symbol in the center of the display, minimalistic interface (no buttons, no joysticks), scrolling around the map by touch, eight consecutively seized stone monuments, summarizing data after each walk in the form of an image with a path and sometimes for each segment of the path;
- Respondents: 60 users, aged 19–24, from a homogenous group of Adam Mickiewicz University Poznan students, participating in the game voluntarily, without any financial gratification;
- Survey: each participant receives the same task instructions, which are to move on the map using roads only to reach each of the eight stone monuments as quickly as possible; 30 participants do the task on the smartphone in the landscape orientation, and 30 participants in the portrait orientation;
- Software and equipment: graphic and cartographic software (Inkscape, Photoshop, QGIS); game engine (Unity), programming language (C#); survey (smartphone 16×9; eye-tracker Gazepoint GP3 HD, Gazepoint Control, laptop MSI);
- Statistical analysis: total time of gameplay, median, statistical significance;
- Cartographic visualization: visual strategy: heat map and proportional point symbol map; road load intensity: flow map (band cartodiagram);
- Expected research results: List of general and particular similarities and differences for landscape and portrait orientation.

3.2. Creating a 2D Map for Smartphone

Simplified content of the topographic map was created in six perceptual colors (gray roads, red—buildings, blue—lakes/rivers, green—forests/parks, black—stone monuments, and yellow—location of the user. A yellow circle is used to symbolize the location of the user; initially is located in the center of map. For designing the map, the landscape and portrait frames (which would provide the initial views) were placed in the center of the area (Figure 2). The location of the eight stone monuments is significant. The first three locations were also placed beside roads but outside the initial view. They had a similar distance from the center of the map and kept well inside the frame of the map extent.

New cartographic elements were created in a vector graphical program, Inkscape, as an SVG file. The basic cartographic symbols are easily distinguishable, both perceptively and intuitively, by using conventional colors and styles. The fundamental content (roads, rivers, lakes, forests-parks, buildings) was then supplemented with the monument point symbols of high contrast (Figure 2). The symbol of the user's location, always located on the road exactly in the middle of the map displayed, was the focus of visual attention [10]. To increase the contrast, the yellow circle has a black outline [7]. Following the rules of map design, the background of the map was made very light yellow. Stone monuments are characterized by the shape from topographic maps [34], although each of them received a unique white number (1–8). The finished graphics of the map was saved as a PNG file at a high resolution.

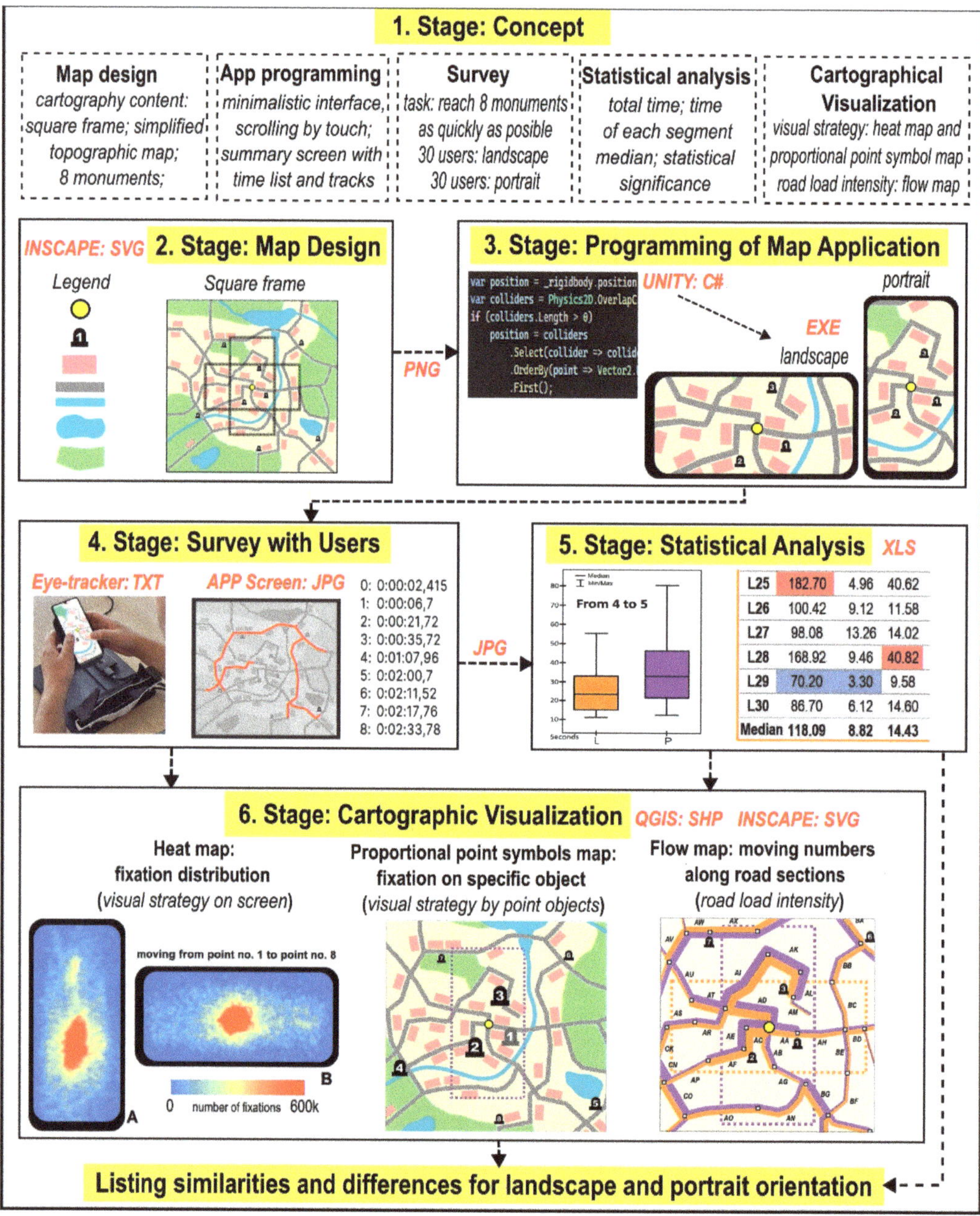

Figure 1. Graphical scheme with six research stages.

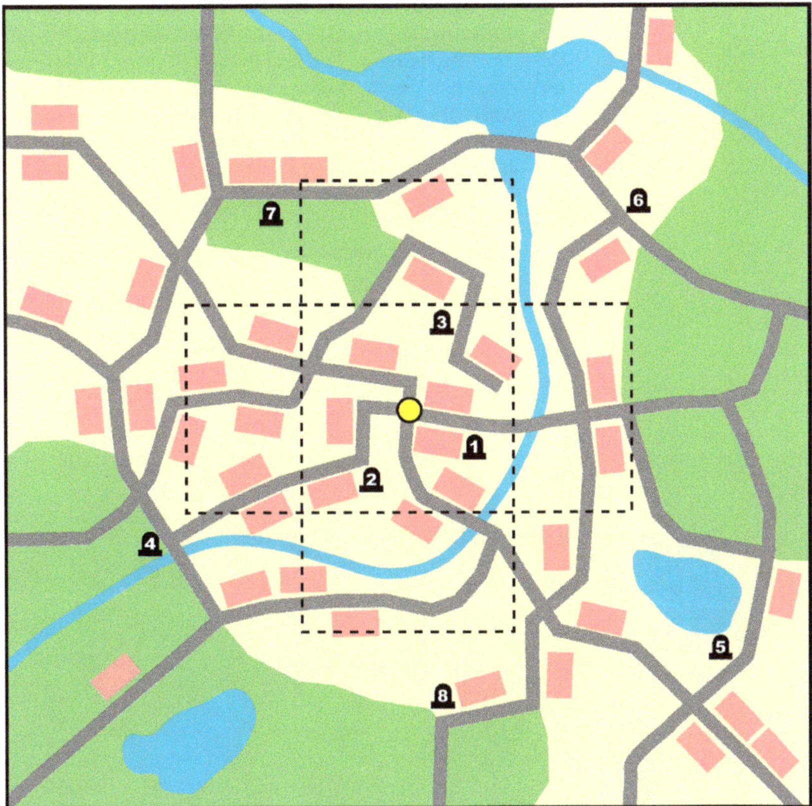

Figure 2. Simplified content of the designed topographic map with initial view frames in landscape and portrait orientations.

3.3. Creating an Application for Smartphone

App development began with preparing a scene in which the map, the player, monuments, the camera, and a summary screen were located. Then, the behavior of these elements, their interactions, and reactions to user input were implemented. The graphic of the map was directly placed on the scene and included roads that had been reflected in the physics system by roads made of *Edge Colliders*. The player was represented in physics as *Circle Collider* but also had a component of *Rigidbody* component that is necessary for moving objects. The player moves in the direction indicated by the finger that touches the screen. While the player is moving, contact with the nearest road is constantly checked (the collision of *Colliders* of the player and roads), and the player is required to be in the middle of the road to prevent them from leaving the path.

Monuments had their own *Circle Colliders*, which were placed on the adjacent road. Only the *Collider* for the monument being searched for is active. When the player *Collider* meets the *Collider* for active monument information about successfully seizing that control point. The action of saving this data changes the active monument to the next one in the sequence. In the beginning, only monument 1 is active, and after monument 8 is seized, the results are summed up. The order of collecting stones was not voluntary. After stone no. 1 is captured, stone no. 2 would activate in black. After stone no. 2 was reached, stone no. 3 would activate, etc.

The camera on the scene is constantly following the player, which means that the player remains in the center of the view throughout the activity. Initially, the script adjusts the size of the camera view to the selected portrait/landscape mode. The summary screen includes a list of times when individual stones were seized and the tracks that the player completed. The data on tracks are collected, starting with the moment the app is started, and *LineRenderer* is used for displaying them on screen.

Figure 3 presents the code to implement moving toward the touch by the user's finger. The Update method is activated approximately 60 times per second and serves to interpret the user's touch. The *FixedUpdate* method is activated exactly 50 times per second and is used for moving the symbol of the player. Placing the player on the road is crucial for moving on the roads on the map. Out of all roads located nearby the touch point, the nearest road is picked, and the new position of the player becomes the contact point with this road. Furthermore, obtaining the number of walks that users took down individual stretches of the roads was implemented (stretches marked with two letters are presented in a flow map. The load of the road stretch is measured by the number of walks taken down the specific stretch by 30 respondents.

```csharp
private void Update()
{
    if (Input.touchCount > 0)
    {
        var pointerPosition = Input.touches[0].position;
        _movementDirection = Camera.main.ScreenToWorldPoint(pointerPosition) - transform.position;
        _movementDirection = _movementDirection.normalized;
    }
    else if (Input.GetMouseButton(0))
    {
        var pointerPosition = Input.mousePosition;
        _movementDirection = Camera.main.ScreenToWorldPoint(pointerPosition) - transform.position;
        _movementDirection = _movementDirection.normalized;
    }
    else
    {
        _movementDirection = Vector2.zero;
    }
}

private void FixedUpdate()
{
    var position = _rigidbody.position + _speed * Time.fixedDeltaTime * _movementDirection;
    var colliders = Physics2D.OverlapCircleAll(position, 1f, LayerMask.GetMask("Paths"));
    if (colliders.Length > 0)
        position = colliders
            .Select(collider => collider.ClosestPoint(position))
            .OrderBy(point => Vector2.Distance(point, position))
            .First();

    _rigidbody.position = position;
}
```

Figure 3. The implementation of moving toward the touch by the user's finger.

3.4. Surveys among Map Users

The researcher would invite a participant into the room, and the respondent would sit down at the table in front of a smartphone mounted on a handle. Then, the researcher explained the research objective in general terms and presented the way the smartphone worked with an eye-tracker. The participant selected the required orientation of the smartphone and then the eye-tracker calibrated eyesight (Figure 4). After the player entered the game, the researcher informed the participant about his/her location with a yellow circle in the middle of the map. Next, the participant would receive a task to reach all black stone moments, starting with 1 to the last one, as quickly as possible. The authors of this study decided that the time between the game entering and taking monument no. 1 will be time for the user to familiarize themselves with the application and its gameplay. After collecting monument no. 1, the game time was officially counted, which was later analyzed in the study. When the participant reached stone monument no. 8, the researcher thanked the participant for completing the task and recorded the data of that completed walk.

In the research, all the respondents held the smartphone in two hands and used their thumbs to interact with the screen, which is in keeping with how mobile devices are typically held [35–37]. Such use of smartphones also resulted from the ability of the user's marker to move down the roads, i.e., from smooth scrolling down the map. The settings resulting from the eye-tracking technology supported this solution too. The data were recorded on the smartphone after each completed gameplay. Figure 5 presents an example of data collected from participant no. 18 (see Figure 6B).

4. Results

4.1. Statistical Analysis

The time data of the 60 research participants were divided into two main categories: the total time of the walk and the time of individual paths (Figure 6). The shortest time in portrait orientation was achieved by user P12 (83.96 s) and the longest time—by P29 (185.42 s) (Figure 6B). The median value of the total game time for the entire portrait group was 116.70 s, and the standard deviation was 22.37, which confirms the temporal discrepancies between players. The shortest time of going through the entire application in landscape orientation (Figure 6A) was achieved by user L29 (70.20 s), and the longest time was by user L25 (182.70 s). The median of the total game time for the entire landscape group was 118.09 s, and the standard deviation was 25.45, which also confirms time intervals between users. The non-parametric Mann–Whitney U test for independent samples was carried out in PQStat (v 1.8.). This is an appropriate test to compare two groups when the data distribution does not fit the criteria of a normal distribution [31,32]. To evaluate differences for portrait and landscape orientation groups, the medians for individual categories are compared (Figure 7). Users that went through the application in portrait orientation would complete the search faster by 1.39 s than those in landscape orientation. The statistical test results did not confirm the significance of this difference.

The Mann–Whitney U test on these median times is shown in Figure 8. For moving between monuments 4–5 (*p*-value = 0.04) and for 5–6 (*p*-value = 0.009), the test demonstrated the differences in parameters between the two orientations. For paths 4–5, users in landscape orientation would find the point faster than in portrait orientation. The difference may result from the fact that points 4 and 5 were more visible for users in horizontal orientation, which allowed users to find the right route to the destination faster. The difference may be corroborated by eye-tracking analyses and route traffic loads. For paths 5–6, users in portrait orientation would find the point faster than in landscape orientation. Although while seizing point 3, the landscape players would see the location of monument 6 (considering the width of the field of vision), this knowledge failed to translate into moving from 5–6 faster than portrait users.

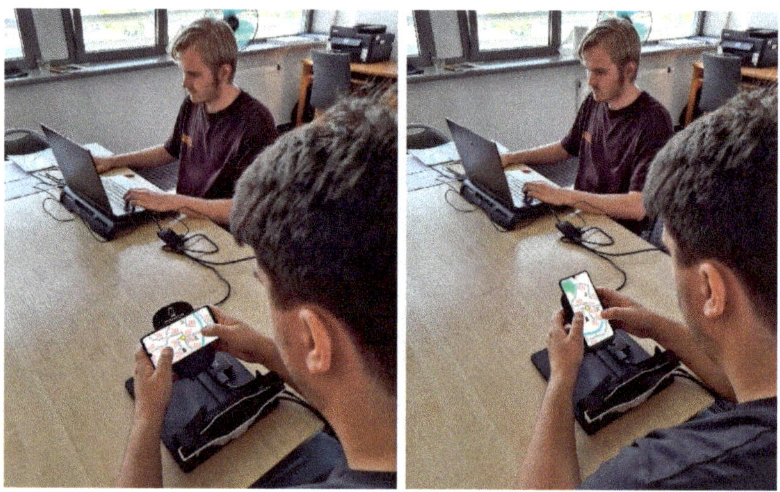

Figure 4. Researcher and participant with a smartphone in landscape and portrait orientation.

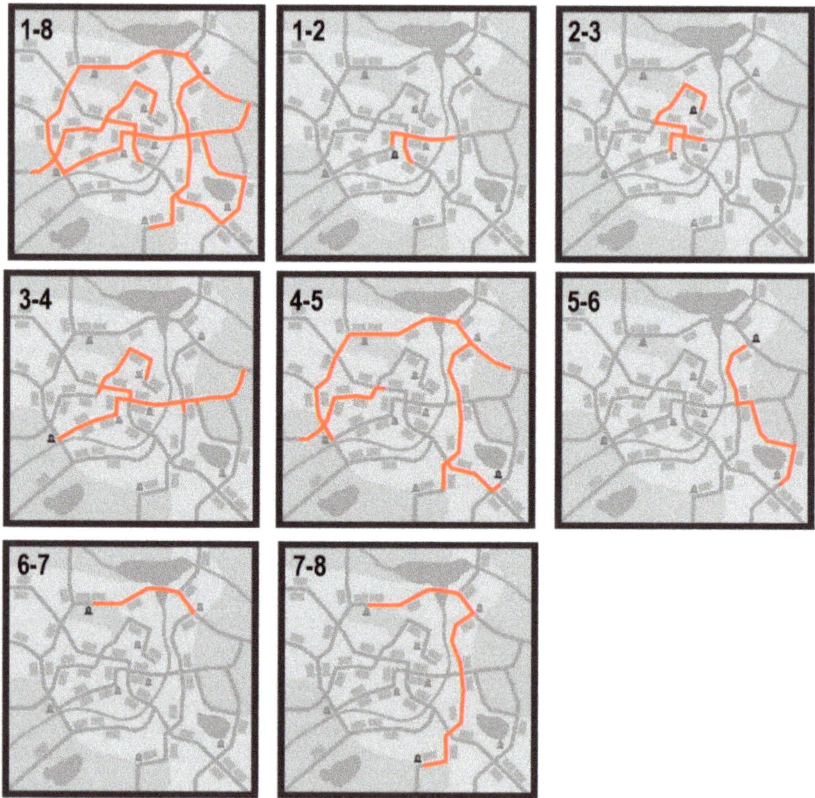

Figure 5. Path data of participant no. P18 (see Figure 6).

A — Total time (s) — LANDSCAPE — Time of each segment (s)

Id	T1-8:	T1-2:	T2-3:	T3-4:	T4-5:	T5-6:	T6-7:	T7-8:
L01	118.86	5.44	20.52	14.34	31.06	12.04	8.56	26.90
L02	99.28	6.08	26.48	15.98	14.94	12.18	7.90	15.72
L03	126.48	13.20	14.02	49.46	13.62	11.20	6.24	18.74
L04	93.36	6.76	11.56	8.82	24.04	8.68	10.02	23.48
L05	120.38	9.90	17.70	16.16	43.20	10.62	6.68	16.12
L06	133.06	21.60	9.82	13.76	45.40	11.34	6.12	25.02
L07	84.12	6.96	13.46	10.40	15.26	10.86	5.92	21.26
L08	128.10	8.84	14.26	28.32	19.32	19.78	11.42	26.16
L09	129.54	8.80	27.06	17.46	20.36	16.28	10.72	28.86
L10	141.90	36.96	15.68	13.74	33.60	11.70	6.22	24.00
L11	100.54	4.04	11.46	12.62	25.70	13.52	7.10	26.10
L12	117.32	9.80	31.90	16.38	18.62	11.80	6.46	22.36
L13	128.66	9.14	29.56	10.84	40.50	10.80	8.50	19.32
L14	83.50	3.50	10.16	15.10	13.78	10.74	6.88	23.34
L15	114.12	9.86	13.44	19.46	28.96	12.76	6.34	23.30
L16	124.34	12.86	17.18	16.50	23.62	23.76	11.08	19.34
L17	137.86	5.20	17.54	17.92	34.94	28.28	6.16	27.82
L18	130.04	31.82	12.44	13.72	14.70	12.10	15.86	29.40
L19	147.38	9.24	22.38	16.32	43.46	11.36	6.88	37.74
L20	115.74	7.70	21.60	13.32	24.82	12.90	6.86	28.54
L21	127.50	7.56	10.32	40.34	13.14	10.28	30.24	15.62
L22	74.68	6.48	9.38	9.60	11.96	9.88	5.92	21.46
L23	104.36	12.28	21.24	11.22	21.92	9.96	6.50	21.24
L24	99.68	4.24	10.02	14.70	28.56	9.04	6.28	26.84
L25	182.70	4.96	40.62	30.82	47.80	19.92	7.10	31.48
L26	100.42	9.12	11.58	14.10	15.22	12.42	8.08	29.90
L27	98.08	13.26	14.02	14.18	22.74	10.26	6.06	17.56
L28	168.92	9.46	40.82	15.10	55.06	13.72	8.30	26.46
L29	70.20	3.30	9.58	8.86	11.18	10.22	5.64	21.42
L30	86.70	6.12	14.60	10.20	16.34	12.38	9.22	17.84
Median	118.09	8.82	14.43	14.52	23.18	11.75	6.88	23.41

B — Total time (s) — PORTRAIT — Time of each segment (s)

Id	T1-8:	T1-2:	T2-3:	T3-4:	T4-5:	T5-6:	T6-7:	T7-8:
P01	118.18	20.16	14.88	11.62	30.44	10.64	7.82	22.62
P02	137.56	10.02	26.94	18.36	34.84	13.32	12.06	22.02
P03	105.24	6.84	10.16	13.36	38.86	9.20	5.76	21.06
P04	115.22	6.42	15.32	11.84	47.08	10.00	6.52	18.04
P05	110.84	10.66	18.54	12.52	18.92	10.76	6.18	33.26
P06	144.94	4.24	11.08	10.66	43.06	9.14	6.02	60.74
P07	147.64	4.38	11.18	50.70	13.52	15.10	15.38	37.38
P08	121.54	10.72	13.54	17.10	12.56	9.88	19.66	38.08
P09	119.76	10.76	15.14	12.32	32.14	16.14	7.72	25.54
P10	136.66	39.90	23.80	12.64	13.40	9.18	8.76	28.98
P11	133.30	4.94	10.48	12.38	55.70	11.90	21.20	16.70
P12	83.96	6.36	12.20	10.30	13.16	10.96	6.26	24.72
P13	92.40	3.38	11.34	12.38	14.54	10.12	26.92	13.72
P14	92.58	9.78	11.12	10.82	30.56	10.38	6.06	13.86
P15	101.74	6.48	10.82	9.80	39.28	12.22	5.74	17.40
P16	118.58	23.58	11.16	11.76	30.42	8.88	14.42	18.36
P17	89.72	5.22	10.86	15.78	27.26	9.50	6.32	14.78
P18	147.08	15.02	14.00	32.24	52.74	10.82	6.24	16.02
P19	113.58	4.44	9.28	15.82	52.18	8.80	5.70	17.36
P20	105.96	2.84	10.26	14.86	39.82	9.76	5.90	22.52
P21	114.24	8.98	15.84	15.16	28.90	10.36	7.08	27.92
P22	118.74	11.06	10.64	16.26	34.56	13.34	6.08	26.80
P23	123.26	3.28	10.24	12.50	49.66	11.88	7.82	27.88
P24	151.20	12.20	12.58	11.52	80.42	12.22	6.26	16.00
P25	100.54	8.18	12.50	15.26	17.34	10.66	10.24	26.36
P26	97.04	10.66	12.28	11.56	19.78	9.30	6.32	27.14
P27	143.38	15.04	14.20	12.22	59.12	11.18	5.80	25.82
P28	97.12	7.48	15.22	10.02	28.02	10.40	5.76	20.22
P29	185.42	13.18	21.82	59.68	54.26	9.14	6.58	20.76
P30	104.28	5.00	9.86	9.52	29.80	9.96	6.52	33.62
Median	116.70	8.58	12.24	12.44	31.35	10.39	6.42	22.57

Figure 6. (**A**): Time data of 30 participants with landscape orientation; (**B**): Time data of 30 participants with portrait orientation (red is the longest time, and blue is the shortest time in each column).

Median for Total time (s) — B. Median for time of each segment (s)

	T1-8:	T1-2:	T2-3:	T3-4:	T4-5:	T5-6:	T6-7:	T7-8:
P	116.70	8.58	12.24	12.44	31.35	10.39	6.42	22.57
L	118.09	8.82	14.43	14.52	23.18	11.75	6.88	23.41

Figure 7. Comparison of the median time for landscape and portrait orientation.

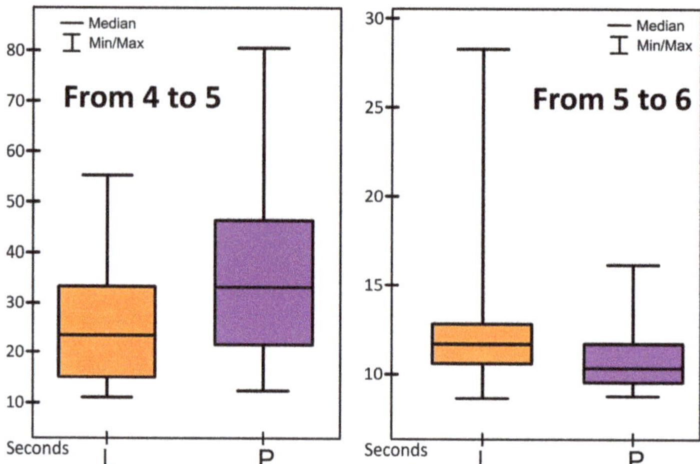

Figure 8. Box plots of statistical significance for paths 4–5 and 5–6 in portrait (P) and landscape (L) orientation.

4.2. Cartographic Visualization

A cartographic visualization of the research results included three quantitative mapping techniques: a heat map, a comparative flow map (band cartodiagram), and a proportional point symbol map [38] (Figure 1).

The first visualization in the form of six heat maps was based on fixation distribution (Figure 9). This cartographic image presents a visual strategy for completing tasks in portrait and landscape orientation. The results presented are related to the entire task, i.e., the user's moving from monument 1 toward monument 8. In the portrait version, the users' gaze was focused on the central part of the smartphone screen, slightly below the geometric center (Figure 9A). Fixations are distributed significantly more along the longer vertical axis of the screen. Therefore, the point of the highest density of fixations (app. 600,000) had an ellipsoidal shape. The further away from the center, the lower the number of fixations. The place located above the main concentration of fixations was the only exception, as an area that was smaller but with a high number of fixations. In the landscape version, the users' gaze also focused on the central part of the screen and slightly below the geometric center of the display. Even though the distribution of the red area with the highest number of fixations has a semi-circular shape, one can still observe a slight elongation towards the ellipse (Figure 9B).

Due to significant differences in the time of completing paths 4–5 (Figure 7), we have analyzed the differences in visual strategy during moving between those two stone monuments (Figure 9C,D). The concentration of fixations on the choropleth map C is close to the total gameplay visible in visualization 9A. The deviation of the significant part of fixations toward the right side of the screen, both in portrait and in landscape orientation, constitutes the main difference, the reason for it being surely the fact that the user was moving from monument 4 (located in the most left part of the map, in the west) to monument 5 (located in the most right part of the map, in the east) (Figure 2). The places where concentration was the greatest oscillated at around 110,000 fixations, which constituted around 18% of fixations from the entire task.

The differences were also outlined in the analysis of the visual strategy research participants adopted for moving from monument 5 to monument 6 (Figure 7). For portrait orientation, fixations are concentrated slightly below the center of the screen along the vertical axis, forming the ellipsoidal shape (Figure 9E). There are extra hot spots of visual activity closer to the upper part of the screen, which may be related to moving toward monument 6 at the top (the north) of the map. On the other hand, in landscape orientation, extra hot spots are located closer to the bottom and the right part of the screen (Figure 9F).

The proportional point symbols map in Figure 10 represents the number of fixations on individual stone monuments while they were active in black, and the participant was moving toward that point. No. 2 was the most frequently observed point in portrait orientation (4.4 fixations on average) and point 3 in landscape orientation (4.9 fixations on average). In this aspect, the greatest difference between orientations occurs for monument 2. In portrait, while point 2 was active, it had 4.4 fixations on average, whereas in landscape, it had 3.1 fixations. Interestingly, monuments 1–4 had more than three fixations on average in both orientations, whereas other points had less than three fixations of the value of 2–3.

A comparative flow map (band cartodiagram) is the last statistical map, presenting the numerical load of individual road stretches (Figure 11). The number of walks down a specific stretch taken by 30 research participants in landscape orientation and by 30 participants in portrait orientation was presented from monument 1 to monument 8 in six classes. The maximum load for a single road stretch for one user is 7 and for 30 users is 210. To obtain the number of walks that participants took, stretches were labeled with two letters, e.g., AA. Each two-letter description is placed in the middle of the distance between nodes. We adapted this method of placing the description for sections from the description of the kilometers value in road maps. The roads have been divided into small sections according to road crossings, section length, and the location of the monument.

Based on the comparative flow map, a significant load of roads is visible in both orientations in the center of the area, between monuments 1-2-3 (Figure 11). Moreover, a similar load of road stretches between monuments 5–7 in the north and east of the area is visible in both versions. In landscape orientation stretches, BB, BC, BD, and BV report a greater load. Whereas, when it comes to portrait, it is typical to have a greater load on BU, BW, and BX. In the western part, in landscape orientation, the AP stretch has a significantly greater load than in portrait orientation. CI and CF stretches have a greater load in portrait orientation.

5. Discussion

The visual strategy of research participants that were doing a topographic task on the smartphone with portrait orientation is highly similar to the one applied in task-oriented monster elimination during the experience with mobile games [39]. Supposedly, it is also a strategy similar to observing one's position and the elements that occur in a mobile car navigation. On the other hand, tasks related to navigation in topographic space in landscape orientation demonstrate a visual strategy focused on the central part of the screen. Certain similarities may also be observed in the research by Hejtmánek et al. [40], in which during navigation in a virtual city, the central part of the monitor in landscape orientation was the spot that the gaze was most focused on.

The difference in fixation distribution between tasks 1–8 and 4–5 may be explained by the fact that point 5 was located in the right part of the map. Hence, the greater the density of fixations on the right side of the smartphone screen. The situation is similar for the track between points 5 and 6 but only for landscape orientation. When it comes to the average number of fixations on individual points of the map, the smaller number of fixations on points 5 to 8 may result from having familiarized oneself with topographic space participants to a satisfactory degree. Time results seem to corroborate this, not showing any drop in the amount of time necessary to complete the task, whereas participants do not need to look at sequentially activating monuments that often anymore.

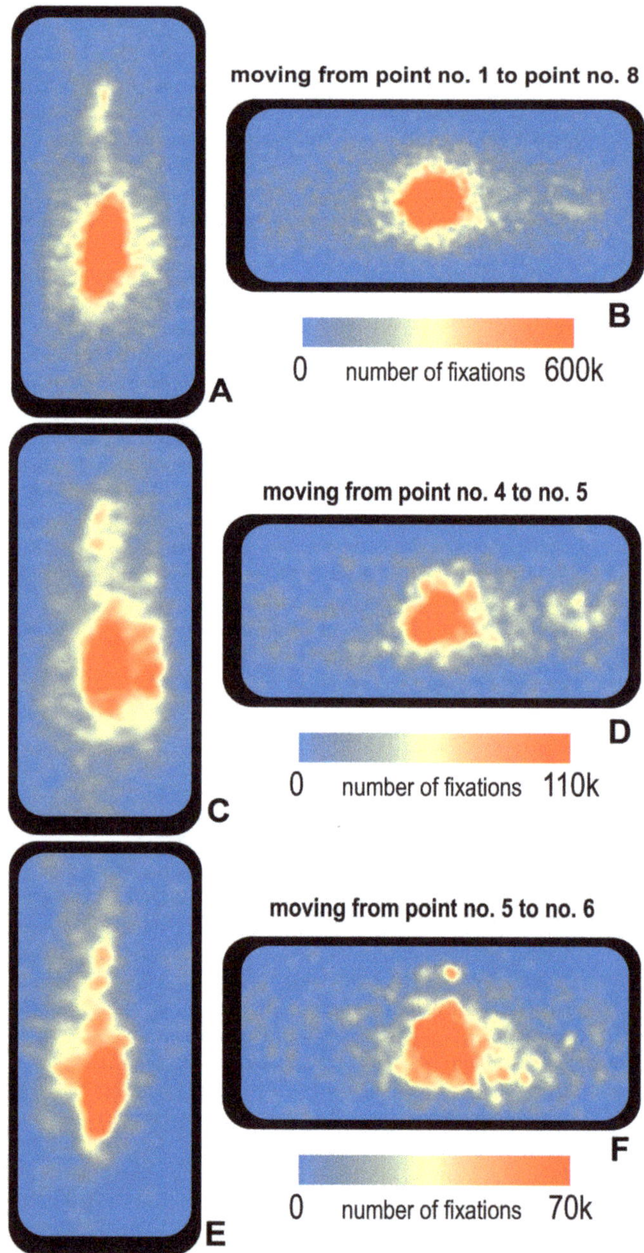

Figure 9. Heat map with participant fixation locations on smartphone: (**A**)—portrait orientation for the movement from monuments 1 to 8; (**B**)—landscape orientation for the movement from monuments 1 to 8; (**C**)—portrait orientation for the movement from monuments 4 to 5; (**D**)—landscape orientation for the movement from monuments 4 to 5; (**E**)—portrait orientation for the movement from monuments 5 to 6; (**F**)—landscape orientation for the movement from monuments 5 to 6.).

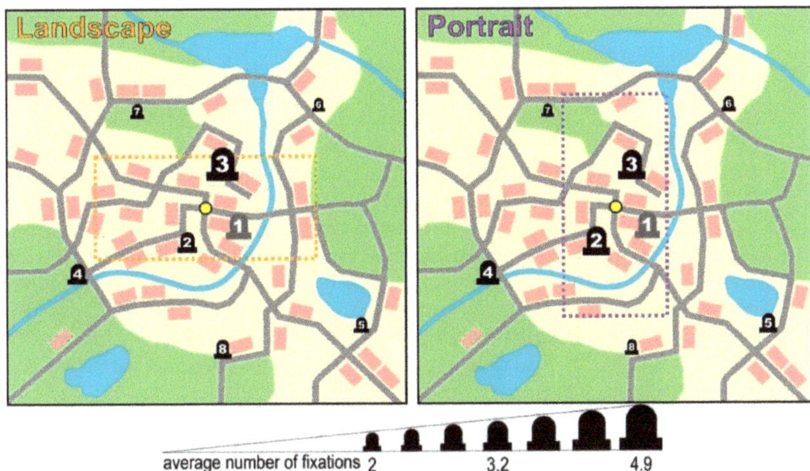

Figure 10. Proportional point symbols map with the presentation of the fixation on specific stone monuments.

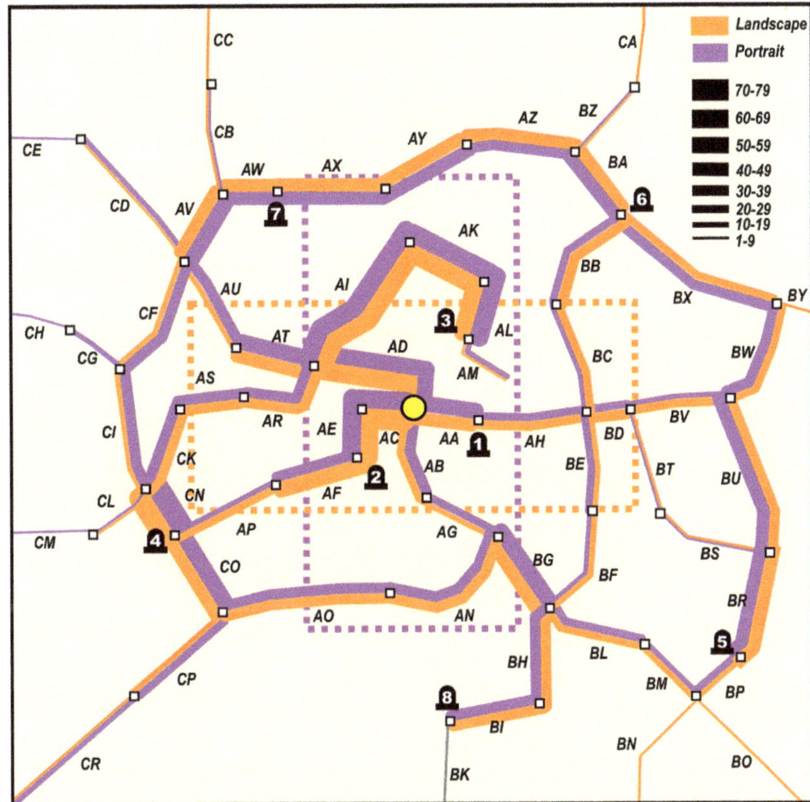

Figure 11. Comparative flow map with the presentation of the moving number along individual road sections (two-letter description of sections from AA to CR).

Each form of result analysis presents different similarities and differences but what is important is to use them in a complementary way [19]. In analyzing the fixation field deviation to the right from the center of the display in moving from monument 4 to 5, the heat map is not enough because it is only the comparison with the basic topographic map and the flow map that reveals the cause of the deviation. Hence, we can conclude that the area of high fixations is located slightly below the geometric center of the smartphone in the shape of the ellipse, significantly elongated vertically for portrait orientation, whereas for landscape orientation, there is little elongation. However, the direction of moving around on the map has a great impact on the deviation of the high fixations field. The effect of perceiving the symbol exactly in the optical center of the graphical image is one of the rules of map design that recommends placing the symbol in the optical center of the image, i.e., slightly above the geometric center. Therefore, it is worth considering whether one may apply this rule to influence the distribution of fixations.

The analysis of the load of individual stretches of the road made by means of a flow map proves that there are differences in the choice of the road between landscape and portrait orientation in the eastern part of the map. The track that is chosen more often in portrait orientation from monument 5 to monument 6 is a consequence of the screen orientation. The change of the screen orientation would lead to east–west road stretches being chosen more often than north–south stretches, which translated into slower times. In the western part of the field, a similar tendency occurs. In portrait orientation, the north–south stretches (CI and CF) have a definitely higher traffic load, whereas in landscape orientation, east–west stretches (AP and AF) were used more frequently. It has also been observed that individual research participants were wandering around vertically in landscape orientation because they reached CC, CA, BN, and BO stretches. It may result from the smartphone's orientation. On the other hand, in portrait orientation, individual participants were wandering around horizontally, reaching CE, CH, and CM stretches in the west. In both orientations, research participants would wander around CR and CP stretch. Interestingly, in landscape orientation, no participant would enter stretches that were directed leftwards, i.e., CE, CH, or CM. What is convincing is the fact that, when moving around on frequented roads AV-CF-CI-CN, the participant would already see the edge of the frame enclosing the content of the map. The users of portrait orientation had an easier situation when it came to noticing the edge of the map at the top and at the bottom of the screen.

The implementation of the minimized touch interface that follows current tendencies in app design turned out to be a practical solution because no research participant would report any problems with moving around on the map by means of touch [41]. The elimination of joysticks and other buttons resulted in the participant's focus solely on cartographic content. However, in some cases, it was necessary to remind participants that they were allowed to move around only on roads, which was not obvious to players. Probably a car or walker symbol would be more obvious to the player, but we assumed that we would use very simplified graphics.

The number of cartographic visualizations of results and their graphic complexity of form remains a topic to be discussed [38]. The opportunities of complementary partial analysis and full synthetic presentation of map use features have both effective and impressive aspects [42]. From the example of six heat maps, which can be analyzed separately but without being interpreted as a whole, it is not possible to draw more general conclusions, proving that fact. Comparative cartodiagrams have great advantages, but it is necessary to use a maximum of six ranges of the range scaling. The use of absolute scaling for the map of proportional point symbols may complicate the adjustment of the symbol's size to the factual numerical value. The authors used uncomplicated graphic result mapping techniques because methods that are graphically attractive often fail to be effective enough for the proper interpretation of phenomena in geographical space.

6. Conclusions

To sum up the research, we can say that the concept of research suggested resulted in the demonstration of similarities and differences in usability on a simplified topographical 2D map on the smartphone in the landscape and portrait orientation:

Main similarity:

- regardless of the orientation of the smartphone, participants completed the whole task at a similar time (this is the recommendation that both device orientations are adequate for presenting topographical space on a mobile phone);

Main differences:

- both orientations implied different path selections (the recommendation is that the orientation of the smartphone has great importance to the better orientation on the north–south and the west–east axis);
- both orientations implied different participants' visual strategies (the recommendation is that the smartphone orientation should support the visibility of crucial information on the map).

Answering the questions raised at the beginning, we can conclude that interactive map design for smartphones should take place according to the rules of cartographic design and current IT trends to create a map application with gameplay elements. Such an approach applies both to designing the movement on the topographic graphics and obtaining data on movement in two smartphone orientations.

The presentation of data on how users move requires the employment of complementary cartographic methods. The application of a single representation method shows only one aspect of the issue, whereas using a few methods allows us to link the results together and demonstrate the broader spectrum of the problem of moving around on the map on the smartphone. The flow map, the proportional point symbols, and the heat map method turned out to be particularly useful.

In our study, we were attempting to solve a few problems related to the use of a traditional cartographic medium in the form of a 2D map in its geomedia extension on smartphones. We understand such a geomedia extension of the map's functionality as the user's opportunity to interactively move around the map's content on the smartphone screen. 'The voyage of a finger across the map' has a new geomedia dimension nowadays. On a printed static map, a finger moves around on the roads, whereas in the research, the user's fingers move around on the roads as the user is 'anchored' to the road in the middle of the smartphone display.

While users feel more comfortable in a portrait orientation of a smartphone, most vehicles have their screens in landscape format. This fact makes the development of research in the context of the usability of applications or cartographic products inevitable. The authors of this article hope that the obtained results may contribute to the development of research in this area and, consequently, lead to improvements in maps on various devices, not only smartphones. Future studies should consider conducting an experiment with both device orientations for both groups so that there would be feedback if the different orientation improves previous user experience.

Author Contributions: Conceptualization, Beata Medyńska-Gulij; data curation, Paweł Cybulski, Krzysztof Zagata, Jakub Zawadzki and Tymoteusz Horbiński; formal analysis, Beata Medyńska-Gulij, Paweł Cybulski, Krzysztof Zagata and Jakub Zawadzki; investigation, Paweł Cybulski, Krzysztof Zagata and Jakub Zawadzki; methodology, Beata Medyńska-Gulij, Jacek Gulij, Paweł Cybulski, Krzysztof Zagata and Tymoteusz Horbiński; project administration, Beata Medyńska-Gulij; resources, Beata Medyńska-Gulij, Jacek Gulij, Paweł Cybulski, Krzysztof Zagata, Jakub Zawadzki and Tymoteusz Horbiński; software, Beata Medyńska-Gulij, Jacek Gulij, Paweł Cybulski, Krzysztof Zagata, Jakub Zawadzki and Tymoteusz Horbiński; validation, Beata Medyńska-Gulij, Paweł Cybulski, Krzysztof Zagata and Tymoteusz Horbiński; visualization, Beata Medyńska-Gulij and Krzysztof Zagata; writing—original draft, Beata Medyńska-Gulij, Jacek Gulij, Paweł Cybulski, Krzysztof Zagata

ISPRS Int. J. Geo-Inf. **2022**, *11*, 577

and Tymoteusz Horbiński; writing—review and editing, Beata Medyńska-Gulij. All authors have read and agreed to the published version of the manuscript.

Funding: This research received no external funding.

Data Availability Statement: Not applicable.

Conflicts of Interest: The authors declare no conflict of interest.

References

1. Muehlenhaus, I. Internet mapping. In *The Routledge Handbook of Mapping and Cartography*; Kent, A.J., Vujakovic, P., Eds.; Routledge: London, UK, 2018; pp. 375–387, ISBN 9780367581046.
2. Robinson, A.H.; Morrison, J.L.; Muehrecke, P.C.; Kimerling, A.J.; Guptil, S.C. *Elements of Cartography*, 6th ed.; Wiley: New York, NY, USA, 1995; ISBN 9788126524549.
3. Kent, A.; Vujakovic, P. Cartographic Language: Towards a New Paradigm for Understanding Stylistic Diversity in Topographic Maps. *Cartogr. J.* **2011**, *48*, 21–40. [CrossRef]
4. Medyńska-Gulij, B. Point Symbols: Investigating Principles and Originality in Cartographic Design. *Cartogr. J.* **2008**, *45*, 62–67. [CrossRef]
5. Wood, C.H.; Keller, C.P. *Cartographic Design: Theoretical and Practical Perspectives*; Wiley: Chicherster, UK, 1996; pp. 1–10, ISBN 9780471965879.
6. Spiess, E. Map compilation. In *Basic Cartography: Volume Two. For Students and Technicians*, 2nd ed.; Anson, R.W., Ormeling, F.J., Eds.; Butterworth-Heinemann: Oxford, UK, 2002; pp. 25–84, ISBN 9781851662326.
7. Dent, B.D. *Cartography: Thematic Map Design*, 5th ed.; McGraw-Hill: Boston, CA, USA, 1999; ISBN 9780697384959.
8. Lloyd, R. Attention on Maps. *Cartogr. Perspect.* **2005**, *52*, 28–57. [CrossRef]
9. Medyńska-Gulij, B. How the Black Line, Dash and Dot Created the Rules of Cartographic Design 400 Years Ago. *Cart. Graph. J.* **2013**, *50*, 356–368. [CrossRef]
10. Horbiński, T.; Zagata, K. View of Cartography in Video Games: Literature Review and Examples of Specific Solutions. *KN J. Cartogr. Geogr. Inf.* **2022**, *72*, 117–128. [CrossRef]
11. Kapenekakis, I.; Chorianopoulos, K. Citizen science for pedestrian cartography: Collection and moderation of walkable routes in cities through mobile gamification. *Hum. Cent. Comput. Inf. Sci.* **2017**, *7*, 10. [CrossRef]
12. Lammes, S.; Wilmott, C. The map as playground: Location-based games as cartographical practices. *Convergence* **2018**, *24*, 648–665. [CrossRef]
13. Koulaxidis, G.; Xinogalos, S. Improving Mobile Game Performance with Basic Optimization Techniques in Unity. *Modelling* **2022**, *3*, 201–223. [CrossRef]
14. Edler, D.; Keil, J.; Dickmann, F. From Na Pali to Earth—An 'Unreal' Engine for Modern Geodata? In *Modern Approaches to the Visualization of Landscapes*; Edler, D., Jenal, C., Kühne, O., Eds.; Springer: Wiesbaden, Germany, 2020; ISBN 9783658309558.
15. Zaina, L.A.M.; de Mattos Fortes, R.P.; Casadei, V.; Nozaki, L.S.; Paiva, D.M.B. Preventing accessibility barriers: Guidelines for using user interface design patterns in mobile applications. *J. Syst. Softw.* **2022**, *186*, 111213. [CrossRef]
16. Giurgiu, L.; Gligorea, I. Responsive Web Design Techniques. *Int. Conf. Knowl. -Based Organ.* **2017**, *23*, 37–42. [CrossRef]
17. Horbiński, T.; Cybulski, P.; Medyńska-Gulij, B. Web Map Effectiveness in the Responsive Context of the Graphical User Interface. *ISPRS Int. J. Geo-Inf.* **2021**, *10*, 134. [CrossRef]
18. Horbiński, T.; Cybulski, P. Similarities of global web mapping services functionality in the context of responsive web design. *Geod. Cartogr.* **2018**, *67*, 159–177. [CrossRef]
19. Halik, Ł.; Kent, A.J. Measuring user preferences and behaviour in a topographic immersive virtual environment (TopoIVE) of 2D and 3D urban topographic data. *Int. J. Digit. Earth* **2021**, *14*, 1835–1867. [CrossRef]
20. Korycka-Skorupa, J.; Gołębiowska, I. Multivariate mapping for experienced users: Comparing extrinsic and intrinsic maps with univariate map. *Misc. Geogr.* **2021**, *25*, 259–271. [CrossRef]
21. Medyńska-Gulij, B.; Zagata, K. Experts and Gamers on Immersion into Reconstructed Strongholds. *ISPRS Int. J. Geo-Inf.* **2020**, *9*, 655. [CrossRef]
22. Wielebski, Ł.; Medyńska-Gulij, B.; Halik, Ł.; Dickmann, F. Time, Spatial, and Descriptive Features of Pedestrian Tracks on Set of Visualizations. *ISPRS Int. J. Geo-Inf.* **2020**, *9*, 348. [CrossRef]
23. Cybulski, P. An Empirical Study on the Effects of Temporal Trends in Spatial Patterns on Animated Choropleth Maps. *ISPRS Int. J. Geo-Inf.* **2022**, *11*, 273. [CrossRef]
24. Horbiński, T.; Zagata, K. Interpretation of Map Symbols in the Context of Gamers' Age and Experience. *ISPRS Int. J. Geo-Inf.* **2022**, *11*, 150. [CrossRef]
25. Edler, D.; Keil, J.; Tuller, M.-C.; Bestgen, A.-K.; Dickmann, F. Searching for the 'Right' Legend: The Impact of Legend Position on Legend Decoding in a Cartographic Memory Task. *Cartogr. J.* **2020**, *57*, 6–17. [CrossRef]
26. Zagata, K.; Gulij, J.; Halik, Ł.; Medyńska-Gulij, B. Mini-Map for Gamers Who Walk and Teleport in a Virtual Stronghold. *ISPRS Int. J. Geo-Inf.* **2021**, *10*, 96. [CrossRef]

27. Lee, E.C.; Park, M.W. A New Eye Tracking Method as a Smartphone Interface. *KSII Trans. Internet Inf. Syst.* **2013**, *7*, 834–848. [CrossRef]
28. Qu, Q.X.; Zhang, L.; Chao, W.Y.; Duffy, V. User Experience Design Based on Eye-Tracking Technology: A Case Study on Smartphone APPs. In *Advances in Applied Digital Human Modeling and Simulation*; Duffy, V., Ed.; Springer: Cham, Switzerland, 2017; ISBN 9783319416267.
29. Chynał, P.; Szymański, J.M.; Sobecki, J. Using Eyetracking in a Mobile Applications Usability Testing. In *Intelligent Information and Database Systems*; Pan, J.S., Chen, S.M., Eds.; Springer: Berlin/Heidelberg, Germany, 2012; ISBN 9783642284922.
30. Calka, B.; Nowak Da Costa, J.; Bielecka, E. Fine scale population density data and it's application in risk assessment. *Geomat. Nat. Hazards Risk* **2017**, *8*, 1440–1455. [CrossRef]
31. Dong, W.; Wang, S.; Chen, Y.; Meng, L. Using Eye Tracking to Evaluate the Usability of Flow Maps. *ISPRS Int. J. Geo-Inf.* **2018**, *7*, 281. [CrossRef]
32. Keskin, M.; Ooms, K.; Dogru, A.O.; De Maeyer, P. Exploring the Cognitive Load of Expert and Novice Map Users Using EEG and Eye Tracking. *ISPRS Int. J. Geo-Inf.* **2020**, *9*, 429. [CrossRef]
33. Smaczyński, M.; Medyńska-Gulij, B.; Halik, Ł. The Land Use Mapping Techniques (Including the Areas Used by Pedestrians) Based on Low-Level Aerial Imagery. *ISPRS Int. J. Geo-Inf.* **2020**, *9*, 754. [CrossRef]
34. Kent, A.J. Topographic Maps: Methodological Approaches for Analyzing Cartographic Style. *J. Map Geogr. Libr.* **2009**, *5*, 131–156. [CrossRef]
35. Jelizakov, D. Mobile Usability Made Simple. 2019. Available online: https://uxplanet.org/mobile-usability-made-simple-945e106e23eb (accessed on 5 July 2022).
36. Gao, Y.; Li, A.; Zhu, T.; Liu, X.; Liu, X. How smartphone usage correlates with social anxiety and loneliness. *PeerJ* **2016**, *4*, e2197. [CrossRef]
37. Hoober, S. How Do Users Really Hold Mobile Devices. 2013. Available online: https://www.uxmatters.com/mt/archives/2013/02/how-do-users-really-hold-mobile-devices.php (accessed on 5 July 2022).
38. Forrest, D. Thematic Maps in Geography. In *International Encyclopedia of the Social & Behavioral Sciences*, 2nd ed.; Wright, J.D., Ed.; Elsevier: Oxford, UK, 2015; ISBN 9780080970875.
39. Jiang, J.-Y.; Guo, F.; Chen, J.-H.; Tian, X.-H.; Lyu, W. Applying Eye-Tracking Technology to Measure Interactive Experience Towards the Navigation Interface of Mobile Games Considering Different Visual Attention Mechanisms. *Appl. Sci.* **2019**, *9*, 242. [CrossRef]
40. Hejtmánek, L.; Oravcová, I.; Motýla, J.; Horáček, J.; Fajnerováa, I. Spatial knowledge impairment after GPS guided navigation: Eye-tracking study in a virtual town. *Int. J. Hum.-Comput. Stud.* **2018**, *116*, 15–24. [CrossRef]
41. Christopoulou, E.; Xinogalos, S. Overview and Comparative Analysis of Game Engines for Desktop and Mobile Devices. *Int. J. Serious Games* **2017**, *4*, 21–36. [CrossRef]
42. Wielebski, Ł.; Medyńska-Gulij, B. Graphically supported evaluation of mapping techniques used in presenting spatial accessibility. *Cartogr. Geogr. Inf. Sci.* **2019**, *46*, 311–333. [CrossRef]

International Journal of
Geo-Information

Article

Mini-Map Design Features as a Navigation Aid in the Virtual Geographical Space Based on Video Games

Krzysztof Zagata * and Beata Medyńska-Gulij

Adam Mickiewicz University, Poznań, Poland; Department of Cartography and Geomatics,
61-712 Poznań, Poland
* Correspondence: krzysztof.zagata@amu.edu.pl; Tel.: +48-61-829-6242

Abstract: The main objective of this study is to identify features of mini-map design as a navigational aid in the virtual geographical space in 100 popular video games for a computer platform. The following research methods were used: visual comparative analysis, classification and selection of cartographic material, comparison of specific parameters for selected features of design elements, and application of cartographic design rules and popularity of design solutions in video games. The study revealed eight features of mini-map design and their popular parameters and attributes in video games, with only one game meeting all conditions of popularity: projection: orthographic; centring: player-centred; base layers: artificial; shape: circle; orientation: camera view; position: bottom left; proportions: 2.1–3%; additional navigational element: north arrow. The key attributes of the mini-map's features were captured, which, when considered separately, complementarily and potentially holistically, confirm the possibility of designing the mini-map according to traditional cartographic design principles. The identified parameters of the mini-map can be useful not only in the design of the game cartography interface, but also for other geomedia products.

Keywords: mini-map; map design; virtual geographical space; navigation aid; video games; game cartography interface; graphical features

Citation: Zagata, K.; Medyńska-Gulij, B. Mini-Map Design Features as a Navigation Aid in the Virtual Geographical Space Based on Video Games. *ISPRS Int. J. Geo-Inf.* **2023**, *12*, 58. https://doi.org/10.3390/ijgi12020058

Academic Editors: Florian Hruby and Wolfgang Kainz

Received: 12 December 2022
Revised: 24 January 2023
Accepted: 6 February 2023
Published: 8 February 2023

1. Introduction

A video game is a complex technological system that combines many programming, geographical, cartographic, psychological and cognitive aspects [1,2]. Many video games are based on real-world concepts to represent virtual geographical space and present a series of phenomena linked by spatial information [3]. Year by year, game developers are building increasingly complex geographical spaces with the help of growing computer game engines (e.g., Unity, Unreal Engine, CryEngine, Frostbite) [4]. The subject of "cartography and games" is taken up in research in the context of designing cartographic products, game production, as well as aspects of gamification [5,6].

Maps are an integral part of cartographic research in the broad field of geomedia [7]. Maps are very often used in video games as a representation of a fictional geographical space [8]. Thanks to the basic content and the cartographic elements used, players can obtain spatial information about current events in the virtual world [9]. The design of a map under traditional conditions must be based on conscious rules and decisions due to the large variety of available methods and visual variables [10–12]. Therefore, in contrast to video game developers who do not rely on cartographic principles to design a virtual world, the authors believe it is worthwhile to investigate the relationship between the user and cartographic products in games [13,14]. Maps play an important role in common navigation both in video games and in real life by helping the user find their way and get directly from A to B [15,16].

Game cartography interfaces are practically ubiquitous in games that involve exploration and navigation of virtual geographical space [17]. As video games evolve, game developers are implementing newer and newer cartographic elements such as mini-maps, compasses, arrows, cartographic symbols and others to better understand space [18]. Toups et al. [19] investigated how the player interacts with the maps during the game. Based on the analysis, they referred to the design of the cartographic interface, but also to its functionality in terms of manipulating the map and its navigational elements.

A very important element of the user interface in topographic orientation (comparison of space with a map) in the virtual geographical space is the mini-map. The mini-map, sometimes called a 'radar screen' or 'corner map', shows a miniature version of the game world, or part of it, from the perspective of the top view [20]. The mini-map usually represents a larger space than the area visible from the player's main camera [21]. The element in the game cartography interface is a kind of navigational aid that provides detailed information about the position, orientation, objects nearby or terrain around the player. The mini-map, usually created as one of the non-diegetic parts of the HUD (heads-up display), provides important information to make the right decisions during the game and plays an important role in navigation in different game genres (e.g., role-playing games, multiplayer online role-playing games, racing games, shooter games, battle royale games, survival games and strategy games), and especially in open-world games [22,23].

Indeed, geographical space is closely linked to different societies of people who relate to each other through the natural environment that surrounds them [24]. On the other hand, virtual geographical space is an environment that, through rules derived from real geographical space, allows the reflection and shaping of existing geographical space or the creation of fictitious spaces based on real-world rules [25]. The basic activity in video games is to navigate through three-dimensional and two-dimensional virtual geographical spaces in order to find a way to new destinations or to freely explore the game in conjunction with the game's open world [26]. One of the tasks of cartography that has found its place in video game research over the years is to understand the definition and role of "space" [27–30]. Almost all video games with programmed gameplay mechanics allow players to memorise and assimilate new spatial knowledge, which they use to transparently employ interactive navigational elements as they move through the game world [31,32]. In the article by Si et al. [33], the term 'spatial exploration' is mentioned as a fundamental element of the game, where players increase their knowledge of geographical space and improve navigation between important topographical objects.

The main problem addressed in the research concerns the development and the selection of basic features of mini-maps design for a navigation aid in the virtual geographical space. The main objective of the study is to identify mini-map design features for a navigation aid in the virtual geographical space in video games. To achieve the main objective of the research, the specific aims are: to identify and arrange features of mini-maps according to their individual parameters and to capture key attributes for mini-map features that are considered separately, complementarily and as holistically as possible.

2. Materials and Methods

The fundamental methods of the research were: visual comparative analysis, classification and selection of cartographic material, comparison and classification of specific parameters for selected features of design elements and the application of cartographic design rules, and popularity of design solutions in video games [34–38].

To meet the research objectives, the stages of research were:

- Selection of mini-map features: shape, position, orientation, centring, projection, base layers, proportions, additional navigational elements (Section 2.1);
- Selection of the most popular and top rated video games that had a mini-map element in the game cartography interface (Section 2.2);
- Identification of the parameters and attributes for classifying mini-map features (Section 2.3);
- Analysis and juxtaposition of the mini-map features expressed as a percentage of each specific parameter or attribute (Sections 3.1–3.8);

- Graphical representation of mini-map design features (Section 3.9).

2.1. Selection of Mini-Map Features

The study proposes the following eight features of the mini-map: shape, position, orientation, centring, projection, base layers, proportions and additional navigation elements. After a careful analysis of the scientific literature, the authors noticed the lack of other scientific studies on the research topic under consideration. In selecting mini-map features, we were guided by the principles of the traditional cartographic map design and its components, as well as the design solution in video games [39,40]. Traditional cartographic map design principles helped enforce mini-map features such as shape, position, orientation, proportions and projection. The design of user interface elements in video games, on the other hand, helped organise features such as centring, base layers and additional navigational elements. Many of these mini-map features overlap between these two design worlds, and the combination of interdisciplinary principles allows for an objective evaluation of the design of the mini-map.

The first feature that can be used to classify the parameters of mini-maps used in video games is their shape. When creating interface elements, game developers adapt them to the commonly understood style of the game, especially its graphic design [41]. The second premise that determines the choice of the optimal shape of the mini-map is the willingness to adapt it to other elements of the user interface, which is the most important part of the navigation system due to its practical aspect [42]. The choice of the appropriate shape allows proper use of space on the user's screen (e.g., no unused spaces). The last factor considered when designing a user interface is the experience of video game developers, who sometimes choose to duplicate patterns that have worked well in other projects [43].

Another feature that characterizes the mini-map is the position of the interface elements on the screen. According to current trends in Western culture, interface elements should be located in the upper or left part of the screen, which is due to the user's reflex regarding the reading direction [44]. In the habits of Eastern cultures, on the other hand, user interface segments tend to be located at the top or right edge of the screen [45]. The relationship described above is used especially in games that require players to react quickly during the game. Placing important navigational elements in an intuitive position allows the user to significantly reduce reaction time and also avoid unnecessary clicks and deceptions [46]. In cartography, the creators of cartographic visualisations face a similar problem concerning the position of inset maps and in the map layout [47].

An important feature in the design of a mini-map is an orientation and, thus, the question in which direction the mini-map rotates in interaction with the camera of the player looking at virtual space. Considering this feature of mini-maps, it should be pointed out that it fulfils the role of a traditional compass, confirming that the mini-map should be perceived as a navigational element [48]. The choice of an appropriate orientation by video game developers facilitates the user's topographical orientation in the virtual geographical space, which affects the comfort of the game [49].

Video game developers have a choice between two types of mini-maps, depending largely on the level of detail of the world created. The mini-map can represent the entire game world, but it is possible, if not necessary, to represent many elements on the mini-map that make up the synthetic appearance of the virtual space [50]. However, some games require the representation of numerous elements that have a significant impact on the gameplay, which can affect the readability of the mini-map [51]. Here, a larger scale of the map is recommended and, thus, a stronger focus on the player's current position in the virtual world [52]. The above considerations show the importance of choosing the appropriate scale of the mini-map for the needs of a particular video game, which also argues for recognising the mini-map as a cartographic element in the user interface.

Another cartographic feature that video game developers must consider when designing a mini-map is how it is projected onto the plane. The vast majority of virtual visualisations use orthographic projection, which is perpendicular to the displayed space [53,54].

This allows the player to have a "bird's eye view" of the virtual space surrounding them, which allows them to estimate the distance more accurately between important points in the game. Perspective projection is the second type of projection, which is much less common in mini-maps. In this case, it is more difficult for the user to estimate the actual distances, as the virtual world is displayed from a certain angle and is, thus, distorted. However, this projection enables the representation of a larger field of view during the game, which allows the player to quickly develop initial ideas about the next fragments of the virtual geographical space.

An essential feature of the mini-map is the issue of the graphic design of the virtual geographical space. This aspect concerns the features of the mini-map, which, on the one hand, should fit the overall aesthetics of the interface and the game world and, on the other hand, should aim to increase navigational efficiency during the game. Choosing the right gameplay background and basic texture layers makes it easier to find one's way around the virtual space [55]. Game developers undertake to create a kind of graphic-aesthetic scheme that allows defining a closed set of markers always associated with the same topographic objects represented on the map and the mini-map (artificial layers), or on the use of unique analogies consisting in the reflection of the mini-map textures used in the game (in-game layers) [56,57]. In this way, the player has the possibility of constantly analysing the totality of the space surrounding him. Video game developers can also choose transparent layers, which allows the mini-map to focus only on the representation of cartographic symbols that are central to topographical orientation in an unfamiliar space.

The proportions of the individual interface elements on the displayed screen is another important design feature of the UI (user interface), which directly affects the appearance and readability of the game during play. The proportions of the individual elements are understood as the percentage of the area they occupy in relation to the total available screen [58]. The appropriate size of the cartographic element of the mini-map is extremely important, as too small an element size may result in insufficient visibility of the elements displayed on it, making it only a decorative element of the user interface. On the other hand, a mini-map that is too large may obscure other important elements of the video game's graphic design that may prove important during gameplay. As suggested by game designers, mini-maps should generally not exceed 10% of the available viewing area [50].

The final optional element of graphical user interface design in video games can be all kinds of navigational components designed to increase navigational efficiency when navigating virtual geographical space. A significant distinguishing feature of the criterion discussed here is the fact that it is not necessary to commit to a single additional navigational element [59]. However, when deciding on additional elements, video game designers should think about the functionality of the chosen solutions, because too much complexity in the number of elements may not make it easier or even more difficult for the user to find his way around the virtual space during the game [60,61].

2.2. Selection of Video Games

The number of professional video games released grows every year, and it is impossible to describe all available games in one article. Modern technologies of computer game engines offer increasingly better ways to create a graphic design and integrate a programming environment to develop a video game. The research focuses on games published for a computer platform (PC). Another criterion for selecting a game for analysis was whether the game had a navigation element in the user interface, i.e., a mini-map.

From the large number of online portals that provide video game ratings and publish lists of popular video games in history, 14 online portals were selected. The websites were selected by the Google search engine and ranked based on the phrases "top 100 video games" and "best 100 PC video games" (17 January 2022) in order of their display on the first and second page of popularity. According to Su et al. [62], a higher position in the internet search engine enables more frequent visits by internet users.

The 100 most popular and highest rated video games on video game websites were selected and categorised. The video games were rated by website users (gamers) and video game critics based on their subjective opinion of the game. The selection of games did not take into account that the same video games were duplicated from different websites. For each criterion of the mini-map features, a game was analysed only once.

The list of web portals analysed includes portals related to:

- Information services such as IGN [63–71];
- websites that collect critics' ratings, e.g., Metacritic [72–74];
- gaming magazines, such as, e.g., PC Gamer [75,76];
- websites that collect surveys on different fields, such as, e.g., Ranker [77].

On the other hand, the list of 100 games collected includes such titles as: The Witcher 3: Wild Hunt, Dead Rising 4, Grand Theft Auto V, Assassin's Creed IV: Black Flag, Forza Horizon 5, Red Dead Redemption 2, Guild Wars 2, World of Warcraft, Counter- Strike: Global Offensive, Warcraft III, Call of Duty: Warzone, Cyberpunk 2077, World of Tanks or League of Legends (Figure 1). All 100 games were installed on a computer and tested during a gameplay [Appendix A].

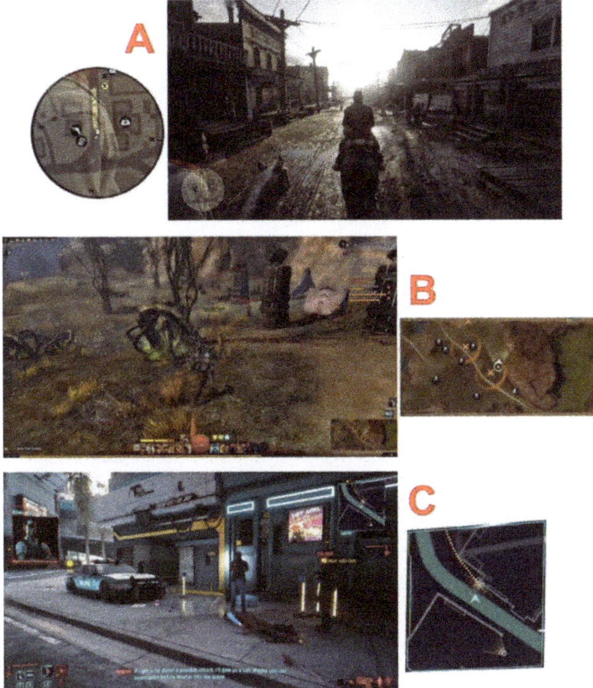

Figure 1. Examples of video games with a mini-map in the game cartography interface (**A**) Red Dead Redemption 2; (**B**) Guild Wars 2; (**C**) Cyberpunk 2077.

2.3. Parameters and Attributes for Classifying Mini-Map Features

Based on the visual analysis of the chosen 100 video game interfaces, the following parameters and attributes were classified for the selected 8 features of the mini-map (Figure 2):

- Shape divided into 7 parameters: circle, irregular, rectangle, diamond, parallelogram, square and ellipse (Figure 2A).
- Position of the mini-map on the user interface in 5 basic parameters of placement: top left, top right, bottom left, bottom centre, bottom right (Figure 2B).

- Orientation of the mini-map to the game environment: camera view (rotation of the mini-map relative to the player's camera during the game), north view (rotation of the mini-map relative to the north direction in the game) and static (no rotation of the mini-map) (Figure 2C).
- Mini-map centring: player-centred (only a limited field of view is represented, covering the player's immediate surroundings) and world-centred (miniature map of the world) (Figure 2D).
- Projection: orthographic (top view) and perspective (oblique view) (Figure 2E).
- Base layers: artificial (with different textures than in the virtual geographical space), transparent (without spatial background or with artificial textures for important objects in the game) and in-game (textures as in virtual space) (Figure 2F).
- Proportions of the size of the mini-map and the whole game screen: 0–1%, 1.1–2%, 2.1–3%, 3.1–4%, 4.1–5%, 5.1–6%, 6.1–10% (Figure 2G).
- Additional navigational elements that support the mini-map to navigate in the virtual geographical space: chessboard, compass, directional cues, peripheral arrows, text and north arrow. This criterion took into account that a single video game may use several additional navigation elements (Figure 2H).

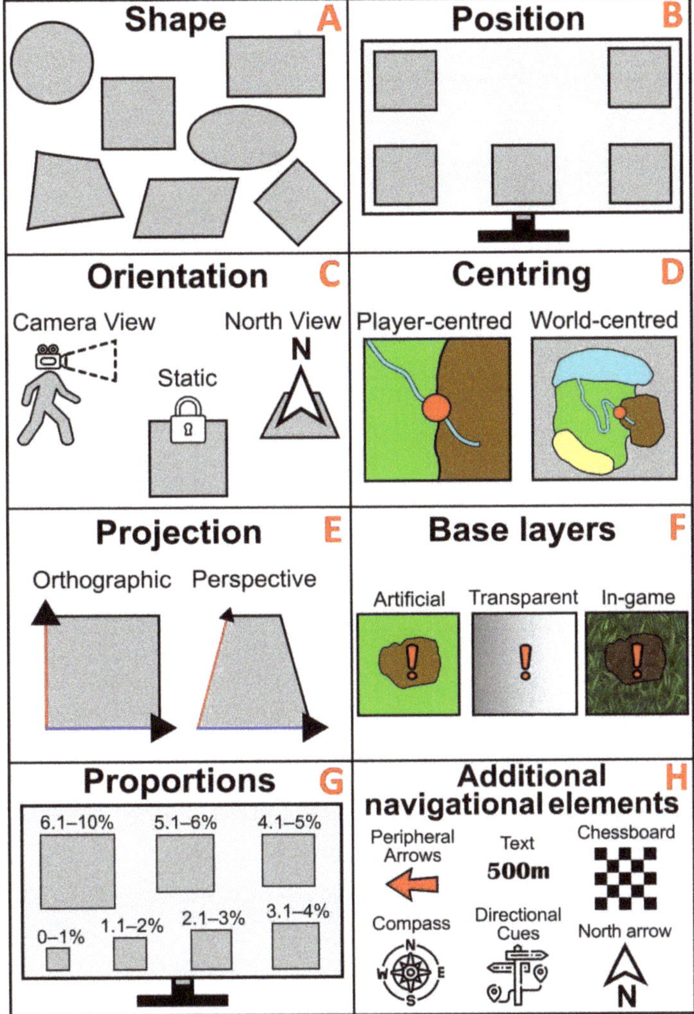

Figure 2. Parameters and attributes for 8 mini-map features according to the chosen 100 video games ((**A**) shape, (**B**) position, (**C**) orientation, (**D**) centring, (**E**) projection, (**F**) base layers, (**G**) proportions, (**H**) additional navigational elements).

3. Results

The analysis result in the juxtaposition form of mini-map features expressed as a percentage for each specific parameter or attribute is presented in Figure 3. The design features are listed in order from the first with the clearest indication of the preferred parameter (projection: orthographic 97%) to the last features with an unclear indication of the popularity of the parameters in the selected 100 video games.

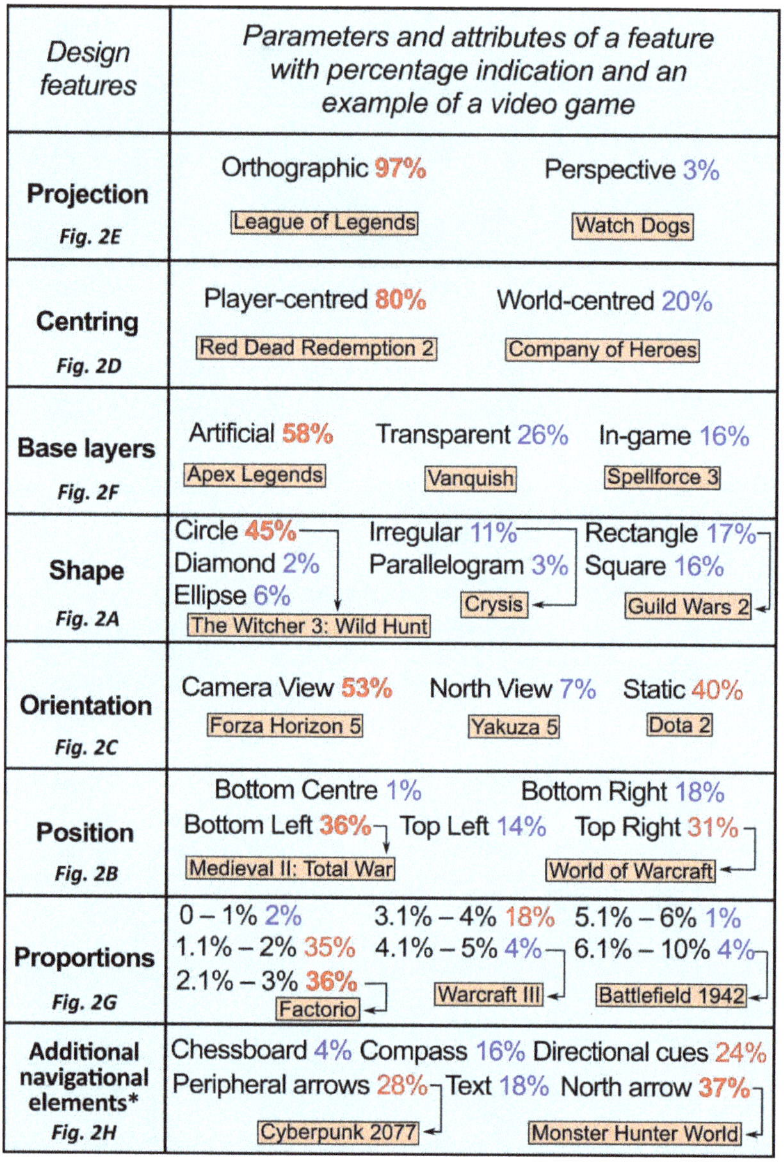

Design features	Parameters and attributes of a feature with percentage indication and an example of a video game		
Projection *Fig. 2E*	Orthographic **97%** League of Legends		Perspective 3% Watch Dogs
Centring *Fig. 2D*	Player-centred **80%** Red Dead Redemption 2		World-centred 20% Company of Heroes
Base layers *Fig. 2F*	Artificial **58%** Apex Legends	Transparent 26% Vanquish	In-game 16% Spellforce 3
Shape *Fig. 2A*	Circle **45%** Diamond 2% Ellipse 6% The Witcher 3: Wild Hunt	Irregular 11% Parallelogram 3% Crysis	Rectangle 17% Square 16% Guild Wars 2
Orientation *Fig. 2C*	Camera View **53%** Forza Horizon 5	North View 7% Yakuza 5	Static 40% Dota 2
Position *Fig. 2B*	Bottom Centre 1% Bottom Left **36%** Medieval II: Total War	Top Left 14%	Bottom Right 18% Top Right 31% World of Warcraft
Proportions *Fig. 2G*	0 − 1% 2% 1.1% − 2% **35%** 2.1% − 3% **36%** Factorio	3.1% − 4% **18%** 4.1% − 5% 4% Warcraft III	5.1% − 6% 1% 6.1% − 10% 4% Battlefield 1942
Additional navigational elements* *Fig. 2H*	Chessboard 4% Compass 16% Directional cues 24% Peripheral arrows 28% Text 18% North arrow **37%** Cyberpunk 2077		Monster Hunter World

Figure 3. The percentage of use of a specific parameter of the mini-map with examples of video games according to their majority unambiguity. Bold red stands for the highest percentage, solid red for a value close to the highest, blue colour for values far from the highest. * One video game can use many additional navigational elements.

3.1. Projection

Almost all video games used the mechanism of orthogonal projection in the mini-map to visualise the virtual world of the game (97%). Only 3% of the mini-maps used perspective projection to visualise the game space (Grand Theft Auto V, Forza Horizon 5, Watch Dogs).

3.2. Centring

The vast majority of mini-maps use the mechanism of a limited field of view covering the immediate surroundings of the player (80%). The predominant option is used especially in the genres RPG (role-playing game), MMORPG (massively multiplayer online role-playing game) and FPS (first-person shooter). Only 20% of the games choose the variant of a miniature map of the virtual world. This type of mini-map was mainly found in the game genres MOBA (multiplayer online battle arena) and RTS (real-time strategy).

3.3. Base Layers

This feature is dominated by mini-maps in which artificial layered textures were used (58%). Secondly, video game developers opted for a transparent mini-map that only highlights the key elements of the game (26%). Most rarely, game developers used in-game textures of the virtual geographical space that reflected the actual view of the player's camera (16%).

3.4. Shape

Most of the mini-maps (45%) are in the shape of a circle. They are followed by a rectangle (17%) and a square (16%). The least popular are unusual geometric shapes such as: irregular (11%), ellipse (6%), parallelogram (3%) and diamond (2%).

3.5. Orientation

In the case of the analysed video games, more than half chose to orient themselves to the player's camera (53%). Static orientation is second (40%) and orientation to the north of the virtual geographical space is the least common (7%).

3.6. Position

Holzinger et al. [78] noted that user interface elements embedded on the left side of the screen performed better than those on the right side. On the other hand, Edler et al. [79] showed in their study that the legend of the thematic map placed on the right side is read and decoded much faster than the legend on the left side. The analysis shows that the most frequently chosen position for a mini-map in the user interface is "bottom left" (36%). The next choice of developers was "top right" (31%), while the option "bottom centre" was only used in one game (1%). It can be seen that the split between the left side of the mini-map (50%) and the right side (49%) is insignificant.

3.7. Proportions

For the criterion of the size of the mini-map, the range of 2.1–3% (36%) in relation to the entire screen displayed was chosen most frequently. The size options 1.1–2% (35%) and 3.1–4% (18%) were also frequently chosen. The smallest size range 0–1% (2%) and mini-maps covering a larger area of the screen: 4.1–5% (4%), 5.1–6% (1%), 6.1–10% (4%) were used least frequently.

3.8. Additional Navigational Elements

Each game can have more than one additional navigational element on the user interface (mini-map is the main navigational element). Four-fifths of the video games have additional navigational elements besides the mini-maps to help the player find his way around the virtual geographical space (80%). The most popular navigational element is the north arrow, which points north in space (37%). Slightly less often, developers used peripheral arrows to indicate the destination (28%). The implementation of directional cues in the game was seen comparatively often (24%). The use of text elements was seen in 18% of games to increase the accuracy of navigation, while a compass element was introduced in 16% of games. Occasionally, a chessboard mechanism was added to the mini-map to better distinguish individual areas of the virtual world (4%).

3.9. Mini-Map Composition

Figure 4 shows a graphical representation of the most popular attributes from eight mini-map design features. The composition is the result of the percentage rating of the parameters calculated separately for each feature from Figure 3. Therefore, it can be assumed that the features should meet the following popular parameters: projection: orthographic; centring: player-centred; base layers: artificial; shape: circle; orientation: camera view; position: bottom left; proportions: 2.1–3%; additional navigational element: north arrow. It is particularly noteworthy that only one mini-map from Assassin's Creed IV: Black Flag meets the above-mentioned parameters of all eight mini-map features (Figure 4).

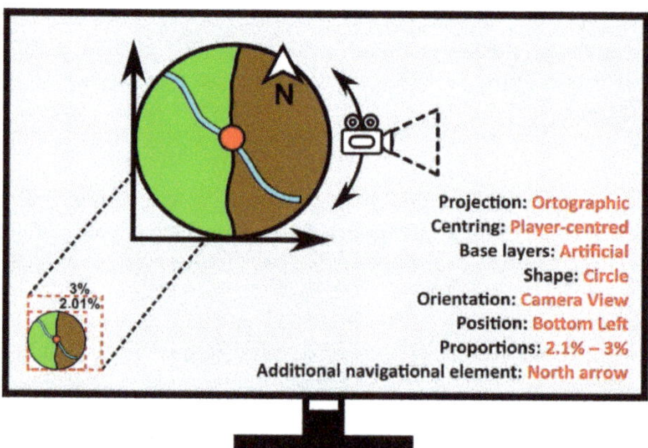

Figure 4. Mini-map composition on the display view (16:9) with the most popular parameters of 8 design features—by separate analysis for each parameter.

4. Discussion

Basic geometric shapes, such as a rectangle and a square, are usually used in designing shapes for traditional maps for navigation. The mini-map as a cartographic interface element in video games breaks with the basic principles of cartographic design by preferred use of the circle shape. Such a design of the mini-map by the game developers could be due to the fact that the shape of the circle often fits better with the graphic aesthetics of the game. The authors note an additional functionality of the shape described, as it allows symmetrical distance to be shown in any direction of the displayed space.

With regard to the position feature, the lower left corner of the screen proved to be the most frequent choice of the developers in designing the position of the mini-map. As a result of the analysis, a slight tendency to place the mini-map element at the bottom of the screen can be identified. Traditional cartographic design principles also suggest placing location maps below the main cartographic content.

The orientation, centring and projection features support the interaction between the player and the user interface. The continuous rotation of the mini-map in relation to the movement of the virtual character's camera, the display of the player's current position together with the immediate surroundings and the orthogonal projection onto the plane of the geographical space support the player in permanent topographical orientation.

As with the design of traditional topographic maps, game developers usually opt for an artificial texture layer that is distinct from the referenced geographical space. The graphic process aims to develop a pattern of navigational behaviour that enhances the interpretation of topographic objects in virtual geographic space.

The vast majority of video games favour smaller mini-maps with an aspect ratio of 1.1% to 3% of the total game screen. It can be assumed that the creators of these games intended the mini-map to play the role of an additional navigation option that enables orientation in virtual space. Even in traditional cartography, the location map is relatively small compared to the frame of the main content of the cartography.

When designing a complex navigation system, game developers do not limit themselves to individual interface elements. As the results of the analysis show, in addition to the mini-map, the developers often use other additional elements that are familiar to users from everyday life (north arrow or compass). It is worth mentioning that the above-mentioned elements are always a fixed part of the interface during the game. However, many games also use dynamic navigational elements that depend on current events or phenomena in the virtual geographical space (different types of lines or highlighted symbols).

The unexpected fact of the research that all eight parameters are fulfilled with the highest percentages by only one game gives rise to a discussion about the complementarity and holistic nature of the design process in the context of the analysis of mini-map features [80]. Therefore, the authors draw attention to the possibility of grouping several selected features instead of treating all eight features of the mini-map together. Based on the comparison in Figure 3, four design features can be identified that have clearly defined priority parameters. Figure 5 shows the four most preferred parameters, which 22 out of 100 games fulfil: projection (orthographic), centring (player-centred), base layers (artificial) and shape (circle). The four features with the addition of camera view orientation implement mini-maps from 16 games, while the addition of the north arrow implements 15 selected video games. For the position feature, an unchanged trend of slightly more mini-maps in the lower left corner of the screen was noted (10 games). However, when identifying the four priority features, the proportion of smaller mini-maps is surprisingly much larger (13 games 1.1–2%).

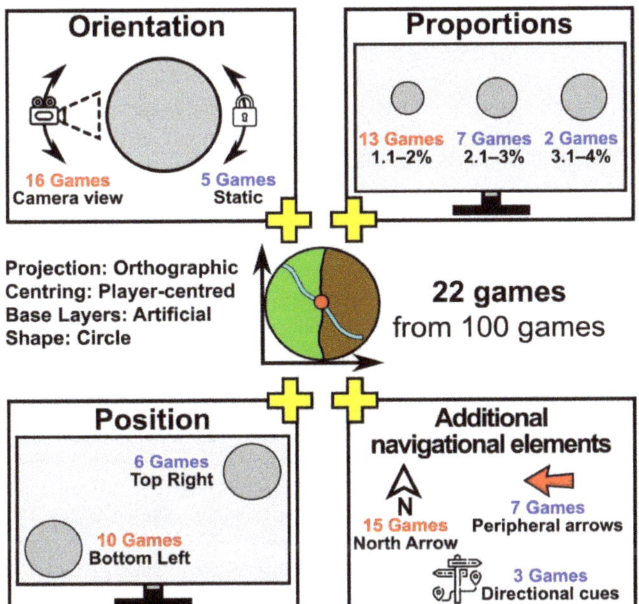

Figure 5. Grouping of the most popular and preferred parameters of the mini-map features: projection, centring, base layers and shape with separate addition of other popular attributes or parameters from four other features (see Figure 3).

5. Conclusions

In summary, the identification of mini-map design features in video games is very valuable for a navigational aid in the virtual geographical space. The authors focused on the selection and classification of the cartographic features of the mini-map element in the graphical user interface of video games in terms of the proposed parameters and attributes. The proposed research approach resulted in revealing eight features of mini-map design in video games, according to their most popular parameters and attributes: projection: orthographic; centring: player-centred; base layers: artificial; shape: circle; orientation: camera view; position: bottom left; proportions: 2.1–3%; additional navigational element: north arrow (Figure 3). If we interpret the mini-map features individually according to the popular parameters and attributes, we find that only one game fulfils all conditions (Figure 4). Based on the identification and arrangement of the eight mini-map features according to their individual parameters and attributes, we can conclude that the four most commonly used attributes are: orthographic projection, player-centred view, artificial base layer, circle shape (Figure 5). The feature of mini-map position on the screen is the most difficult to define, as the bottom left and bottom right corners are similarly preferred and the other two corners also have high usability.

In response to the additional research objectives of the article, it can be stated that the fundamental elements of the mini-map most commonly found among video game designers were indicated (Figure 5). They form the core of the cartographic design of the mini-map, on which further research on other evaluation criteria can be based.

The authors captured key attributes for mini-map features that are considered separately, complementarily and as holistically as possible, confirming the nature of mini-map design as part of an individual creative process that can be supported by traditional cartographic design principles [81,82]. These studies also confirm the importance of map design in the context of usability, which goes hand in hand with the popularity of cartographic products in new geomedia, such as mini-maps in video games [83].

The features analysed make the mini-map an important cartographic element in the visualisation of geographical space, providing detailed spatial information in a useful form. In general, the mini-map as the main element of the game cartography interface, is designed in accordance with the principles of cartographic design. The only significant divergency from the rules of map design is the use of a circle shape instead of a traditional rectangle. However, such a divergency is related to the specificity of navigation in virtual geographic space, which is the result of the necessary adaptation of cartographic design to new geomedia.

A navigation in spaces is not limited to virtual worlds, but takes place in all activities that have to do with moving around in our world. Cartographic elements in games, such as a mini-map, fulfil many conditions that can be compared to satellite applications on mobile phones [84]. The design of these elements usually varies according to the needs of presenting and exploring the geographical space. Fundamentally, they have a common task, which is to help people navigate through real or virtual spaces. Further research should include the possibility of comparing the individual design processes to identify differences in functions and tasks.

The authors hope that the identified most popular parameters of the eight features can be useful not only in the design of video games, but also in other cartographic products dealing with virtual and augmented reality. User-centred design focuses on the needs of the player [85] and the needs of the map users [57,86]. Future projects will also need an evaluation of popular solutions by gamers and map users to determine which solutions actually work best and are not just the most popular. Given the growing interest in the video game industry worldwide, more research is needed on the optimisation and effectiveness of the user interface and cartographic products.

Author Contributions: Conceptualization, Krzysztof Zagata and Beata Medyńska-Gulij; methodology, Krzysztof Zagata; formal analysis, Krzysztof Zagata; resources, Krzysztof Zagata; writing—original draft preparation, Krzysztof Zagata and Beata Medyńska-Gulij; writing—review and editing, Krzysztof Zagata and Beata Medyńska-Gulij. All authors have read and agreed to the published version of the manuscript.

Funding: This research received no external funding.

Data Availability Statement: Not applicable.

Acknowledgments: This paper is the result of research on the mini-map in the interface based on the analysis of video games carried out within statutory research on the doctorate at the Department of Cartography and Geomatics, Faculty of Geographical and Geological Sciences, Doctoral School, Adam Mickiewicz University Poznań, Poland.

Conflicts of Interest: The authors declare no conflict of interest.

Appendix A

Alphabetical list of the 100 games with a mini-map by *Name*-**Year**-Genre:

Age of Empires II: HD Edition-**2013**-Strategy; *Albion Online*-**2017**-MMORPG; *Anno 1404*-**2009**-Strategy; *Apex Legends*-**2019**-Action; *Assassin's Creed IV: Black Flag*-**2013**-Action; *Battlefield 1942*-**2002**-Action; *Battlefield 2*-**2005**-Action; *Battlefield 2042*-**2021**-Action; *Black Desert Online*-**2014**-MMORPG; *Borderlands 2*-**2012**-Action; *Call of Duty 4: Modern Warfare*-**2007**-Action; *Call of Duty: Black Ops*-**2010**-Action; *Call of Duty: Warzone*-**2020**-Action; *Company of Heroes*-**2006**-Strategy; *Company of Heroes 2*-**2013**-Strategy; *Cossacks 3*-**2016**-Strategy; *Counter-Strike: Global Offensive*-**2012**-Action; *Counter-Strike: Source*-**2005**-Action; *Crackdown 3*-**2019**-Action; *Crysis*-**2007**-Action; *Cyberpunk 2077*-**2020**-Action; *Darksiders 2*-**2012**-Action; *Days Gone*-**2021**-Action; *Dead Cells*-**2018**-Platform; *Dead Island*-**2011**-Action; *Dead Rising 4*-**2016**-Action; *Dekaron*-**2007**-MMORPG; *Destiny 2*-**2017**-Action; *Deus EX: Human Revolution*-**2011**-RPG; *Divinity: Original Sin II*-**2017**-RPG; *Dota 2*-**2013**-MOBA; *Dragon Age: Origins*-**2009**-RPG; *DRAGON QUEST XI S: Echoes of an Elusive Age—Definitive Edition*-**2018**-RPG; *Dying Light*-**2015**-Action; *Elex*-**2017**-RPG; *Factorio*-**2020**-Strategy; *Far Cry 3*-**2012**-Action; *Far Cry 4*-**2014**-Action; *Fifa 22*-**2021**-Sports; *Final Fantasy XII: The Zodiac Age*-**2018**-RPG; *Final Fantasy XIV*-**2013**-MMORPG; *For Honor*-**2017**-Action; *Fortnite*-**2017**-Action; *Forza Horizon 5*-**2021**-Racing; *Genshin Impact*-**2020**-RPG; *Grand Theft Auto III*-**2002**-Action; *Grand Theft Auto V*-**2015**-Action; *Grand Theft Auto: San Andreas*-**2005**-Action; *Grand Theft Auto: Vice City*-**2003**-Action; *Grim Dawn*-**2016**-RPG; *Guild Wars*-**2005**-MMORPG; *Guild Wars 2*-**2012**-MMORPG; *Heroes of Might and Magic III*-**2004**-Strategy; *Hitman 3*-**2021**-Action; *L.A. Noire*-**2011**-Adventure; *League of Legends*-**2009**-MOBA; *Lost Ark*-**2022**-MMORPG; *Mad Max*-**2015**-Action; *Medal of Honor*-**2010**-Action; *Medieval II: Total War*-**2006**-Strategy; *Middle-earth: Shadow of War*-**2017**-Action; *Monster Hunter World*-**2018**-RPG; *Need for Speed Payback*-**2017**-Racing; *Nioh*-**2017**-RPG; *Northgard*-**2018**-Strategy; *Orcs Must Die! 2*-**2012**-Action; *Path of Exile*-**2013**-MMORPG; *PlayerUnknown's Battlegrounds*-**2017**-Action; *Red Dead Redemption 2*-**2019**-Action; *Rome: Total War*-**2004**-Strategy; *RuneScape*-**2003**-MMORPG; *S.T.A.L.K.E.R.: Call of Pripyat*-**2009**-Action; *Saints Row 4*-**2013**-Action; *Sleeping Dogs*-**2012**-Action; *Sniper Elite 4*-**2017**-Action; *Sniper: Ghost Warrior 3*-**2017**-Action; *Spellforce 3*-**2017**-Strategy; *Star Wars: Knights of the Old Republic*-**2003**-RPG; *Star Wars: The Old Republic*-**2011**-MMORPG; *Starcraft II: Wings of Liberty*-**2010**-Strategy; *The Ascent*-**2021**-RPG; *The Binding of Isaac: Afterbirth*-**2015**-Action; *The Crew 2*-**2018**-Racing; *The Riftbreaker*-**2021**-Action; *The Witcher 2: Assassins of Kings*-**2011**-RPG; *The Witcher 3: Wild Hunt*-**2015**-RPG; *Titanfall 2*-**2016**-Action; *Tom Clancy's Ghost Recon: Wildlands*-**2017**-Action; *Torchlight 2*-**2012**-RPG; *Total War: Shogun 2*-**2011**-Strategy; *Valheim*-**2021**-Action; *Valorant*-**2020**-Action; *Vanquish*-**2017**-Action; *Warcraft III: The Frozen Throne*-**2003**-Strategy; *Warframe*-**2013**-Action; *Warhammer 40,000: Dawn of War—Dark Crusade*-**2006**-Strategy; *Watch Dogs*-**2014**-Action; *World of Tanks*-**2010**-Action; *World of Warcraft*-**2004**-MMORPG; *Yakuza 5 Remastered*-**2021**-Action.

References

1. Ash, J.; Gallacher, L.A. Cultural Geography and Videogames. *Geogr. Compass* **2011**, *5*, 351–368. [CrossRef]
2. Behnke, M.; Gross, J.J.; Kaczmarek, L.D. The role of emotions in esports performance. *Emotion* **2022**, *22*, 1059–1070. [CrossRef] [PubMed]
3. Wolf, M. 3 Space in the Video Game. In *The Medium of the Video Game*; Wolf, M., Ed.; University of Texas Press: New York, NY, USA, 2021.
4. Coulton, P.; Huck, J.; Gradinar, A.; Salinas, L. Mapping the beach beneath the street: Digital cartography for the playable city. In *Playable Cities. The City as Digital Playground*; Nijholt, A., Ed.; Springer: Singapore, 2017.
5. Horbiński, T.; Zagata, K. View of Cartography in Video Games: Literature Review and Examples of Specific Solutions. *KN J. Cartogr. Geogr. Inf.* **2022**, *72*, 117–128. [CrossRef]
6. Lammes, S. Spatial Regimes of the Digital Playground: Cultural Functions of Spatial Practices in Computer Games. *Space Cult.* **2008**, *11*, 260–272. [CrossRef]
7. Medyńska-Gulij, B. *Kartografia I Geomedia*; Wydawnictwo Naukowe PWN: Warsaw, Poland, 2021; ISBN 978-83-01-21554-5.
8. Eskidemir, K.; Kubat, A.S. The visual structure of fictional space in video games: A wayfinding research in games based on real world cities. In *Annual Conference Proceedings of the XXVIII International Seminar on Urban Form, Glasgow, UK, 29 June–3 July 2021*; University of Strathclyde Publishing: Glasgow, UK, 2022.
9. Johanson, C.; Gutwin, C.; Mandryk, R.L. The Effects of Navigation Assistance on Spatial Learning and Performance in a 3D Game. In Proceedings of the Annual Symposium on Computer-Human Interaction in Play (CHI PLAY 17), Amsterdam, The Netherlands, 15–18 October 2017; Association for Computing Machinery: New York, NY, USA, 2017; pp. 341–353.
10. Ratajski, L. Logiczno-semiotyczne zasady porządkowania i standaryzacji znaków kartograficznych. *Pol. Przegląd Kartogr.* **1971**, *3*, 106–116.
11. Bertin, J. *Semiology of Graphics: Diagrams*; University of Wisconsin Press: Madison, WI, USA, 1983.
12. Medyńska-Gulij, B. Map compiling, map reading and cartographic design in "Pragmatic pyramid of thematic mapping". *Quaest. Geogr.* **2010**, *29*, 57–63. [CrossRef]
13. Horbiński, T.; Zagata, K. Map Symbols in Video Games: The Example of "Valheim". *KN J. Cartogr. Geogr. Inf.* **2021**, *71*, 269–283. [CrossRef]
14. Lammes, S. Terra Incognita: Computer Games, Cartography and Spatial Stories. In *Digital Material: Tracing New Media in Everyday Life and Technology*; Van Den Boomen, M., Lammes, S., Eds.; Amsterdam University Press: Amsterdam, The Netherlands, 2009; pp. 223–238.
15. Ouriques, L.; Xexéo, G.; Mangeli, E. Analyzing Space Dimensions in Video Games. In Proceedings of the XVIII SBGames 2019, Rio de Janeiro, Brazil, 28–31 October 2019; IEEE: New York, NY, USA, 2019.
16. Britta, N. World and Place: Map and Territory. In Proceedings of the DiGRA 2009 'Breaking New Ground: Innovation in Games, Play, Practice and Theory', London, UK, 1–4 September 2009.
17. Mat Zain, N.H.; Abdul Razak, F.H.; Jaafar, A.; Zulkipli, M.F. Eye Tracking in Educational Games Environment: Evaluating User Interface Design through Eye Tracking Patterns. In Proceedings of the Visual Informatics: Sustaining Research and Innovations, Selangor, Malaysia, 9–11 November 2011; Springer: Berlin/Heidelberg, Germany, 2011; pp. 64–73.
18. Khan, N.; Rahman, A.U. Rethinking the Mini-Map: A Navigational Aid to Support Spatial Learning in Urban Game Environments. *Int. J. Hum. Comput. Interact.* **2018**, *34*, 1135–1147.
19. Toups, Z.O.; Lalone, N.; Alharthi, S.A.; Sharma, H.N.; Webb, A.M. Making Maps Available for Play: Analyzing the Design of Game Cartography Interfaces. *ACM Trans. Comput. Hum. Interact.* **2019**, *26*, 1–43. [CrossRef]
20. Edler, D.; Husar, A.; Keil, J.; Vatter, M.; Dickmann, F. Virtual Reality (VR) and Open Source Software: A Workflow for Constructing an Interactive Cartographic VR Environment to Explore Urban Landscapes. *J. Cartogr. Geogr. Inf.* **2018**, *68*, 5–13. [CrossRef]
21. Mahalil, I.; Yusof, A.M.; Ibrahim, N.; Mahidin, E.M.M.; Rusli, M.E. Virtual Reality Mini Map Presentation Techniques: Lessons and experience learned. In Proceedings of the 2019 IEEE Conference on Graphics and Media (GAME), Pulau Pinang, Malaysia, 19–21 November 2019; IEEE: New York, NY, USA, 2019; pp. 26–31.
22. Peacocke, M.; Teather, R.J.; Carette, J.; MacKenzie, I.S.; McArthur, V. An empirical comparison of first-person shooter information displays: Huds diegetic displays and spatial representations. *Entertain. Comput.* **2018**, *26*, 41–58.
23. Marre, Q.; Caroux, L.; Sakdavong, J.C. Video game interfaces and diegesis: The impact on experts and novices' performance and experience in virtual reality. *Int. J. Hum. Comput. Interact.* **2021**, *37*, 1089–1103. [CrossRef]
24. Couclelis, H. Space, Time, Geography. *Geogr. Inf. Syst.* **1999**, *1*, 29–38.
25. Lin, H.; Jianhua, G. Exploring Virtual Geographic Environments. *Geogr. Inf. Sci.* **2001**, *7*, 1–7. [CrossRef]
26. Mukherjee, S. The Playing Fields of Empire: Empire and Spatiality in Video Games. *J. Gaming Virtual Worlds* **2015**, *7*, 299–315. [CrossRef] [PubMed]
27. Batty, M. Virtual geography. *Futures* **1997**, *29*, 337–352. [CrossRef]
28. Konecny, M. Review: Cartography: Challenges and potential in the virtual geographical environments era. *Ann. GIS* **2011**, *17*, 135–146.
29. Kraak, M.J.; Ormeling, F. *Cartography: Visualization of Geospatial Data*, 4th ed.; CRC Press: Boca Raton, FL, USA, 2020.
30. Gekker, A. (Mini) Mapping the Game-Space. In *Playful Mapping in the Digital Age (Theory on Demand Series)*; The Playful Mapping Collective, Ed.; Institute of Network Cultures: Amsterdam, The Netherlands, 2018.
31. McGregor, G.L. Situations of Play: Patterns of Spatial Use in Videogames. In Proceedings of the 2007 DiGRA International Conference: Situated Play, Tokyo, Japan, 24–28 September 2007; Digital Games Research Association: Helsinki, Finland, 2007.

32. Di Tore, P. Perception of Space, Empathy and Cognitive Processes: Design of a Video Game for the Measurement of Perspective Taking Skills. *Int. J. Emerg. Technol. Learn.* **2014**, *9*, 23–29.
33. Si, C.; Pisan, Y.; Tan, C.T.; Shen, S. An initial understanding of how game users explore virtual environments. *Entertain. Comput.* **2017**, *19*, 13–27. [CrossRef]
34. Davis, M.; Kent, A.J. Soviet City Plans and OpenStreetMap: A comparative analysis. *Int. J. Cartogr.* **2022**. Available online: https://www.tandfonline.com/doi/pdf/10.1080/23729333.2022.2047396 (accessed on 30 September 2022).
35. Schobesberger, D. Integrating User and Usability Research in Web-Mapping Application Design. In *Modern Trends in Cartography*; Brus, J., Vondrakova, A., Vozenilek, V., Eds.; Springer: Cham, Switzerland, 2014.
36. Kuveždić Divjak, A.; Đapo, A.; Pribičević, B. Cartographic Symbology for Crisis Mapping: A Comparative Study. *ISPRS Int. J. Geo. Inf.* **2020**, *9*, 142.
37. Cybulski, P. Design rules and practices for animated maps online. *J. Spat. Sci.* **2016**, *61*, 461–471.
38. Medyńska-Gulij, B.; Wielebski, Ł.; Halik, Ł.; Smaczyński, M. Complexity Level of People Gathering Presentation on an Animated Map—Objective Effectiveness Versus Expert Opinion. *ISPRS Int. J. Geo.Inf.* **2020**, *9*, 117. [CrossRef]
39. Medyńska-Gulij, B. Educating tomorrow's cartographers. In *The Routledge Handbook of Mapping and Cartography*; Kent, A.J., Vujakovic, P., Eds.; Routledge: Milton Park, UK, 2018.
40. Gazzard, A. Gaming maps and virtual words. In *The Routledge Handbook of Mapping and Cartography*; Kent, A.J., Vujakovic, P., Eds.; Routledge: Milton Park, UK, 2018.
41. Johnson, D.; Wiles, J. Effective affective user interface design in games. *Ergonomics* **2003**, *46*, 1332–1345. [CrossRef] [PubMed]
42. Toups, Z.O.; LaLone, N.; Spiel, K.; Hamilton, B. Paper to Pixels: A Chronicle of Map Interfaces in Games. In Proceedings of the 2020 ACM Designing Interactive Systems Conference (DIS 20), Eindhoven, The Netherlands, 6–10 July 2020; Association for Computing Machinery: New York, NY, USA, 2020.
43. Juul, J.; Norton, M. Easy to use and incredibly difficult: On the mythical border between interface and gameplay. In Proceedings of the 4th International Conference on Foundations of Digital Games (FDG 2009), Orlando, FL, USA, 26–30 April 2009; Association for Computing Machinery: New York, NY, USA, 2009; pp. 107–112.
44. Miraz, M.H.; Excell, P.S.; Ali, M. User interface (UI) design issues for multilingual users: A case study. *Univers. Access Inf. Soc.* **2016**, *15*, 431–444.
45. Callahan, E. Interface design and culture. *Annu. Rev. Inf. Sci. Technol.* **2006**, *39*, 255–310.
46. Horbiński, T.; Cybulski, P.; Medyńska-Gulij, B. Graphic Design and Button Placement for Mobile Map Applications. *Cartogr. J.* **2020**, *57*, 196–208. [CrossRef]
47. Slocum, T.; McMaster, R.B.; Kessler, F.C.; Howard, H.H. *Thematic Cartography and Geographic Visualization*, 2nd ed.; Pearson Prentice Hall: Upper Saddle River, NJ, USA, 2005.
48. Hayatpur, D.; Xia, H.; Wigdor, D. DataHop: Spatial Data Exploration in Virtual Reality. In Proceedings of the 33rd Annual ACM Symposium on User Interface Software and Technology (UIST 2020), virtual event, USA, 20–23 October 2020; Association for Computing Machinery: New York, NY, USA, 2020; pp. 818–828.
49. Okada, A.; Rocha, K.; Fuchter, S.; Zucchi, S.; Wortley, D. Formative assessment of inquiry skills for Responsible Research and Innovation using 3D Virtual Reality Glasses and Face Recognition. In Proceedings of the Technology Enhanced Assessment Conference (TEA 2018), Hong Kong, 8–9 February 2018; Springer: Berlin, Germany, 2019; pp. 91–101.
50. Adams, E. *Fundamentals of Game Design*, 3rd ed.; New Riders Publishing: Indianapolis, IN, USA, 2014.
51. Horbiński, T.; Zagata, K. Interpretation of Map Symbols in the Context of Gamers' Age and Experience. *ISPRS Int. J. Geo-Inf.* **2022**, *11*, 150. [CrossRef]
52. Zagata, K.; Gulij, J.; Halik, Ł.; Medyńska-Gulij, B. Mini-Map for Gamers Who Walk and Teleport in a Virtual Stronghold. *ISPRS Int. J. Geo-Inf.* **2021**, *10*, 96.
53. Nesbitt, K.; Sutton, K.; Wilson, J.; Hookham, G. Improving player spatial abilities for 3D challenges. In Proceedings of the Sixth Australasian Conference on Interactive Entertainment (IE 2009), Sydney, Australia, 17–19 December 2009; Association for Computing Machinery: New York, NY, USA, 2009.
54. Edler, D.; Keil, J.; Wiedenlübbert, T.; Sossna, M.; Kühne, O.; Dickmann, F. Immersive VR Experience of Redeveloped Post-industrial Sites: The Example of "Zeche Holland" in Bochum-Wattenscheid. *KN J. Cartogr. Geogr. Inf.* **2019**, *69*, 267–284.
55. Edler, D.; Keil, J.; Dickmann, F. Varianten interaktiver Karten in Video-und Computerspielen—Eine Übersicht. *KN J. Cartogr. Geogr. Inf.* **2018**, *68*, 57–65. [CrossRef]
56. Fadaeddini, A.; Majidi, B.; Eshghi, M. A Case Study of Generative Adversarial Networks for Procedural Synthesis of Original Textures in Video Games. In Proceedings of the 2nd National and 1st International Digital Games Research Conference: Trends, Technologies, and Applications (DGRC 2018), Tehran, Iran, 29–30 November 2018; Institute of Electrical and Electronics Engineers: New York, NY, USA, 2018; pp. 118–122.
57. Medyńska-Gulij, B.; Zagata, K. Experts and Gamers on Immersion into Reconstructed Strongholds. *ISPRS Int. J. Geo-Inf.* **2020**, *9*, 655. [CrossRef]
58. Wiemer, S. Interface Analysis: Notes on the "Scopic Regime" of Strategic Action in Real-Time Strategy Games. In *Computer Games and New Media Cultures*; Fromme, J., Unger, A., Eds.; Springer: Dordrecht, The Netherlands, 2012.
59. Burigat, S.; Chittaro, L. Navigation in 3D virtual environments: Effects of user experience and location-pointing navigation aids. *Int. J. Hum. Comput. Stud.* **2007**, *65*, 945–958. [CrossRef]

60. Novak, J. *Game Development Essentials: An Introduction*, 3rd ed.; Cengage Learning: Boston, MA, USA, 2011; ISBN 9781111307653.
61. Targett, S.; Verlysdonk, V.; Hamilton, H.J.; Hepting, D. A study of user interface modifications in World of Warcraft. *Game Stud.* **2012**, *12*. Available online: https://www.researchgate.net/publication/290732545_A_Study_of_User_Interface_Modifications_in_World_of_Warcraft (accessed on 30 September 2022).
62. Su, A.J.; Hu, Y.C.; Kuzmanovic, A.; Koh, C.K. How to improve your search engine ranking: Myths and reality. *ACM Trans. Web* **2014**, *8*, 1–25. [CrossRef]
63. IGN. Top 100 Video Games of All Time. Available online: https://www.ign.com/lists/top-100-games/100 (accessed on 17 January 2022).
64. GryOnline. Najlepsze Gry Wszech Czasów. Available online: https://www.gry-online.pl/ranking-gier.asp?CZA=2 (accessed on 17 January 2022).
65. Slant Magazine. The 100 Best Video Games of All Time. Available online: https://www.slantmagazine.com/games/the-100-best-video-games-of-all-time/ (accessed on 17 January 2022).
66. GAMINGbible. The Greatest Video Games Of All Time. Available online: https://www.gamingbible.co.uk/features/games-the-greatest-video-games-of-all-time-100-81-20201130 (accessed on 17 January 2022).
67. Screenage Wasteland. The 500 Greatest Video Games of All Time (25-1). Available online: https://screenagewasteland.com/the-500-greatest-video-games-of-all-time-25-1/ (accessed on 18 January 2022).
68. Rock Paper Shotgun Part 1. The RPS 100: Our Top PC Games of All Time (100-51). Available online: https://www.rockpapershotgun.com/the-rps-100-2021-part-one (accessed on 18 January 2022).
69. Rock Paper Shotgun Part 2. The RPS 100: Our Top PC Games of All Time (50-1). Available online: https://www.rockpapershotgun.com/the-rps-100-2021-part-two (accessed on 18 January 2022).
70. Tom's Guide. The Best PC Games of 2023. Available online: https://www.tomsguide.com/best-picks/best-PC-games (accessed on 18 January 2022).
71. ProFanboy. 25 Best PC Games of All Time. Available online: https://profanboy.com/best-pc-games/ (accessed on 18 January 2022).
72. Metacritic. Available online: https://www.metacritic.com/browse/games/score/userscore/all/all/filtered?sort=desc (accessed on 18 January 2022).
73. OpenCritic. Available online: https://opencritic.com/browse/all (accessed on 18 January 2022).
74. Giant Bomb. Available online: https://www.giantbomb.com/profile/dantebk/lists/game-informers-top-200-games-of-all-time/32009/ (accessed on 19 January 2022).
75. PC Gamer. The Top 100 PC Games. Available online: https://www.pcgamer.com/top-100-pc-games-2021/ (accessed on 19 January 2022).
76. Popular Mechanics. The 100 Greatest Video Games of All Time. Available online: https://www.popularmechanics.com/culture/gaming/g134/the-100-greatest-video-games-of-all-time/ (accessed on 19 January 2022).
77. Ranker. Available online: https://www.ranker.com/list/most-popular-pc-games-today/ranker-games (accessed on 19 January 2022).
78. Holzinger, A.; Scherer, R.; Ziefle, M. Navigational User Interface Elements on the Left Side: Intuition of Designers or Experimental Evidence? In *Human-Computer Interaction—INTERACT 2011*; Campos, P., Graham, N., Jorge, J., Nunes, N., Palanque, P., Winckler, M., Eds.; Springer: Berlin/Heidelberg, Germany, 2011.
79. Edler, D.; Keil, J.; Tuller, M.C.; Bestgen, A.K.; Dickmann, F. Searching for the 'Right' Legend: The Impact of Legend Position on Legend Decoding in a Cartographic Memory Task. *Cartogr. J.* **2018**, *57*, 6–17.
80. Medyńska-Gulij, B. Geomedia Attributes for Perspective Visualization of Relief for Historical Non-Cartometric Water-Colored Topographic Maps. *ISPRS Int. J. Geo-Inf.* **2022**, *11*, 554. [CrossRef]
81. Robinson, A.H.; Morrison, J.L.; Muehrecke, P.C.; Kimerling, A.J.; Guptil, S.C. *Elements of Cartography*, 6th ed.; Wiley: New York, NY, USA, 1995.
82. Dent, B.D. *Cartography: Thematic Map Design*, 5th ed.; McGraw-Hill: Boston, CA, USA, 1998.
83. Medyńska-Gulij, B.; Gulij, J.; Cybulski, P.; Zagata, K.; Zawadzki, J.; Horbiński, T. 2022, Map Design and Usability of a Simplified Topographic 2D Map on the Smartphone in Land-scape and Portrait Orientations. *ISPRS Int. J. Geo-Inf.* **2022**, *11*, 577.
84. Chesher, C. Navigating Sociotechnical Spaces: Comparing Computer Games and Sat Navs as Digital Spatial Media. *Converg. Int. J. Res. Into New Media Technol.* **2012**, *18*, 315–330. [CrossRef]
85. Chammas, A.; Quaresma, M.; Mont'Alvão, C. A closer look on the user centred design. *Procedia Manuf.* **2015**, *3*, 5397–5404.
86. Kramers, E. Interaction with Maps on the Internet—A User Centred Design Approach for The Atlas of Canada. *Cartogr. J.* **2008**, *45*, 98–107.

International Journal of
Geo-Information

Article

Interpretation of Map Symbols in the Context of Gamers' Age and Experience

Tymoteusz Horbiński * and Krzysztof Zagata

Department of Cartography and Geomatics, Faculty of Geographical and Geological Sciences,
Adam Mickiewicz University, 61-712 Poznań, Poland; krzysztof.zagata@amu.edu.pl
* Correspondence: tymoteusz.horbinski@amu.edu.pl; Tel.: +48-61-829-6307

Abstract: In this article researchers examined the differences that may characterise selected groups of gamers with regard to age and time spent on playing a survival game, Valheim, confronted with their interpretation of map symbols used in the game. The Valheim video game, which was released in early 2021, is a survival game set in a gameplay world inspired by Norse mythology. The authors of the article noted that the age factor and gaming experience may have different results in terms of the interpretation of cartographic products. The obtained data came from an online questionnaire. In the statistical analysis the authors employed the Pearson's chi-squared test and the Benjamini–Hochberg procedure to find statistically significant differences between selected groups of respondents and their subjective interpretation of map symbols. Statistical analysis showed several differences and interesting relationships in the interpretation of symbols in relation to the age of the players and in the interpretation of symbols in relation to the time spent in the game. For future research, it is worth continuing towards researching new video games with existing cartographic products in order to investigate how games and players influence the culture in which they participate.

Keywords: map symbols; video games; statistical analysis; gamers' experience; cartographic design; multimedia cartography

Citation: Horbiński, T.; Zagata, K. Interpretation of Map Symbols in the Context of Gamers' Age and Experience. *ISPRS Int. J. Geo-Inf.* **2022**, *11*, 150. https://doi.org/10.3390/ijgi11020150

Academic Editors: Wolfgang Kainz, Beata Medynska-Gulij, David Forrest and Thomas P. Kersten

Received: 16 December 2021
Accepted: 20 February 2022
Published: 21 February 2022

Publisher's Note: MDPI stays neutral with regard to jurisdictional claims in published maps and institutional affiliations.

1. Introduction

Video games are one of the most popular and common forms of entertainment every day. Most video games are spatial in nature, i.e., they present a number of phenomena related to each other by coherent spaces [1–3]. Game developers often provide in-game maps that help players navigate the virtual world [4,5]. Many games have more and more complex geographic areas and the need for easier understanding of the presented world requires improvement and adding newer and newer cartographic elements: minimap, compass, arrows, world maps, cartographic symbols and others [6–8]. The inherent spatiality of video games allows the creation of maps in any video game genre. From maps depicting the game world (World of Warcraft), to interactive maps in computer games (Civilization series), to mobile games that support GPS (Pokémon GO), as well as using satellite maps (Zombies, Run!) that have interaction with the player at all time.

Video games constitute complex technological systems that combine numerous spatial, physical, and social issues. Geographers and cartographers are dealing with the topic of "cartography and games" [9–12] more and more often, describing the technical aspects of game and component production. The discussion about a constantly increasing impact of games on cartographic visualizations manifests itself through increasing popularity of virtual reality and augmented reality environments [13–17].

The problem of the evaluation of cartographic products is quite wide. Survey methods used in cartography appear in various forms: direct survey, internet survey, eye-tracking methods, GPS mobile device task, direct user observation in digital tasks [18–21]. When creating a survey, we must shift the emphasis to which population of participants we are targeting, which is limited by factors that are important for the presented purpose of the study, such as: age, gender, culture, country of origin, experience, factor confirming the purpose of the study or psychological motivations [22]. One of the most popular methods in cartography is the online survey, which allows you to collect data from a wide range of users and allows you to obtain subjective opinions [23,24]. In studies devoted to cartography, the number of respondents varies from 20 to 100 questionnaires concerning one version of the map [25–27], the number of which is proportionally greater depending on the number of versions of cartographic products [28,29].

The space in video games is represented by map symbols that convey information using visual variables (Figure 1). Visual variables were introduced by Bertin [30] by identifying seven categories of variables, incl. shape or color. This classification was based on selectivity, i.e., the ability to distinguish one category while ignoring others. Graphical information, such as the shape and color of the symbol, and the context of use, can influence the users' interpretation of the symbols. Keil et al. [31] noted that in many cases, the design of cartographic symbols is not based on the visualization of representative features of a given object, but on focusing on redefined design standards related to the function of the object.

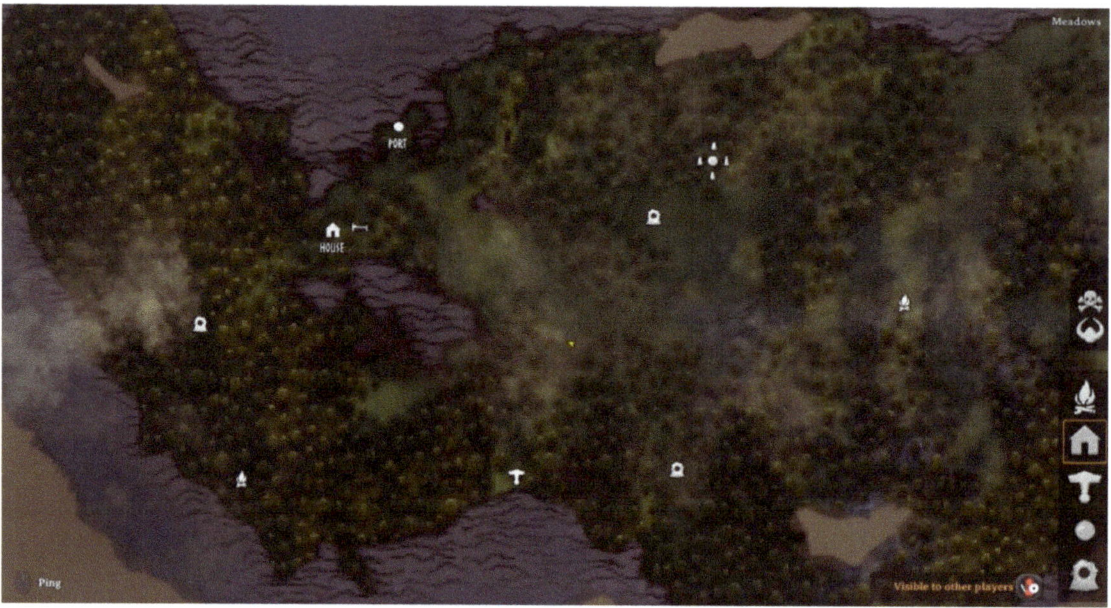

Figure 1. Map from the Valheim game.

Maps are an important integral element used in broadly understood geomedies [32]. Maps are often used in video games as a representation of a fictional geographical space, while the interpretation of cartographic symbols by various groups of users becomes an important problem [33]. The best recognizable are cartographic symbols that directly represent a specific object or refer to associations relating to the features of phenomena [34]. Another aspect studied in cartography is the experience of users in interpreting symbols on maps by experts or novices [35]. In the case of experts and novices, very large differences in the valorization of cartographic design are noted for the same visualization [36,37]. There are also differences in studies in which the length of time in which users use the application is important and they acquire the ability to use symbols and their functions [38]. In this research, we touched upon the importance of the length of experience in using cartographic symbols in one video game, which is important for the differences between first contact with the map and after longer use.

2. Research Goals and Questions

The data obtained through questionnaires are usually the nominal data. In this case, researchers can use a pool of nonparametric tests in the process of testing statistical hypotheses. Everything depends on what research objectives are, what researchers intend to achieve and what kind of variables and categories they want to juxtapose. In this article, the authors want to use the data from the questionnaire to examine the correlations in interpretations of map symbols from the "Valheim" video game. The authors looked for correlation or differences between independent groups of respondents (gamers) in the context of their age and time spent in the virtual game world. The authors concluded that the two research questions posed could help to obtain answers for understanding the interpretation of cartographic symbols:

- Is there any correlation between the age of gamers and their interpretation of symbols?
- Is there any correlation between the time spent on playing Valheim and the interpretation of symbols?

Questions about the age and time spent in a particular game are a universal factor in the questionnaire when it comes to video game literature. In the literature, the age of players is examined psychologically, how it affects psychosocial behavior among players and sets the boundary of life experiences that can be reflected in in-game behavior [39]. In contrast, time spent in a video game is often combined in psychology in research on mental health and social skills, as well as in cognitive research and interpretation of in-game phenomena that directly affect the player [40,41]. The authors, through the analysis and experience of scientific literature, concluded that these two factors will be the best reflection of the interpretation of cartographic symbols for various groups of respondents.

3. Material and Methods

The Valheim computer game was created by Iron Gate studio, which became one of the most popular games in 2021. According to the Steam platform, at the beginning of February 2021, the game was downloaded by about 4 million people within 20 days [42]. It is a survival game set in a world based on Norse mythology with a sandbox concept. The game world has a very simple and understandable user interface that includes the map of the game's world. The game map is a simple cartographic product that contains a designed sketch of the game world and 12 cartographic symbols occurring in various locations (Figure 2) [43].

Questionnaire data have been collected through the LimeSurvey platform [43–45] (Figure 3). A link to online survey was distributed internationally through gaming-related fora (Reddit, Gamespot, IGN, NeoGAF, Facebook groups). Data were collected between 19 and 31 March 2021. The number of respondents who participated in the survey was 1043. Out of 1043 respondents who filled out the entire survey, we choose only the 513 players of Valheim. According to temporal data, players had to spend 6 min on average for the survey (the maximum of 30 min and a minimum of 1 min) [43]. In this study, the opinions of gamers were taken into consideration. Respondents had full subjectivity of answers in the questionnaire; hence, to separate individual groups, the process of categorization was suggested. All the answers to the specified questions received their separate categories during the categorization. The authors developed a separate data categorization for each of the previously discussed symbols, due to their diversity and meaning. In the next stage, the number of responses from all the identified categories for each question was analyzed and it was determined that the category will be correct in the data analysis when the number of responses exceeds 1% of all responses in a particular group of respondents. This study does not address the problem of interpreting the symbols used in the game Valheim. This aspect was analyzed in the article [43].

Figure 2. Screenshot from the game of Valheim and map symbols occurring on the map.

As far as questionnaire data are concerned, the verification of hypotheses including qualitative variables takes place through nonparametric tests. The Pearson's chi-squared test (χ^2) is one of the nonparametric tests that are most frequently used in numerous research disciplines. The χ^2 test may be used for examining the correlation between two variables. It was introduced by Karl Pearson and was an object of many discussions. In a series of articles, the χ^2 test was analysed [46,47] and its problems were discussed [48,49]. The chi-square test has been applied in all research areas [50]. Its main uses are the following: goodness-of-fit [51–53], association/independence [54,55], homogeneity [56,57], classification [58–62], etc.

The authors of the study used the Pearson test in the context of the collected data from the article. It is commonly known that the Pearson's chi-squared test belongs to the family of tests for which the following assumptions have been made [50,63,64]:

I.   The data are collected at random from the population;

II.   The size of the sample is large enough. The employment of the χ^2 test with a research sample that is too small may lead to the unacceptable indicator of the type II error (adopting the null hypothesis, whereas it is, in fact, false) [65]. There is no arbitrarily established acceptable size of the sample, hence, the minimal value of the sample oscillates between 20 and 50 people;

III.   The value in individual cells is right when no more than 1/5 of the expected values are lower than 5 and there are no cells with the number zero [66].

Figure 3. Online questionnaire on LimeSurvey platform.

W. G. Cochran is the source of the above rules [67] and, as Bolboacă et al. [50] write, they have been arbitrarily selected. When, out of the abovementioned assumptions, the third one is not fulfilled, the Yates's correction for continuity is applied [68]. The Fisher's exact test is an alternative when the Yates's correction for continuity is unacceptable [69]. As Bolboacă et al. [50] highlight, the Fisher's exact test turns out to be "the golden test" in the analysis of correlations whereas the Yates's correction for continuity and the Mantel-Haenszel test may be used as alternative tests. It is worth emphasising that the Fisher's exact test [70] is preferred when the data with the small-expected value are analysed.

The Pearson's χ^2 test, like other nonparametric tests, is not contingent upon the distribution of the examined set. Thus, it may be applied both for the normal distribution and for all other cases [65]. The χ^2 test may be particularly useful in situations, in which two variables obtained by means of the questionnaire are measured in the nominal scale, and the results are presented in the matrix with the freely chosen number of verses and columns. The Pearson's chi-square test is used mainly in studying correlations between nominal variables [65]. The variable is treated as nominal when its values represent categories, without any internal ranging, i.e., it is impossible to establish which case is larger, superior or more relevant than the other [71].

The chi-square test provides the information whether or not a given correlation occurs (a possibility of rejecting the null hypothesis); however, it fails to inform one about the significance of the indicated correlation. In the chi-square test, one compares the observed value with the expected value. Using the Fisher's exact test helps one eliminate the failure to meet the third assumption of the χ^2 test. To find significant differences between individual variables, one should apply a testing correction. In the context of the authors' research, among the numerous corrections, the Benjamini–Hochberg treatment is the strongest. The correction demonstrates all the (statistically significant) differences and minimises the expected false positive rate among them [72].

4. Results

Statistical research has been conducted in the PQStat (v 1.8.0) software. The authors considered only the data concerning gamers that play Valheim to examine the differences in their interpretation of symbols included in the game. The differences will be studied between independent groups related to the age of respondents and the total time they had spent on playing the game before the research. Due to the fact that, as a part of statistical analyses that the authors carry out, they were dealing with the data coming from the questionnaire and they meet the criterion of nominal data, the authors selected the Pearson's chi-square test to identify statistically significant differences in symbol interpretations. For all the analysed data, PQStat failed to show that the Cochran's theorem (third assumption of the test) [66] was met; hence, the Fisher's exact test was additionally performed [50,70]. To be able to specify significant differences between groups, the Benjamini–Hochberg procedure was implemented [72].

4.1. Age and Symbol Interpretation

The question about age is one of the most frequently asked questions in question-naires [73]. It is also a way to verify whether or not respondents are a homogenous group agewise [74]. In this case, respondents were asked to reveal their age directly. After the questionnaire data had been collected, the authors distinguished five age groups based on the published statistical data on the global sector of computer games [75,76]. Then, the results were grouped into age cohorts:

- More than 46 years old (8.8% of players);
- From 31–45 years old (38.4% of players);
- From 26–30 years old (28.4% of players);
- From 20–25 years old (19.1% of players);
- Less than 19 years old (3.9% of players);
- None (1.4% of players).

Each gamer was free not to give their age and seven people used that opportunity. The largest age group consisted of gamers aged 31–45, and the smallest one of those below 19. Based on the data, one can conclude that gamers that participated in the research do not create a homogenous group of respondents.

One of the crucial objectives of the article was to examine whether the age of gamers directly affected the interpretation of symbols in the game of Valheim (the research question). Significant results of the Pearson's χ^2 test allowed researchers to find two correlations (Table 1):

Table 1. Pearson's test for the age of players.

Symbol	1	2	3	4	5	6
Pearson χ^2 test (p-value)	0.7753	0.0526	0.2255	0.0447	0.4464	0.4360
Symbol	7	8	9	10	11	12
Pearson χ^2 test (p-value)	0.7238	0.3941	0.8696	0.0351	0.4229	0.0604

4.1.1. Age and the Interpretation of Symbol 4

For this comparison, the p-value of the Pearson's chi-square test was 0.044737. The Cochran's theorem was not met (the third assumption of the chi-square test) [66]; therefore, the Fisher's exact test (p-value = 0.01999) was performed. As it turns out from previous studies, gamers (questionnaire respondents) interpreted the symbol in the following way [43]:

- Cat. 1—Bed/sleep—54%;
- Cat. 2—Current spawn point/save point—42.5%;
- Cat. 3—Main base/home—3.3%;
- Other—0.2%.

The Benjamin–Hochberg procedure demonstrated two significant differences for the interpretation of symbol 4 between categories 1 and 3 (p-value = 0.032695), as well as 2 and 3 (p-value = 0.032695) (Figure 4B).

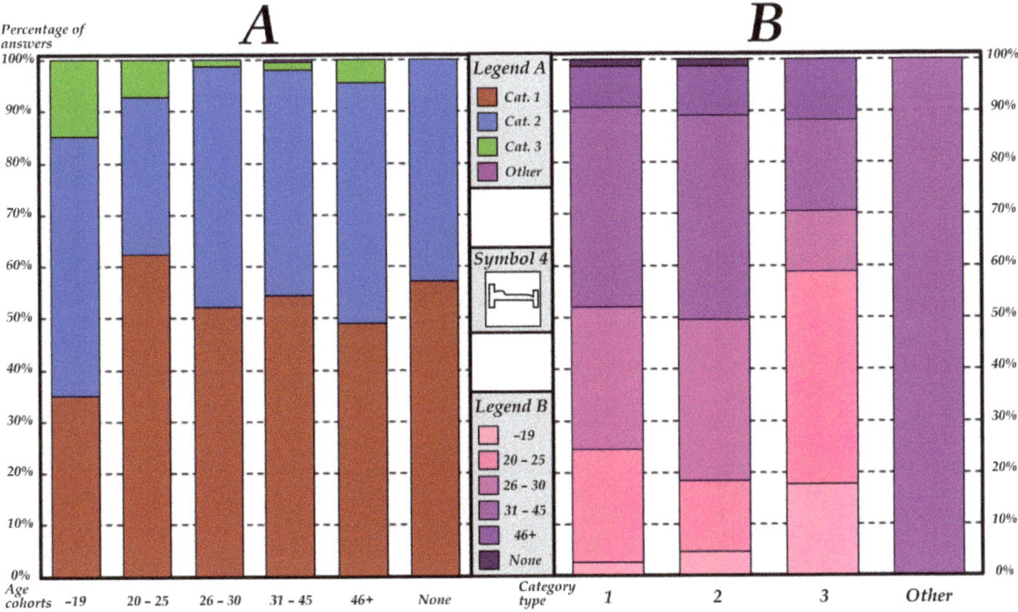

Figure 4. Cumulative charts presenting a percentage of individual categories of symbol 4 interpretation, divided according to age (**A**) and including individual age groups, divided according to interpretation category (**B**).

As far as part A Figure 4 is concerned, the Benjamin–Hochberg procedure failed to show any statistically relevant results (p-value > 0.05) concerning the differences between categories of interpretation of symbol 4.

4.1.2. Age and Symbol 10 Interpretation

Symbol 10 was interpreted by gamers in 10 different ways (plus the *None* and the *Other* group) [43]:

- Cat. 1—POI/waypoint—18.9%;
- Cat. 2—Dungeons/crypts/ruins—15.6%;
- Cat. 3—Altar/shrine/monument—4.7%;
- Cat. 4—Mining/ore/resources—9.2%;
- Cat. 5—Ship location/port/boat/anchor—8.0%;
- Cat. 6—Mjolnir—7.8%;
- Cat. 7—Hammer/forge/anvil/weapon Smith/armor—2.9%;
- Cat. 8—Enemy camp/enemy/monsters—4.1%;
- Cat. 9—Fight/challenge/combat/quest—1.9%;

- Cat. 11—Celtic scroll/Maypole/Norse myth—1.9%;
- None—17.0%;
- Other—8%.

The Pearson's chi-square test confirms that the comparison made between age groups and symbol 10 interpretation groups is statistically significant (p-value = 0.035139; the Cochran's theorem is not met). Unfortunately, the Fisher's exact test fails to show a statistically significant result. Analysing the results of the Benjamin–Hochberg procedure, researchers noticed that it demonstrated statistically significant results, indicating differences between category 6 and category 2 (p-value = 0.048093) and *Other* (p-value = 0.048093) (Figure 5B). As far as part A Figure 5 is concerned, the Benjamini–Hochberg procedure demonstrated no statistically significant differences.

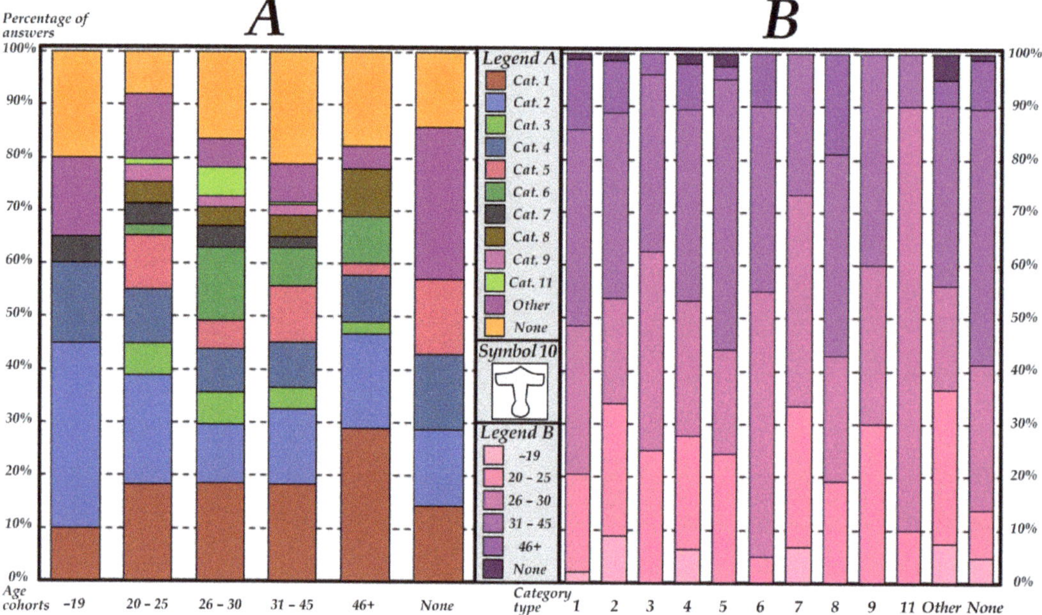

Figure 5. Cumulative charts presenting a percentage of individual categories of symbol 10 interpretation, divided according to age (**A**) and including individual age groups, divided according to interpretation category (**B**).

4.2. Time Spent on Playing the Game of Valheim and Symbol Interpretation

Time spent on playing a given game/each game objectively demonstrates the gamer's commitment. It may directly translate into the correct understanding of some aspects of the game. In some video games the time needed for a single game is limited, e.g. a single game in the League of Legends takes between 25:02 to 32:30 min in the 5 vs. 5 mode on Summoner Rifts [77]. The gamer may play many games, but it is difficult to quit a certain time scheme in which certain amount of time is necessary to complete a single game. It looks different for survival games, such as the game analysed in the research, Valheim. In such games the time that the gamer may devote to a single game is unlimited. As long hours of playing the game go by, the gamer discovers new places or new mechanics of the game that was difficult to spot at the beginning. Therefore, the observation whether there are statistically significant differences between the time spent on playing Valheim and the interpretation of map symbols used in the game is a significant aspect of this article. Like for the age groups, the authors divided gamers into several groups with regard to time spent on playing the game. Considering the data that come from independent websites [78,79], which analysed the time spent in the game of Valheim based on time data from the Steam platform, the authors determined the following five groups in terms of the time spent on playing the game:

- 1—less than 30 h (86.0% gamers);
- 2—30–60 h (15.0% gamers);
- 3—60–90 h (21.8% gamers);
- 4—90–120 h (16.2% gamers);
- 5—120 and more hours (38.4% gamers).

Based on the statistical analyses carried out, researchers managed to achieve one statistically significant results of the Pearson's chi-square test and one statistically significant difference of the Benjamini–Hochberg procedure (Table 2):

Table 2. Pearson's test for time spent in-game.

Symbol	1	2	3	4	5	6
Pearson χ^2 test (p-value)	0.0842	0.6506	<0.000001	0.0684	0.7413	0.4352
Symbol	7	8	9	10	11	12
Pearson χ^2 test (p-value)	0.3414	0.5664	0.9001	0.1883	0.1464	0.1430

4.2.1. Time Spent on Playing the Game and Symbol 3 Interpretation

The statistically significant result of the Pearson's chi-square test (p-value < 0.000001) was achieved for the comparison of groups of gamers divided according to the time spent on playing Valheim and symbol 3 interpretation (the Cochran's theorem unfulfilled: lack of statistical significance for the Fisher's exact test). The symbol 3 itself was interpreted by gamers in the following way:

- Cat. 1—Trader/Merchant—80.7%;
- Cat. 2—Loot/Bag of goods—8%;
- Cat. 3—Coin purse/Money—5.3%;
- Other—4.9%;
- None—1.2%.

The Benjamini–Hochberg procedure confirmed significant differences in interpretations of symbol 3 by groups of gamers selected in terms of the time they had spent on playing Valheim between category 2 and 1 (p-value < 0.000001), 3 (p-value = 0.000211) and *Other* (p-value = 0.020922) (Figure 6B). In the juxtaposition of the group of gamers (time spent on playing the game) with the category of symbol 3 interpretation the Benjamin–Hochberg procedure confirmed four differences between the <30 h of playing group and other groups. For all the differences statistical significance (p-value) was <0.00001.

4.2.2. Time Spent on Playing and Symbol 4 Interpretation

For groups selected on the basis of the time spent on playing Valheim, the Pearson's χ^2 test failed to achieve a statistically significant result for symbol 4 (p-value > 0.05; the Cochran's theorem unfulfilled) but the Fisher's exact test had a significant result (p-value = 0.019258). Unfortunately, the Benjamini–Hochberg's multiple testing correction showed no differences, either for groups (time spent on playing the game) or the symbol 4 interpretation category.

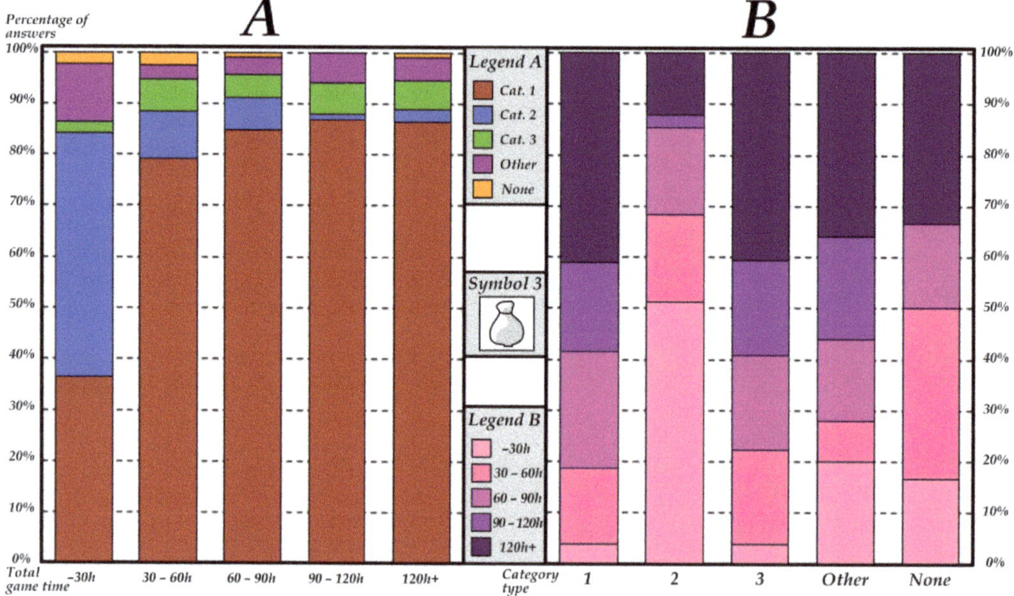

Figure 6. Cumulative charts presenting a percentage of individual categories of symbol 3 interpretation, divided according to the time spent on playing Valheim (**A**) and including individual groups (time spent on playing), divided according to interpretation categories (**B**).

4.2.3. Time Spent on Playing the Game and Symbol 11 Interpretation

Symbol 11, presenting a black circle filled with the white colour, is the most universal map symbol in the game of Valheim, due to the fact that there is no graphic or color relationship with any of the places in the game. Each player could mark each place by distinguishing them with a signature. It was interpreted by gamers in the following way:

- Cat. 1—POI/waypoint—48.3%;
- Cat. 2—Resources/veins/berries—19.5%;
- Cat. 3—Interesting places/checkpoints/reference points—7.8%;
- Cat. 4—Dot/cicrle/pin/ball—8.2%;
- Cat. 7—Dungeon/crypt/tomb/cave/tunnel—1.6%;
- None—10.9%;
- Other—3.7%.

Even though it failed to achieve statistically significant results in the Pearson's χ^2 test (unfulfilled Cochran's theorem) and in the Fisher's exact test (p-value > 0.05), the Benjamini–Hochberg procedure was conducted, providing two significant differences. One of them occurred between the <30 h and 120 h+ hour of gaming groups (time spent on gaming) (p-value = 0.037632) (Figure 7A) and the other one between category 2 of the symbol 11's interpretation and the group *Other* (p-value = 0.038049) (Figure 7B).

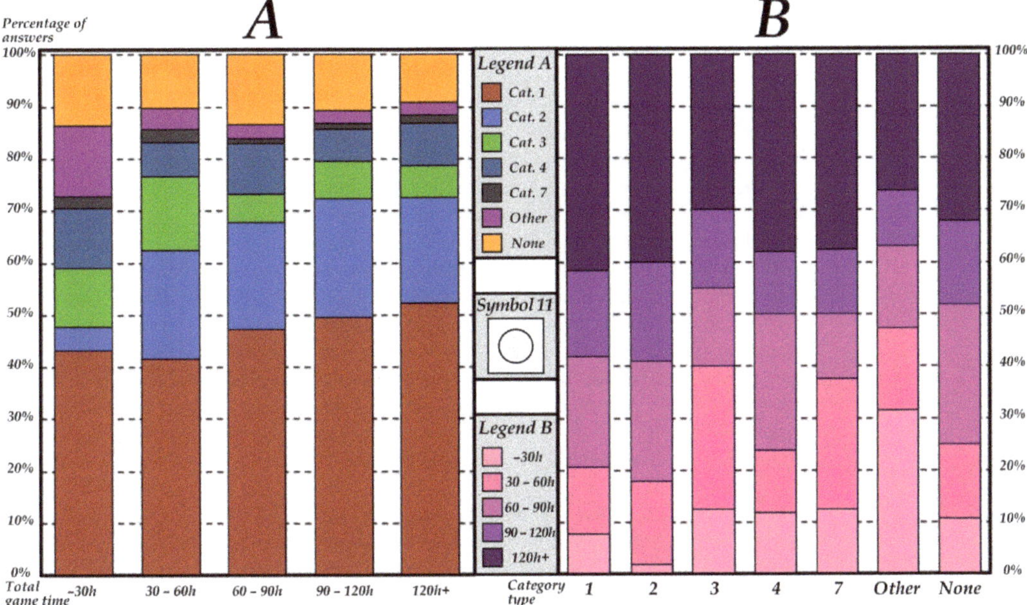

Figure 7. Cumulative charts presenting a percentage of individual categories of symbol 11 interpretation, divided according to the time spent on playing Valheim (**A**) and including individual groups (time spent on playing), divided according to interpretation categories (**B**).

5. Discussion and Conclusions

The aim of the presented research was to investigate the correlations that can characterize the interpretations of individual cartographic symbols used in the game "Valheim" in the context of the age of the players and the time spent in the virtual world. In addition, the authors also asked two research questions. The research is based on the obtained survey data, applied statistical methods and developed conclusions in the context of cartography, which were obtained on the basis of the analysis of cartographic symbols in games, taking into account two parameters of the players (age and time spent in the game).

In the context of symbol 3, analysing Figure 4B, one can conclude that the fundamental difference between categories 3, as well as 1 and 2, lies in a significant disproportion between the younger and the older gamers. The younger gamers (the -19 and the 20–25 group) to the greater extent associated the symbol with main base/home, compared to the older gamers (26–30 and 31–45). On the other hand, older gamers had a greater tendency to interpret symbol 4 as bed/sleep or current spawn point/save point. It constitutes a significant difference in the interpretation of the symbol because the age groups 26–30 and 31–45 associated the symbol mostly with the function that the bed fulfils in the game, whereas groups -19 and 20–25 with the actual position in the game (in the game the bed needs to be roofed; therefore, it may be also easily associated with the constructed house or even the main headquarters of the gamer). It is worth highlighting that all the *Other* answers were added by gamers from the group aged 31–45.

In the juxtaposition of age groups and symbol 10 interpretation it may be observed that category 6 of the 26–30 age group has a greater contribution than of the 20–25 age group, and that there is neither -19 nor *None* group there (Figure 5B). What it directly means is that the youngest gamers (aged -19 and 20–25) failed to interpret symbol 10 as Mjolnir (Mjolnir is symbolically represented as a reversed symbol 10). It is possible, then, that the majority of younger respondents are not interested in Norse mythology. Additionally, the movie saga about Thor and his hammer Mjolnir were created in 2011 so the age groups analysed were at a very young age.

As far as Figure 6 symbol 3 is concerned, for some gamers it may actually be very distant from the starting point in the game; therefore, it frequently happens that many hours of playing the game are necessary to be able to reach that symbol on the map. Symbol 3, representing a bag (money) in the game of Valheim, denotes the location of the only NPC in the game and may occur in multiple places at the same time. Some gamers may be extremely lucky and encounter this symbol on their initial island or during their first travel but the results of our research related to the interpretation of symbol 3 prove clearly that the gamers that played less (below 30 h), constitute the greatest percentage of those who interpreted the symbol as category 2 (Loot/Bag of goods) (Figure 6B). Those gamers interpreted the symbol literally. Their answers differ significantly from interpretation 1, 3 and *Other*, as well as from other groups selected on the basis of the time spent on playing the game (Figure 6A). Category 1 refers directly to the function that NPC performs in the game, whereas category 3 may be strongly associated with the fact that in the location of symbol 3 the gamer may only spend the jewels and gold that they have gained. A small percentage of the <30 h and 30–60 h groups in the interpretation categories 1 and 3 corroborates the statement that gamers need a lot of time to discover that symbol in the game.

When groups (time spent on playing the game) and symbol 11 interpretations were juxtaposed, researchers noticed that the cases observed were characterized by the larger percentage of the *Other* group for gamers with the shortest gaming experience (<30 h) and by the smaller percentage of interpretation of number 2 symbol 11 (particularly in relation to gamers who had the longest gaming experience—120 h). The difference may be caused by the fact that gamers, who had spent more time playing the game, marked all the places they had visited or planned to visit again and associated them directly with things like resources, veins, and berries. The gamers with the shortest gaming experience (<30 h) came up with so many interpretations (which is related to the universality of the symbol) that those interpretations failed to be classified as one unique category.

The first research question asked in the introduction of the article: is there any correlation between the age of gamers and their interpretation of symbols? To answer, one needs to conclude on the basis of statistical analyses that, indeed, there is a correlation, and it is corroborated by two verified cases. Both cases were related to symbol 4 and 10. As far as symbol 4 is concerned, the 26–30 and 31–45 age groups interpreted the symbol, mostly associating it with the function it performs in the game, whereas the -19 and 20–25 age groups interpreted the symbol as the actual ranking in the game. For symbol 10, the youngest gamers (-19 and 20–25) failed to interpret the symbol as Mjolnir (Mjolnir is symbolically represented as a reversed symbol 10).

The second research question was related to the correlation between the time spent on playing the game of Valheim and the interpretation of symbols. In the comparison of these two groups of respondents the authors achieved significant results of statistical tests related to symbol 3. The results of the research leave no doubts: the gamers that had played less (below 30 h) constitute the largest percentage in the interpretation of the symbol as category 2 (loot/bag of goods) (Figure 6B). Those gamers interpreted the symbol literally. A small percentage of the <30 h and 30–60 h groups in the category of interpretation 1 and 3 corroborates the assumption that gamers need a lot of time to be able to discover that symbol during the game.

The methods applied (the Pearson's chi-square test, the Fisher's exact test, the Benjamini–Hochberg procedure) fulfilled their task. For the questionnaire data (nominal data), the employment of the nonparametric test was obligatory. In the analyses presented, the authors confronted the data about symbol interpretation with the data related to the age and time spent on playing that are usually used solely for characterizing the group of respondents. The authors also emphasise that the data have not been selected at random because they intended to confront the data with symbol interpretation, showing both the gamers' commitment (time spent on playing the game) regardless of their age (age of gamers). The map symbols that occur in video games with specific graphic features also play a significant part in multimedia cartography [80], and the results obtained in the research may be taken into consideration in the process of map symbol design in geomedia [33].

The authors would like to point out that the obtained statistically significant results concern only three symbols (two symbols for the age and one symbol for the time spent in the game). It is not a substantial majority for the realized set of 12 symbols. Therefore, it is worth noting that the correlation between the age of the players and the time spent in the game, and the interpretations of symbols exists for individual symbols, not for the entire set. As a consequence, a question arises for subsequent research and the possibility of checking subsequent game titles and trying to find a title in which such dependencies exist for a larger number of symbols.

The procedure of verification of statistical hypotheses requires a set of specific skills, such as selectivity and the employment of suitable statistical tools, from the researcher. To a large extent, such skills decide upon the reliability of the results obtained and correctness of their interpretation. From the researcher's point of view, it is interesting to discover correlations between the data/variables, e.g., similarities and differences. In this article, the authors verified and drew interesting conclusions on the correlation between the interpretation of symbols from the game of Valheim and the time spent on playing this game, as well as the age of gamers.

Along with new video games being launched, the authors are going to undertake new studies related to map symbols that appear in them. Video games are perceived as products of culture [81]; hence, the authors would like to develop their studies in this direction as games allow one to visualise how gamers influence and shape the culture they participate in [82], with culture being reproduced by symbols and elements presented, distributed, taught, and perpetuated in video games [83].

Author Contributions: Conceptualization, Tymoteusz Horbiński and Krzysztof Zagata; methodology, Tymoteusz Horbiński; formal analysis, Tymoteusz Horbiński and Krzysztof Zagata; resources, Tymoteusz Horbiński and Krzysztof Zagata; writing—original draft preparation, Tymoteusz Horbiński and Krzysztof Zagata; writing—review and editing, Tymoteusz Horbiński All authors have read and agreed to the published version of the manuscript.

Funding: This research received no external funding.

Institutional Review Board Statement: Not applicable.

Informed Consent Statement: Not applicable.

Data Availability Statement: Not applicable.

Acknowledgments: This paper is the result of research on cartography in video games and statistical analysis carried out within statutory research in the Department of Cartography and Geomatics, Faculty of Geographical and Geological Sciences, Adam Mickiewicz University Poznań, Poland.

Conflicts of Interest: The authors declare no conflict of interest.

References

1. Ash, J.; Gallacher, L.A. Cultural Geography and Videogames. *Geogr. Compass* **2011**, *5*, 351–368. [CrossRef]
2. Álvarez, R.; Duarte, F. Spatial Design and Placemaking: Learning from Video Games. *Space Cult.* **2017**, *21*, 208–232. [CrossRef]
3. Wolf, M. 3 Space in the Video Game. In *The Medium of the Video Game*; Wolf, M., Ed.; University of Texas Press: New York, NY, USA, 2021; pp. 51–76. [CrossRef]
4. Chądzyńska, D.; Gotlib, D. Maps in video games—Range of applications. *Pol. Cartogr. Rev.* **2015**, *47*, 137–145. [CrossRef]
5. Toups, Z.O.; Lalone, N.; Alharthi, S.A.; Sharma, H.N.; Webb, A.M. Making Maps Available for Play. *ACM Trans. Comput. Interact.* **2019**, *26*, 43. [CrossRef]
6. Johanson, C.; Gutwin, C.; Mandryk, R.L. The Effects of Navigation Assistance on Spatial Learning and Performance in a 3D Game. In Proceedings of the Annual Symposium on Computer-Human Interaction in Play (CHI PLAY '17). Association for Computing Machinery, New York, NY, USA, 15–18 October 2017; pp. 341–353. [CrossRef]

7.   Peacocke, M.; Teather, R.J.; Carette, J.; MacKenzie, I.S.; McArthur, V. An empirical comparison of first-person shooter information displays: HUDs, diegetic displays, and spatial representations. *Entertain. Comput.* **2018**, *26*, 41–58. [CrossRef]

8.   Caroux, L.; Isbister, K. Influence of head-up displays' characteristics on user experience in video games. *Int. J. Hum. Comput. Stud.* **2016**, *87*, 65–79. [CrossRef]

9.   Garfield, S. *On the Map: Why the World Looks the Way It Does*; Profile Books: London, UK, 2012.

10.   Coulton, P.; Huck, J.; Gradinar, A.; Salinas, L. Mapping the beach beneath the street: Digital cartography for the playable city. In *Playable Cities. The City as Digital Playground*; Nijholt, A., Ed.; Springer: Singapore, 2017; pp. 137–162.

11.   Edler, D.; Dickmann, F. The Impact of 1980s and 1990s Video Games on Multimedia Ca;rtography. *Cartogr. Int. J. Geogr. Inf. Geovis.* **2017**, *52*, 168–177. [CrossRef]

12.   Edler, D. Where Spatial Visualization Meets Landscape Research and "Pinballology": Examples of Landscape Construction in Pinball Games. *KN-J. Cartogr. Geogr. Inf.* **2020**, *70*, 55–69. [CrossRef]

13.   Hruby, F.; Ressl, R.; Valle, G.D.L.B.D. Geovisualization with immersive virtual environments in theory and practice. *Int. J. Digit. Earth* **2018**, *12*, 123–136. [CrossRef]

14.   Büyüksalih, G.; Kan, T.; Özkan, G.E.; Meriç, M.; Isın, L.; Kersten, T.P. Preserving the Knowledge of the Past Through Virtual Visits: From 3D Laser Scanning to Virtual Reality Visualisation at the Istanbul Çatalca İnceğiz Caves. *PFG–J. Photogramm. Remote Sens. Geoinf. Sci.* **2020**, *88*, 133–146. [CrossRef]

15.   Keil, J.; Korte, A.; Ratmer, A.; Edler, D.; Dickmann, F. Augmented Reality (AR) and Spatial Cognition: Effects of Holographic Grids on Distance Estimation and Location Memory in a 3D Indoor Scenario. *PFG–J. Photogramm. Remote Sens. Geoinf. Sci.* **2020**, *88*, 165–172. [CrossRef]

16.   Walmsley, A.P.; Kersten, T.P. The Imperial Cathedral in Königslutter (Germany) as an Immersive Experience in Virtual Reality with Integrated 360° Panoramic Photography. *Appl. Sci.* **2020**, *10*, 1517. [CrossRef]

17.   Zagata, K.; Gulij, J.; Halik, Ł.; Medyńska-Gulij, B. Mini-Map for Gamers Who Walk and Teleport in a Virtual Stronghold. *ISPRS Int. J. Geo-Inf.* **2021**, *10*, 96. [CrossRef]

18.   Cybulski, P. Spatial distance and cartographic background complexity in graduated point symbol map-reading task. *Cartogr. Geogr. Inf. Sci.* **2020**, *47*, 244–260. [CrossRef]

19.   Popelka, S.; Brychtova, A. Eye-tracking Study on Different Perception of 2D and 3D Terrain Visualisation. *Cartogr. J.* **2013**, *50*, 240–246. [CrossRef]

20.   Czepkiewicz, M.; Jankowski, P.; Młodkowski, M. Geo-questionnaires in urban planning: Recruitment methods, participant engagement, and data quality. *Cartogr. Geogr. Inf. Sci.* **2016**, *44*, 551–567. [CrossRef]

21.   Kapenekakis, I.; Chorianopoulos, K. Citizen science for pedestrian cartography: Collection and moderation of walkable routes in cities through mobile gamification. *Hum. Cent. Comput. Inf. Sci.* **2017**, *7*, 111. [CrossRef]

22.   Slocum, T.A.; Blok, C.; Jian, B.; Koussoulakou, A.; Montello, D.R.; Fuhrmann, S. Cognitive and usability issues in geovisualization. *Cartogr. Geogr. Inf. Sci.* **2001**, *28*, 61–75. [CrossRef]

23.   Pirani, N.; Ricker, B.A.; Kraak, M.J. Feminist cartography and the United Nations Sustainable Development Goal on gender equality: Emotional responses to three thematic maps. *Can. Geogr. Géographe Can.* **2020**, *64*, 184–198. [CrossRef]

24.   Poplin, A. How user-friendly are online interactive maps? Survey based on experiments with heterogeneous users. *Cartogr. Geogr. Inf. Sci.* **2015**, *42*, 358–376. [CrossRef]

25.   Medyńska-Gulij, B.; Zagata, K. Experts and Gamers on Immersion into Reconstructed Strongholds. *ISPRS Int. J. Geo-Inf.* **2020**, *9*, 655. [CrossRef]

26.   Çöltekin, A.; Heil, B.; Garlandini, S.; Fabrikant, S.I. Evaluating the effectiveness of interactive map interface designs: A case study integrating usability metrics with eye-movement analysis. *Cartogr. Geogr. Inf. Sci.* **2009**, *36*, 5–17. [CrossRef]

27.   Golebiowska, I.; Opach, T.; Rød, J.K. Breaking the Eyes: How Do Users Get Started with a Coordinated and Multiple View Geovisualization Tool? *Cartogr. J.* **2020**, *57*, 235–248. [CrossRef]

28.   Halik, Ł.; Medyńska-Gulij, B. The Differentiation of Point Symbols using Selected Visual Variables in the Mobile Augmented Reality System. *Cartogr. J.* **2016**, *54*, 147–156. [CrossRef]

29.   Cybulski, P. Effectiveness of Memorizing an Animated Route—Comparing Satellite and Road Map Differences in the Eye-Tracking Study. *ISPRS Int. J. Geo-Inf.* **2021**, *10*, 159. [CrossRef]

30.   Bertin, J. *Semiology of Graphics: Diagrams, Networks, Maps, Trans*; Berg, W., Ed.; University of Wisconsin Press: Madison, WI, USA, 1983.

31.   Keil, J.; Edler, D.; Dickmann, F.; Kuchinke, L. Meaningfulness of landmark pictograms reduces visual salience and recognition performance. *Appl. Ergon.* **2018**, *75*, 214–220. [CrossRef]

32.   Medyńska-Gulij, B. *Cartography and Geomedia*; Wydawnictwo Naukowe PWN: Warszawa, Poland, 2021; p. 286. (In Polish)

33.   Medyńska-Gulij, B.; Tegeler, T.; Bauer, H.; Zagata, K.; Wielebski, Ł. Filming the Historical Geography: Story from the Realm of Maps in Regensburg. *ISPRS Int. J. Geo-Inf.* **2021**, *10*, 764. [CrossRef]

34.   MacEachren, A.M. *How Maps Work: Representation, Visualization and Design*; Guilford Press: New York, NY, USA, 1995.

35.   Halik, Ł. The analysis of selected point-symbol visual variables of the Augmented Reality system for smartphone-type mobile devices. *Geod. Cartogr.* **2012**, *61*, 19–30.

36.   Robinson, A.H.; Morrison, J.L.; Muehrecke, P.C.; Kimerling, A.J.; Guptil, S.C. *Elements of Cartography*, 6th ed.; Wiley: New York, NY, USA, 1995.

37. Dent, B.D. *Cartography: Thematic Map Design*, 5th ed.; McGraw-Hill: Boston, MA, USA, 1999.
38. Cybulski, P.; Horbiński, T. User experience in using graphical user interfaces of web maps. *ISPRS Int. J. Geo. Inf.* **2020**, *9*, 412. [CrossRef]
39. Przybylski, A.K. Electronic Gaming and Psychosocial Adjustment. *Pediatrics* **2014**, *134*, e716–e722. [CrossRef]
40. Kovess-Masfety, V.; Keyes, K.; Hamilton, A.; Hanson, G.; Bitfoi, A.; Golitz, D.; Koç, C.; Kuijpers, R.; Lesinskiene, S.; Mihova, Z.; et al. Is time spent playing video games associated with mental health, cognitive and social skills in young children? *Soc. Psychiatry Psychiatr. Epidemiol.* **2016**, *51*, 349–357. [CrossRef] [PubMed]
41. Shen, C.; Williams, D. Unpacking Time Online: Connecting Internet and Massively Multiplayer Online Game Use with Psychosocial Well-Being. *Commun. Res.* **2010**, *38*, 123–149. [CrossRef]
42. Available online: https://steamcommunity.com/games/892970/announcements/detail/3058478454611894476 (accessed on 12 February 2021).
43. Horbiński, T.; Zagata, K. Map Symbols in Video Games: The Example of "Valheim". *KN-J. Cartogr. Geogr. Inf.* **2021**, *71*, 269–283. [CrossRef]
44. Cybulski, P.; Wielebski, Ł. Effectiveness of dynamic point symbols in quantitative mapping. *Cartogr. J.* **2019**, *56*, 146–160. [CrossRef]
45. Horbiński, T.; Cybulski, P.; Medyńska-Gulij, B. Graphic Design and Button Placement for Mobile Map Applications. *Cartogr. J.* **2020**, *57*, 196–208. [CrossRef]
46. Mirvaliev, M. The components of chi-squared statistics for goodness-of-fit tests. *J. Math. Sci.* **1987**, *38*, 2357–2363. [CrossRef]
47. Plackett, R.L. Karl Pearson and the Chi-Squared Test. *Int. Stat. Rev.* **1983**, *51*, 59. [CrossRef]
48. Baird, D. The fisher/pearson Chi-squared controversary: A turning point for inductive inference. *Br. J. Philos. Sci.* **1983**, *34*, 105–118. [CrossRef]
49. Cochran, W.G. Some methods for strengthening the common chi-square tests. *Biometrics* **1954**, *10*, 417–451. [CrossRef]
50. Bolboacă, S.D.; Jäntschi, L.; Sestraş, A.F.; Sestraş, R.E.; Pamfil, D.C. Pearson-Fisher Chi-Square Statistic Revisited. *Information* **2011**, *2*, 528–545. [CrossRef]
51. Li, G.; Doss, H. Generalized pearson-fisher Chi-square goodness-of-fit tests, with applications to models with life history data. *Ann. Stat.* **1993**, *21*, 772–797. [CrossRef]
52. Moore, D.S.; Stubblebinea, J.B. Chi-square tests for multivariate normality with application to common stock prices. *Commun. Stat. Theory Methods* **1981**, *10*, 713–738. [CrossRef]
53. Joe, H.; Maydeu-Olivares, A. A general family of limited information goodness-of-fit statistics for multinomial data. *Psychometrika* **2010**, *75*, 393–419. [CrossRef]
54. Nathan, G. On the asymptotic power of tests for independence in contingency tables from stratified samples. *J. Am. Stat. Assoc.* **1972**, *67*, 917–920. [CrossRef]
55. O'Brien, P.C.; Fleming, T.H. A multiple testing procedure for clinical trials. *Biometrics* **1979**, *35*, 549–556. [CrossRef]
56. Cox, M.K.; Key, C.H. Post hoc pair-wise comparisons for the Chi-square test of homogeneity of proportions. *Key Educ. Psychol. Meas.* **1993**, *53*, 951–962. [CrossRef]
57. Pardo, L.; Martín, N. Omogeneity/heterogeneity hypotheses for standardized mortality ratios based on minimum power-divergence estimators. *Biom. J.* **2009**, *51*, 819–836. [CrossRef] [PubMed]
58. Andrés, A.M.; Tejedor, I.H. Comments on 'Tests for the homogeneity of two binomial proportions in extremely unbalanced 2×2 contingency tables'. *Stat. Med.* **2009**, *28*, 528–531.
59. Baker, S.; Cousins, R.D. Clarification of the use of Chi-square and likelihood functions in fits to histograms. *Nucl. Instrum. Methods Phys. Res.* **1984**, *221*, 437–442. [CrossRef]
60. Elmore, K.L. Alternatives to the Chi-Square Test for Evaluating Rank Histograms from Ensemble Forecasts. *Weather Forecast.* **2005**, *20*, 789–795. [CrossRef]
61. Zhang, J.-T. Approximate and Asymptotic Distributions of Chi-Squared–Type Mixtures with Applications. *J. Am. Stat. Assoc.* **2005**, *100*, 273–285. [CrossRef]
62. Gagunashvili, N.D. Chi-square tests for comparing weighted histograms. *Nucl. Instrum. Methods Phys. Res. Sect. A* **2010**, *614*, 287–296. [CrossRef]
63. Agresti, A. *Introduction to Categorical Data Analysis*; John Wiley and Sons: New York, NY, USA, 1996; pp. 231–236.
64. Levin, I.P. *Relating Statistics and Experimental Design*; Sage Publications: Thousand Oaks, CA, USA, 1999.
65. Neyman, J.; Pearson, E.S. (Eds.) *The Testing of Statistical Hypotheses in Relation to Probabilities a Priori*; Cambridge University Press: Cambridge, UK, 1967; pp. 186–202.
66. Rosner, B. *Fundamentals of Biostatistics. Chapter 10. Chi-Square Goodness of Fit*, 6th ed.; Thomson Learning Academic Resource Center: Duxbury, MA, USA, 2006; pp. 438–441.
67. Cochran, W.G. The χ^2 test of goodness of fit. *Ann. Math. Stat.* **1952**, *25*, 315–345. [CrossRef]
68. Yates, F. Contingency table involving small numbers and the χ^2 test. *Suppl. J. R. Stat. Soc.* **1934**, *1*, 217–235. [CrossRef]
69. Fisher, R.A. The logic of inductive inference. *J. R. Stat. Soc.* **1935**, *98*, 39–54. [CrossRef]
70. Scholz, F.W.; Stephens, M.A. K-sample anderson-darling tests. *J. Am. Stat. Assoc.* **1987**, *82*, 918–924.
71. Słowińska, M. Wykorzystanie testu chi-kwadrat w badaniach preferencji żywieniowych konsumentów. *Eng. Sci. Technol.* **2019**, *1*, 24–38. [CrossRef]

72. Benjamini, Y.; Hochberg, Y. Controlling the False Discovery Rate: A Practical and Powerful Approach to Multiple Testing. *J. R. Stat. Soc. Ser. B* **1995**, *57*, 289–300. [CrossRef]
73. Healey, B.; Gendall, P. Asking the Age Question in Mail and Online Surveys. *Int. J. Mark. Res.* **2007**, *50*, 309–317. [CrossRef]
74. Wielebski, Ł.; Medyńska-Gulij, B. Graphically supported evaluation of mapping techniques used in presenting spatial accessibility. *Cartogr. Geogr. Inf. Sci.* **2019**, *46*, 311–333. [CrossRef]
75. Available online: https://www.statista.com/statistics/189582/age-of-us-video-game-players/ (accessed on 12 February 2021).
76. Available online: https://www.isfe.eu/data-key-facts/key-facts-about-europe-s-video-games-sector/ (accessed on 12 February 2021).
77. Available online: https://www.leagueofgraphs.com/pl/rankings/game-durations (accessed on 12 February 2021).
78. Available online: https://howlongtobeat.com/game?id=87873 (accessed on 12 February 2021).
79. Available online: https://howlongis.io/app/892970/Valheim (accessed on 12 February 2021).
80. Medyńska-Gulij, B.; Forrest, D.; Cybulski, P. Modern Cartographic Forms of Expression: The Renaissance of Multimedia Cartography. *ISPRS Int. J. Geo-Inf.* **2021**, *10*, 484. [CrossRef]
81. Martin, C.B.; Deuze, M. The Independent Production of Culture: A Digital Games Case Study. *Games Cult.* **2009**, *4*, 276–295. [CrossRef]
82. Steinkuehler, C.A. Why game (culture) studies now? *Games Cult.* **2006**, *1*, 97–102. [CrossRef]
83. Shaw, A. What is video game culture? Cultural studies and game studies. *Games Cult.* **2010**, *5*, 403–424. [CrossRef]

International Journal of
Geo-Information

Article

Aesthetics and Cartography: Post-Critical Reflections on Deviance in and of Representations

Dennis Edler [1,*] and Olaf Kühne [2]

[1] Department of Geography, Ruhr University Bochum, Universitätsstr. 150, 44801 Bochum, Germany
[2] Department of Geosciences, Eberhard Karls University of Tübingen, Rümelinstr. 19-23, 72070 Tübingen, Germany
* Correspondence: dennis.edler@ruhr-uni-bochum.de

Abstract: Cartographic representations are subject to sensory perception and rely on the translation of sensory perceptions into cartographic symbols. In this respect, cartography is closely related to aesthetics, as it represents an academic discipline of sensory perceptions. The scholarly concern with cartographic aesthetics, by today, has strongly been focused on the aesthetic impact of cartographic representations. The consideration of the philosophical sub-discipline of aesthetics however is rather restrained. This is also true for the connection between sociological questions and the social construction of aesthetic judgments. We address both topics in this article. We refer to post-critical cartographic theory. It accepts the socially constructed nature and power-bound nature of maps but does not reject "traditional" and widely established positivist cartography. Drawing on the theory of deviant cartographies related to this, we understand cartography designed according to aesthetic criteria as meta-deviant, as it makes the contingency of world interpretations clear. Especially augmented and virtual environments show a great potential to generate aesthetically constructed cartographic representations. Participatory cartography enables many people to reflect on the contingency of their spatial experiences and spatial abstractions without expert-like special knowledge. A prerequisite, however, is the greatest possible openness to topics and representations. This is not subject to a moral restriction.

Keywords: aesthetics; cartography; deviant cartography; post-critical cartography; neopragmatism; modernism; postmodernism

Citation: Edler, D.; Kühne, O. Aesthetics and Cartography: Post-Critical Reflections on Deviance in and of Representations. *ISPRS Int. J. Geo-Inf.* **2022**, *11*, 526. https://doi.org/10.3390/ijgi11100526

Academic Editors: Wolfgang Kainz, Beata Medynska-Gulij, David Forrest and Thomas P. Kersten

Received: 17 August 2022
Accepted: 15 October 2022
Published: 18 October 2022

Publisher's Note: MDPI stays neutral with regard to jurisdictional claims in published maps and institutional affiliations.

1. Introduction

Cartography can be understood as the translation of sensory perception into representations based on (changeable) professional conventions. In the past, these representations were mainly two-dimensional approaches, before three-dimensional forms of representations became more and more established [1–4]. Today, approaches of virtual and augmented realities broaden the spectrum of 3D cartography [5–9].

If we follow the approach of cartography, as it results from the first part of the presented conceptual understanding, aesthetics would be a central access mode of cartography. Until the 18th century, the old European notion of the unity of the true, the beautiful, and the good dominated (e.g., in [10,11]). Aesthetics, as an independent philosophical discipline, emerged only in the modern era [12,13]. The work 'Aesthetica' by Alexander Gottlieb Baumgarten [14] introduced a new way of thinking about the aesthetic that "increasingly displaced the paradigm of an ontologically based theory of the beautiful that had survived from antiquity and the Middle Ages" [15]. Baumgarten's reflections on aesthetics can be understood as a fundamental turn in the approach to aesthetics, since he upgraded sensuality to the medium of cognition [16].

This obvious mode of attention is overlaid by the second aspect of our conceptual understanding of cartography: the professional conventions. These (changeable) professional conventions have led to a hegemony of positivistic thinking, which aims at a simple mapping of facts (especially evident in topographic cartography). This is criticized by representatives of critical cartography. However, this approach of map-making, both in the form of the underlying conventions and the (extensive) state monopolization of (topographic) cartography, is strongly associated with the binding of power. These are discussed to be revealed, criticized and overcome (among many: [17–20]). The "traditional cartography" was even declared "dead" in this context [21]. In contrast to a cartography (powerfully) executed by trained cartographers, a 'counter-mapping' driven by inequalities and that affected people themselves was suggested and discussed (e.g., [22–26]).

In recent years, Alexander Kent has elaborated on the importance of aesthetics for cartography, addressing critical aspects, especially concerning the power of representations. This is also increasingly taken into account in web cartography [27–30]. Web cartography in particular not only opens up increasing possibilities for the design of cartographic representations. It also enables an increasing number of people (without a special education in cartography and related sciences) to create and distribute cartographic representations [31,32]. Our work follows a similar path. We acknowledge the power-bound nature of maps as aesthetic objects. However, we extend this consideration in the direction of philosophical and sociological research on aesthetics.

The existing literature on cartographic aesthetics is strongly focused on the aesthetic impact of cartographic representations. However, the connection to the philosophical sub-discipline of aesthetics is rather unexplored. The same is true for the connection to sociological questions of the social construction of aesthetic judgments. We would like to address this research gap in this paper. In our approach, we especially draw on the tradition of aesthetics published in German-speaking countries, as this has led to the formation of aesthetics as an independent sub-discipline of philosophy. In the international negotiation of cartographic aesthetics, it has hardly found resonance so far, and we intend to broaden the spectrum of philosophical and sociological theory in cartographic aesthetic research with this article. For this purpose, we make use of elaborate considerations on another aesthetic synthesis of spatially arranged material objects, i.e., the research field of landscape aesthetics. This field is also based on the exploration of social patterns of interpretation, categorization, and evaluation, and it leads to a spatial synthesis (among others: [33–38]).

In our contribution, we follow up on the development of a post-critical cartography [39,40], considering the rejection of dysfunctional aspects of critical cartography, such as its criticism of a power-bound nature of maps. Our aim with this essay is to clarify the contingency of cartographic interpretations of the world, in the form of functional and meta-functional deviances, by means of a return to the aesthetic dimension of cartographic representations. In our article, we deal with deviations from the current cartographic tradition in two ways: Firstly, with deviancies that produce innovative forms of cartographic representations, and secondly, with deviancies that provide opportunities for developing and testing contingent thinking (here in relation to cartographic representations).

This article will first deal with a brief introduction to the approach of deviant cartographies, which emerged from the theoretical operationalization of post-critical cartographies. Followed by this, we will deal with central (and for cartography relevant) aspects of philosophical aesthetics as well as the importance of aesthetic judgments for social processes of distinction. We will then apply these considerations to the development of cartographic representations and illustrate them using case examples the cartographic representation of post-industrial objects (in virtual reality representations). In the conclusion, we will again highlight the importance of aesthetics for cartography and elaborate perspectives from the chosen theoretical framework.

2. Basic Theoretical Considerations: Deviant Cartographies

The development of the deviant cartographies approach [40,41] resulted from the critical reflection of critical cartography pursued since the 1980s. Critical cartography claimed to represent a new paradigm against the positivist tradition of cartography. This was classified as outdated [39]. Instead of the complete replacement of one paradigm by another, the authors rather prefer a neo-pragmatic approach (the philosophical foundations: [42,43]; to transfer to the concern with spaces: [44–46]). The basis for an evaluation of the development of theories, research questions, methods, etc. is not going back to moral preferences or ideological principles, but to the questions of whether they are suitable for the progress of the discipline or social understandings of space. As a measure of suitability, we built on the concept of life chances.

This concept is originally based on Max Weber [47] and it is further differentiated by Ralf Dahrendorf [48,49]. Dahrendorf understands life chances, figuratively speaking, as "the baking forms of human life in society; they determine how far people can develop" [48] (p. 24). Dahrendorf [48] (p. 98), who followed Max Weber [47] considers that a chance is a "structurally based [...] probability of behavior" as well as "something that the individual can have, something as an opportunity to satisfy interests" [48] (p. 98). These chances are "socially shaped. Social structures order opportunities" [48] (p. 98). Life chances originate in options (as choices of action) and ligatures (as a socially predetermined structure of norms, roles, and social positions). Moreover, ligatures give meaning to options. However, they also restrict options, since they morally influence the development of options. An example is the attainment of higher education through religious restrictions [50] (see in more details: [51–53]).

The basis for this study of functionality implies deviance, an established term and concept in sociology. Deviance describes the deviation of behavior from an existing (often unquestioned) norm [54]. This deviation from a norm can have functional and dysfunctional effects. Deviation from a norm, for example, can highlight confidence in the ability of a society (or parts of it) to regulate itself. However, it can also serve to clarify norms or lead to innovation [55]. Especially the scientific process of knowledge is constitutively bound to deviation from existing conventions. However, these deviances have to prove their suitability [56].

In order to conceptualize this, four types of functional deviance have been distinguished ([40]; see also: [57–60]):

Deviance that proves to be suitable in relation to the professional, further: scientific or social development, as they expand life chances, can be described as functional.

Deviance that limits life chances are described as dysfunctional. It proves to be unsuitable for the scientific and social progress.

Deviance can be described as a-functional, if there is no reception of any resonance from a scientific to a social point of view.

Such deviance can be considered meta-functional if it is suitable to reflect the social construction processes of the world and makes people aware of the contingency of world interpretations.

Against the backdrop of this terminology, critical cartography has achieved a meta-functional impact because it has initiated a reflection on the cartographic construction of the world, including its power-bound nature. However, an important point of criticism, i.e., the securing of power through the governmental monopolization of geospatial data, has become irrelevant. Geospatial data is now made freely available by official (national and/or regional) agencies in many countries [61]. In the meantime, it has also become possible to use digital maps and virtual and augmented reality applications by means of commercial offers. Therefore, it is hardly possible to speak of a isolation of expert knowledge based on commercial interests [62–64]. The persistence of this argument seems dysfunctional in view of the achievements of a (positivistically pursued) cartography established by public authorities. It is not least the foundation for spatial orientation on the Earth's surface (and beyond). This is widely ignored by critical cartography. The expansion of access to low-threshold software tools has greatly increased the possibilities for creative production of cartographic representations. As a result, the call by critical cartographers for artistic creation and participatory involvement has proven largely a-functional. While this has greatly expanded, it is widely based on commercially developed tools rather than on activist counter-mapping [40].

An artistic but also a participatory cartography updates the question of the design of cartographic representations. For a long time, it seemed to be pushed into the background as a result of standardization efforts [30,65]. The increased importance of design, in turn, raises the question of aesthetic reflection and classification. As a basis for such a reflection and classification, the recourse to philosophical and sociological aesthetics becomes important. On the one hand, this enables the connection of cartographic-aesthetic considerations to a centuries-old tradition of the science of sensual perception. On the other hand, cartographic-aesthetic communication can be grasped in its social meanings. This is what we will deal with in the following section.

3. Aesthetics between Philosophical Foundations and Social Distinctions

In his work 'Aesthetica',Alexander Gottlieb Baumgarten [14] introduces a new form of thinking about the aesthetic which "replaces the paradigm of an ontic paradigm of an ontologically based theory of the beautiful that has come down to us from antiquity and the Middle Ages." [15] (p. 7). The doctrine of sensual perception (Aisthetike Episteme) thus becomes complementary to the doctrines of thinking (Logike Episteme) and morality (Ethike Episteme; [12,66]). In the following, we will briefly introduce essential basic terms of philosophical and sociological aesthetics. Later, we apply them to questions of cartography.

The central concept of aesthetic reflections is formed by the concept of beauty. Thus, the development of aesthetics can be understood as a history of "a constant reinterpretation of the concept of beauty" [67] (p. 100). Beauty can be characterized as "unity in multiplicity" [68] (p. 63). However, the ability to perceive beauty is linked to conditions. Following Kant [69], beauty is something that is generally pleasing without interest. An interest, for example, can refer to the economic use of something. For the perception of beauty, it is fundamental that no immediate (e.g., economic) individual or social interest would exist in the object described as "beautiful". This view of the separation between aesthetic and practical world is contradicted by John Dewey [70,71]. Thus, "beauty" can be seen as being guided by interests. It could certainly have a consumptive meaning, which is indicated not least by the consumptive meaning of material spaces described as landscapes influenced by tourism [72].

A complement to the concept of beauty is that of sublimity. A fundamental distinction between the two concepts can be described by Edward Burke [73]: the beautiful inspires love, the sublime generates admiration. Thus, the experience of the sublime is bound to large, impressive or terrible material objects. Material objects from the realm of the 'natural' become dominant here. Berleant (1997: 28) refers to this, such as volcanoes, dark forests, or high mountains. Kant moves away from Burke's strong object-centeredness by founding the beautiful "in the harmonious interplay of understanding and sensual imagination ('Einbildungskraft')" ([74] (p. 38); see also [75,76]). In contrast, he traces "the 'sublime' to a disharmonious interplay of reason and sensuous imagination." [74] (p. 38). Compared to the beautiful, the access to understanding the sublime is more difficult. It belongs "to the feeling of the sublime the powerlessness and questioning of the subject in the face of the overpowering nature, of the invading 'too much'." [77] (p. 10). The sublime resists the effort of intellectual control ([77] following Kant). With this non-ambiguity and the deprivation of intelligible control, the sublime resists the efforts of modernity in its striving for unambiguity and cognitive domestication [36,77,78]. In a nutshell, the project of modernity can be understood as an attempt to anesthetize the world, i.e., to withdraw it from aesthetic perception, whereby it "does not evoke an aesthetic experience" ([79] (p. 28); in more detail in [80]). With the emergence of the postmodern discussion since the 1970s, the sublime has experienced a renaissance ([81]; cf. also [82]), while the beautiful entered a crisis in the second half of the 20th century: after all, the "aesthetics of the beautiful [...] had uncritically degenerated into mere 'design' and had been reduced to a commodity in the consumer society" [83] (p. 90). The "threatening oversaturation, anesthetization, and social desensitization" [84] (p. 229) became the subject of conservative as well as (Neo-)Marxist social criticism.

The picturesque arises from a tension that is particularly widespread in landscape painting. It arises from the combination of objects and object constellations in the foreground that are connoted with "beauty" (often: flowers) with objects and object constellations in the background that are considered "sublime" (mountains or storm clouds; for example: [85,86]). This motif arrangement is still common in landscape photography today.

The ugly, in turn, like the beautiful, compared to the sublime, evokes "no too strong emotions. One pleases, the other does not; one produces pleasure, the other aversion, which as a sensation is arguably stronger than its positive counterpart, may also give rise to immediate reactions, but is rarely felt as dramatic" [87] (p. 72). In his 'Aesthetics of the Ugly' Karl Rosenkranz conceptualizes [88] the ugly as the "negative beautiful," assigning to the ugly a "secondary existence" [88] (pp. 14–15). He assigns three manifestations to the ugly:

1. Amorphia, the shapelessness or an indeterminacy of the shape. Beauty is not achieved in the absence of a 'being-appropriate' limitation or of the unity in the necessary difference (cf: [89]).
2. Asymmetry, the imbalance of opposites. He also describes this form of ugliness as unshaped.
3. Disharmony, the disproportion between the part to the whole. In place of the agreement, a disunity arises here. This is produced by the generation of "wrong" contrasts.

The ugly, according to Rosenkranz [88], is not trapped in itself, but can, like the beautiful, experience an aesthetic revaluation through a transformation towards 'das Komische' (in English approximately: the comic, the funny, the strange, the weird). It unites 'das Komische', "the beautiful and the ugly by liberating both from their respective (pseudo-ideal) one-sidednesses." [90] (p. 61). The highest form of transformation of the ugly (as well as of the beautiful or picturesque) into 'das Komische' is the caricature accompanied by means of exaggeration and disproportion.

A central question of aesthetics is formed by whether, for instance, the beautiful (but also the ugly, comic, etc.) is a property of an object or the attribution by the observer. The former, naturalistic position of an 'objective aesthetics' is already represented by Plato [91]. He attributes an underlying idea to every object. The degree of beauty is measured by the degree to which the object succeeds in (materially) expressing the idea that characterizes it. This essentialist conception is contrasted with the subjectivist conception of aesthetics. This was made prominent by Francis Hutcheson (1694–1747; [92]): beauty arises on the basis of the ability to feel beauty. Because of this ability, people are able to have an aesthetic perception by combining uniformity and multiplicity of material objects. In this understanding, beauty is always linked to the process of perception. Theodor Vischer (1807–1887; [93] (p. 438)) succinctly states: "beauty is not a thing, but an act." The subjectivist understanding of the beautiful (the ugly, the picturesque, etc.) has far-reaching consequences. If beauty becomes a "product of the subject and his mental dispositions and faculties" [94] (p. 3), the question of the nature and condition of these "dispositions and faculties" becomes the focus of the study of aesthetics. Thus, the aesthetic also becomes an object of sociological research. Kant had already pointed out the social conditionality of aesthetic judgments. On the one hand, aesthetic patterns of interpretation and evaluation have to be learned (a question of socialization). On the other hand, there is the question of how aesthetic judgments are media of stratification in society. For example, this is pointed out by Pierre Bourdieu [95] in his study on distinction mechanisms in society. In particular, "legitimate taste" distinguishes itself from "middle taste" through the production of new aestheticized objects or the aesthetic charging of familiar objects. The latter does not have these aesthetic access patterns but tries to imitate them. This causes the "legitimate taste" to develop new aesthetic attributions. The pejorative aesthetic judgment of the superior taste towards the subordinate tastes is kitsch. The associated judgment of the "bad taste" implies the (distinctive) attribution of a lack of self-realization. This lack of self-realization is measured by an aesthetic standard that is declared to be valid universally [96]. This difference between trivial and high culture, in turn, is subject to leveling in the context of postmodern individualization and further differentiation [97,98]. The patterns of distinction, which were formerly applied to society as a whole, are now found at best in milieu-specific discourse communities [99].

Aesthetic access to the world by humans is multisensory and not limited to the (nevertheless dominant) sense of vision. While the sense of vision, as a sense of distance, suggests an abstracted access to the external world. The perception of the media of the world is directly connected to other senses through material interaction with the body (see already: [100]). Through this bodily involvement, a synesthetic turning to the world takes place. Synesthetics is understood as the synthesis of different sensual perceptions in an aesthetic mode. The preoccupation with synesthetics is strongly oriented toward atmospheres. These can be described, following Thibaud [101], as a medium of sensual relations between the sensing human being and his environment. This experience of the world in the form of atmospheres in turn leads from a cognitive preoccupation with aesthetic stimuli to a preoccupation with emotional attentions to the world.

Regardless of the aesthetic concept, it becomes clear: the imprinted patterns of aesthetic access to the world, according to Seel [102], are tied to experience. Aesthetic experience has become a "self-reflexive lifeworld experience, that is, an 'experience of experience'" [103] (p. 31). An exception is the atmospheric turning to the world. Here, experiencing is practiced and not a reflecting.

4. The Operationalization of Aesthetic References in Cartography

The cartographic representation of "something" represents the result of a threefold construction process [104]:

a. On the basis of socially mediated and individually updated patterns of interpretation, categorization, and valuation, (mostly material) objects are subjected to a synthesis (for example, as a landscape).
b. These syntheses, in turn, are translated into a cartographic symbol (usually professionally mediated) based on relevance criteria into a cartographic representation.
c. The interpreter of these cartographic representations constructs them on the basis of her or his internalized patterns of interpretation, categorization, and valuation. These are often based on common sense understandings of the mapped world (a) and of cartographic representations (b).

These relationships become more complex when, as a result of increasing participation processes, the producer and recipient sides are no longer clearly separated. This results in a network structure of production and reception. This connection also becomes more complex when it is taken into consideration that all steps of world observation, production and interpretation of cartographic representations are subject to sensory cognition. Thus, aesthetic judgments are always included as well. This shows the importance of addressing the relationship between aesthetics and cartography. We address this following basic concept of philosophical and sociological aesthetics introduced in the previous section:

A functional deviance produced by critical cartography is the assertion of the social (and individual) constructedness of maps. The finding of the social constructedness of cartographic representations can be understood as a functional deviance in the history of cartography. This functionality arises not least from the fact that it enables a connection to research in social and cultural sciences. Social constructivist thinking, as it has become prevalent in social science landscape research (among many: [105–108]), is also connected to the interpretation of the process of map production presented above. From this perspective, also in relation to cartographic representations, the subject (in recourse to the learned and updated aesthetic patterns of interpretation, categorization and evaluation) is the constitutive instance of aesthetic judgments. This also determines aesthetic inscriptions in cartographic representations. In principle, this concerns the attribution of aesthetic judgments, for instance in relation to the beautiful, the sublime, the picturesque, the ugly, kitsch, and 'das Komische', but in a thoroughly differentiated manner.

Beyond the presented basic considerations on the topic of aesthetic construction of cartographic representations, the aspect of beauty has some peculiarities: It is rarely found on the side of the represented; after all, the largest number of objects described as "beautiful" is below the threshold of what can be mapped for reasons of scale (such as individual species of vegetation). The objects attributed as "beautiful" are accordingly subject to generalization in a special way, which can take place in two ways. The first way refers to the process of generalization. Aesthetic judgments about "beautiful" individual material objects are generalized (e.g., the material arrangement of various individual grass, trees and other vegetation species into "green spaces"). The second way is to transpose the beautiful with the sublime into the picturesque. Thus, aspects of material space that can be experienced as sublime are translated cartographic representations, such as mountains. Topographic objects that are commonly evaluated as beautiful fall below the threshold of representation because they are commonly considered too small. On the side of representation, the "beautiful" can be inscribed in cartographic representations, for example, through the use of "harmonious" color choices and proportioning of symbols and ornaments. The evaluation of these style aspects, however, is dependent on the zeitgeist, as we will discuss later. The challenge of cartographic translation of the sublime, however, is somewhat different: 'traditional' positivist cartography has played a significant role in 'disenchanting' the sublime(in the sense of: [109]). It transfers material objects formerly experienced as sublime into the realm of a calculable, manageable, and abstracted cognitive construction of the world; the sublime has been domesticated by means of cartographic representation [110]. On the side of cartographic representations, the sublime is indicated, for example, by the shading in the representation of mountains, such as in the Official Cartography of Switzerland. While the scale reduction (just from the dimension) makes it difficult to experience maps in the mode of sublimity in classical cartography, VR-based cartography (incl. immersion) can certainly help to create such an experience. The synthesis of beauty and sublimity in picturesqueness unites the statements about the two modes of aesthetic construction. In this respect, we now turn to ugliness. In ugliness, there is a translation to the anesthetic means of abstract symbology. On the side of cartographic representation, ugliness is definitely present, as it cannot be avoided even through abstraction, generalization and formalization. This is especially obvious when the depicted objects themselves or their distribution (after translation into the language of cartography) evoke a representation characterized by amorphousness, asymmetry, and disharmony. Cartographers have a greater influence on the avoidance of "ugliness" in the design of cartographic representations when it comes to the design and arrangement of the legend (unless these are standardized).

Generalizing these remarks, we can speak of an attempt by modernity to withdraw the world in general and cartographic representations in particular from aesthetic reference. Thus, the attempt is made to transfer the references into the realm of the anesthetic. This transfer succeeded with artifacts, such as maps (or in a larger dimension: functionalist urban planning), rather than this was possible with more naturally formed objects and object constellations. On the one hand, this is due to the inherent logic of natural objects (a mountain range can only be removed with great effort), on the other hand, it is also due to aesthetic persistence that is not least rooted in Romanticism [111]. However, even with artifacts, success remained incomplete. Even in topographic cartography, attempts to completely displace aesthetic representations were ultimately unsuccessful, as Alexander Kent [65] impressively illustrates with the arrangement and design of cartographic symbols.

The theme of multisensory technology in and of cartographic representations can also be interpreted with the theme of fundamental scientific and social changes: In spatial research before the 20th century, non-visual stimuli were quite present [112]. Raab understands the extensive fading out of the non-visual senses in modern science [113] (p. 16) as immanent to the quality criteria of this science (such as freedom of value, general validity and comprehensibility): for example, "visual perception could be assigned both optical qualities (colors) of a physically measurable dimension (wavelength of light)". Furthermore, a manageable subjective category system for the classification of these qualities is available (e.g., basic colors). In contrast, "in the olfactory domain, neither consistent relationships between the chemical–physical characteristics of scents and their sensations are discernible, nor are systematic classification criteria, according to which subjective scent qualities can be ordered, available" [113] (p. 16). Olfactory and acoustic stimuli are also challenging for mapping for another reason: they are usually fleeting. Accordingly, the preoccupation with them has been pushed back in favor of the preoccupation with visual stimuli. Here, after all, there is a simpler translation from an optical stimulus by means of conveying subject-specific conventions to a medium that is related to the sense of vision. The restriction is broken in the course of new technical possibilities (e.g., VR and AR cartography). Using these opportunities, the hematic fixation is not reduced to only exactly measurable and locatable physical objects (for example: [114–116]; more on this in detail in the following sections).

The triad of pre-modernity, modernity and post-modernity is also connected to the cartographic meaning of kitsch. The modernization of cartography took place not least in the form of standardizations and the pushing out of individual (socially shaped) aesthetic preferences in the design of maps. Thus, the map was transformed in large parts from an object designed according to aesthetic ideas to an anesthetic object. This was accompanied by the distinctive devaluation of the map and its authors. The map was no longer a medium of a creative practice (from which pride and recognition could be generated), but the processing of technical routines. A map was designed following the modernist urge for order, categorization, and the generation of monovalences (one object, one function). The principle of "form follows function" offered no room for ornamentation on the one hand, nor for outdated aesthetic standards that could have been described as "kitsch" by subsequent generations of cartographers and users. With the de-standardization and individualization that characterize post-modernization, new possibilities for the return of aesthetic cartographic representations opened up, in conjunction with the development and spread of technical innovations. Irony is considered an essential stylistic element of postmodern communication [43] to highlight the contingency of world constructions. This can also be used in cartographic representations, with the aim of visually emphasizing the contingency of spatial constructs [45,46]. In the sense of aesthetics, 'das Komische' can be used as an example: By means of exaggeration and disproportion, irony can be symbolized here, and contingency can be made clear. Concrete symbols (with their particular openness to exaggeration), but also unusual combinations of symbols, are particularly suitable for such a metafunctional-deviant cartography. However, an ironic distance to the usual norms of cartographic representations can also be created by an unusual selection of orientation indications in the design of topographic bases of thematic maps. The postmodern re-entry of the aesthetic into cartography means that cartographic representations (again) become polyvalent. They receive an explicit aesthetic charge in addition to the function of the thematic statement. Even if general distinctive aesthetic standards are hardly socially widespread anymore, distinctive aesthetic standards continue to be milieu-specific. Accordingly, cartographic representations designed with artistic approaches not only become outdated in terms of content, but their form also runs the risk of being subjected to a distinctive aesthetic judgment by a subsequent generation of mapmakers: Kitsch!

5. Possibilities of Reintegrating Aesthetic References by Means of Virtual and Augmented Realities

The current possibilities of cartographic integration of virtual and augmented realities (VR/AR) means a considerable expansion of representation possibilities. These can be used not least for a design of cartographic representations that focus on creativity and the testing of new aesthetic approaches. These possibilities of new cartographic representations are not limited to the visual symbol language, but also allow, for example, the integration of soundscape elements (Figure 1). When using appropriate hardware, tactile and occasional olfactory elements can also be integrated.

Figure 1. A multisensory urban traffic situation for immersive VR experience. (Source: Marco Weißmann & Dennis Edler).

This enables a multisensory experience of cartographic representations instead of a solely visual cognitive attention. The multisensory approach in turn can contribute to strengthening the aesthetic attention to spaces and their representations. One possibility for this is already known from computer game cartography. The gradual transition from an abstract cartographic 2D representation to an elevated and detailed 3D landscape representation, accompanied by the transition of a musical background, up to concrete sound elements. These are usually connoted with landscape experiences and can also be supported with the action of avatars (e.g., navigation) by means of cartographic representations [117–123]. However, immersive 3D representations have other functions in the (aesthetic) construction of spaces. By applying a reduction of complexity, repeatability is made possible. This is based on the fact that there is a limited selection of possible elements in a spatial arrangement, but also a reduced number of options for action. They can also be related to the extent of virtual spaces, as this is usually limited as well. This contributes to the experience of aesthetic contingency. Thus, it can have a meta-functional effect [124,125].

The development of virtual cartographic representations not only imply hybridizations and the creation of smooth transitions (for example, as shown above, of an abstract 2D map and elevation representations), they also enable the consideration of social hybridizations. It refers to augmented forms of representation that enable the integration of different (cartographic) elements into the representation of material spaces. The degree of integration can range from sparse to intensive, from an addition of virtual elements to the integration of complex, even multisensory information [126–130]. Social hybridizations have an effect on the differentiation of commonsense and 'expert special knowledge stocks' regarding to spatial questions. They also influence the cartographic representation of spatial information. This is based on the availability of low-threshold (mostly commercial) offers for the generation of cartographic representations [40,131]. This is also where the possibilities of creative representations arise, beyond classical standardized procedures which characterized cartography as a discipline for a long time.

The possibilities that virtual and augmented cartographic representations offer in terms of aesthetics have not yet been fully explored and investigated. They range from the adaptation of representations according to current user interests, to the user-oriented individual design of cartographic representations according to the user's ability to understand complex representations. In addition, they contain sensory capabilities to process cartographic representations (e.g., in case of color vision deficiencies), and they have great potentials for user-oriented real-time visualization, e.g., by using modern techniques and methodologies of artificial intelligence [132,133].

With post-industrialization (e.g., [134]), there is a change from "industrial-space" to "post-industrial-space" ([135] (p. 193)). The transformation of places of work into residual spaces is accompanied by two symbolic patterns of attribution. On the one hand, they represent symbols of failure. On the other hand, they are objects of aestheticization [136,137]. Aestheticization is essential in our context, and it takes place through the transfer of patterns of interpretation and aestheticization originating from the times of industrialization (Figure 2).

Figure 2. An oblique bird's eye view on a post-industrial area in the Ruhr Valley represented in immersive VR. Buildings of the industrial past are preserved and integrated into the landscape architectural setting of the present. (Source: Timo Wiedenlübbert & Dennis Edler).

Old industrial (urban)landscapes combine "baroque aesthetics of ruin with decaying blast furnaces and memories of the picturesque garden of the eighteenth century" ([138] (p. 154)). In analogy to pre-modern agricultural life, old industrial objects become symbols of the "simple, hard worker's life" [139] (p. 231). However, they can be interpreted as a reaction to the de-standardization and fragmentation of post-industrial society (cf. [140–142]). This aestheticization, in turn, is initiated by the bearers of "legitimate taste" by (almost arbitrarily) turning objects from the realm of the worthless into media of social distinction, for instance by staging them using out-of-context illumination (as, for example, in the photographs of Bernd and Hilla Becher; [143]). Based on modern VR and AR techniques and methodologies, these old industrial landscapes can be represented in a multisensory way. As these VR applications support immersion, the process and results of aestheticization is supported based modern media of 3D cartographic communication (e.g., VR environments).

6. Conclusions

The critique by critical cartography concerning 'traditional' cartography relates in some parts explicitly and in others implicitly to aspects of the aesthetic construction of the world through cartographic representations. It is explicit in the call for artistic cartography. It is less explicit (with exceptions) in the secondary consequence of anesthetization. This anesthetization of cartography occurred with the enforcement of the positivist paradigm and the principle of 'form follows function', which was essentially related to the standardization of cartographic symbols in abstract rather than concrete symbol language. However, this anesthetization also occurs to a certain canonization of what can be represented. While the enforcement and maintenance of the positivist paradigm is understood as a powerful process, the distinctive function that aesthetic interpretations and evaluations contribute to this process is only hardly considered. With the canonization of representations, not only a canonization of the representable takes place, but also a contingency reduction of the modes of representation. The artistic design of cartographic representations is however associated with an activist mission. This is made clear by the rejection of design. This makes maps a commodity. Here, the link to a (somewhat naive) search for the authentic, not professionalized, also becomes obvious. The critical rigorism of Critical Cartography towards "traditional" cartography can accordingly be characterized as dysfunctional, as its abandonment goes along with the reduction of life chances. At the same time, the power-bound nature of "traditional" cartography is reflected (similarly: [144]). It is only acceptable if more life chances are created by positivist cartography than life chances are lost by power-boundness. However, the same is true for critical cartography. It also strives for a hegemonic power of interpretation: who is when and in which context is allowed to create cartographic representations of certain socially defined issues and who just is not. In this respect, we suggest neopragmatic considerations of evaluating cartographic representations according to the criteria of usefulness, suitability, and usefulness for cognition. We continue to recognize great functional-deviant potentials for positivist cartography, while the functionality of critical cartography has some dysfunctional implications, as we will explain in the following.

Moreover, the demands for artistic and participatory scenarios of cartography, as suggested by critical cartography, remain a-functional to a large extend. Maps are classified as morally acceptable if they are designed out of a critical mindset. Corresponding developments outside the critical discourse are—accordingly rejected, especially if they can be outlined as commercial products [24]. Activist mapping has largely remained a marginal social phenomenon. In particular, it has the function of achieving gains in distinction when different people prefer different forms of creating and designing cartographic representations. These individuals, who exhibit a different worldview on cartography, represent "a pathological case" [145] (p. 47) (see in detail: [50]). Central concerns of critical cartography, such as an artistic design and a greater participation of people without professional backgrounds, have become more important in recent years and decades. Activist counter-mapping, however, has remained a marginal phenomenon in these fields. This contradicts the claim of critical cartography in particular, of critical science in general, to be avant-garde of social change (here as an expression of cartography). Participatory cartography thus shows potentials of realizing contingency and of expanding individual life chances as a result of reflecting on one's own spatial constraints on the one hand and one's own spatial constructions on the other hand. However, it also bears the danger of dysfunctional deviance, through the overemphasis of (moral) ligatures and thus the limitation of options. Based on the norm of expanding life chances, artistic approaches in cartography can generally be attributed a great meta-functional potential. They can support becoming aware of spatial contingency. Moreover, they provide an occasion to reflect on contingency in the transformation of physical space into cartographic representations, their reduction of complexity and intricacy. Any reduction of complexity and intricacy, however, runs the risk of freezing into ligatures. For example, if the rules on which the reduction is based are no longer subject to critical reflection but are solidified into unquestioned instructions.

The re-aestheticization of cartography, however, is not solely carried by people without appropriate professional training, including artists. Even in professional cartography, new forms of translating the beautiful, the sublime, the picturesque, the ugly, etc. into the language of cartography are emerging. Going beyond this, an increasing openness to innovative symbol languages as well as the use of new technologies facilitates a design that is aligned with aesthetic ideas. We have also indicated this in this article. Cartography conducted in this way is not only aware of the contingency of its 'world-making', but also highlights it through the use of (self-)irony. This stands out from the persistence that characterizes cartographic representations generated based on the 'jargon of moral ligatures'. This (self-)irony (in the sense of [43]) is distinguished by its self-distance from a critical cartographic determination of the will to deconstruct the other [146–148].

In our article, we have shown how the concepts of philosophical and sociological aesthetics can be profitably introduced for a reflexive engagement with cartography. We have thus outlined an approach that can be used to investigate the development of maps, virtual and augmented representations, complementary to content and technical issues.

Author Contributions: Conceptualization, Dennis Edler and Olaf Kühne; Writing—Original draft preparation, Dennis Edler and Olaf Kühne; Writing—review and editing, Dennis Edler and Olaf Kühne; Visualization, Dennis Edler and Olaf Kühne. All authors have read and agreed to the published version of the manuscript.

Funding: This research received no external funding.

Data Availability Statement: Not applicable.

Conflicts of Interest: The authors declare no conflict of interest.

References

1. Buchroithner, M. *True-3D in Cartography: Autostereoscopic and Solid Visualisation of Geodata*; Springer: Berlin/Heidelberg, Germany, 2012, ISBN 9783642122729.
2. Fernández, A.; Iván, P.; Buchroithner, M.F. *Paradigms in Cartography: An Epistemological Review of the 20th and 21st Centuries*; Springer: Berlin/Heidelberg, Germany, 2014.
3. Vetter, M. 3D-Visualisierung von Landschaft–Ein Ausblick auf zukünftige Entwicklungen. In *Handbuch Landschaft*; Kühne, O., Weber, F., Berr, K., Jenal, C., Eds.; Springer VS: Wiesbaden, Germany, 2019; pp. 559–573.
4. Bröhmer, K.; Knust, C.; Dickmann, F.; Buchroithner, M.F. Z-axis Based Visualization of Map Elements–Cartographic Experiences with 3D Monitors Using Lenticular Foil Technology. *Cartogr. J.* **2013**, *50*, 211–217. [CrossRef]
5. Liu, B.; Ding, L.; Wang, S.; Meng, L. Designing Mixed Reality-Based Indoor Navigation for User Studies. *KN–J. Cartogr. Geogr. Inf.* **2022**, *72*, 129–138. [CrossRef]
6. Lochhead, I.; Hedley, N. Designing Virtual Spaces for Immersive Visual Analytics. *KN–J. Cartogr. Geogr. Inf.* **2021**, *71*, 223–240. [CrossRef]
7. Çöltekin, A.; Lochhead, I.; Madden, M.; Christophe, S.; Devaux, A.; Pettit, C.; Lock, O.; Shukla, S.; Herman, L.; Stachoň, Z.; et al. Extended Reality in Spatial Sciences: A Review of Research Challenges and Future Directions. *Int. J. Geo-Inf.* **2020**, *9*, 439. [CrossRef]
8. Hruby, F.; Ressl, R.; de La Borbolla Valle, G. Geovisualization with immersive virtual environments in theory and practice. *Int. J. Digit. Earth* **2019**, *12*, 123–136. [CrossRef]
9. Hruby, F.; Castellanos, I.; Ressl, R. Cartographic Scale in Immersive Virtual Environments. *KN–J. Cartogr. Geogr. Inf.* **2020**, *71*, 45–51. [CrossRef]
10. Augustinus, A. *Theologische Frühschriften. Vom Freien Willen (De Libero Arbitrio) u. Von der Wahren Religion (De Vera Religione)*; Artemis-Verlag: Zürich, Switzerland; Stuttgart, Germany, 1962; p. 390.
11. Areopagita, P.-D. *Über die Göttlichen Namen*; BGL: Stuttgart, Germany, 1988; p. 500.
12. Gilbert, K.E.; Kuhn, H. *A History of Esthetics*; Indiana University Press: Bloomington, IN, USA, 1953.
13. Majetschak, S. *Ästhetik zur Einführung*; Junius Verlag: Hamburg, Germany, 2007.
14. Baumgarten, A.G. *Ästhetik*; Meiner: Hamburg, Germany, 2009; pp. 1750–1758. ISBN 9783787318995.
15. Schneider, N. *Geschichte der Ästhetik von der Aufklärung bis zur Postmoderne: Eine Paradigmatische Einführung, 4. Auflage*; Reclam: Stuttgart, Germany, 2005, ISBN 3150094577.
16. Franke, U. Alexander Gottlieb Baumgarten. In *Ästhetik und Kunstphilosophie: Von der Antike bis zur Gegenwart in Einzeldarstellungen*; Nida-Rümelin, J., Betzler, M., Eds.; Krömer: Stuttgart, Germany, 1998; pp. 72–79.
17. Harley, B.J. Deconstructing the map. In *The New Nature of Maps. Essays in the History of Cartography*; Laxton, P., Ed.; John Hopkins University Press: Baltimore, MD, USA, 2002; pp. 149–168.
18. Wood, D.; Fels, J. Design on Signs/Myth and Meaning in Maps. *Cartogr. Int. J. Geogr. Inf. Geovis.* **1986**, *23*, 54–103. [CrossRef]
19. Crampton, J.; Krygier, J. An Introduction to Critical Cartography. *ACME Int. J. Crit. Geogr.* **2005**, *4*, 11–33.
20. Wood, D.; Krygier, J. Cartography: Critical Cartography. In *The International Encyclopedia of Human Geography*; Kitchin, R., Thrift, N., Eds.; Elsevier: Oxford, UK, 2009; pp. 345–357.
21. Wood, D. Cartography is Dead (Thank God!). *CP* **2003**, *45*, 4–7. [CrossRef]
22. Kim, A.M. Critical cartography 2.0: From "participatory mapping" to authored visualizations of power and people. *Landsc. Urban Plan.* **2015**, *142*, 215–225. [CrossRef]
23. Fellner, A.M. Counter-Mapping Corporeal Borderlands: Border Imaginaries in the Americas. In *Geographien der Grenzen: Räume–Ordnungen–Verflechtungen*; Weber, F., Wille, C., Caesar, B., Hollstegge, J., Eds.; Springer VS: Wiesbaden, Germany, 2020.
24. Schranz, C. Shifts in Mapping: Two Concepts have Changed the World View. In *Shifts in Mapping: Maps as a Tool of Knowledge*; Schranz, C., Ed.; Transcript Verlag: Bielefeld, Germany, 2021; pp. 23–37.

25. Counter Cartographies Collective; Craig Dalton; Liz Mason-Deese. Counter (Mapping) Actions: Mapping as Militant Research. *ACME* **2012**, *11*, 439–466.
26. Thatcher, J.; Bergmann, L.; Ricker, B.; Rose-Redwood, R.; O'Sullivan, D.; Barnes, T.J.; Barnesmoore, L.R.; Imaoka, L.B.; Burns, R.; Cinnamon, J.; et al. Revisiting critical GIS. *Environ. Plan. A Econ. Space* **2016**, *48*, 815–824. [CrossRef]
27. Kent, A.J. Aesthetics: A Lost Cause in Cartographic Theory? *Cartogr. J.* **2005**, *42*, 182–188. [CrossRef]
28. Kent, A.J. Maps, Materiality and Tactile Aesthetics. *Cartogr. J.* **2019**, *56*, 1–3. [CrossRef]
29. Kent, A. Cartographic Style and the Aesthetic Fix. *Cartogr. J.* **2017**, *54*, 1–4. [CrossRef]
30. Kent, A. From a dry statement of facts to a thing of beauty:understanding aesthetics in the mappingand counter-mapping of place. *Cartogr. Perspect.* **2013**, *73*, 39–60.
31. Griffin, A.L.; Fabrikant, S.I. More Maps, More Users, More Devices Means More Cartographic Challenges. *Cartogr. J.* **2012**, *49*, 298–301. [CrossRef]
32. Fabrikant, S.I.; Christophe, S.; Papastefanou, G.; Maggi, S. Emotional Response to Map Design Aesthetics. In Proceedings of the GIScience 2012: Seventh International Conference on Geographic Information Science, Columbus, OH, USA, 18–21 September 2012.
33. Bourassa, S.C. *The Aesthetics of Landscape*; Belhaven Press: London, UK, 1991, ISBN 978-1852930714.
34. Winchester, H.P.M.; Kong, L.; Dunn, K. *Landscapes: Ways of Imagining the World*; Routledge: London, UK; New York, NY, USA, 2003, ISBN 0-582-28878-9.
35. Wylie, J. *Landscape*; Routledge: Abingdon, UK, 2007, ISBN 9780203480168.
36. Kühne, O. *Landscape Theories: A Brief Introduction*; Springer VS: Wiesbaden, Germany, 2019.
37. Brook, I. Aesthetic appreciation of landscape. In *The Routledge Companion to Landscape Studies, 2. Auflage*; Howard, P., Thompson, I., Waterton, E., Atha, M., Eds.; Routledge: London, UK; New York, NY, USA, 2019; pp. 39–50, ISBN 9781315195063.
38. Porteous, J.D. *Environmental Aesthetics: Ideas, Politics and Planning*; Routledge: Abingdon, UK; New York, NY, USA, 2013, ISBN 9780203437322.
39. Kühne, O. Contours of a 'Post-Critical' Cartography–A Contribution to the Dissemination of Sociological Cartographic Research. *KN–J. Cartogr. Geogr. Inf.* **2021**, *71*, 133–141. [CrossRef]
40. Edler, D.; Kühne, O. Deviant Cartographies: A Contribution to Post-critical Cartography. *KN–J. Cartogr. Geogr. Inf.* **2022**, *72*, 103–116. [CrossRef]
41. Kühne, O.; Koegst, L. Cartographic Representations of Coastal Land Loss in Louisiana: An Investigation Based on Deviant Cartographies. *KN-J. Cartogr. Geogr. Inf.* **2022**, *2022*, 1–12. [CrossRef]
42. Rorty, R. *Consequences of Pragmatism: Essays: 1972–1980*; University of Minnesota Press: Minneapolis, MN, USA, 1982, ISBN 0816610649.
43. Rorty, R. *Contingency, Irony, and Solidarity, Reprint*; Cambridge University Press: Cambridge, UK, 1997, ISBN 9780521353816.
44. Kühne, O.; Koegst, L. *Land Loss in Louisiana: A Neopragmatic Redescription*; Springer VS: Wiesbaden, Germany, 2022.
45. Kühne, O.; Jenal, C. Baton Rouge–A Neopragmatic Regional Geographic Approach. *Urban Sci.* **2021**, *5*, 17. [CrossRef]
46. Kühne, O.; Jenal, C. Baton Rouge (Louisiana): On the Importance of Thematic Cartography for 'Neopragmatic Horizontal Geography'. *KN–J. Cartogr. Geogr. Inf.* **2020**, *71*, 23–31. [CrossRef]
47. Weber, M. *Wirtschaft und Gesellschaft: Grundriss der verstehenden Soziologie, 5., Revidierte Auflage*; Mohr, J.C.B., Ed.; Paul Siebeck: Tübingen, Germany, 1972; ISBN 3165336318.
48. Dahrendorf, R. *Lebenschancen: Anläufe zur Sozialen und Politischen Theorie*; Suhrkamp: Frankfurt, Germany, 1979; ISBN 3518370596.
49. Dahrendorf, R. *Der Moderne Soziale Konflikt: Essay zur Politik der Freiheit*; Dtv: München, Germany, 1994; ISBN 3-423-04628-7.
50. Kühne, O.; Berr, K.; Jenal, C. *Die Geschlossene Gesellschaft und Ihre Ligaturen: Eine Kritik am Beispiel 'Landschaft'*; Springer VS: Wiesbaden, Germany, 2022.
51. Kühne, O. Religion und Landschaft–Überlegungen zu einem komplexen Verhältnis aus sozialwissenschaftlicher Perspektive. In *Religion und Landschaft*; Bund Heimat und Umwelt in Deutschland, Ed.; Selbstverlag: Bonn, Germany, 2013; pp. 7–15, ISBN 9783925374319.
52. Kühne, O.; Leonardi, L. *Ralf Dahrendorf: Between Social Theory and Political Practice*; Palgrave Macmillan: London, UK, 2020.
53. Kühne, O. Landscape Conflicts: A Theoretical Approach Based on the Three Worlds Theory of Karl Popper and the Conflict Theory of Ralf Dahrendorf, Illustrated by the Example of the Energy System Transformation in Germany. *Sustain. Sci. Pract. Policy* **2020**, *12*, 6772. [CrossRef]
54. Berger, P.L.; Luckmann, T. *The Social Construction of Reality: A Treatise in the Sociology of Knowledge*; Anchor Books: New York, NY, USA, 1966, ISBN 0-385-05898-5.
55. Durkheim, É. *Die Regeln der Soziologischen Methode*; Luchterhand: Neuwied, Germany, 1961.
56. Popper, K.R. *The Logic of Scientific Discovery*; Harper & Row: New York, NY, USA, 1959.
57. Goode, E. *Deviant Behavior, 12. Auflage*; Routledge: New York, NY, USA; London, UK, 2019; ISBN 978-0-367-19317-1.
58. Becker, H.S. *Outsiders: Studies in the Sociology of Deviance*; The Free Press: New York, NY, USA, 1963.
59. Peuckert, R. Abweichendes Verhalten und soziale Kontrolle. In *Einführung in Hauptbegriffe der Soziologie, 6. Auflage*; Korte, H., Schaefers, B., Eds.; VS Verlag für Sozialwissenschaften: Wiesbaden, Germany, 2006; pp. 105–125, ISBN 978-3-531-90032-2.
60. Stehr, J. Normalität und Abweichung. In *Soziologische Basics: Soziologische Basics: Eine Einführung für Pädagogen und Pädagoginnen*; Scherr, A., Ed.; VS Verlag für Sozialwissenschaften: Wiesbaden, Germany, 2013; pp. 191–197.

61. Coetzee, S.; Ivánová, I.; Mitasova, H.; Brovelli, M. Open Geospatial Software and Data: A Review of the Current State and A Perspective into the Future. *ISPRS Int. J. Geo-Inf.* **2020**, *9*, 90. [CrossRef]
62. Keil, J.; Edler, D.; Schmitt, T.; Dickmann, F. Creating Immersive Virtual Environments Based on Open Geospatial Data and Game Engines. *KN–Ournal Cartogr. Geogr. Inf.* **2021**, *71*, 53–65. [CrossRef]
63. Lindner, C.; Ortwein, A.; Staar, K.; Rienow, A. Different Levels of Complexity for Integrating Textures Extra-terrestrial Elevation Data in Game Engines for Educational Augmented and Virtual Reality Applications. *KN-J. Cartogr. Geogr. Inf.* **2021**, *71*, 253–267. [CrossRef]
64. Könen, D.; Gryl, I.; Pokraka, J.; Zwischen, W. Interessenvertretung und Verantwortung: Bürger*innenbeteiligung am Beispiel Windkraft im Spiegel von Neocartography und Spatial Citizenship. In *Bausteine der Energiewende*; Kühne, O., Weber, F., Eds.; Springer VS: Wiesbaden, Germany, 2018; pp. 207–230. ISBN 978-3-658-19508-3.
65. Kühne, O. Urban nature between modern and postmodern aesthetics: Reflections based on the social constructivist approach. *Quaest. Geograph.* **2012**, *31*, 61–70. [CrossRef]
66. Reicher, M.E. *Einführung in die Philosophische Ästhetik, 3., Überarbeitete Auflage*; WBG: Darmstadt, Germany, 2015; ISBN 9783534739776.
67. Borgeest, C. *Das Sogenannte Schoene: Ästhetische Sozialschranken*; S. Fischer: Frankfurt, Germany, 1977; ISBN 3-10-007602-8.
68. Schweppenhäuser, G. *Ästhetik: Philosophische Grundlagen und Schlüsselbegriffe*; Campus-Verlag: Frankfurt, Germany, 2007; ISBN 3593383470.
69. Kant, I. *Kritik der Urteilskraft*; Unveränd. Neudr. der Ausg. von 1924; Meiner: Hamburg, Germany, 1959; ISBN 39a.
70. Dewey, J. *Experience and Nature*; Dover Publications: New York, NY, USA, 1958.
71. Dewey, J. *Kunst als Erfahrung*; Suhrkamp: Frankfurt, Germany, 1988, ISBN 3-518-28303-0.
72. Aschenbrand, E. *Die Landschaft des Tourismus: Wie Landschaft von Reiseveranstaltern Inszeniert und von Touristen Konsumiert Wird*; Springer VS: Wiesbaden, Germany, 2017.
73. Burke, E. *Philosophische Untersuchung über den Ursprung Unserer Ideen vom Erhabenen und Schönen, 2. Aufl.*; Meiner: Hamburg, Germany, 1989; ISBN 3787309446.
74. Peres, C. Philosophische Ästhetik: Eine Standortbestimmung. In *Kunst, Ästhetik, Philosophie: Im Spannungsfeld der Disziplinen*; Friesen, H., Wolf, M., Eds.; Mentis: Münster, Germany, 2013; pp. 13–69, ISBN 9783897857384.
75. Graham, G. *Philosophy of the Arts: An Introduction to Aesthetics*, 3rd ed.; Routledge: London, UK; New York, NY, USA, 2005; ISBN 9780415349796.
76. Lothian, A. Landscape and the philosophy of aesthetics: Is landscape quality inherent in the landscape or in the eye of the beholder? *Landsc. Urban Plan.* **1999**, *44*, 177–198. [CrossRef]
77. Pries, C. Einleitung. In *Das Erhabene: Zwischen Grenzerfahrung und Größenwahn*; Pries, C., Ed.; VCH Acta Humaniora: Weinheim, Germany, 1989; pp. 1–30, ISBN 3527176640.
78. Fayet, R. *Reinigungen: Vom Abfall der Moderne zum Kompost der Nachmoderne*; Passagen-Verlag: Wien, Austria, 2003; ISBN 9783851656046.
79. Linke, S. Ästhetik, Werte und Landschaft–eine Betrachtung zwischen philosophischen Grundlagen und aktueller Praxis der Landschaftsforschung. In *Landschaftsästhetik und Landschaftswandel*; Kühne, O., Megerle, H., Weber, F., Eds.; Springer VS: Wiesbaden, Germany, 2017; pp. 23–40.
80. Welsch, W. Das Ästhetische–eine Schlüsselkategorie unserer Zeit? In *Die Aktualität des Ästhetischen*; Welsch, W., Ed.; Fink: München, Germany, 1993; pp. 13–47. ISBN 9783770528967.
81. Lyotard, J.-F. Das Erhabene und die Avantgarde. In *Verabschiedung von der (Post-)Moderne?* Le Rider, J., Raulet, G., Eds.; Narr: Tübingen, Germany, 1987; pp. 251–269.
82. Welsch, W. *Unsere Postmoderne Moderne*; VCH Acta Humaniora: Weinheim, Germany, 1987; ISBN 3527175849.
83. Friesen, H. Philosophische Ästhetik und die Entwicklung der Kunst. In *Kunst, Ästhetik, Philosophie: Im Spannungsfeld der Disziplinen*; Friesen, H., Wolf, M., Eds.; Mentis: Münster, Germany, 2013; pp. 71–160, ISBN 9783897857384.
84. Recki, B. Stil im Handeln Oder Die Aufgaben der Urteilskraft. In *Kunst, Ästhetik, Philosophie: Im Spannungsfeld der Disziplinen*; Friesen, H., Wolf, M., Eds.; Mentis: Münster, Germany, 2013; pp. 221–244, ISBN 9783897857384.
85. Büttner, N. *Geschichte der Landschaftsmalerei*; Hirmer: München, Germany, 2006.
86. Carlson, A. *Nature and Landscape: An Introduction to Environmental Aesthetics*; Columbia University Press: New York, NY, USA, 2009; ISBN 9780231518550.
87. Liessmann, K.P. *Schönheit*; Facultas Verlags-und Buchhandels AG: Wien, Austria, 2009; ISBN 9783825230487.
88. Rosenkranz, K. *Ästhetik des Häßlichen, 2., überarbeitete Auflage*; Reclam: Leipzig, Germany, 1996; ISBN 3379015555.
89. Pöltner, G. *Philosophische Ästhetik*; Kohlhammer: Stuttgart, Germany, 2008; ISBN 9783170169760.
90. Hauskeller, M. *Was ist Kunst?: Positionen der Ästhetik von Platon bis Danto, 8*; Auflage; Beck: München, Germany, 2005; ISBN 3406459994.
91. Platon. *Werke in 8 Bänden*; WBG: Darmstadt, Germany, 2005.
92. Hutcheson, F. *Eine Untersuchung über den Ursprung Unserer Ideen von Schönheit und Tugend: Über Moralisch Gutes und Schlechtes*; Meiner: Hamburg, Germany, 1986; ISBN 3-7873-0632-3.
93. Vischer, F.T.V. *Kritische Gänge, 2., Verm. Auflage*; Meyer & Jessen: München, Germany, 1922.
94. Hartmann, E.V. *Philosophie des Schönen, 2. Aufl*; Wegweiser-Verlag: Berlin, Germany, 1924.

95.   Bourdieu, P. *Distinction: A Social Critique of the Judgement of Taste*; Harvard University Press: Cambridge, MA, USA, 1984; ISBN 0674212800.

96.   Illing, F. *Kitsch, Kommerz und Kult: Soziologie des Schlechten Geschmacks*; UVK Verlagsgesellschaft: Konstanz, Germany, 2006; ISBN 389669541X.

97.   Liessmann, K.P. *Kitsch! Oder Warum der Schlechte Geschmack der Eigentlich Gute Ist*; Brandstätter: Wien, Austria, 2002; ISBN 9783854981701.

98.   Welsch, W. *Ästhetisches Denken*; Reclam: Stuttgart, Germany, 2006.

99.   Kühne, O. *Landscape and Power in Geographical Space as a Social-Aesthetic Construct*; Springer International Publishing: Dordrecht, The Netherlands, 2018; ISBN 3319729020.

100.   Simmel, G. Soziologie der Sinne. *Die Neue Rundsch.* **1907**, *18*, 1025–1036.

101.   Thibaud, J.-P. Die sinnliche Umwelt von Städten.: Zum Verständnis urbaner Atmosphären. In *Die Kunst der Wahrnehmung: Beiträge zu Einer Philosophie der Sinnlichen Erkenntnis*; Hauskeller, M., Ed.; SFG-Servicecenter Fachverlage: Kusterdingen, Germany, 2003; pp. 280–297.

102.   Seel, M. *Die Kunst der Entzweiung: Zum Begriff der Ästhetischen Rationalität*; Suhrkamp: Frankfurt, Germany, 1985; ISBN 3518577301.

103.   Lehmann, H. *Ästhetische Erfahrung: Eine Diskursanalyse*; Wilhelm Fink: Paderborn, Germany, 2016; ISBN 3846761397.

104.   Kühne, O. Potentials of the Three Spaces Theory for Understandings of Cartography, Virtual Realities, and Augmented Spaces. *KN-J. Cartogr. Geogr. Inf.* **2021**, *71*, 297–305. [CrossRef]

105.   Koegst, L. Potentials of Digitally Guided Excursions at Universities Illustrated Using the Example of an Urban Geography Excursion in Stuttgart. *KN-J. Cartogr. Geogr. Inf.* **2022**, *72*, 59–71. [CrossRef] [PubMed]

106.   Kühne, O. *Landschaftstheorie und Landschaftspraxis: Eine Einführung aus Sozialkonstruktivistischer Perspektive*; 3., Aktualisierte und Überarbeitete Auflage; Springer VS: Wiesbaden, Germany, 2021.

107.   Aschenbrand, E. Einsamkeit im Paradies. Touristische Distinktionspraktiken bei der Aneignung von Landschaft. *Berichte. Geogr. Und Landeskd.* **2016**, *90*, 219–234.

108.   Greider, T.; Garkovich, L. Landscapes: The Social Construction of Nature and the Environment. *Rural. Sociol.* **1994**, *59*, 1–24. [CrossRef]

109.   Weber, M. *Wissenschaft als Beruf, 11. Auflage*; Duncker & Humblot: Berlin, Germany, 2011; ISBN 9783428135097.

110.   Stichweh, R. Raum und moderne Gesellschaft: Aspekte der sozialen Kontrolle des Raumes. In *Die Gesellschaft und ihr Raum: Raum als Gegenstand der Soziologie*; Krämer-Badoni, T., Ed.; Leske+Budrich: Opladen, Germany, 2003; pp. 93–102, ISBN 3810040223.

111.   Kirchhoff, T.; Trepl, L. Landschaft, Wildnis, Ökosystem: Zur kulturbedingten Vieldeutigkeit ästhetischer, moralischer und theoretischer Naturauffassungen: Einleitender Überblick. In *Vieldeutige Natur: Landschaft, Wildnis und Ökosystem als Kulturgeschichtliche Phänomene*; Kirchhoff, T., Trepl, L., Eds.; Transcript: Bielefeld, Germany, 2009; pp. 13–68, ISBN 3899429443.

112.   Faure, P. *Magie der Düfte: Eine Kulturgeschichte der Wohlgerüche von den Pharaonen zu den Römern*; Artemis & Winkler: München, Germany, 1993; ISBN 3760819230.

113.   Raab, J. Die Soziale Konstruktion Olfaktorischer Wahrnehmung: Eine Soziologie des Geruchs. Available online: http://kops.uni-konstanz.de/bitstream/handle/123456789/11429/260_1.pdf?sequence=1&isAllowed=y (accessed on 20 December 2018).

114.   Kühne, O.; Edler, D. Multisensorische Landschaften–die Bedeutung des Nicht-Visuellen bei der sozialen und individuellen Konstruktion von Landschaft und Herausforderungen für ihre Erfassung und Wiedergabe. *Berichte. Geogr. Und Landeskd.* **2018**, *92*, 27–45.

115.   Meenar, M.; Kitson, J. Using Multi-Sensory and Multi-Dimensional Immersive Virtual Reality in Participatory Planning. *Urban Sci.* **2020**, *4*, 34. [CrossRef]

116.   Edler, D.; Kühne, O. Nicht-visuelle Landschaften. In *Handbuch Landschaft*; Kühne, O., Weber, F., Berr, K., Jenal, C., Eds.; Springer VS: Wiesbaden, Germany, 2019; pp. 599–612.

117.   Horbiński, T.; Zagata, K. Map Symbols in Video Games: The Example of "Valheim". *KN–J. Cartogr. Geogr. Inf.* **2021**, *71*, 269–283. [CrossRef]

118.   Kühne, O.; Edler, D.; Jenal, C. The Abstraction of an Idealization: Cartographic Representations of Model Railroads: Die Abstraktion der Idealisierung–über kartographische Repräsentationen von Modellbahnlandschaften. *KN–J. Cartogr. Geogr. Inf.* **2021**, *71*, 207–217. [CrossRef]

119.   Zagata, K.; Gulij, J.; Halik, Ł.; Medynska-Gulij, B. Mini-Map for Gamers Who Walk and Teleport in a Virtual Stronghold. *ISPRS–Int. J. Geo-Inf.* **2021**, *10*, 96. [CrossRef]

120.   Kühne, O. Representations of landscape in the strategy game Civilization. In *The Social Construction of Landscape in Games*; Edler, D., Kühne, O., Jenal, C., Eds.; Springer: Wiesbaden, Germany, 2022; pp. 261–272.

121.   Kaltman, E. The Construction of Civilization. *Kinephanos* **2014**, *1*, 105–118.

122.   Fontaine, D. Landscape in Computer Games–The Examples of GTA V and Watch Dogs 2. In *Modern Approaches to the Visualization of Landscapes*; Edler, D., Jenal, C., Kühne, O., Eds.; Springer VS: Wiesbaden, Germany, 2020; pp. 293–306.

123.   Pietsch, S.M. Landscape as Frontier–Experiencing the *Wild West in Red Dead Redemption 2* (2018). In *The Social Construction of Landscape in Games*; Edler, D., Kühne, O., Jenal, C., Eds.; Springer: Wiesbaden, Germany, 2022; pp. 273–288.

124.   Kühne, O.; Edler, D.; Jenal, C. Landschaften und Spiele: Von Virtualisierungen, Hybridisierungen und der Steigerung von Kontingenz. *Berichte. Geogr. Und Landeskd.* **2022**, *96*, 246–267. [CrossRef]

125. Schuster, K. The Social Psychological Function of Play. In *The Social Construction of Landscape in Games*; Edler, D., Kühne, O., Jenal, C., Eds.; Springer: Wiesbaden, Germany, 2022; pp. 39–57.
126. Koegst, L. Über drei Welten, Räume und Landschaften: Digital geführte Exkursionen an Hochschulen aus der Perspektive der drei Welten Theorie im Allgemeinen und der Theorie der drei Landschaften im Speziellen. *Ber. Geogr. Und Landeskd.* **2022**, *69*, 288–308. [CrossRef]
127. Stintzing, M.; Pietsch, S.; Wardenga, U. How to Teach "Landscape" Through Games? In *Modern Approaches to the Visualization of Landscapes*; Edler, D., Jenal, C., Kühne, O., Eds.; Springer VS: Wiesbaden, Germany, 2020; pp. 333–349.
128. Pietsch, S.M.; Stintzing, M.; Heyer, I. SpielRäume–Entdeckungs- und Erlebnisraum Landschaft. *GW Unterr.* **2020**, *1*, 37–49. [CrossRef]
129. Dickmann, F.; Keil, J.; Dickmann, P.L.; Edler, D. The Impact of Augmented Reality Techniques on Cartographic Visualization. *KN-J. Cartogr. Geogr. Inf.* **2021**, *71*, 285–295. [CrossRef]
130. Edler, D.; Dickmann, F. Interaktive Multimediakartographie in frühen Videospielwelten–Das Beispiel „Super Mario World". *KN–J. Cartogr. Geogr. Inf.* **2016**, *66*, 51–58. [CrossRef]
131. Fish, C.S. Elements of Vivid Cartography. *Cartogr. J.* **2021**, *58*, 150–166. [CrossRef]
132. Edler, D.; Edler, S.; Dickmann, F. Eine empirische Studie zu Effekten von simulierter Gründblindheit (Deuteranopie) auf das kartenbasierte Positionsgedächtnis. *Kartogr. Nachr.* **2015**, *65*, 183–194.
133. Schnürer, R.; Sieber, R.; Schmid-Lanter, J.; Öztireli, A.C.; Hurni, L. Detection of Pictorial Map Objects with Convolutional Neural Networks. *Cartogr. J.* **2021**, *58*, 50–68. [CrossRef]
134. Bell, D. *The Coming of Post-Industrial Society: A Venture in Social Forecasting*; Basic Books: New York, NY, USA, 1999.
135. Lash, S.; Urry, J. *Economies of Signs and Space*; SAGE Publications: London, UK, 1994; ISBN 0803984723.
136. Kühne, O. Soziale Akzeptanz und Perspektiven der Altindustrielandschaft: Ergebnisse einer empirischen Untersuchung im Saarland. *RaumPlanung* **2007**, *3*, 156–160.
137. Jenal, C. (Alt)Industrielandschaften. In *Handbuch Landschaft*; Kühne, O., Weber, F., Berr, K., Jenal, C., Eds.; Springer VS: Wiesbaden, Germany, 2019; pp. 831–841.
138. Hauser, S. Industrieareale als urbane Räume. In *Die Europäische Stadt*; Siebel, W., Ed.; Suhrkamp: Frankfurt, Germany, 2004; pp. 146–157.
139. Vicenzotti, V. Kulturlandschaft und Stadt-Wildnis. In *Kulturen der Landschaft: Ideen von Kulturlandschaft Zwischen Tradition und Modernisierung*; Kazal, I., Voigt, A., Weil, A., Zutz, A., Eds.; Technische Universität Berlin: Berlin, Germany, 2006; pp. 221–236.
140. Höfer, W. *Natur als Gestaltungsfrage: Zum Einfluß Aktueller Gesellschaftlicher Veränderungen auf die Idee von Natur und Landschaft als Gegenstand der Landschaftsarchitektur. Dissertation*; Herbert Utz Verlag GmbH: München, Germany, 2001.
141. Fischer, L. Das Erhabene und die ‚feinen Unterschiede': Zur Dialektik in den sozio-kulturellen Funktionen von ästhetischen Deutungen der Landschaft. In *Natur–Kultur: Volkskundliche Perspektiven auf Mensch und Umwelt*; Brednich, R.W., Schneider, A., Werner, U., Eds.; Waxmann: Münster, Germany, 2001; pp. 347–356, ISBN 978-3-8309-1100-5.
142. Eisel, U. *Landschaft und Gesellschaft: Räumliches Denken im Visier*; Westfälisches Dampfboot: Münster, Germany, 2009; ISBN 3896917722.
143. Sander, A.; Blossfeldt, K.; Renger-Platzsch, A.; Becher, B.; Becher, H. *Vergleichende Konzeptionen*; Schirmer/Mosel: München, Germany; Paris, France; London, UK, 1997.
144. Kent, A.J. Trust Me, I'm a Cartographer: Post-truth and the Problem of Acritical Cartography. *Cartogr. J.* **2017**, *54*, 193–195. [CrossRef]
145. Grau, A. *Hypermoral: Die neue Lust an der Empörung, 2. Auflage*; Claudius: München, Germany, 2017; ISBN 9783532628034.
146. Kühne, O.; Berr, K.; Jenal, C.; Schuster, K. *Liberty and Landscape: In Search of Life-Chances with Ralf Dahrendorf*; Palgrave Macmillan: Basingstoke, UK, 2021.
147. Sofsky, W. *Verteidigung des Privaten: Eine Streitschrift*; Beck: München, Germany, 2007; ISBN 3-406-56298-1.
148. Kersting, W. *Verteidigung des Liberalismus*; Murmann: Hamburg, Germany, 2009; ISBN 9783867740739.